This book is the first complete presentation in English of the combinatorial theory of species, introduced by A. Joyal in 1980. It gives a unified understanding of the use of generating functions for both labeled and unlabeled structures as and also provides a tool for the specification and analysis of these structures. Of particular importance is the capacity of combinatorial species to transform recursive definitions of tree-like structures into functional or differential equations, and conversely.

Some of the features of this book are:

– A detailed analysis of classes of combinatorial structures such as permutations, total orders, graphs, trees, rooted trees, data structures, etc., including the basic constructions and operations that can be performed on them and on their various generating series.
– A thorough study of data structures defined by functional equations: binary trees, (2,3)-, (a,b)- and B-trees, AVL trees, PQ-trees, leftist trees, ordered rooted trees and many other classes of enriched rooted trees.
– Extensive discussions of various links between combinatorics and other parts of mathematics: orthogonal polynomials, Lagrange inversion formula, implicit function theorem, Newton–Raphson iteration, differential equations, symmetric functions, and asymptotic analysis.
– An extension of Pólya theory to classes of structures defined by combinatorial operations or by functional equations and to asymmetric structures.

This book will be a valuable reference to graduate students and researchers in combinatorics, analysis, probability theory, and theoretical computer science.

ENCYCLOPEDIA OF MATHEMATICS AND ITS APPLICATIONS

EDITED BY G.-C. ROTA

Editorial Board

R. Doran, M. Ismail, T.-Y. Lam, E. Lutwak

Volume 67

Combinatorial Species and Tree-like Structures

ENCYCLOPEDIA OF MATHEMATICS AND ITS APPLICATIONS

6   H. Minc   *Permanents*
18   H. O. Fattorini   *The Cauchy Problem*
19   G. G. Lorentz, K. Jetter, and S. D. Riemenschneider   *Birkhoff Interpolation*
22   J. R. Bastida   *Field Extensions and Galois Theory*
23   J. R. Cannon   *The One-Dimensional Heat Equation*
24   S. Wagon   *The Banach-Tarski Paradox*
25   A. Salomaa   *Computation and Automata*
26   N. White (ed.)   *Theory of Matroids*
27   N. H. Bingham, C. M. Goldie, and J. L. Teugels   *Regular Variation*
28   P. P. Petrushev and V. P. Popov   *Rational Approximation of Real Functions*
29   N. White (ed.)   *Combinatorial Geometries*
30   M. Pohst and H. Zassenhaus   *Algorithmic Algebraic Number Theory*
31   J. Aczel and J. Dhombres   *Functional Equations in Several Variables*
32   M. Kuczma, B. Choczewski, and R. Ger   *Iterative Functional Equations*
33   R. V. Ambartzumian   *Factorization Calculus and Geometric Probability*
34   G. Gripenberg, S.-O. Londen and O. Staffans   *Volterra Integral and Functional Equations*
35   G. Gasper and M. Rahman   *Basic Hypergeometric Series*
36   E. Torgersen   *Comparison of Statistical Experiments*
37   A. Neumaier   *Interval Methods for Systems of Equations*
38   N. Korneichuk   *Exact Constants in Approximation Theory*
39   R. A. Brualdi and H. Ryser   *Combinatorial Matrix Theory*
40   N. White (ed.)   *Matroid Applications*
41   S. Sakai   *Operator Algebras in Dynamical Systems*
42   W. Hodges   *Basic Model Theory*
43   H. Stahl and V. Totik   *General Orthogonal Polynomials*
45   G. Da Prato and J. Zabczyk   *Stochastic Equations in Infinite Dimensions*
46   A. Björner et al.   *Oriented Matroids*
47   G. Edgar and L. Sucheston   *Stopping Times and Directed Processes*
48   C. Sims   *Computation With Finitely Presented Groups*
49   T. Palmer   *Banach Algebras and the General Theory of *-Algebras*
50   F. Borceux   *Handbook of Categorical Algebra 1*
51   F. Borceux   *Handbook of Categorical Algebra 2*
52   F. Borceux   *Handbook of Categorical Algebra 3*
54   A. Katok and B. Hasselblatt   *Introduction to the Modern Theory of Dynamical Systems*
55   V. N. Sachkov   *Combinatorial Methods in Discrete Mathematics*
56   V. N. Sachkov   *Probabilistic Methods in Discrete Mathematics*
57   P. M. Cohn   *Skew Fields*
58   Richard Gardner   *Geometric Tomography*
59   George A. Baker Jr. and Peter Graves-Morris   *Pade Approximants*
60   Jan Krajicek   *Bounded Arithmetic, Propositional Logic and Complexity Theory*
61   H. Groemer   *Geometric Applications of Fourier Series and Spherical Harmonics*
62   H. O. Fattorini   *Infinite Dimensional Optimization and Control Theory*
63   A. C. Thompson   *Minkowski Geometry*

ENCYCLOPEDIA OF MATHEMATICS AND ITS APPLICATIONS

Combinatorial Species and Tree-like Structures

F. BERGERON
Université du Québec à Montréal

G. LABELLE
Université du Québec à Montréal

P. LEROUX
Université du Québec à Montréal

Translated from French by Margaret Readdy

PUBLISHED BY THE PRESS SYNDICATE OF THE UNIVERSITY OF CAMBRIDGE
The Pitt Building, Trumpington Street, Cambridge CB2 1RP, United Kingdom

CAMBRIDGE UNIVERSITY PRESS
The Edinburgh Building, Cambridge CB2 2RU, United Kingdom
40 West 20th Street, New York, NY 10011-4211, USA
10 Stamford Road, Oakleigh, Melbourne 3166, Australia

© F. Bergeron G. Labelle P. Leroux 1998
French edition:
Théroie des espèces et combinatoire des structures arborescentes,
Publications du LACIM, vol. 19, 1994
© LACIM, UQAM, octobre 1994.

This book is in copyright. Subject to statutory exception
and to the provisions of relevant collective licensing agreements,
no reproduction of any part may take place without
the written permission of Cambridge University Press.

First published in English in 1998

Printed in the United States of America

Typeset in Times

Library of Congress Cataloging-in-Publication Data
Bergeron, F.
 [Théorie des espèces et combinatoire des structures arborescentes. English]
 Combinatorial species and tree-like structures / F. Bergeron, G. Labelle, P. Leroux: translated from the French by Margaret Readdy.
   p. cm. – (Encyclopedia of mathematics and its applications; v. 67)
 Includes bibliographical references (p. –) and index.
 ISBN 0-521-57323-8 (hardback)
 1. Combinatorial enumeration problems. I. Labelle, Gilbert, 1944–   . II. Leroux, P. (Pierre). 1942–   . III. Title. IV. Series.
QA164.8.B4713   1997   96-46136
511′.5—dc20   CIP

*A catalog record for this book is available from
the British Library*

ISBN 0 521 57323 8 hardback

Foreword

by Gian-Carlo Rota

Advances in mathematics occur in one of two ways.

The first occurs by the solution of some outstanding problem, such as the Bieberbach conjecture or Fermat's conjecture. Such solutions are justly acclaimed by the mathematical community. The solution of every famous mathematical problem is the result of joint effort of a great many mathematicians. It always comes as an unexpected application of theories that were previously developed without a specific purpose, theories whose effectiveness was at first thought to be highly questionable.

Mathematicians realized long ago that it is hopeless to get the lay public to understand the miracle of unexpected effectiveness of theory. The public, misled by two hundred years of Romantic fantasies, clamors for some "genius" whose brain power cracks open the secrets of nature. It is therefore a common public relations gimmick to give the entire credit for the solution of famous problems to the one mathematician who is responsible for the last step.

It would probably be counterproductive to let it be known that behind every "genius" there lurks a beehive of research mathematicians who gradually built up to the "final" step in seemingly pointless research papers. And it would be fatal to let it be known that the showcase problems of mathematics are of little or no interest for the progress of mathematics. We all know that they are dead ends, curiosities, good only as confirmation of the effectiveness of theory. What mathematicians privately celebrate when one of their showcase problems is solved is Polya's adage: "no problem is ever solved directly."

There is a second way by which mathematics advances, one that mathematicians are also reluctant to publicize. It happens whenever some commonsense notion that had heretofore been taken for granted is discovered to be wanting, to need clarification or definition. Such foundational advances produce substantial dividends, but not right away. The usual accusation that is leveled against mathematicians who dare propose overhauls of the obvious is that of being "too abstract." As if one piece of mathematics could be "more abstract" than another, except in the eyes of the beholder (it is time to raise a cry of alarm against the misuse of the word "abstract," which has become as meaningless as the word "Platonism.")

An amusing case history of an advance of the second kind is uniform convergence, which first made headway in the latter quarter of the nineteenth century. The late Herbert Busemann told me that while he was a student, his analysis teachers admitted their inability to visualize uniform convergence, and viewed it as the outermost limit of abstraction. It took a few more generations to get uniform convergence taught in undergraduate classes.

The hostility against groups, when groups were first "abstracted" from the earlier "group of permutations" is another case in point. Hadamard admitted to being unable to visualize groups except as groups of permutations. In the thirties, when groups made their first inroad into physics via quantum mechanics, a staunch sect of reactionary physicists, repeatedly cried "Victory!" after convincing themselves of having finally rid physics of the "Gruppenpest." Later, they tried to have this episode erased from the history of physics.

In our time, we have witnessed at least two displays of hostility against new mathematical ideas. The first was directed against lattice theory, and its virulence all but succeeded in wiping lattice theory off the mathematical map. The second, still going on, is directed against the theory of categories. Grothendieck did much to show the simplifying power of categories in mathematics. Categories have broadened our view all the way to the solution of the Weil conjectures. Today, after the advent of braided categories and quantum groups, categories are beginning to look downright concrete, and the last remaining anticategorical reactionaries are beginning to look downright pathetic.

There is a common pattern to advances in mathematics of the second kind. They inevitably begin when someone points out that items that were formerly thought to be "the same" are not really "the same," while the opposition claims that "it does not matter," or "these are piddling distinctions." Take the notion of species that is the subject of this book. The distinction between "labeled graphs" and "unlabeled graphs" has long been familiar. Everyone agrees on the definition of an unlabeled graph, but until a while ago the notion of labeled graph was taken as obvious and not in need of clarification. If you objected that a graph whose vertices are labeled by cyclic permutations – nowadays called a "fat graph" – is not the same thing as a graph whose vertices are labeled by integers, you were given a strange look and you would not be invited to the next combinatorics meeting.

The correct definition of a labeled graph turned out to be more sophisticated than the definition of an unlabeled graph. A labeled graph – or any "labeled" combinatorial construct – is a functor from the groupoid of finite sets and bijections to itself. This definition of a labeled object is not "abstract": on the contrary, it expresses in precise terms the commonsense idea of "being able to label the vertices of a graph either by integers or by colors, it does not matter," and it is the only way of making this commonsense idea precise. The notion of groupoid, which is one of the key ideas of contemporary mathematics, makes it possible to withhold the assignement of a specific set of labels to the vertices of a graph without making the graph unlabeled.

Joyal's definition of "labeled object" as a species discloses a vast horizon of new combinatorial constructions, which cannot be seen if one holds on to the reactionary view that "labeled objects" need no definition. The simplest, and the most remarkable, application of the definition of species is the rigorous combinatorial rendering of functional composition, which was formerly dealt with by handwaving – always a bad sign. But it is just the beginning.

Species are related to generating functions in much the same way as random variables are related to probability distributions. Those probabilists of the thirties who held on to distributions, while rejecting random variables as "superfluous," were eventually wiped out, and their results are not even acknowledged today.

I dare make a prediction on the future acceptance of this book. At first, the old fogies will pretend the book does not exist. This pretense will last until sufficiently many younger combinatorialists publish papers in which interesting problems are solved using the theory of species. Eventually, a major problem will be solved in the language of species, and from that time on everyone will have to take notice. The rewriting, copying and imitating will start, and mathematicians who capitulate to the new theory will begin to tell us what species really are. Considering the speed at which mathematics progresses in our day, that time is more likely to come sooner than later.

The present book is the first thorough treatment in English of the theory of species. It is lucidly and clearly written, and it should go a long way to making this fundamental chapter of combinatorial mathematics available to the entire spectrum of mathematicians, computer scientists and cultivated scientists generally.

Cambridge, April 27, 1997

Table of Contents

Foreword		*page* v
Preface		xi
1	**Introduction to Species of Structures**	**1**
	1.0 Introduction	1
	1.1 Notion of Species of Structures	1
	1.2 Associated Series	12
	1.3 Addition and Multiplication	26
	1.4 Substitution and Differentiation	40
2	**Complements on Species of Structures**	**59**
	2.0 Introduction	59
	2.1 Pointing and Cartesian Product	60
	2.2 Functorial Composition	70
	2.3 Weighted Species	79
	2.4 Extension to the Multisort Context	100
	2.5 Virtual Species	120
	2.6 Molecular and Atomic Species	139
3	**Combinatorial Functional Equations**	**162**
	3.0 Introduction	162
	3.1 Lagrange Inversion	164
	3.2 Implicit Species Theorem	191
	3.3 Quadratic Iterative Methods	222
	3.4 Elements of Asymptotic Analysis	247
4	**Complements on Unlabeled Enumeration**	**277**
	4.0 Introduction	277
	4.1 The Dissymmetry Theorem for Trees	278

	4.2	Connected Graphs and Blocks	297
	4.3	Proof of the Substitution Formulas	309
	4.4	Asymmetric Structures	321

5 Species on Totally Ordered Sets — **341**

	5.0	Introduction	341
	5.1	$\mathbb{L}$-Species	342
	5.2	Combinatorial Differential Equations	358

Appendix 1: Group Actions and Pólya Theory	393
Appendix 2: Miscellaneous Tables	408
Bibliography	434
Notation Indexx	447
Index	449

Preface

During the last decades considerable progress has been made in clarifying and strengthening the foundation of enumerative combinatorics. A number of useful theories, especially to explain algebraic techniques, have emerged. We mention, among others, *Möbius inversion* (Rota [284], Rota–Smith [289], Rota–Sagan [288]), *partitional composition* (Cartier–Foata [45], Foata [106, 112]), *prefabs* (Bender–Goldman [11]), *reduced incidence algebras* (Mullin–Rota [253], Doubilet, Rota, and Stanley [80], Dür [84]), *binomial posets* and *exponential structures* (Stanley [301, 300]), *Möbius categories* (Content–Lemay–Leroux [66], Leroux [212, 214]), *umbral calculus* and *Hopf algebras* (Rota [286], Joni–Rota [157]), *Pólya theory* (Pólya [263], Redfield [275], de Bruijn [68], Robinson [282]), and *species of structures* (Joyal [158]). Many authors have also underlined the importance of these methods to solve problems of enumeration, in particular, Bender–Williamson [12], Berge [13], Comtet [58], Flajolet [91], Goulden–Jackson [133], Graham–Knuth–Patashnik [136], Kerber [169], Harary–Palmer [144], Knuth [172], Liu [222], Riordan [281], Moon [251], Sagan [290], Stanley [304, 302], Stanton–White [306], van Lint–Wilson [316], Wehrhahn [324], and Wilf [326].

In addition, during this same period, the subject has been greatly enriched by its interaction with theoretical computer science as a source of application and motivation. The importance of combinatorics for the analysis of algorithms and the elaboration of efficient data structures is established in the fundamental book of Knuth [172]. A good knowledge of combinatorics is now essential to the computer scientist. Of particular importance are the following areas: formal languages, grammars, and automata theory (see for instance Berstel–Reutenauer [30], Eilenberg [88], Greene [137], Lothaire [227], Reutenauer [276, 277], and the work of Schützenberger, Cori, Viennot and the Bordeaux School); asymptotic analysis and average case complexity (see Bender [9], Bender–Canfield [10], Flajolet–Odlyzko [97], Flajolet–Salvy–Zimmermann [99], Knuth [172], and Sedgewick–Flajolet [295]); and combinatorics of data structures (see Aho–Hopcroft–Ullman

[1], Baeza-Yates–Gonnet [5], Brassard–Bratley [38], Mehlhorn [244], and Williamson[329]).

The combinatorial theory of species, introduced by Joyal in 1980, is set in this general framework. It provides a unified understanding of the use of generating series for both labeled and unlabeled structures, as well as a tool for the specification and analysis of these structures. Of particular importance is its capacity to transform recursive definitions of (tree-like) structures into functional or differential equations, and conversely. Encompassing the description of structures together with permutation group actions, the theory of species conciliates the calculus of generating series and functional equations with Pólya theory, following previous efforts to establish an algebra of cycle index series, particularly by de Bruijn [68] and Robinson [282]. This is achieved by extending the concept of group actions to that of functors defined on groupoids, in this case the category of finite sets and bijections. The functorial concept of species of structures goes back to Ehresmann [87]. The functorial property of combinatorial constructions on sets is also pointed out in a paper of Mullin and Rota [253] in the case of reluctant functions, a crucial concept for the combinatorial understanding of Lagrange inversion. There are also links between the algebra of operations on species and category theory. For example, the partitional composition of species can be described in the general settings of doctrines (see Kelly [166]), operads (see May [233] and Loday [224]), and analytic functors (see Joyal [163]).

Informally, a species of structures is a rule, F, associating with each finite set U, a finite set $F[U]$ which is "independent of the nature" of the elements of U. The members of the set $F[U]$, called F-structures, are interpreted as combinatorial structures on the set U given by the rule F. The fact that the rule is independent of the nature of the elements of U is expressed by an invariance under relabeling. More precisely, to any bijection $\sigma : U \longrightarrow V$, the rule F associates a bijection $F[\sigma] : F[U] \longrightarrow F[V]$ which transforms each F-structure on U into an (isomorphic) F-structure on V. It is also required that the association $\sigma \mapsto F[\sigma]$ be consistent with composition of bijections. In this way the concept of species of structures puts as much emphasis on isomorphisms as on the structures themselves. In categorical terms, a species of structures is simply a functor from the category $\mathbb{B}$ of finite sets and bijections to itself.

As an example, the class $\mathcal{G}$ of simple (finite) graphs and their isomorphisms, in the usual sense, give rise to the species of graphs, also denoted $\mathcal{G}$. For each set U, the elements of $\mathcal{G}[U]$ are just the simple graphs with vertex set U. For each $\sigma : U \longrightarrow V$, the bijection $\mathcal{G}[\sigma] : \mathcal{G}[U] \longrightarrow \mathcal{G}[V]$ transforms each simple graph on U into a graph on V by relabeling via σ. Similarly, any class of discrete structures closed under isomorphisms gives rise to a species.

Furthermore, species of structures can be combined to form new species by using set theoretical constructions. There results a variety of combinatorial operations on species, including addition, multiplication, substitution, derivation, etc., which extend the familiar calculus of formal power series. Indeed to each species of structures, we can associate various formal power series designed to treat enumeration problems of a specific kind (labeled, unlabeled, asymmetric, weighted, etc.). Of key importance is the fact that these associated series are "compatible" with operations on species. Hence each (algebraic, functional, or differential) identity between species implies identities between their associated series. This is in the spirit of Euler's method of generating series.

For example, let $\mathfrak{a}$ denote the species of *trees* (acyclic connected simple graphs) and $\mathcal{A}$, that of *rooted trees* (trees with a distinguished vertex). Then the functional equation

$$\mathcal{A} = XE(\mathcal{A}), \tag{1}$$

expresses the basic fact that any rooted tree on a finite set U can be naturally described as a root (a vertex $x \in U$) to which is attached a set of disjoint rooted trees (on $U \setminus \{x\}$); see Figure 1.4.4. Equation (1) yields immediately the following equalities between generating series

$$A(x) = xe^{A(x)}, \qquad T(x) = x \exp\left(\sum_{k \geq 0} \frac{T(x^k)}{k}\right). \tag{2}$$

These formulas go back to Cayley [46] and Pólya [263]. The first refers to the exponential generating series $A(x) = \sum_{n \geq 0} a_n x^n / n!$, where a_n is the number of rooted trees on a set of n elements (labeled rooted trees), and yields Cayley's formula $a_n = n^{n-1}$ via the Lagrange inversion formula. The second refers to the ordinary generating series $T(x) = \sum_{n \geq 0} T_n x^n$, where T_n is the number of isomorphism types of rooted trees (unlabeled rooted trees) on n elements, and yields a recurrence formula for these numbers (see 4.1.44).

Analogously, the identity

$$2(n-1)n^{n-2} = \sum_{k=1}^{n-1} \binom{n}{k} k^{k-1}(n-k)^{n-k-1}, \tag{3}$$

and Otter's formula [259]

$$t(x) = T(x) + \frac{1}{2}(T(x^2) - T^2(x)), \tag{4}$$

where $t(x) = \sum_{n \geq 1} t_n x^n$ is the ordinary generating series of the number t_n of unlabeled trees on n elements, both follow from the species isomorphism

$$\mathcal{A} + E_2(\mathcal{A}) = \mathfrak{a} + \mathcal{A}^2, \tag{5}$$

allowing us to express the species $\mathfrak{a}$ of trees as a function of the species of rooted trees. We call this identity the *dissymmetry* theorem for trees (see Leroux [213], Leroux and Miloudi [215]). It is inspired from the dissimilarity formula of Otter

[259] and the work of Norman [257] and Robinson [282] on the decomposition of graphs into 2-connected components.

Since its introduction, the theory of species of structures has been the focus of considerable research by the Montréal school of combinatorics as well as numerous other researchers. The goal of this book is to present the basic elements of the theory and to give a unified account of some of it's developments and applications.

Chapter 1 contains the first key ideas of the theory. A general discussion on the notion of discrete structures leads naturally to the formal definition of species of structures. Some of the basic formal power series associated to a species F are introduced: the (exponential) generating series $F(x)$ for labeled enumeration, the type generating series $\widetilde{F}(x)$ for unlabeled enumeration, and the cycle index series $Z_F(x_1, x_2, x_3, \ldots)$ as a general enumeration tool. Finally, we introduce the combinatorial operations of addition, multiplication, substitution (partitional composition), and derivation of species of structures. These operations extend and interpret in the combinatorial context of species the corresponding operations on formal power series.

The second chapter begins with an introduction to three other operations: pointing, Cartesian product and functorial composition. Pointing is a combinatorial analogue of the operator $x(d/dx)$ on series. The Cartesian product, consisting of superposition of structures, corresponds to the *Hadamard product* of series (coefficientwise multiplication). Functorial composition, not to be confused with substitution, is the natural composition of species considered as functors (see Décoste–Labelle–Leroux [74]). Many species of graphs and multigraphs can be expressed easily by this operation.

The theory is then extended to weighted species where structures are counted according to certain parameters, and to multisort species in analogy with functions of several variables. These generalizations broaden the range of applications to more refined enumeration problems.

For example, the generating series for Laguerre polynomials,

$$\sum_{n\geq 0} \mathcal{L}_n^{(\alpha)}(t) \frac{x^n}{n!} = \left(\frac{1}{1-x}\right)^{\alpha+1} \exp\left(\frac{-tx}{1-x}\right), \qquad (6)$$

suggests a combinatorial "model" for these polynomials, consisting of permutations (with cycle counter $\alpha + 1$) and oriented chains (each with weight $-t$). This model gives rise to a combinatorial theory of Laguerre polynomials, where identities appear as consequences of elementary constructions on discrete structures. The same approach can be applied to many other families of polynomials. See, for example, Bergeron [20], Dumont [83], Foata [107, 109], Foata–Labelle [110], Foata–Leroux [111], Foata–Schützenberger [112], Foata–Strehl [113, 114], Foata–Zeilberger [115], Labelle–Yeh [202, 204, 208], Leroux–Strehl [216], Strehl [308, 309, 310], Viennot [319], and Zeng [340]. Following those lines, the book contains a combinatorial treatment of Eulerian, Hermite, Laguerre, and Jacobi polynomials.

Finally Chapter 2 introduces *virtual* species due to Joyal [162, 161] and Yeh [333, 334], making possible the subtraction of species, and develops an analysis of *molecular* (indecomposable with respect to addition) and *atomic* (indecomposable with respect to both addition and multiplication) species. These tools allow the construction of a multiplicative inverse $1/F$, for species F such that $F(0) = 1$, as well as the construction of an inverse for substitution $G^{\langle -1\rangle}$, for species G such that $G(0) = 0$ and $G'(0) = 1$.

Chapter 3 is concerned with a deeper study of the combinatorial functional equation

$$Y = XR(Y), \tag{7}$$

related to Lagrange inversion, as well as of more general equations of the form

$$Y = H(X, Y), \tag{8}$$

or even $Y = H(X, Y \circ G)$, where R, H, and G are given species, and Y is the unknown. The combinatorial resolution of these equations gives rise to tree-like structures that can be described recursively. In particular, the solution of the functional equation $Y = XR(Y)$ is a species of structures consisting of rooted trees whose fibers are "enriched" (or "structured") by the species R. The enumeration of these structures by combinatorial means leads to a combinatorial proof of the classical Lagrange inversion formulas (see Labelle [178] and Joyal [158]). Another proof of Lagrange inversion, also based on enriched rooted trees, is given by Chen [48, 51]. The enumeration of enriched rooted trees also leads to binomial type sequences naturally associated with the series $R(x)$ (see Rota [286]).

The more general equation $Y = H(X, Y)$, $Y(0) = 0$, is then considered. We give sufficient conditions for the existence and "uniqueness" of combinatorial solutions of systems of equations of this type. In particular, a multidimensional Lagrange inversion formula is established, following the approach of Gessel [127], which is then applied to the calculation of the index series of R-enriched rooted trees (see Décoste–Labelle–Leroux [73] and Labelle [181]).

The iterative method of Newton–Raphson for the solution of functional equations has a natural analogue in the context of species of structures (see Décoste–Labelle–Leroux [73] and Labelle [181]). It yields an iterative process with "quadratic convergence" for calculating species of structures defined by equations of the form $Y = XR(Y)$ or of the form

$$Y = X + G(Y), \tag{9}$$

where G is a given species, or more generally of the form $Y = H(X, Y)$. Functional equations of type

$$Y = H(X, Y \circ G), \tag{10}$$

which we call *Read–Bajraktarević* equations, lend themselves to an iterative approach and are also treated here (see Bergeron–Labelle–Leroux [23]). This work is motivated, from the point of view of computer science, by the complexity analysis of update algorithms for various kind of "balanced trees" for which equation (10) arises naturally.

This chapter concludes with an overview of various techniques of asymptotic analysis applied to the enumeration of F-structures on $\{1, 2, \ldots, n\}$ as $n \to \infty$. We give particular attention to the case of tree-like structures. The main methods considered are the analysis of dominant singularities, the method of Hayman (singularity at infinity), a theorem of Meir and Moon [234, 241] on the equation $y = xR(y)$, and one of Bender [9] on the equation $y = H(x, y)$.

Chapter 4 is devoted to a more thorough analysis of unlabeled enumeration. For species of structures, this often involves calculation of their cycle index series and the establishment of functional equations between them. The case of (enriched) trees and rooted trees is treated in detail, including the dissymmetry theorem for trees and its corollary, Otter's formula. We also study classes of simple graphs, for example, those whose *blocks* (2-connected components) belong to a given family (see Hanlon–Robinson [141]).

An objective of Chapter 4 is the detailed proof of the substitution formulas for weighted species, which are essential tools for unlabeled enumeration:

$$\widetilde{F_v \circ G_w}(x) = Z_{F_v}(\widetilde{G_w}(x), \widetilde{G_{w^2}}(x^2), \ldots), \tag{11}$$

$$Z_{F_v \circ G_w} = Z_{F_v}(Z_{G_w}(x_1, x_2, x_3, \ldots), Z_{G_{w^2}}(x_2, x_4, x_6, \ldots), \ldots). \tag{12}$$

This last formula generalizes both the theorem of Pólya for the wreath product of group actions (see Pólya [263], Section 27) and the composition theorem of Robinson for graphs (see Harary–Palmer [144], Chapter 8). The proofs given here of these formulas are based on the notion of wreaths of G-structures (see Joyal [158]) and on the algebraic independence of *power sum* symmetric functions

$$p_n = t_1^n + t_2^n + t_3^n + \cdots, \tag{13}$$

following the approach of Pólya. Substitution formula (12) describes the relation between "partitional composition" of species and plethysm of symmetric functions. Indeed, to any species of structures F corresponds a family of set-theoretical (linear) representations of the symmetric groups $\mathcal{S}_n$, $n \geq 0$. The symmetric function $Z_F(p_1, p_2, p_3, \ldots)$ is then the Frobenius characteristic of these representations and formula (12) reflects the fact that the characteristic of the composite of two representations is the plethysm of their characteristics (see Macdonald [228], Appendix A).

The last section of Chapter 4 is devoted to the study of *asymmetric* structures, that is, structures whose stabilizer reduces to the identity permutation. This problem has been solved, in the context of group actions, by Rota (in 1968) who introduced an acyclic indicator polynomial using Möbius inversion in the lattice of periods of the group action (see [285, 288, 289]). Asymmetric structures can also be studied in the general context of enumeration according to stabilizers, using Möbius inversion in the lattice of subgroups of a permutation group (see Stockmeyer [307], White [325], Kerber–Thürlings [171]). Following Pólya [263, Section 23], Harary–Prins [146], and Rota, our main tool for the enumeration of asymmetric structures is the *asymmetry index series* $\Gamma_F(x_1, x_2, x_3, \ldots)$ of a species F, which plays an analogous role for this problem as the cycle index series Z_F for the enumeration of isomorphism types. In particular, as shown by G. Labelle [186], the assignment $F \mapsto \Gamma_F$ transforms combinatorial identities on species into corresponding functional and differential equations on asymmetry index series. For a unified presentation in the context of *Möbius species*, see Yang [331]. The series Z_F and Γ_F also give rise to canonical q-series induced by principal specialization (see Décoste [70, 71], Décoste–Labelle [72]).

In enumerative and algebraic combinatorics, structures often involve a given total (linear) order on the underlying set. Good examples are provided by alternating permutations or Young standard tableaux. Similarly, in computer science, data structures often make use of some explicit order on the set of data, as in the case of searching or sorting algorithms. To better deal with such constructions, we develop in Chapter 5 a variant of the theory of species of structures, the $\mathbb{L}$-*species*, where structures are supported by sets with a given total order. In this context, the differential equation

$$Y' = R(Y), \qquad Y(0) = 0, \tag{14}$$

admits a natural combinatorial solution which can be expresssed in terms of *increasing* enriched rooted trees.

We develop, in the context of $\mathbb{L}$-species, a combinatorial theory of autonomous and nonautonomous differential equations. See Leroux and Viennot [217, 218, 219, 220], and also Bergeron–Flajolet–Salvy [22], Bergeron–Reutenauer [26, 27], Rawlings [270], and Unger [315]. In particular, classical results (Lie–Gröbner–Taylor expansions, linear differential equations, the method of separation of variables) are analyzed from a combinatorial viewpoint. An application is given in the exercises to the generic solution (with computer algebra techniques) of differential equations occurring in control theory, for example, Duffing equation.

This book has two intended audiences: graduate students and researchers in pure and applied mathematics and computer science. It can be used as a graduate text in enumerative and algebraic combinatorics. The only prerequisites are a good knowledge of basic algebra and analysis and some familiarity with discrete mathematics. It offers a modern introduction to the use of various generating functions in labeled and unlabeled enumeration. The professional researcher in the areas of combinatorics, algebra, analysis, computer science, and probability theory will find combinatorial tools to solve problems occurring in their field.

This book has the following special features:
- A comprehensive presentation of the combinatorial theory of species of structures.
- A clear explanation of when to use the different types of generating series (ordinary, exponential, multivariate, cycle index, asymmetry index, molecular, etc.).
- An extensive study of combinatorial functional and differential equations and their relations with tree-like structures.
- Efficient tools for construction, analysis, classification, and enumeration of various species of discrete structures.
- A review of the connections between the theory of species and algebra via Möbius inversion; symmetric functions; finite group actions; molecular, atomic, and virtual species; and Pólya theory.
- Detailed discussion of various links between combinatorics and classical analysis: families of (orthogonal) polynomials, q-series, Lagrange inversion, implicit function theorem, Newton–Raphson iteration, functional and differential equations, asymptotic analysis, etc.
- Extended reference tools: numerous tables (see Appendix 2), a notation index, a general index, and a bibliography.

Finally, the book contains more than 350 exercises ranging from very easy (e.g., routine computations based on definitions, direct applications of theorems, ...) to very elaborate or difficult (e.g., step by step development of a topic not covered in the text, delicate analysis of hidden properties of complex species of stuctures, ...). They are found at the end of each section and serve as a complement to the theory as well as a "hands on" practice. Accordingly, the statements in most exercises are intentionally very explicit and can often be thought of as "theorems" to be proved. For instance, instead of asking to find a formula in a given context, an exercise may ask to prove (or verify) that an "explicitly stated" formula is true. The most difficult exercises are not really meant to be solved but should be read in order to have a more complete view of the theory and its applications.

Due to space and time limitations, some of the recent and interesting work related to species theory has not been covered in this book. The most notable omissions concern
- Connections with symmetric functions, λ-rings, and binomial rings. See, for instance, Bergeron [17, 18], Bonetti–Rota–Senato–Venezia [32, 33], Flores de Chela–Méndez [105], Joyal [159, 160], Kerber [170], Méndez [246], Nava [255], Senato–Venezia [296], and Yeh [333, 334].
- Extensions to tensorial species and analytic functors, and connections with group representations. See Bergeron–Yeh [29], Joyal [163], Joyal–Street [164], Macdonald [228], and Méndez [245].
- Colored and Möbius species. See Ehrenborg–Méndez [86], Méndez–Nava [247], Méndez–Yang [248], and Yang [331, 332].

- Connections with category theory, Hopf algebras, and umbral calculus. See Chen [49, 50, 51], Dür [84], Loeb [225], Nava–Rota [256], Rajan [267, 268], Ray [272], and Schmitt [292].
- Further developments and applications of species theory. See, for example, Bergeron [14, 19], Bergeron–Sattler [28], Chen [48, 51], Chiricota [53], Chiricota–Labelle [55], Constantineau [59, 61], Constantineau–Labelle [65], Gessel [128], Gessel–Labelle [129], Labelle [182, 184, 185], Labelle–Labelle–Pineau [190], Labelle–Laforest [192], Labelle–Yeh [203, 206], Longtin [226], Pineau [262], Strehl [311, 312], Unger [315], and Yeh [335].

Acknowledgments

We wish to warmly thank all the people who helped us during the preparation of this work and in particular, at the risk of forgetting someone, the following people:
- André Joyal, for having founded the combinatorial theory of species of structures and shown its importance in enumerative and algebraic combinatorics.
- Jean Berstel, for having realized the interest of species of structures in computer science and encouraged the creation of this book.
- Gian-Carlo Rota, for his constant interest in the theory of species and for inviting us to publish the English version of the book in this collection.
- The researchers and collaborators who have been interested in the theory of species of structures and who have participated in its development, including P. Bouchard, Ph. Flajolet, I. Gessel, A. Kerber, J. Labelle, L. Laforest, A. Longtin, D. Rawlings, C. Reutenauer, R.W. Robinson, G.-C. Rota, B. Salvy, S. Schanuel, R.P. Stanley, V. Strehl, G.X. Viennot, T. Walsh, K.H. Wehrhahn, Y.N. Yeh, and D. Zeilberger; as well as D. Foata, A. Garsia, J. Françon, G. Nicoletti, R.C. Read, B. Sagan, D. Stanton, D. White, H.S. Wilf, and A. Zvonkin.
- Young researchers whose dissertations dealt with species of structures, including W.C. Chen, Y. Chiricota, I. Constantineau, H. Décoste, R. Ehrenborg, M. Mendez, O. Nava, K. Pineau, B. Unger, and J. Yang.
- All the students and other colleagues who have read various versions of this book and have made comments and suggestions, including P. Auger, N. Bergeron, M. Bousquet, A. de Medicis, P. Dumais, L. Favreau, J.F. Gagné, A. Goupil, J. Labelle, J.B. Lévesque, G. Melançon, B. Miloudi, A. Nickel, S. Poirier, M. Poulin, E. Roblet, J. Roger, L. Shapiro, and J. Turgeon.
- Margaret Readdy for translating the book from French to English.
- For their more specific aid for
    typing and linguistic corrections: G. Hetyei, L. Remon, and A. Robitaille
    preparation of tables: Y. Chiricota, H. Décoste, M. Ikollo, J. Labelle, and K. Pineau
    preparation of index: A. Robitaille and F. Strasbourg
    preparation of figures: B. Girard and S. Plouffe.

- For their financial support, the Université du Québec à Montréal and the funding councils Fonds pour la formation de chercheurs et l'aide à la recherche (FCAR), Québec, and National Science and Engineering Research Council (NSERC), Canada.
- And above all, for their patience, understanding, and constant encouragement, our life companions Rosamaria, Hélène, and Madeleine.

1

Introduction to Species of Structures

1.0. Introduction

This chapter contains the basic concepts of the combinatorial theory of species of structures. It is an indispensable starting point for the developments and applications presented in the subsequent chapters. We begin with some general considerations on the notion of structure, everywhere present in mathematics and theoretical computer science. These preliminary considerations lead us in a natural manner to the fundamental concept of species of structures.

The definition puts the emphasis on the transport of structures along bijections and is due to C. Ehresmann [87], but it is A. Joyal [158] who showed its effectiveness in the combinatorial treatment of formal power series and for the enumeration of labeled structures as well as unlabeled (isomorphism types of) structures.

We introduce in Section 1.2 some of the first power series that can be associated to species: generating series, types generating series, cycle index series. They serve to encode all the information concerning labeled and unlabeled enumeration.

Sections 1.3 and 1.4 form an introduction to *the algebra* of species of structures. Various combinatorial operations on species of structures are used to produce new ones, in general more complex. The operations introduced here are addition, multiplication, substitution, and differentiation of species of structures. They constitute a combinatorial lifting of the corresponding operations on formal power series. The problems of specification, classification, and enumeration of structures are then greatly simplified, using this algebra of species. Also, this approach reveals a remarkable link between the composition of functions and the plethystic substitution of symmetric functions, in the context of Pólya theory.

1.1. Notion of Species of Structures

The concept of structure is fundamental, recurring in all branches of mathematics, as well as in computer science. From an informal point of view, a *structure s*

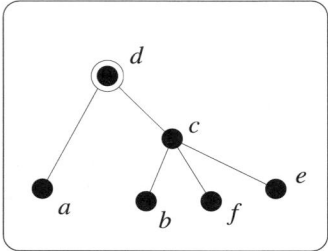
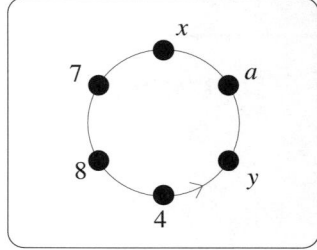

Fig. 1.

is a *construction* γ which one performs on a set U (of data). It consists of a pair

$$s = (\gamma, U). \tag{1}$$

It is customary to say that U is the underlying set of the structure s or even that s is a structure constructed from (or labeled by) the set U. Figure 1 depicts two examples of structures: a rooted tree and an oriented cycle. In a set theoretical fashion, the tree in question can be described as $s = (\gamma, U)$, where

$$U = \{a, b, c, d, e, f\},$$
$$\gamma = (\{d\}, \{\{d, a\}, \{d, c\}, \{c, b\}, \{c, f\}, \{c, e\}\}). \tag{2}$$

The singleton $\{d\}$ which appears as the first component of γ indicates the *root* of this rooted tree. As for the oriented cycle, it can be put in the form $s = (\gamma, U)$, where

$$U = \{x, 4, y, a, 7, 8\},$$
$$\gamma = \{(4, y), (y, a), (a, x), (x, 7), (7, 8), (8, 4)\}. \tag{3}$$

The abuse of notation $s = \gamma$, which consists of identifying a structure $s = (\gamma, U)$ with the construction γ, will be used if it does not cause any ambiguity with regard to the nature of the underlying set U. Here is an example which could give rise to such an ambiguity:

$$s = (\gamma, U) \quad \text{where} \quad U = \{c, x, g, h, m, p, q\}, \quad \gamma = \{x, m, p\}. \tag{4}$$

Here the structure is that of a-subset-of-a-set U. The knowledge of γ alone does not enable one to recover the set U.

A traditional approach to the concept of structure consists of generalizing the preceding examples within axiomatic set theory. However, in the present work we adopt a *functorial* approach which puts an emphasis on the *transport of structures* along bijections. Here is an example which illustrates the concept of transport of structures.

Example 1. Consider the rooted tree $s = (\gamma, U)$ of Figure 1 whose underlying set is $U = \{a, b, c, d, e, f\}$. Replace each element of U by those of

1.1. Notion of Species of Structures

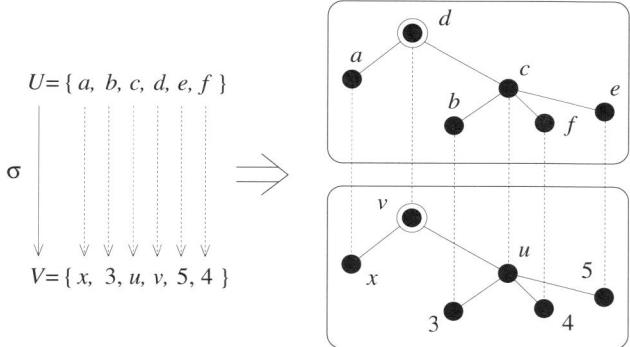

Fig. 2.

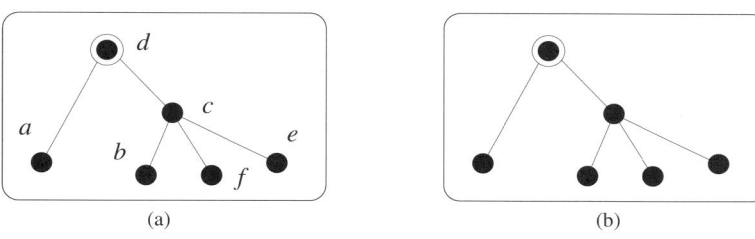

Fig. 3.

$V = \{x, 3, u, v, 5, 4\}$ via the bijection $\sigma : U \longrightarrow V$ described by Figure 2. This figure clearly shows how the bijection σ allows the transport of the rooted tree s onto a corresponding rooted tree $t = (\tau, V)$ on the set V, simply by replacing each vertex $u \in U$ by the corresponding vertex $\sigma(u) \in V$. We say that the rooted tree t has been obtained by transporting the rooted tree s along the bijection σ, and we write

$$t = \sigma \cdot s. \qquad (5)$$

From a purely set theoretical point of view, this amounts to replacing simultaneously each element u of U appearing in γ by the corresponding element $\sigma(u)$ of V in the expression of γ. The rooted trees s and t are said to be *isomorphic*, and σ is called an *isomorphism* of s to t.

Intuitively two isomorphic structures can be considered as identical if the nature of the elements of their underlying sets is ignored. This "general form" that isomorphic structures have in common is their *isomorphism type*. It often can be represented by a diagram (see, for example, Figure 3(b)) in which the elements of the underlying set are represented by "indistinguishable" points. The structure is then said to be *unlabeled*. Figure 4 illustrates a rooted tree *automorphism*. In this case, the sets U and V coincide, the bijection $\sigma : U \longrightarrow U$ is a permutation

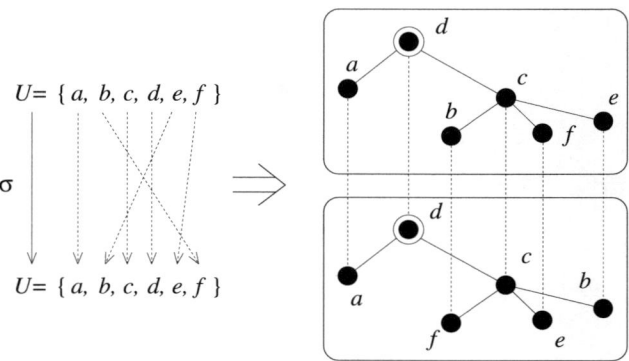

Fig. 4.

of U, and the transported rooted tree $\sigma \cdot s$ is identical to the tree s, that is to say, $s = \sigma \cdot s$.

The preceding examples show that the concept of transport of structures is of prime importance since it enables one to define the notions of isomorphism, isomorphism type, and automorphism. In fact, the transport of structures is at the very base of the general concept of species of structures.

Example 2. As an introduction to the formal definition of species of structures, here is a detailed description of the *species* $\mathcal{G}$ of all *simple graphs* (i.e., undirected graphs without loops or multiple edges). For each finite set U, we denote by $\mathcal{G}[U]$ the set of all structures of simple graph on U. Thus

$$\mathcal{G}[U] = \{g \mid g = (\gamma, U), \gamma \subseteq \wp^{[2]}[U]\}, \qquad (6)$$

where $\wp^{[2]}[U]$ stands for the collection of (unordered) pairs of elements of U. In the simple graph $g = (\gamma, U)$, the elements of U are the *vertices* and γ is the set of *edges*. Clearly $\mathcal{G}[U]$ is a finite set. The following three expressions are considered to be equivalent:
– g is a simple graph on U;
– $g \in \mathcal{G}[U]$;
– g is a $\mathcal{G}$-structure on U.
Moreover, each bijection $\sigma : U \longrightarrow V$ induces, by transport of structure (see Figure 5), a function

$$\mathcal{G}[\sigma] : \mathcal{G}[U] \longrightarrow \mathcal{G}[V], \qquad g \mapsto \sigma \cdot g, \qquad (7)$$

describing the transport of graphs along σ. Formally, if $g = (\gamma, U) \in \mathcal{G}[U]$, then $\mathcal{G}[\sigma](g) = \sigma \cdot g = (\sigma \cdot \gamma, V)$, where $\sigma \cdot \gamma$ is the set of pairs $\{\sigma(x), \sigma(y)\}$ of elements of V obtained from pairs $\{x, y\} \in \gamma$. Thus each edge $\{x, y\}$ of g finds itself relabeled $\{\sigma(x), \sigma(y)\}$ in $\sigma \cdot g$.

1.1. Notion of Species of Structures

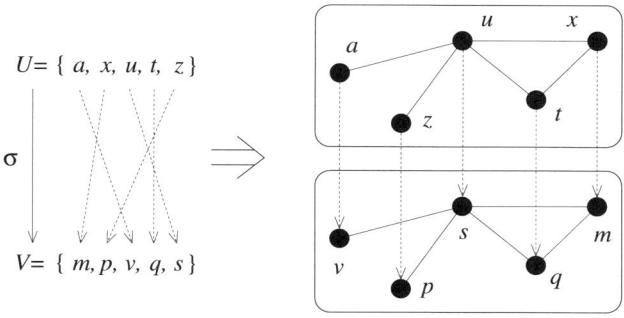

Fig. 5.

Since this transport of graphs along σ is only a relabeling of the vertices and edges by σ, it is clear that for bijections $\sigma : U \longrightarrow V$ and $\tau : V \longrightarrow W$, one has

$$\mathcal{G}[\tau \circ \sigma] = \mathcal{G}[\tau] \circ \mathcal{G}[\sigma], \tag{8}$$

and that, for the identity map $\mathrm{Id}_U : U \longrightarrow U$, one has

$$\mathcal{G}[\mathrm{Id}_U] = \mathrm{Id}_{\mathcal{G}[U]}. \tag{9}$$

These two equalities express the *functoriality* of the transports of structures $\mathcal{G}[\sigma]$. It is this property which is abstracted in the definition of species of structures.

Definition of Species of Structures

Definition 3. A *species of structures* is a rule F which
 i) *produces*, for each finite set U, a finite set $F[U]$,
 ii) *produces*, for each bijection $\sigma : U \longrightarrow V$, a function

$$F[\sigma] : F[U] \longrightarrow F[V]. \tag{10}$$

The functions $F[\sigma]$ should further satisfy the following functorial properties:
a) for all bijections $\sigma : U \longrightarrow V$ and $\tau : V \longrightarrow W$,

$$F[\tau \circ \sigma] = F[\tau] \circ F[\sigma], \tag{11}$$

b) for the identity map $\mathrm{Id}_U : U \longrightarrow U$,

$$F[\mathrm{Id}_U] = \mathrm{Id}_{F[U]}. \tag{12}$$

An element $s \in F[U]$ is called an F-*structure on* U (or even a *structure of species F on U*). The function $F[\sigma]$ is called the *transport* of F-structures *along σ*.

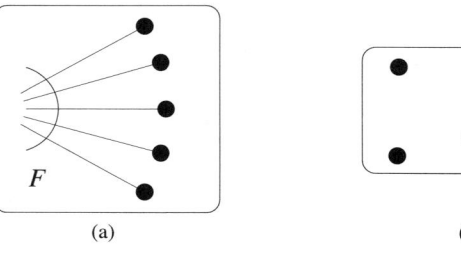

Fig. 6.

As for graphs, the three following expressions are considered to be equivalent:
- s is a structure of species F on U;
- $s \in F[U]$;
- s is an F-structure on U.

It immediately follows from the functorial properties that each transport function $F[\sigma]$ is necessarily a bijection (see Exercise 2). We use the notation $\sigma \cdot s$, or sometimes $\sigma \cdot_F s$ to avoid ambiguity, to designate $F[\sigma](s)$.

In order to represent a generic F-structure, we often utilize drawings like those of Figures 6(a) and 6(b). The black dots in these figures represent the (distinct) elements of the underlying set. The F-structure itself is represented in 6(a) by a circular arc labeled F, and in 6(b) by the superposition of the symbol F.

Observe that the notions of isomorphism, isomorphism type, and automorphism of F-structures are implicitly contained in the definition of the species F.

Definition 4. Consider two F-structures $s_1 \in F[U]$ and $s_2 \in F[V]$. A bijection $\sigma : U \longrightarrow V$ is called an *isomorphism* of s_1 to s_2 if $s_2 = \sigma \cdot s_1 = F[\sigma](s_1)$. One says that these structures have the same *isomorphism type*. Moreover, an isomorphism from s to s is said to be an *automorphism* of s.

The advantage of this definition of species is that the rule F which produces the structures $F[U]$ and the transport functions $F[\sigma]$ can be described in any fashion, provided that the functoriality conditions (11) and (12) are satisfied. For example, one can either use axiomatic systems, explicit constructions, algorithms, combinatorial operations, functional equations, or even simple geometric figures to specify a species. We will illustrate each of these approaches with some examples.

Examples of Species Defined Using Set Theoretic Axioms

A species F can be defined by means of a system $\mathfrak{A}$ of well-chosen axioms by requiring

$$s = (\gamma, U) \in F[U] \quad \text{if and only if} \quad s = (\gamma, U) \text{ is a model of } \mathfrak{A}.$$

The transport $F[\sigma]$ is carried out in the natural fashion illustrated earlier. Clearly one can introduce in this manner a multitude of species, including the following (see Exercise 4):
- the species $\mathcal{A}$, of *rooted trees*;[1]
- the species $\mathcal{G}$, of *simple graphs*;
- the species $\mathcal{G}^c$, of *connected simple graphs*;
- the species $\mathfrak{a}$, of *trees* (connected simple graphs without cycles);[1]
- the species $\mathcal{D}$, of *directed graphs*;
- the species Par, of *set partitions*;
- the species $\wp$, of *subsets*, i.e.,

$$\wp[U] = \{S \mid S \subseteq U\}; \tag{13}$$

- the species End, of *endofunctions*, i.e.,

$$\text{End}[U] = \{\psi \mid \psi : U \longrightarrow U\}; \tag{14}$$

- the species Inv, of *involutions*, i.e., those endofunctions ψ such that $\psi \circ \psi = \text{Id}$;
- the species $\mathcal{S}$, of *permutations* (i.e., bijective endofunctions);
- the species $\mathcal{C}$, of *cyclic permutations* (or oriented cycles);
- the species L, of *linear* (or *total*) *orders*, etc.

For example, identifying each endofunction ψ with its (γ, U), we can describe the species End of all endofunctions as follows: $\psi = (\gamma, U) \in \text{End}[U]$ if and only if

$$\gamma \subseteq U \times U \quad \text{and} \quad (\forall x)[(x \in U) \implies (\exists ! y)[(y \in U) \quad \text{and} \quad ((x, y) \in \gamma)]]. \tag{15}$$

Directed graphs γ satisfying (15) are called *functional digraphs*. We also say that γ is the *sagittal graph* of the endofunction ψ.

Note that the transport $\text{End}[\sigma] : \text{End}[U] \longrightarrow \text{End}[V]$ along the bijection $\sigma : U \longrightarrow V$ is given by the formula

$$\text{End}[\sigma](\psi) = \sigma \circ \psi \circ \sigma^{-1}, \tag{16}$$

for each $\psi \in \text{End}[U]$. Indeed, upon setting $\theta = \text{End}[\sigma](\psi) \in \text{End}[V]$, the pairs $(u, \psi(u))$ run over the functional digraph determined by ψ if and only if the pairs $(\sigma(u), \sigma(\psi(u)))$ run over the sagittal graph determined by θ. Moreover, the relation $v = \sigma(u)$ is equivalent to the relation $u = \sigma^{-1}(v)$. We then deduce that the functional digraph of θ is given by pairs of the form $(v, \sigma \circ \psi \circ \sigma^{-1}(v))$ with $v \in V$.

Examples of Explicit Constructions

When the structures of a species F are particularly simple or not numerous, it can be advantageous to define the species by an explicit description of the sets $F[U]$

---

[1] Translator's note: *arbre* and *arborescence* are the French words for tree and rooted tree, respectively; hence the notation $\mathfrak{a}$ and $\mathcal{A}$.

and transport functions $F[\sigma]$. The following species fall under this category. In each case the transport of structures $F[\sigma]$ is obvious.
- The species E, of *sets*, defined by $E[U] = \{U\}$. For each finite set U, there is a unique E-structure, namely the set U itself.
- The species ε, of *elements*, defined by $\varepsilon[U] = U$, where the structures on U are the elements of U.
- The species X, characteristic of *singletons*, defined by

$$X[U] = \begin{cases} \{U\}, & \text{if } |U| = 1, \\ \emptyset, & \text{otherwise.} \end{cases} \tag{17}$$

Here $\emptyset$ denotes the empty set. As a consequence, there are no X-structures on a set U when $|U| \neq 1$.
- The species 1, characteristic of the *empty set*, defined by

$$1[U] = \begin{cases} \{U\}, & \text{if } U = \emptyset, \\ \emptyset, & \text{otherwise.} \end{cases} \tag{18}$$

- The *empty* species, denoted by 0, defined by $0[U] = \emptyset$ for all U.
- The species E_2, characteristic of sets of cardinality 2, defined by

$$E_2[U] = \begin{cases} \{U\}, & \text{if } |U| = 2, \\ \emptyset, & \text{otherwise.} \end{cases} \tag{19}$$

Example of Algorithmic Construction

One can specify structures in an algorithmic fashion. For instance, an algorithm can be given which generates all the binary rooted trees on a given set of vertices. If we designate this algorithm by $\mathcal{B}$, and by $\mathcal{B}[U]$ the set of structures produced by $\mathcal{B}$ for a given U, then we have

Algorithm 5. INPUT: A FINITE SET U; OUTPUT: THE SET $\mathcal{B}[U]$ OF BINARY ROOTED TREES ON U.
1) $\mathcal{B}[U] := \{\emptyset\}$, if the set U is empty;
2) $\mathcal{B}[U] :=$ the set of triples (g, x, d),
   obtained by choosing in all possible ways:
  a) x an element of U;
  b) S a subset of $U \setminus \{x\}$;
     then let $T :=$ the complement of S in $U \setminus \{x\}$;
  c) g an element of $\mathcal{B}[S]$;
  d) d an element of $\mathcal{B}[T]$.
End.

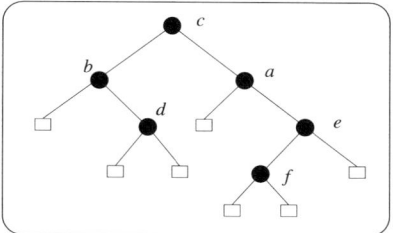

Fig. 7.

For each bijection $\sigma: U \longrightarrow V$ and each binary rooted tree $(g, x, d) \in \mathcal{B}[U]$, algorithm $\mathcal{B}$ also produces a transport of structures, by setting
1) $\mathcal{B}[\sigma](\emptyset) := \emptyset$, if $U = \emptyset$;
2) $\mathcal{B}[\sigma](g, x, d) := (l, y, r)$ with:
   a) $y := \sigma(x)$;
   b) $l := \mathcal{B}[\sigma](g)$;
   c) $r := \mathcal{B}[\sigma](d)$.
End.

For example, one of the structures produced by algorithm $\mathcal{B}$ on the set

$$U = \{a, b, c, d, e, f\}$$

is

$$((\emptyset, b, (\emptyset, d, \emptyset)), c, (\emptyset, a, ((\emptyset, f, \emptyset), e, \emptyset)))$$

(see Figure 7 for a representation of this binary rooted tree). The transport of this structure along the bijection

$$\sigma : \{a, b, c, d, e, f\} \longrightarrow \{A, B, C, D, E, F\},$$

which replaces each letter by its corresponding capital letter, clearly gives

$$\mathcal{B}[\sigma]((\emptyset, b, (\emptyset, d, \emptyset)), c, (\emptyset, a, ((\emptyset, f, \emptyset), e, \emptyset)))$$
$$= ((\emptyset, B, (\emptyset, D, \emptyset)), C, (\emptyset, A, ((\emptyset, F, \emptyset), E, \emptyset))).$$

In general, species of structures satisfying a "functional equation" can be defined in an algorithmic or recursive manner. See below and in Chapter 3.

Examples of Species Defined Using Combinatorial Operations

Another way of producing species of structures is by applying operations to known species. These operations (addition, multiplication, substitution, differentiation, etc.) will be described in detail in Sections 1.3, 1.4, 2.1, and 2.2. Here are some examples:

- The species E^3, of *tri-colorings*, $\qquad E^3 = E \cdot E \cdot E.$ (20)
- The species E_+, of *nonempty sets*, $\qquad 1 + E_+ = E.$ (21)
- The species $\mathcal{H}$, of *hedges* (or lists) *of rooted trees*, $\qquad \mathcal{H} = L(\mathcal{A}).$ (22)
- The species Der, of *derangements*, $\qquad E \cdot \text{Der} = \mathcal{S}.$ (23)
- The species Bal, of *ballots* (ordered partitions), $\qquad \text{Bal} = L(E_+).$ (24)

Examples of Functional Equations

It frequently happens that a species of structures is described or characterized recursively by a functional equation. Here are some examples:

- The species $\mathcal{A}$, of *rooted trees*, $\qquad \mathcal{A} = X \cdot E(\mathcal{A}).$ (25)
- The species L, of *linear orders*, $\qquad L = 1 + X \cdot L.$ (26)
- The species $\mathcal{A}_L$, of *ordered rooted trees*, $\qquad \mathcal{A}_L = X \cdot L(\mathcal{A}_L).$ (27)
- The species $\mathcal{B}$, of *binary rooted trees*, $\qquad \mathcal{B} = 1 + X \cdot \mathcal{B}^2.$ (28)
- The species $\mathcal{P}$, of *commutative parenthesizations*, $\qquad \mathcal{P} = X + E_2(\mathcal{P}).$ (29)

The interpretation and analysis of these equations is the object of Chapter 3. Describing species of structures with the help of functional equations plays a central role in this book.

Geometric Descriptions

One can sometimes gain in simplicity or clarity by describing a species F with the help of one (or several) figure(s) which schematically represents a typical F-structure. Figure 8 represents a typical structure belonging to the species P of polygons (i.e., nonoriented cycles) on a set of cardinality 5. By definition, $P[U]$ is the set of polygons on U.

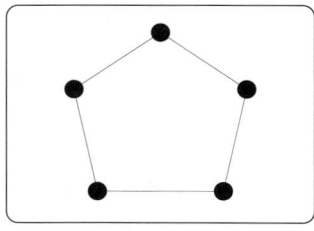

Fig. 8.

Remark 6. The reader who is familiar with category theory will have observed that a species is simply a functor

$$F : \mathbb{B} \longrightarrow \mathbb{E} \tag{30}$$

from the category $\mathbb{B}$ of finite sets and bijections to the category $\mathbb{E}$ of finite sets and functions. Although knowledge of category theory is not necessary in order to read this book, the interested reader is encouraged to consult a basic text on category theory, such as Maclane [229].

Exercises for Section 1.1

1. Verify the *functoriality* of the transport of graphs along bijections, i.e., show that the transport functions $\mathcal{G}[\sigma]$ satisfy Equations (8) and (9).
2. Let F be a species of structures and let $\sigma : U \longrightarrow V$ be any bijection between finite sets. Use functoriality to show that the transport function $F[\sigma] : F[U] \longrightarrow F[V]$ is necessarily a bijection.
   HINT: Show that $F[\sigma]^{-1} = F[\sigma^{-1}]$.
3. Describe the transport functions for the following species: $E, \varepsilon, X, 1$, and 0.
4. Show how to define, with the help of axioms, the following species: $\mathcal{G}^c$, $\mathcal{D}$, $\mathfrak{a}$, $\wp$, Par, Inv, $\mathcal{S}, \mathcal{C}, L$. In each case describe the transport of structures.
5. Figure 9 describes a structure belonging to the species Cha of *chains* (non-oriented). Describe rigorously this species.
6. a) For all integers $n \geq 0$, designate by $\mathcal{S}_n$ the symmetric group formed of permutations (bijections) of $[n] = \{1, 2, \ldots, n\}$, under the operation of composition. Show that every species of structures F induces, for each $n \geq 0$, an action

   $$\mathcal{S}_n \times F[n] \longrightarrow F[n], \tag{31}$$

   of the group $\mathcal{S}_n$ on the set $F[n]$ of F-structures on $[n]$, by setting $\sigma \cdot s = F[\sigma](s)$ for $\sigma \in \mathcal{S}_n$ and $s \in F[n]$. The definition of group action is reviewed in Appendix 1.
   b) Conversely, show that any family of set actions

   $$(\mathcal{S}_n \times F_n \longrightarrow F_n)_{n \geq 0} \tag{32}$$

   allows the definition of a species of structures F for which the families of actions in (31) and (32) are isomorphic.

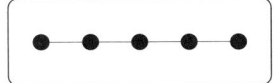

Fig. 9. A chain.

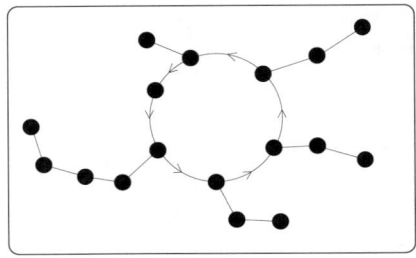

Fig. 10. An octopus.

7. The geometric Figure 10 describes a structure belonging to the species Oct of *octopuses* (see Bergeron [20]). Describe rigorously this species.
8. Let F be a species of structures. Prove that the relation "is isomorphic to" is an equivalence relation on the totality of F-structures. More precisely, prove that for all F-structures s, t, and r, one has
   a) $s \sim s$
   b) $s \sim t \implies t \sim s$
   c) $s \sim t$ and $t \sim r \implies s \sim r$,
   where $s \sim t$ signifies that there exists an isomorphism from s to t.
9. a) List all trees on each of the following sets:

   $$\emptyset, \{1\}, \{1, 2\}, \{1, 2, 3\}, \text{ and } \{1, 2, 3, 4\}.$$

   b) List all isomorphism types of trees (i.e., unlabeled trees) having at most 7 vertices.
10. Let L be the species of linear orders and $\mathcal{S}$ be that of permutations.
    a) Show that for all $n \geq 0$ the number of L-structures on a set of cardinality n and the number of $\mathcal{S}$-structures on a set of cardinality n are both equal to $n!$.
    b) Show that for all $n \geq 2$ the number of isomorphism types of L-structures on a set of cardinality n is strictly less than the number of types of $\mathcal{S}$-structures on a set of cardinality n.

1.2. Associated Series

We will now associate to each species of structures F three important formal power series related to the enumeration of F-structures. An F-structure $s \in F[U]$ on a set U is often refered to as a *labeled* structure, whereas an *unlabeled* structure is an isomorphism class of F-structures. The three series are
– the (exponential) *generating series* of F, denoted $F(x)$, for labeled enumeration,
– the *type generating series* of F, denoted $\tilde{F}(x)$, for unlabeled enumeration,
– the *cycle index series* of F, denoted $Z_F(x_1, x_2, x_3, \ldots)$, as a general enumeration tool.

1.2. Associated Series

These series serve to "encode" all the information concerning the enumeration of labeled or unlabeled F-structures (i.e., up to isomorphism). Note that for all finite sets U, the number of F-structures on U only depends upon the number of elements of U (and not on the elements of U). In other words,

$$|F[U]| \text{ only depends upon } |U|.$$

This property immediately follows from the earlier observation (see Exercise 1.2) that transport functions $F[\sigma]$ are always bijections. Hence, the cardinalities $|F[U]|$ are completely characterized by the sequence of values $f_n = |F[\{1, 2, \ldots, n\}]|$, $n \geq 0$.

For ease in notation, we write
- $[n]$ to designate the set $\{1, 2, 3, \ldots, n\}$,
- $F[n]$ to designate the set $F[\{1, 2, \ldots, n\}]$, rather than $F[[n]]$.

Generating Series of a Species of Structures

Definition 1. The *generating series* of a species of structures F is the formal power series

$$F(x) = \sum_{n=0}^{\infty} f_n \frac{x^n}{n!}, \tag{1}$$

where $f_n = |F[n]|$ = the number of F-structures on a set of n elements (labeled structures).

Note that this series is of exponential type in the indeterminate x in the sense that $n!$ appears in the denominator of the term of degree n. The series $F(x)$ is also called the *exponential generating series* of the species F.

The following notation is used to designate the coefficients of formal power series. For an ordinary formal power series

$$G(x) = \sum_{n \geq 0} g_n x^n, \tag{2}$$

we set

$$[x^n] G(x) = g_n. \tag{3}$$

For a formal power series of exponential type, of the form (1), we then have

$$n! \, [x^n] F(x) = f_n. \tag{4}$$

Taking the Taylor expansion (at the origin) of $F(x)$ shows that

$$n! \, [x^n] F(x) = \left. \frac{d^n F(x)}{dx^n} \right|_{x=0}. \tag{5}$$

More generally, for a formal power series in any number of variables expressed in the form

$$H(x_1, x_2, x_3, \ldots) = \sum_{n_1,n_2,n_3,\ldots} h_{n_1,n_2,n_3,\ldots} \frac{x_1^{n_1} x_2^{n_2} x_3^{n_3} \cdots}{c_{n_1,n_2,n_3,\ldots}}, \tag{6}$$

where $c_{n_1,n_2,n_3,\ldots}$ is a given family of nonzero scalars, we have

$$c_{n_1,n_2,n_3,\ldots} \left[x_1^{n_1} x_2^{n_2} x_3^{n_3} \cdots \right] H(x_1, x_2, x_3, \ldots) = h_{n_1,n_2,n_3,\ldots}. \tag{7}$$

Example 2. Referring to species described earlier, it is easy to verify by direct enumeration the following identities:

$$\begin{aligned}
&\text{a)} \quad L(x) = \frac{1}{1-x}, &&\text{b)} \quad S(x) = \frac{1}{1-x}, \\
&\text{c)} \quad C(x) = -\log(1-x), &&\text{d)} \quad E(x) = e^x, \\
&\text{e)} \quad \epsilon(x) = xe^x, &&\text{f)} \quad \wp(x) = e^{2x}, \\
&\text{g)} \quad X(x) = x, &&\text{h)} \quad 1(x) = 1, \\
&\text{i)} \quad 0(x) = 0, &&\text{j)} \quad \mathcal{G}(x) = \sum_{n \geq 0} 2^{\binom{n}{2}} \frac{x^n}{n!}, \\
&\text{k)} \quad \mathcal{D}(x) = \sum_{n \geq 0} 2^{n^2} \frac{x^n}{n!}, &&\text{l)} \quad \text{End}(x) = \sum_{n \geq 0} n^n \frac{x^n}{n!}.
\end{aligned} \tag{8}$$

The computation of the generating series for other species $\mathcal{G}^c$, Par, Inv, $\mathcal{A}$, etc. which have been mentioned earlier is less direct. It will be done after the introduction of combinatorial operations on species of structures.

Type Generating Series

Let us now consider the enumeration of isomorphism types of F-structures. We may restrict ourselves to structures on sets of the form $U = \{1, 2, \ldots, n\} = [n]$. One defines an equivalence relation $\sim$ on the set $F[n]$ by setting, for $s, t \in F[n]$,

$s \sim t$ if and only if s and t have the same isomorphism type.

In other words (see Definition 1.1.4), $s \sim t$ if and only if there exists a permutation $\pi : [n] \longrightarrow [n]$ such that $F[\pi](s) = t$.

By definition, an *isomorphism type of F-structures of order n* is an equivalence class (modulo $\sim$) of F-structures on $[n]$. Such an equivalence class is also called an *unlabeled F-structure* of order n. Denote by $T(F_n)$ the quotient set $F[n]/\sim$ of types of F-structures of order n and let

$$T(F) = \sum_{n \geq 0} T(F_n). \tag{9}$$

1.2. Associated Series

Definition 3. The *(isomorphism) type generating series* of a species of structures F is the formal power series

$$\widetilde{F}(x) = \sum_{n \geq 0} \widetilde{f}_n x^n, \qquad (10)$$

where $\widetilde{f}_n = |T(F_n)|$ is the number unlabeled F-structures of order n.

The notation $F\widetilde{\ }(x)$ will sometimes be used for typographical reasons. Note that this is an ordinary formal power series (i.e., without factorials in the denominators) in one indeterminate x acting as a point counter.

Example 4. Direct calculation yields the following type generating series (see Exercise 3):

a) $\widetilde{L}(x) = \dfrac{1}{1-x}$,   b) $\widetilde{S}(x) = \displaystyle\prod_{k=1}^{\infty} \dfrac{1}{1-x^k}$,   c) $\widetilde{C}(x) = \dfrac{x}{1-x}$,

d) $\widetilde{E}(x) = \dfrac{1}{1-x}$,   e) $\widetilde{\epsilon}(x) = \dfrac{x}{1-x}$,   f) $\widetilde{\wp}(x) = \dfrac{1}{(1-x)^2}$,

g) $\widetilde{X}(x) = x$,   h) $\widetilde{1}(x) = 1$,   i) $\widetilde{0}(x) = 0$.

$$(11)$$

Despite the fact that the generating series of the species L and S coincide, $L(x) = S(x) = 1/(1-x)$, equality does not hold for the type generating series:

$$\widetilde{L}(x) \neq \widetilde{S}(x). \qquad (12)$$

This provides evidence that the species L and S are not the same. Indeed, total orders and permutations are not transported in the same manner along bijections. In particular, a total order only admits a single automorphism, whereas in general a permutation admits many automorphisms. Thus there is an essential difference between permutations π of a set U of cardinality n and lists without repetition $\pi_1 \pi_2 \cdots \pi_n$ of the elements of U. Of course, if the set U happens to be given a fixed order, one can establish a bijection (depending on this order) between permutations and lists (see Example 13).

Cycle Index Series

In general, explicit or recursive calculation of type generating series is difficult. It requires the use of combinatorial operations on species of structures and of a third kind of series associated with each species F, the *cycle index series* of F, denoted by Z_F. This is a formal power series in an infinite number of variables $x_1, x_2, x_3, \ldots$. It contains more information than both series $F(x)$ and $\widetilde{F}(x)$. We first define the *cycle type* of a permutation.

Definition 5. Let U be a finite set and σ, a permutation of U. The *cycle type* of the permutation σ is the sequence $(\sigma_1, \sigma_2, \sigma_3, \ldots)$, where for $k \geq 1$, $\sigma_k =$ is the number of cycles of length k in the decomposition of σ into disjoint cycles.

Observe that σ_1 is the number of fixed points of σ. Moreover, if $|U| = n$ then $\sigma_k = 0$ if $k \geq n$. The cycle type of σ can then be written in the form of a vector with n components, $(\sigma_1, \sigma_2, \ldots, \sigma_n)$. We use the following notation:

$$\text{Fix } \sigma = \{u \in U \mid \sigma(u) = u\},$$
$$\text{fix } \sigma = |\text{Fix } \sigma|. \tag{13}$$

Fix σ denotes the set of fixed points of σ, whereas fix $\sigma = \sigma_1$ denotes the number of fixed points of σ. Figure 1 shows a permutation of type $(3, 4, 0, 3, 2)$. Now let F be any species. Each permutation σ of U induces, by transport of structures, a permutation $F[\sigma]$ of the set $F[U]$ of F-structures on U. Consider, for illustrative purposes, the species Inv of involutions (i.e., the endofunctions ψ such that $\psi \circ \psi = Id$) and the permutation σ of the set $U = \{a, b, c, d, e\}$ given by Figure 2.

This permutation σ induces a permutation Inv$[\sigma]$ on Inv$[U]$ given by Figure 3 in which involutions are represented by simple graphs with each vertex having degree ≤ 1. The permutation σ is of type $(0, 1, 1)$ and permutes 5 points, while the

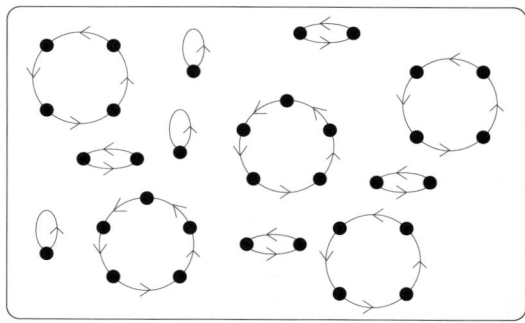

Fig. 1.

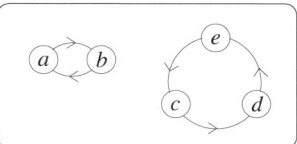

Fig. 2.

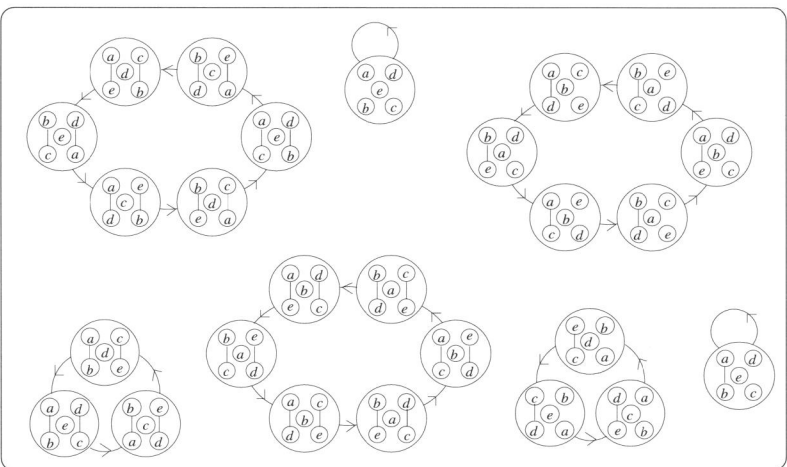

Fig. 3.

permutation $\text{Inv}[\sigma]$ is of type $(2, 0, 2, 0, 0, 3)$ permuting the 26 involutions of U. In particular, we have $\text{fix}\,\text{Inv}[\sigma] = 2$.

Definition 6. The *cycle index series* of a species of structures F is the formal power series (in an infinite number of variables $x_1, x_2, x_3 \cdots$)

$$Z_F(x_1, x_2, x_3, \ldots) = \sum_{n \geq 0} \frac{1}{n!} \left(\sum_{\sigma \in \mathcal{S}_n} \text{fix}\, F[\sigma]\, x_1^{\sigma_1} x_2^{\sigma_2} x_3^{\sigma_3} \cdots \right), \qquad (14)$$

where $\mathcal{S}_n$ denotes the group of permutations of $[n]$ (i.e., $\mathcal{S}_n = \mathcal{S}[n]$) and fix $F[\sigma] = (F[\sigma])_1 = $ is the number of F-structures on $[n]$ fixed by $F[\sigma]$, i.e., the number of F-structures on $[n]$ for which σ is an automorphism.

Example 7. Without the help of various techniques developed in the following sections, direct calculation of cycle index series can only be carried out in very simple cases. For instance, for the species $0, 1, X, L, \mathcal{S}, E, \mathcal{E}$, we have

a) $Z_0(x_1, x_2, x_3, \ldots) = 0$,

b) $Z_1(x_1, x_2, x_3, \ldots) = 1$,

c) $Z_X(x_1, x_2, x_3, \ldots) = x_1$,

d) $Z_L(x_1, x_2, x_3, \ldots) = \dfrac{1}{1 - x_1}$,

e) $Z_\mathcal{S}(x_1, x_2, x_3, \ldots) = \dfrac{1}{(1 - x_1)(1 - x_2)(1 - x_3)\cdots}$, (15)

f) $Z_E(x_1, x_2, x_3, \ldots) = \exp\left(x_1 + \dfrac{x_2}{2} + \dfrac{x_3}{3} + \cdots\right)$,

g) $Z_\varepsilon(x_1, x_2, x_3, \ldots) = x_1 \exp\left(x_1 + \dfrac{x_2}{2} + \dfrac{x_3}{3} + \cdots\right)$.

The notion of cycle index series Z_F gives a simultaneous generalization of both the series $F(x)$ and $\widetilde{F}(x)$. In fact, we have the following fundamental theorem.

Theorem 8. *For any species of structures F, we have*

a) $F(x) = Z_F(x, 0, 0, \ldots)$,

b) $\widetilde{F}(x) = Z_F(x, x^2, x^3, \ldots)$. (16)

Proof. To establish a), proceed as follows. Substituting $x_1 = x$ and $x_i = 0$, for all $i \geq 2$, in Equation (14) gives

$$Z_F(x, 0, 0, \ldots) = \sum_{n \geq 0} \frac{1}{n!} \left(\sum_{\sigma \in \mathcal{S}_n} \operatorname{fix} F[\sigma] x^{\sigma_1} 0^{\sigma_2} 0^{\sigma_3} \cdots \right).$$

Now for each fixed value of $n \geq 0$, $x^{\sigma_1} 0^{\sigma_2} 0^{\sigma_3} \cdots = 0$, except if $\sigma_1 = n$ and $\sigma_i = 0$ for $i \geq 2$. In other words, only the identity permutations $\sigma = \operatorname{Id}_n$ contribute to the sum. Thus

$$Z_F(x, 0, 0, \ldots) = \sum_{n \geq 0} \frac{1}{n!} \operatorname{fix} F[\operatorname{Id}_n] x^n$$

$$= \sum_{n \geq 0} \frac{1}{n!} f_n x^n$$

$$= F(x),$$

since all F-structures are fixed by transport along the identity.

Equality b) is based upon a lemma of Cauchy–Frobenius (alias Burnside, see Appendix 1). Indeed, we have

$$Z_F(x, x^2, x^3, \ldots) = \sum_{n \geq 0} \frac{1}{n!} \sum_{\sigma \in S_n} \text{fix } F[\sigma] x^{\sigma_1} x^{2\sigma_2} x^{3\sigma_3} \cdots$$

$$= \sum_{n \geq 0} \frac{1}{n!} \sum_{\sigma \in S_n} \text{fix } F[\sigma] x^n$$

$$= \sum_{n \geq 0} |F[n]/\sim| x^n$$

$$= \widetilde{F}(x). \qquad \blacksquare$$

Example 9. As an illustration of Theorem 8, consider the case of the species S of permutations. It immediately follows from (15), e) that

$$Z_S(x, 0, 0, \ldots) = \frac{1}{1-x} = S(x),$$

$$Z_S(x, x^2, x^3, \ldots) = \frac{1}{(1-x)(1-x^2)(1-x^3)\cdots} = \widetilde{S}(x),$$
(17)

in agreement with the formulas given earlier for $S(x)$ and $\widetilde{S}(x)$.

Remark 10. In examining Figures 2 and 3, one is easily convinced that the cycle type of $\text{Inv}[\sigma]$ is independent of the nature of the points of U and only depends on the type of σ. This is a general phenomenon. Indeed, for all species F and all permutations σ of U, the cycle type $((F[\sigma])_1, (F[\sigma])_2, \ldots)$ of $F[\sigma]$ only depends on the cycle type $(\sigma_1, \sigma_2, \ldots)$ of σ (see Exercise 5). In particular, the number of fixed points of the permutation $F[\sigma]$, given by

$$\text{fix } F[\sigma] = |\text{Fix } F[\sigma]| = (F[\sigma])_1,$$

only depends on the numbers $\sigma_1, \sigma_2, \ldots$. Hence, in the definition of cycle index series (14), all permutations σ having the same cycle type $(\sigma_1, \sigma_2, \sigma_3, \ldots)$ contribute to the same monomial in the variables $x_1, x_2, x_3, \ldots$. In order to eliminate this redundancy, we regroup the monomials of the index series which correspond to each of these types. Since the number of permutations σ of n elements, of type $(n_1, n_2, n_3, \ldots)$, is given by

$$\frac{n!}{1^{n_1} n_1! \, 2^{n_2} n_2! \, 3^{n_3} n_3! \cdots},$$
(18)

we obtain, after simplification of the $n!$, the following variant for the definition of the index series of any species F:

$$Z_F(x_1, x_2, x_3, \ldots) = \sum_{n_1+2n_2+3n_3+\cdots<\infty} \text{fix } F[n_1, n_2, n_3, \ldots] \frac{x_1^{n_1} x_2^{n_2} x_3^{n_3} \cdots}{1^{n_1} n_1! \, 2^{n_2} n_2! \, 3^{n_3} n_3! \cdots}.$$
(19)

Here fix $F[n_1, n_2, n_3, \ldots]$ denotes the number of F-structures on a set of $n = \sum_{i \geq 1} i n_i$ elements which are fixed under the action of any (given) permutation of type $(n_1, n_2, n_3, \ldots)$. In other words, introducing the compact notation $\mathbf{n} = (n_1, n_2, n_3, \ldots)$,

$$\mathrm{aut}(\mathbf{n}) = 1^{n_1} n_1! \, 2^{n_2} n_2! \, 3^{n_3} n_3! \cdots,$$

and $\mathrm{coeff}_\mathbf{n}$, we have

$$\mathrm{fix}\, F[\mathbf{n}] = \mathrm{coeff}_\mathbf{n} Z_F = \mathrm{aut}(\mathbf{n})\left[x_1^{n_1} x_2^{n_2} x_3^{n_3} \ldots\right] Z_F(x_1, x_2, x_3, \ldots). \tag{20}$$

Combinatorial Equality

To conclude the present section, we discuss various concepts of *equality* which one encounters in the theory of species of structures. Strictly speaking, two species F and G are *equal* or *identical* if they have the same structures and the same transports: For all finite sets U, $F[U] = G[U]$, and for all bijections $\sigma : U \longrightarrow V$, $F[\sigma] = G[\sigma]$. However, this concept of identity is very restrictive. A much weaker version of *equality* between species is that of equipotence. It is obtained by replacing the set equalities $F[U] = G[U]$ by bijections $F[U] \longrightarrow G[U]$.

Definition 11. Let F and G be two species of structures. An *equipotence* α of F to G is a family of bijections α_U, where for each finite set U,

$$\alpha_U : F[U] \xrightarrow{\sim} G[U]. \tag{21}$$

The two species F and G are then called *equipotent*, and one writes $F \equiv G$.

In other words, $F \equiv G$ if and only if there is the same number of F-structures as G-structures on all finite sets U. For example, the species $\mathcal{S}$ of permutations is equipotent to the species L of linear orders since one has $|\mathcal{S}[U]| = |U|! = |L[U]|$ for all finite sets U. Clearly,

$$F \equiv G \quad \Leftrightarrow \quad F(x) = G(x). \tag{22}$$

The concept of equipotence is useful when one is only interested in the enumeration of labeled structures. However, it turns out to be inadequate when one wants to enumerate the isomorphism types of structures. Indeed,

$$\begin{aligned} F \equiv G &\quad \not\Rightarrow \quad \widetilde{F}(x) = \widetilde{G}(x), \\ F \equiv G &\quad \not\Rightarrow \quad Z_F = Z_G, \end{aligned} \tag{23}$$

as has been already observed for the species $\mathcal{S}$ and L.

The "good" notion of equality between species of structures lies halfway between identity and equipotence. It is the concept of isomorphism of species. It requires that the family of bijections $\alpha_U : F[U] \longrightarrow G[U]$ satisfy an additional condition relative to the transport of structures, called the *naturality condition*.

Definition 12. Let F and G be two species of structures. An *isomorphism* of F to G is a family of bijections $\alpha_U : F[U] \longrightarrow G[U]$ which satisfies the following *naturality condition*: For any bijection $\sigma : U \longrightarrow V$ between two finite sets, the following diagram commutes:

$$\begin{array}{ccc} F[U] & \xrightarrow{\alpha_U} & G[U] \\ {\scriptstyle F[\sigma]}\downarrow & & \downarrow{\scriptstyle G[\sigma]} \\ F[V] & \xrightarrow{\alpha_V} & G[V] \end{array} \qquad (24)$$

In other words, for any F-structure $s \in F[U]$, one must have $\sigma \cdot \alpha_U(s) = \alpha_V(\sigma \cdot s)$. The two species F and G are then said to be *isomorphic*, and one writes $F \simeq G$.

Informally, the naturality condition means that, for any F-structure s on U, the corresponding G-structure $\alpha_U(s)$ on U can be described without appealing to the nature of the elements of U. Although much weaker than the concept of identity, the concept of isomorphism is nevertheless compatible with the transition to series (see Exercise 9) in the sense that

$$F \simeq G \quad \Rightarrow \quad \begin{cases} F(x) = G(x), \\ \widetilde{F}(x) = \widetilde{G}(x), \\ Z_F(x_1, x_2, x_3, \ldots) = Z_G(x_1, x_2, x_3, \ldots). \end{cases} \qquad (25)$$

We will have many occasions to verify that two isomorphic species essentially possess the "same" combinatorial properties. Henceforth they will be considered as *equal* in the combinatorial algebra developed in the next sections. Thus we write $F = G$ in place of $F \simeq G$, and say that there is a *combinatorial equality* between the species F and G.

Example 13. There exist many classic bijections showing that the species L and S are equipotents. These bijections $\varphi_U : L[U] \to S[U]$ are all based on a linear order $\leq_U$ given a priori on the underlying set U. The most common, when $U = [n]$, consists of identifying the list

$$(\sigma(1), \sigma(2), \ldots, \sigma(n))$$

with the bijection $i \mapsto \sigma(i)$. Another, called the *fundamental transformation* (see Foata [45] or Knuth [172]), is particularly elegant. Here is the description. Given a list

$$\lambda = (u_1, u_2, \ldots, u_i, \ldots, u_n) \in L[U],$$

let $i_1, i_2, \ldots, i_k$ be the increasing sequence of indices for which the u_{i_j} are the minimum from left to right (records) according to the order $\leq_U$. That is to say

$$u_{i_j} = \min\{u_i \mid i \leq i_j\}, \ j = 1, \ldots, k.$$

In particular, $i_1 = 1$. One then defines $\tau = \varphi_U(\lambda) \in S[U]$ as being the permutation whose disjoint cycle decomposition is

$$\tau = (u_1, \ldots, u_{i_2-1})(u_{i_2}, \ldots, u_{i_3-1}) \ldots (u_{i_k}, \ldots, u_n).$$

For example, for $\lambda = (5, 9, 7, 3, 8, 1, 4, 6, 2) \in L[9]$, the minima from left to right are 5, 3, 1 and $i_1 = 1, i_2 = 4, i_3 = 6$, so that

$$\tau = \varphi_{[9]}(\lambda) = (5, 9, 7)(3, 8)(1, 4, 6, 2),$$

or, rewriting the cycles according to increasing order of their minimum elements,

$$\tau = (1, 4, 6, 2)(3, 8)(5, 9, 7).$$

This is the so-called *standard* form for τ.

Conversely, to recover λ from τ written in standard form, it suffices to write the cycles of τ in decreasing order of their minimum elements, then removing the parentheses.

The fundamental transformation $\lambda \to \tau = \varphi_U(\lambda)$ has the advantage of preserving a large part of the functional digraph of these structures. It is compatible with the transport of structures along increasing bijections $\sigma : (U, \leq_U) \longrightarrow (V, \leq_V)$, in the sense that $\sigma \cdot \varphi_U(\lambda) = \varphi_V(\sigma \cdot \lambda)$. However, this is not the case for an arbitrary bijection $\sigma : U \longrightarrow V$ since the species L and S are not isomorphic (see Exercise 9, d)).

Contact of Order n

Here is a last notion, more topological, of equality, that is extremely useful when constructing species of structures by successive approximations. It is the concept of *contact of order n* between species of structures, for an integer $n \geq 0$.

Recall that given two formal power series $a(x) = \sum_{n \geq 0} a_n x^n$ and $b(x) = \sum_{n \geq 0} b_n x^n$, one says that $a(x)$ and $b(x)$ have *contact of order n*, and one writes

$a(x) =_n b(x)$, if for all $k \le n$, $[x^k]a(x) = [x^k]b(x)$. In other words, letting $a_{\le n}(x) = \sum_{0 \le k \le n} a_k x^k$, one has

$$a(x) =_n b(x) \iff a_{\le n}(x) = b_{\le n}(x). \tag{26}$$

Contact of order n for index series of the form

$$h(x_1, x_2, x_3, \ldots) = \sum_{n_1 + 2n_2 + 3n_3 + \cdots} h_{n_1 n_2 n_3 \ldots} x_1^{n_1} x_2^{n_2} x_3^{n_3} \cdots \tag{27}$$

is defined in a similar fashion by setting

$$h_{\le n}(x_1, x_2, x_3, \ldots) = \sum_{n_1 + 2n_2 + 3n_3 + \cdots \le n} h_{n_1 n_2 n_3 \ldots} x_1^{n_1} x_2^{n_2} x_3^{n_3} \cdots . \tag{28}$$

By analogy, one has the following definition for species of structures.

Definition 14. Let F and G be two species of structures and n, an integer ≥ 0. One says that F and G have *contact of order n*, and one writes

$$F =_n G, \tag{29}$$

if the combinatorial equality

$$F_{\le n} = G_{\le n}$$

is valid, where $F_{\le n}$ denotes the restriction of F to sets of cardinality $\le n$. More precisely, for finite sets U and V, and a bijection $\sigma : U \longrightarrow V$, set

$$\begin{aligned} F_{\le n}[U] &= \emptyset, \text{ if } |U| > n, \\ F_{\le n}[U] &= F[U] \text{ and } F_{\le n}[\sigma] = F[\sigma], \text{ if } |U| \le n. \end{aligned} \tag{30}$$

It is clear that when species F and G have contact of order n, their associated series also have contact of order n:

$$F =_n G \implies \begin{cases} F(x) =_n G(x), \\ \widetilde{F}(x) =_n \widetilde{G}(x), \\ Z_F(x_1, x_2, x_3, \ldots) =_n Z_G(x_1, x_2, x_3, \ldots). \end{cases} \tag{31}$$

Definition 15. LIMIT OF A SEQUENCE OF SPECIES OF STRUCTURES. A sequence $(F_n)_{n \ge 0}$ of species of structures is said to *converge to* a species F, written as

$$\lim_{n \to \infty} F_n = F, \tag{32}$$

if for any integer $N \ge 0$, there exists $K \ge 0$ such that for all $n \ge K$, $F_n =_N F$.

As an example, one trivially has for any species F

$$\lim_{n \to \infty} F_{\leq n} = F. \tag{33}$$

This concept of limit is compatible with passage to the associated series (see Exercise 12).

Exercises for Section 1.2

1. Verify, by direct enumeration, formulas (8) for the generating series of the species $L, S, C, E, \in, \wp, X, 1, 0, \mathcal{G}, \mathcal{D}$, and End.
2. Verify that the type generating series of the species Par and S coincide.
3. Verify, by direct enumeration, formulas (11) for the type generating series for the species $L, S, C, E, \in, \wp, X, 1$, and 0.
4. Consider two finite sets U and V such that $|U| = |V|$.
   a) Show that for all linear orders $s \in L[U]$ and $t \in L[V]$ there exists a unique bijection $\sigma : U \longrightarrow V$ such that $L[\sigma](s) = t$.
   b) Show that two permutations $\alpha \in S[U]$ and $\beta \in S[V]$ are isomorphic if and only if they have the same cycle type.
5. Prove that for any species F and any permutation σ of U, the cycle type

$$((F[\sigma])_1, (F[\sigma])_2, (F[\sigma])_3, \ldots)$$

of $F[\sigma]$ only depends on the type $(\sigma_1, \sigma_2, \sigma_3, \ldots)$ of σ (and does not depend on the nature of the elements of U).

HINT: Use the functoriality of F.

6. Let $(n_1, n_2, n_3, \ldots)$ be a sequence of integers satisfying the condition $\sum_{i \geq 1} i n_i = n$. Prove that the number of permutations σ of type $(n_1, n_2, n_3, \ldots)$ of a set with n elements is given by the expression

$$\frac{n!}{1^{n_1} n_1! 2^{n_2} n_2! 3^{n_3} n_3! \cdots}. \tag{34}$$

HINT: Show that the number of automorphisms (i.e., permutations τ such that $\sigma = \tau^{-1} \sigma \tau$) of a permutation σ of cycle type $\mathbf{n} = (n_1, n_2, n_3, \ldots)$ is aut($\mathbf{n}$) and use formula (A.1.13) of Appendix 1 for the size of an orbit.

7. Starting from the definition of cycle index series of a species, verify the formulas (15) for the index series of the species $0, 1, X, L, S, E$, and ϵ.
8. Verify the formulas $F(x) = Z_F(x, 0, 0, \ldots)$ and $\widetilde{F}(x) = Z_F(x, x^2, x^3, \ldots)$ for the case of the following species: $0, 1, X, L, S, E$, and ϵ.
9. a) Verify that $F \equiv G$ if and only if $F(x) = G(x)$.
   b) Describe two distinct but isomorphic species.
   c) Show that $F \simeq G$ implies $F(x) = G(x), \widetilde{F}(x) = \widetilde{G}(x)$, and $Z_F = Z_G$.
   d) Conclude from this that the species L and S are not isomorphic.

10. a) Show that the extraction of the coefficients in Equations (3), (4), and (7) define linear transformations.
    b) Express the linear transformation defined by Equation (7) in terms of products of differential operators.
11. a) Let $n \geq 0$ and $u(t) \in \mathbb{K}[\![t]\!]$, the ring of formal power series in t with coefficients in $\mathbb{K}$. Show that
$$[t^n]u(t) = 0$$
if and only if there exists $w(t) \in \mathbb{K}[\![t]\!]$ such that
$$u(t) = tw'(t) - nw(t). \tag{35}$$
   b) Let $n \geq k \geq 0$. Also, let $u(t)$ and $v(t)$ be in $K[\![t]\!]$. Show that
$$n! \, [t^n] \, u(t) = k! \, [t^k] \, v(t)$$
if and only if there exists a $w(t) \in \mathbb{K}[\![t]\!]$ such that
$$n_{\langle n-k \rangle} u(t) - t^{n-k} v(t) = tw'(t) - nw(t), \tag{36}$$
where $\lambda_{\langle m \rangle} := \lambda(\lambda - 1) \cdots (\lambda - m + 1)$, if $m > 0$, and $\lambda_{\langle 0 \rangle} = 1$.
   c) Let $n_1, n_2, \ldots, n_k$ be integers ≥ 0, and $h(x_1, x_2, \ldots, x_k) \in \mathbb{K}[\![x_1, x_2 \ldots x_k]\!]$. Show that
$$[x_1^{n_1} x_2^{n_2} \cdots x_k^{n_k}] h(x_1, x_2, \ldots, x_k) = 0$$
if and only if there exists $w(x_1, \ldots, x_k) \in \mathbb{K}[[x_1, \ldots, x_k]]$ such that
$$h(x_1, \ldots, x_k) = \left\{ \left(x_1 \frac{\partial}{\partial x_1} - n_1 \right)^2 + \cdots + \left(x_k \frac{\partial}{\partial x_k} - n_k \right)^2 \right\} w(x_1, \ldots, x_k). \tag{37}$$
   d) State and prove a necessary and sufficient condition analogous to b) for the equality
$$c_{n_1, \ldots, n_k} [x_1^{n_1} x_2^{n_2} \cdots x_k^{n_k}] u(x_1, x_2, \ldots, x_k)$$
$$= c_{m_1, \ldots, m_k} [x_1^{m_1} x_2^{m_2} \cdots x_k^{m_k}] v(x_1, x_2, \ldots, x_k).$$

12. We say that a sequence of formal power series $a_n(x)$ converges to a power series $a(x)$, and we write $\lim_{n \to \infty} a_n(x) = a(x)$ if for any integer $N > 0$, there exists $K > 0$ such that
$$n \geq K \implies a_n(x) =_N a(x).$$
   a) Using the concept of contact of order n for the index series established in (28), define the notion of limit of a sequence of index series.

b) Show that for two species of structures $(F_n)_{n\geq 0}$ and F,

$$\lim_{n\to\infty} F_n = F \implies \begin{cases} \lim_{n\to\infty} F_n(x) = F(x), \\ \lim_{n\to\infty} \widetilde{F}_n(x) = \widetilde{F}(x), \\ \lim_{n\to\infty} Z_{F_n}(x_1, x_2, x_3, \ldots) = Z_F(x_1, x_2, x_3, \ldots). \end{cases} \tag{38}$$

1.3. Addition and Multiplication

We now introduce several operations on species of structures. There results a *combinatorial algebra*, allowing the construction and analysis of a multitude of species, as well as the calculation of associated series (generating series and cycle index series). These operations between species often constitute combinatorial analogs of the usual operations, addition (+), multiplication (·), substitution (∘), and differentiation (′) on their exponential generating functions.

In the algebraic context of formal power series in one variable x, given two series of exponential type

$$f = f(x) = \sum_{n=0}^{\infty} f_n \frac{x^n}{n!} \quad \text{and} \quad g = g(x) = \sum_{n=0}^{\infty} g_n \frac{x^n}{n!}, \tag{1}$$

Table 1 recalls the general coefficient h_n of the series

$$h = h(x) = \sum_{n=0}^{\infty} h_n \frac{x^n}{n!} \tag{2}$$

constructed from f and g in the following cases:

$$h = f + g, \quad h = f \cdot g, \quad h = f \circ g = f(g), \quad h = \frac{d}{dx} f = f'. \tag{3}$$

Table 1

Operation	Coefficient h_n
$h = f + g$	$h_n = f_n + g_n$
$h = f \cdot g$	$h_n = \sum_{i+j=n} \frac{n!}{i!\,j!} f_i g_j$
$h = f \circ g$ $(g(0) = 0)$	$h_n = \sum_{\substack{0 \leq k \leq n \\ n_1 + \cdots + n_k = n}} \frac{n!}{k!\,n_1! \ldots n_k!} f_k\, g_{n_1} \cdots g_{n_k}$
$h = f'$	$h_n = f_{n+1}$

By analogy, let us now consider two species of structures F and G and consider the problem of constructing some other species, denoted by

$$F + G, \quad F \cdot G, \quad F \circ G, \quad \text{and} \quad F',$$

in order to have, for the corresponding generating series,

a) $(F + G)(x) = F(x) + G(x)$,

b) $(F \cdot G)(x) = F(x) G(x)$,

c) $(F \circ G)(x) = F(G(x))$,

d) $F'(x) = \dfrac{d}{dx} F(x)$.

(4)

These equalities between generating series signify that the new species $F + G$, $F \cdot G$, $F \circ G$, and F' should be defined so that the enumeration of their structures depends "solely" on the enumeration of the F- and G-structures, via the following formulas:

1) the number of $(F + G)$-structures on n elements is

$$|(F + G)[n]| = |F[n]| + |G[n]|;$$

2) the number of $(F \cdot G)$-structures on n elements is

$$|(F \cdot G)[n]| = \sum_{i+j=n} \frac{n!}{i!\,j!} |F[i]|\, |G[j]|;$$

3) the number of $(F \circ G)$-structures on n elements is

$$|(F \circ G)[n]| = \sum_{j=0}^{n} \sum_{\substack{n_1+n_2+\cdots+n_j=n \\ n_i > 0}} \frac{1}{j!} \binom{n}{n_1\, n_2\, \cdots\, n_j} |F[j]| \prod_{i=1}^{j} |G[n_i]|;$$

4) the number of F'-structures on n elements is

$$|F'[n]| = |F[n+1]|.$$

There could exist, a priori, many candidates for these definitions. However, there are very natural solutions which are, moreover, compatible with transport of structures. As we will see, this proves to be fundamental, in particular for the calculation of index series. We only consider, in the present section, the operations of addition and multiplication. Substitution and derivation will be treated in Section 1.4.

Sum of Species of Structures

As a motivating example, let us consider the species $\mathcal{G}^c$ of connected simple graphs and the species $\mathcal{G}^d$ of disconnected (i.e., empty or having at least two connected components) simple graphs. Since every graph is either connected or disconnected, the following equality evidently holds for any finite set U:

$$\mathcal{G}[U] = \mathcal{G}^c[U] + \mathcal{G}^d[U] \qquad \text{(set theoretical disjoint union)}, \tag{5}$$

where $\mathcal{G}$ designates the species of all simple graphs. We then say that the species $\mathcal{G}$ is the sum of the species $\mathcal{G}^c$ and $\mathcal{G}^d$, and we write

$$\mathcal{G} = \mathcal{G}^c + \mathcal{G}^d. \tag{6}$$

This example serves as prototype for the general definition of addition of species:

Definition 1. Let F and G be two species of structures. The species $F + G$, called the *sum* of F and G, is defined as follows: an $(F + G)$-structure on U is an F-structure on U or (exclusive) a G-structure on U. In other words, for any finite set U, one has

$$(F + G)[U] = F[U] + G[U] \qquad \text{(disjoint union)}. \tag{7}$$

The transport along a bijection $\sigma : U \longrightarrow V$ is carried out by setting, for any $(F + G)$-structure s on U,

$$(F + G)[\sigma](s) = \begin{cases} F[\sigma](s) & \text{if } s \in F[U], \\ G[\sigma](s) & \text{if } s \in G[U]. \end{cases} \tag{8}$$

In a pictorial fashion, any $(F + G)$-structure can be represented by Figure 2.

Remark 2. In the case where certain F-structures are also G-structures (i.e., $F[U] \cap G[U] \neq \emptyset$), one must at first form distinct copies of the sets $F[U]$ and $G[U]$. A standard way of distinguishing the F-structures from the G-structures is to replace the set $F[U]$ by the isomorphic set $F[U] \times \{1\}$ and $G[U]$ by $G[U] \times \{2\}$, and to set

$$(F + G)[U] = (F[U] \times \{1\}) \cup (G[U] \times \{2\}). \tag{9}$$

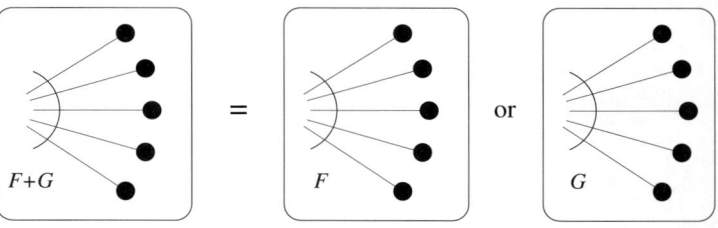

Fig. 2.

The operation of addition is associative and commutative, up to isomorphism. Moreover, the empty species 0 (not having any structure: $0[U] = \emptyset$) is a neutral element for addition:

$$F + 0 = 0 + F = F, \tag{10}$$

for all species F. We leave it to the reader to verify these properties as well as the following proposition.

Proposition 3. *Given two species of structures F and G, the associated series of the species $F + G$ satisfy the equalities*

a) $(F + G)(x) = F(x) + G(x)$,

b) $\widetilde{(F + G)}(x) = \widetilde{F}(x) + \widetilde{G}(x)$, \hfill (11)

c) $Z_{F+G} = Z_F + Z_G$.

Example 4. Let E_{even} (respectively E_{odd}) be the species of sets containing an even number (respectively odd number) of elements. Then $E = E_{\text{even}} + E_{\text{odd}}$ and Proposition 3 takes the form of the equalities (see Exercise 3)

a) $e^x = \cosh(x) + \sinh(x)$,

b) $\dfrac{1}{(1-x)} = \dfrac{1}{(1-x^2)} + \dfrac{x}{(1-x^2)}$, \hfill (12)

c) $\exp\left(x_1 + \dfrac{x_2}{2} + \dfrac{x_3}{3} + \cdots\right) = e^{\left(\frac{x_2}{2} + \frac{x_4}{4} + \cdots\right)}(\cosh(x_1 + x_3/3 + \cdots)$
$\qquad + \sinh(x_1 + x_3/3 + \cdots))$.

The operation of addition can be extended to *summable* families of species as follows:

Definition 5. A family $(F_i)_{i \in I}$ of species of structures is said to be *summable* if for any finite set U, $F_i[U] = \emptyset$, except for a finite number of indices $i \in I$. The sum of a summable family $(F_i)_{i \in I}$ is the species $\sum_{i \in I} F_i$ defined by the equalities

a) $\left(\sum_{i \in I} F_i\right)[U] = \sum_{i \in I} F_i[U] = \bigcup_{i \in I} F_i[U] \times \{i\}$, \hfill (13)

b) $\left(\sum_{i \in I} F_i\right)[\sigma](s, i) = (F_i[\sigma](s), i)$, \hfill (14)

where $\sigma : U \longrightarrow V$ is a bijection and $(s, i) \in (\sum_{i \in I} F_i)[U]$.

We leave to the reader the task of verifying that $\sum_{i \in I} F_i$, defined in this way, is indeed a species of structures, that the families of associated series are summable (see Exercise 2), and that one has

a) $\left(\sum_{i \in I} F_i \right)(x) = \sum_{i \in I} F_i(x),$

b) $\left(\widetilde{\sum_{i \in I} F_i} \right)(x) = \sum_{i \in I} \widetilde{F}_i(x),$ (15)

c) $Z_{(\sum_{i \in I} F_i)} = \sum_{i \in I} Z_{F_i}.$

Example 6. CANONICAL DECOMPOSITION. Each species F gives rise canonically to an enumerable family $(F_n)_{n \geq 0}$ of species defined by setting, for each $n \in \mathbb{N}$,

$$F_n[U] = \begin{cases} F[U], & \text{if } |U| = n, \\ \emptyset, & \text{otherwise}, \end{cases} \quad (16)$$

with the obvious induced transports. We say that F_n is the species F *restricted to cardinality* n. The family $(F_n)_{n \geq 0}$ is clearly summable and we obtain the following *canonical decomposition*:

$$F = F_0 + F_1 + F_2 + \cdots + F_n + \cdots. \quad (17)$$

In the case where $F = F_k$ (i.e., $F_n = 0$ for $n \neq k$), we say that F is *concentrated on the cardinality* k.

Taking, for example, the species P of polygons (introduced in Section 1.1), we obtain $P = P_0 + P_1 + P_2 + \cdots + P_n + \cdots$, where P_n designates the species of all n-gons. In an analogous fashion, E_n is the species of sets of cardinality n (in particular $E_0 = 1$ and $E_1 = X$). One has the combinatorial equality

$$E = E_0 + E_1 + E_2 + \cdots + E_n + \cdots, \quad (18)$$

which is reflected, in terms of the associated series, by the identities

a) $e^x = 1 + \dfrac{x}{1!} + \dfrac{x^2}{2!} + \cdots + \dfrac{x^n}{n!} + \cdots,$

b) $\dfrac{1}{(1-x)} = 1 + x + x^2 + \cdots + x^n + \cdots,$ (19)

c) $\exp\left(x_1 + \dfrac{x_2}{2} + \dfrac{x_3}{3} + \cdots \right) = \displaystyle\sum_{n \geq 0} \sum_{k_1 + 2k_2 + 3k_3 + \cdots = n} \dfrac{x_1^{k_1} x_2^{k_2} x_3^{k_3} \cdots}{1^{k_1} k_1! 2^{k_2} k_2! 3^{k_3} k_3! \cdots}.$

Other examples of infinite sums of species are given by the formulas

$$\mathcal{S} = \sum_{k \geq 0} \mathcal{S}^{[k]} \quad \text{and} \quad \text{Par} = \sum_{k \geq 0} \text{Par}^{[k]}, \tag{20}$$

where $\mathcal{S}^{[k]}$ denotes the species of permutations having exactly k cycles and $\text{Par}^{[k]}$, the species of partitions having exactly k blocks (or classes). The *finite* sum $F + F + \cdots + F$ of n copies of the same F is often denoted by nF. Clearly one has

$$\begin{align} \text{a)} \quad & (nF)(x) = nF(x), \\ \text{b)} \quad & (\widetilde{nF})(x) = n\widetilde{F}(x), \tag{21} \\ \text{c)} \quad & Z_{nF} = nZ_F. \end{align}$$

The particular case where $F = 1$ (the empty set species) gives rise to the species

$$n = \underbrace{1 + 1 + \cdots + 1}_{n} = n \cdot 1 \tag{22}$$

which possesses exactly n structures on the set $U = \emptyset$ and no structure on any set $U \neq \emptyset$. Consequently the natural numbers themselves are embedded in the combinatorial algebra of species of structures.

Product of Species of Structures

Let us examine the permutation described by Figure 3. We can divide this structure into two disjoint structures:
i) a set of fixed points (those having a loop);
ii) a derangement of the remaining elements (i.e., the permutation without fixed points formed by the nontrivial cycles).
Figure 3 illustrates this dichotomy. An analogous decomposition clearly exists for any permutation. We say that the species $\mathcal{S}$ of permutations is the *product* of the

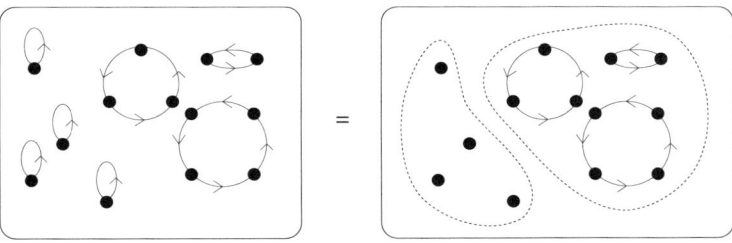

Fig. 3.

species E of sets with the species Der of derangements and we write

$$\mathcal{S} = E \cdot \text{Der}. \tag{23}$$

This is a typical example of the product of species of structures. In general, the operation of *multiplication* is defined as follows.

Definition 7. Let F and G be two species of structures. The species $F \cdot G$ (also denoted FG), called the *product* of F and G, is defined as follows: an $(F \cdot G)$-structure on U is an ordered pair $s = (f, g)$ where
a) f is an F-structure on a subset $U_1 \subseteq U$;
b) g is a G-structure on a subset $U_2 \subseteq U$;
c) (U_1, U_2) is a decomposition of U, i.e., $U_1 \cup U_2 = U$ and $U_1 \cap U_2 = \emptyset$.
In other words, for any finite set U,

$$(F \cdot G)[U] = \sum_{(U_1, U_2)} F[U_1] \times G[U_2], \tag{24}$$

the disjoint sum being taken over all pairs (U_1, U_2) forming a decomposition of U.

The transport along a bijection $\sigma : U \longrightarrow V$ is carried out by setting, for each $(F \cdot G)$-structure $s = (f, g)$ on U,

$$(F \cdot G)[\sigma](s) = (F[\sigma_1](f), G[\sigma_2](g)), \tag{25}$$

where $\sigma_i = \sigma|_{U_i}$ is the restriction of σ on U_i, $i = 1, 2$.

More informally, an $(F \cdot G)$-structure is an ordered pair formed by an F-structure and a G-structure over complementary disjoint subsets. A general $(F \cdot G)$-structure can be represented by Figure 4 or by Figure 5.

The multiplication of species is associative and commutative up to isomorphism, but in general $F \cdot G$ and $G \cdot F$ are not identical. These properties are easily established by constructing appropriate (coherent, see Maclane [230]) isomorphisms. The multiplication admits the species 1 as neutral element, and the species 0 as absorbing element, i.e.,

$$1 \cdot F = F \cdot 1 = F \quad \text{and} \quad F \cdot 0 = 0 \cdot F = 0. \tag{26}$$

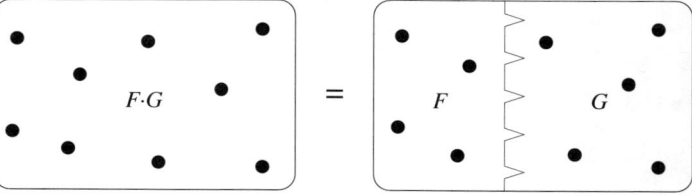

Fig. 4.

1.3. Addition and Multiplication

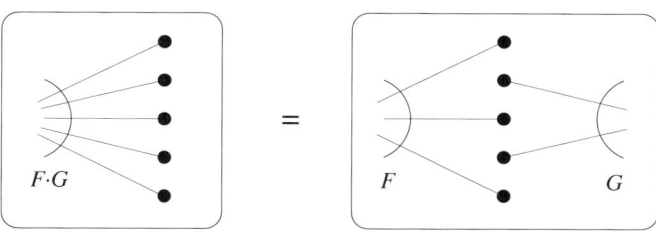

Fig. 5.

Moreover, multiplication distributes over addition. We leave to the reader to prove these properties as well as the following proposition (see Exercise 4).

Proposition 8. *Let F and G be two species of structures. Then the series associated with the species $F \cdot G$ satisfy the equalities*

a) $(F \cdot G)(x) = F(x)G(x)$,

b) $(\widetilde{F \cdot G})(x) = \widetilde{F}(x)\widetilde{G}(x)$, (27)

c) $Z_{F \cdot G}(x_1, x_2, x_3, \ldots) = Z_F(x_1, x_2, x_3, \ldots) Z_G(x_1, x_2, x_3, \ldots)$.

Example 9. The preceding proposition, when applied to combinatorial Equation (23), yields the equalities

a) $\dfrac{1}{1-x} = e^x \operatorname{Der}(x)$,

b) $\displaystyle\prod_{k \geq 1} \dfrac{1}{1-x^k} = \dfrac{1}{1-x} \widetilde{\operatorname{Der}}(x)$, (28)

c) $\displaystyle\prod_{k \geq 1} \dfrac{1}{1-x_k} = \exp\left(x_1 + \dfrac{x_2}{2} + \dfrac{x_3}{3} + \cdots\right) Z_{\operatorname{Der}}(x_1, x_2, x_3, \ldots)$.

One deduces (by simple division) the following expressions for the series associated to the species Der of derangements:

a) $\operatorname{Der}(x) = \dfrac{e^{-x}}{1-x}$,

b) $\widetilde{\operatorname{Der}}(x) = \displaystyle\prod_{k \geq 2} \dfrac{1}{1-x^k}$, (29)

c) $Z_{\operatorname{Der}}(x_1, x_2, x_3, \ldots) = e^{-(x_1 + \frac{x_2}{2} + \frac{x_3}{3} + \cdots)} \displaystyle\prod_{k \geq 1} \dfrac{1}{1-x_k}$.

Note that the classical formula

$$d_n = n!\left(1 - \dfrac{1}{1!} + \dfrac{1}{2!} - \cdots + \dfrac{(-1)^n}{n!}\right),$$ (30)

giving the number of derangements of a set of n elements, is directly obtained from (29), a) by explicitly carrying out the product

$$\sum_{n\geq 0} d_n \frac{x^n}{n!} = \left(\sum_{i\geq 0}(-1)^i \frac{x^i}{i!}\right)\left(\sum_{j\geq 0} x^j\right). \tag{31}$$

The reader can calculate explicitly, starting from (29), c), the numbers

$$\text{fix Der}[n_1, n_2, \ldots],$$

coefficients of the index series Z_{Der} (see Exercise 6). As we see, the simple combinatorial equality $S = E \cdot \text{Der}$ contains structural information which goes well beyond the simple enumeration of the labeled structures.

It is interesting to observe that the species $F + F + \cdots + F$ (n terms), which is denoted nF, is also the product of the species n with the species F. This is to say

$$nF = n \cdot F. \tag{32}$$

Once more, this justifies identifying the integer n with the species n.

Recall the species $\wp$ of subsets of a set, introduced in (1.1.13). We have the combinatorial equality

$$\wp = E \cdot E. \tag{33}$$

Translating this equality into series, we recover the equalities

$$\wp(x) = e^x e^x = e^{2x}, \qquad \widetilde{\wp}(x) = \frac{1}{(1-x)^2}, \tag{34}$$

mentioned in Section 1.2. For the index series $Z_\wp$, we immediately obtain

$$Z_\wp(x_1, x_2, x_3, \ldots) = \left(\exp\left(x_1 + \frac{x_2}{2} + \frac{x_3}{3} + \cdots\right)\right)^2, \tag{35}$$

and deduce the expression

$$\text{fix } \wp[n_1, n_2, \ldots] = 2^{n_1+n_2+\cdots}, \tag{36}$$

which can also be obtained by a direct combinatorial argument.

The species $\wp^{[k]}$ of subsets of cardinality k satisfies the combinatorial equality

$$\wp^{[k]} = E_k \cdot E, \tag{37}$$

where E_k denotes the species of sets of cardinality k. A simple passage to the associated series then yields the equalities

a) $\wp^{[k]}(x) = e^x \dfrac{x^k}{k!}$,

b) $\widetilde{\wp^{[k]}}(k) = \dfrac{x^k}{1-x}$, (38)

c) $Z_{\wp^{[k]}}(x_1, x_2, x_3, \ldots) = \exp\left(x_1 + \dfrac{x_2}{2} + \dfrac{x_3}{3} + \cdots\right)$

$$\times \sum_{n_1 + 2n_2 + \cdots = k} \dfrac{x_1^{n_1} x_2^{n_2} x_3^{n_3} \cdots}{1^{n_1} n_1! 2^{n_2} n_2! 3^{n_3} n_3! \cdots}.$$

The equality $\wp^{[k]}(x) = e^x \dfrac{x^k}{k!}$ gives the well-known combinatorial interpretation of binomial coefficients,

$$|\wp^{[k]}[n]| = \binom{n}{k}, \tag{39}$$

as the number of k-element subsets of a n-element set. Moreover, the explicit formula for the numbers fix $\wp^{[k]}[n_1, n_2, n_3 \cdots]$ given in Exercise 7 constitutes a generalization of the notion of binomial coefficients. Of course, we also have the combinatorial equality

$$\sum_{k \geq 0} \wp^{[k]} = \wp = E^2, \tag{40}$$

which, by passing to generating series, gives the identity

$$\sum_{k \geq 0} \binom{n}{k} = 2^n. \tag{41}$$

By virtue of associativity, the operation of multiplication can be extended to finite families F_i of species, $i = 1, \ldots, k$, by defining the product $F_1 \cdot F_2 \cdot \ldots \cdot F_k$ by

$$(F_1 \cdot F_2 \cdot \ldots \cdot F_k)[U] = \sum_{U_1 + U_2 + \cdots + U_k = U} F_1[U_1] \times F_2[U_2] \times \cdots \times F_k[U_k].$$
(42)

Here the (disjoint) sum is taken over all families $(U_i)_{1 \leq i \leq k}$ of pairwise disjoint subsets of U whose union is U. The transport along a bijection $\sigma : U \longrightarrow V$ is defined in the following *component-wise* manner: For $s_i \in F_i[U_i]$, $i = 1, \ldots, k$, set

$$(F_1 \cdot F_2 \cdot \ldots \cdot F_k)[\sigma]((s_i)_{1 \leq i \leq k}) = (F_i[\sigma_i](s_i))_{1 \leq i \leq k}, \tag{43}$$

where $\sigma_i = \sigma|_{U_i}$ denotes the restriction of σ to U_i. The product $\prod_{i \in I} F_i$ of an infinite family of species $(F_i)_{i \in I}$ can also be defined provided that this family is *multipliable* (see Exercise 9). An example of a multipliable infinite family is given in Example 11.

In the case of a finite family $(F_i)_{1 \leq i \leq k}$ where all of its members are equal to the same species F, the product $F \cdot F \cdots F$ (k factors) is denoted F^k. An F^k-structure on a set U is therefore a k-tuple $(s_1, s_2, \ldots, s_k)$ of disjoint F-structures whose union of underlying sets is U.

Example 10. Taking $F = E_+$, the species of nonempty sets, we obtain the species

$$\mathrm{Bal}^{[k]} = (E_+)^k \tag{44}$$

of all ballots having k levels (i.e., ordered partitions having k blocks; see Figure 6). We therefore have

a) $\mathrm{Bal}^{[k]}(x) = (e^x - 1)^k$,

b) $\widetilde{\mathrm{Bal}^{[k]}}(x) = \left(\dfrac{x}{1-x}\right)^k$, \hfill (45)

c) $Z_{\mathrm{Bal}^{[k]}}(x_1, x_2, x_3, \ldots) = \left(\exp\left(x_1 + \dfrac{x_2}{2} + \dfrac{x_3}{3} + \cdots\right) - 1\right)^k$.

Since the family $(E_+)^k$, $k = 0, 1, 2, \ldots$, is summable, one obtains by summation (see also (1.1.24)) the species Bal of all ballots (independent of the number of levels):

$$\mathrm{Bal} = \sum_{k \geq 0} \mathrm{Bal}^{[k]} = \sum_{k \geq 0} (E_+)^k.$$

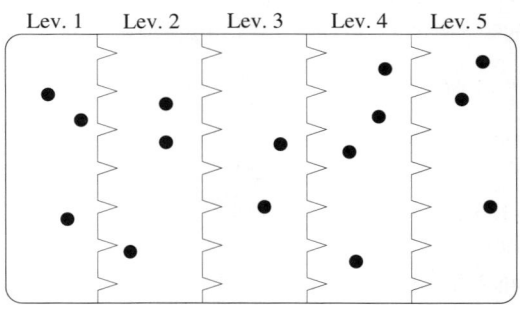

Fig. 6. A Bal$^{[5]}$-structure.

1.3. Addition and Multiplication

A simple summation of the associated series gives

a) $\text{Bal}(x) = \dfrac{1}{2 - e^x}$,

b) $\widetilde{\text{Bal}}(x) = \dfrac{1-x}{1-2x}$, (46)

c) $Z_{\text{Bal}}(x_1, x_2, x_3, \ldots) = \dfrac{1}{2 - \exp\left(x_1 + \dfrac{x_2}{2} + \dfrac{x_3}{3} + \cdots\right)}$.

Example 11. Consider the species L of linear orderings and its restriction L_k to sets of length k. One has the combinatorial equalities

a) $L_k = X^k$, $\quad k = 0, 1, 2, \ldots$;

b) $L = 1 + XL = \displaystyle\sum_{k \geq 0} X^k = \prod_{i \geq 0} (1 + X^{2^i})$, (47)

where X denotes the species of singletons. For a definition of infinite product of species, see Exercise 9.

Exercises for Section 1.3

1. Let F, G, and H be species of structures.
   a) Show, by explicitly describing the isomorphisms, that addition has the following properties:

   i) $(F + G) + H = F + (G + H)$, (associativity)

   ii) $F + G = G + F$, (commutativity) (48)

   iii) $F + 0 = 0 + F = F$. (neutral element)

   b) Show that the passage to generating and cycle index series preserves the operation of addition (see the identities (11)).

2. a) Let $(F_i)_{i \in I}$ be a summable family of species of structures. Show that the sum $\sum_{i \in I} F_i$ of this family, defined by (13) and (14), is indeed a species of structures.
   b) Let $(h_i(x_1, x_2, \ldots))_{i \in I}$ be a family of formal series in the variables $x_1, x_2, \ldots$, expressed in the form

   $$h_i(x_1, x_2, \ldots) = \sum_{n_1, n_2, \ldots} h_{i; n_1, n_2, \ldots} \dfrac{x_1^{n_1} x_2^{n_2} \cdots}{c_{n_1, n_2, \ldots}}, \quad i \in I, \quad (49)$$

where $c_{n_1,n_2,\ldots}$ is a given family of nonzero scalars. By definition, the family

$$(h_i(x_1, x_2, \ldots))_{i \in I}$$

is said to be *summable* if, for each multi-index $n_1, n_2, \ldots$, one has

$$c_{n_1,n_2,\ldots} \left[x_1^{n_1} x_2^{n_2} \cdots\right] h_i(x_1, x_2, \ldots) = h_{i;n_1,n_2,\ldots} = 0, \tag{50}$$

except for a finite number of $i \in I$. The sum of the family is the formal series $h(x_1, x_2, \ldots)$ whose coefficients are given by the (finite) sums

$$c_{n_1,n_2,\ldots} \left[x_1^{n_1} x_2^{n_2} \cdots\right] h(x_1, x_2, \ldots) = \sum_{i \in I} h_{i;n_1,n_2,\ldots}. \tag{51}$$

Show that, if $(F_i)_{i \in I}$ is a summable family of species of structures, then the families of formal series

$$(F_i(x))_{i \in I}, \ (\widetilde{F}_i(x))_{i \in I} \text{ and } (Z_{F_i}(x_1, x_2, \ldots))_{i \in I}$$

are summable and the equalities (15) hold.

3. Let F be a species of structures. The even part and odd part of F are the species defined by the decompositions

$$F_{\text{even}} = F_0 + F_2 + F_4 + \cdots, \tag{52}$$

$$F_{\text{odd}} = F_1 + F_3 + F_5 + \cdots. \tag{53}$$

Show that the following equalities are satisfied:

a) $Z_{F_{\text{even}}} = \dfrac{1}{2} (Z_F(x_1, x_2, x_3, x_4, \ldots) + Z_F(-x_1, x_2, -x_3, x_4, \ldots))$,

b) $Z_{F_{\text{odd}}} = \dfrac{1}{2} (Z_F(x_1, x_2, x_3, x_4, \ldots) - Z_F(-x_1, x_2, -x_3, x_4, \ldots))$. \hfill (54)

In the case where $F = E$, the species of sets, deduce formulas (12).

4. a) Let F, G, and H be species of structures. Describe the isomorphisms establishing the following properties of multiplication:

    i) $(F \cdot G) \cdot H = F \cdot (G \cdot H)$,   (associativity)

    ii) $F \cdot G = G \cdot F$,   (commutativity)

    iii) $F \cdot 1 = 1 \cdot F = F$,   (neutral element)   (55)

    iv) $F \cdot 0 = 0 \cdot F = 0$,   (absorbing element)

    v) $F \cdot (G + H) = F \cdot G + F \cdot H$.   (distributivity)

b) Prove the identities (27) concerning the series associated with the product $F \cdot G$ of two species of structures.

c) Using the definitions of sum and product of species, prove that, for all integers $n \geq 0$, one has the combinatorial equality

$$\underbrace{F + F + \cdots + F}_{n \text{ times}} = n \cdot F, \tag{56}$$

where, by convention, the n on the right-hand side denotes the species having exactly n structures on the empty set and no structure on other sets.

d) Let $n \geq 0$ be an integer and set $F = n! \cdot E_n$ and $G = X^n$. For which values of n are the species F and G isomorphic?

5. a) Show that the species $\wp$, where $\wp[U] = \{B \mid B \subseteq U\}$, is isomorphic to the species $E \cdot E$.

b) Show directly that the number of structures of the species $\wp$ on a set U that are fixed by a permutation of U of cycle type $(n_1, n_2, n_3, \ldots)$ is

$$1^{n_1} n_1! 2^{n_2} n_2! 3^{n_3} n_3! \ldots \left[x_1^{n_1} x_2^{n_2} x_3^{n_3} \cdots \right] Z_\wp(x_1, x_2, x_3, \ldots) = 2^{n_1 + n_2 + n_3 + \cdots}, \tag{57}$$

by partitioning the cycles of the permutation according to whether or not they are contained in the considered subset.

6. Show that, for the species Der of derangements, one has

$$\text{fix Der}[n_1, n_2, \ldots] = n_1! n_2! \ldots \sum_{\substack{0 \leq i_k \leq n_k \\ k \geq 1}} \frac{(-1)^{i_1 + i_2 + \cdots} 1^{n_1 - i_1} 2^{n_2 - i_2} \cdots}{i_1! i_2! \cdots}. \tag{58}$$

7. Show that the species $\wp^{[k]}$, where $\wp^{[k]}[U] = \{B \mid B \subseteq U \text{ and } |B| = k\}$, is isomorphic to the species $E_k \cdot E$. Also show that

$$\text{fix } \wp^{[k]}[n_1, n_2, n_3, \ldots] = \sum_{k_1 + 2k_2 + 3k_3 + \cdots = k} \prod_{i \geq 1} \binom{n_i}{k_i}. \tag{59}$$

8. Let F and G be two species of structures. Show that the family $(F_m \cdot G_n)_{(m,n) \in \mathbb{N} \times \mathbb{N}}$ is summable and that

$$F \cdot G = \sum_{(m,n) \in \mathbb{N} \times \mathbb{N}} F_m \cdot G_n. \tag{60}$$

Deduce the canonical decomposition of $F \cdot G$. (In this problem, F_m denotes the restriction of the species F to the cardinality m.)

9. Let $(F_i)_{i \in I}$ be a family of species of structures. By convention, a $\prod_{i \in I} F_i$-structure on a finite set U is a family $(s_i)_{i \in I}$ where for each $i \in I$ one has $s_i \in F_i[U_i]$ for a subset $U_i \subseteq U$, and the subsets U_i are required to be pairwise disjoint and to satisfy $\bigcup_{i \in I} U_i = U$. The family $(F_i)_{i \in I}$ is said to be *multipliable* if for any finite set U, the set of $\prod_{i \in I} F_i$-structures on U is finite.

a) For a multipliable family $(F_i)_{i \in I}$ of species, define the species product $\prod_{i \in I} F_i$ (do not forget the transports of structures).

b) Show that a nonempty family of species $(F_i)_{i \in I}$ is multipliable if and only if there exists a $J \subseteq I$ such that
   i) $I \setminus J$ is finite,
   ii) $i \in J \iff (F_i)_0 = 1$,
   iii) the family of $((F_i)_+)_{i \in J}$ is summable.
c) Show that if $(F_i)_{i \in I}$ is multipliable and if $J = \{i \in I \mid (F_i)_0 = 1\}$, then $(F_j)_{j \in J}$ is multipliable and one has the combinatorial equalities

$$\text{i)} \quad \prod_{i \in I} F_i = \left(\prod_{i \in I \setminus J} F_i\right) \cdot \left(\prod_{j \in J} F_j\right),$$

$$\text{ii)} \quad \prod_{j \in J} F_j = 1 + \sum_{j \in J} (F_j)_+ + \sum_{\{j_1, j_2\} \in \wp^{[2]}(J)} (F_{j_1})_+ (F_{j_2})_+ + \cdots. \tag{61}$$

10. Denote by $\mathcal{S}^{\langle k \rangle}$ the species of permutations having all cycles of length k. Show that the infinite family $(\mathcal{S}^{\langle k \rangle})_{k \geq 1}$ is multipliable and that the species $\mathcal{S}$ of permutations satisfies

$$\mathcal{S} = \prod_{k \geq 1} \mathcal{S}^{\langle k \rangle}. \tag{62}$$

11. Determine whether the species Bal of ballots is isomorphic to the species

$$\prod_{n \geq 1} (1 + E_n + E_n^2 + E_n^3 + \cdots). \tag{63}$$

12. Prove formulas (46) for the series $\text{Bal}(x)$, $\widetilde{\text{Bal}}(x)$, and $Z_{\text{Bal}}(x_1, x_2, \ldots)$.
    HINT: First establish formulas (45).
13. For the species L of linear orderings, prove the combinatorial equalities

$$L = 1 + XL = \sum_{k \geq 0} X^k = \prod_{i \geq 0} (1 + X^{2^i}). \tag{64}$$

1.4. Substitution and Differentiation

Substitution of Species of Structures

As a motivating example, let us consider an endofunction $\varphi \in \text{End}[U]$ of a set U, determined by its functional digraph, such as that of Figure 1(a). Two kinds of points (elements of U) can be distinguished:
i) the *recurrent points*, i.e., those $x \in U$ for which there exists a $k > 0$ such that $\varphi^k(x) = x$; these are the elements located on cycles;
ii) the nonrecurrent points, i.e., those x for which $\varphi^k(x) \neq x$ for all $k > 0$.

Figure 1(b) shows how the endofunction φ can naturally be identified with a permutation of disjoint rooted trees. The naturality originates from the fact that

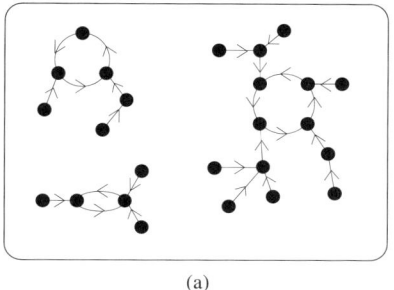 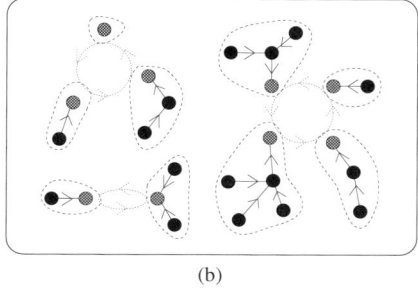

(a)   (b)

Fig. 1.

we need not use the specific nature of the underlying points in order to pass from Figure 1(a) to (b). Clearly, such an analysis can be carried out no matter which endofunction is given. Thus, every End-structure can naturally be identified with an $\mathcal{S}$-structure placed on a set of disjoint $\mathcal{A}$-structures, where, as previously, End denotes the species of endofunctions, $\mathcal{S}$, the species of permutations, and $\mathcal{A}$, the species of rooted trees. In a more concise manner, we say that every End-structure is an $\mathcal{S}$-assembly of $\mathcal{A}$-structures. This situation is summarized by the combinatorial equation

$$\text{End} = \mathcal{S} \circ \mathcal{A}, \quad \text{or} \quad \text{End} = \mathcal{S}(\mathcal{A}). \tag{1}$$

This simple equality expresses the fact that every endofunction is essentially a permutation of (disjoint) rooted trees. It is a typical example of substitution of species, also called the (partitional) composition of species, which can be defined in general as follows.

Definition 1. Let F and G be two species of structures such that $G[\emptyset] = \emptyset$ (i.e., there is no G-structure on the empty set). The species $F \circ G$, also denoted $F(G)$, called the (partitional) *composite* of G in F, is defined as follows: An $(F \circ G)$-structure on U is a triplet $s = (\pi, \varphi, \gamma)$, where
 i) π is a partition of U,
 ii) φ is an F-structure on the set of classes of π,
 iii) $\gamma = (\gamma_p)_{p \in \pi}$, where for each class p of π, γ_p is a G-structure on p.
In other words, for any finite set U, one has

$$(F \circ G)[U] = \sum_{\pi \text{ partition of } U} F[\pi] \times \prod_{p \in \pi} G[p], \tag{2}$$

the (disjoint) sum being taken over the set of partitions π of U (i.e., $\pi \in \text{Par}[U]$).

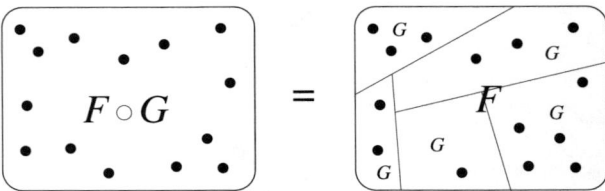

Fig. 2.

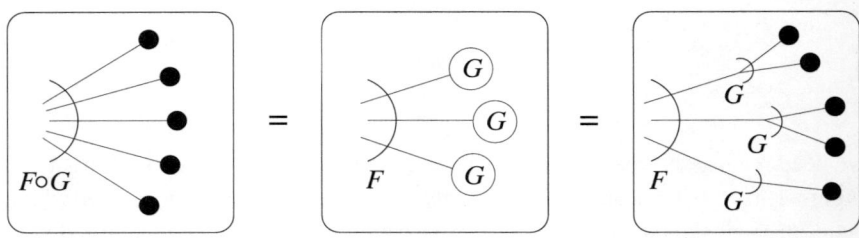

Fig. 3.

The transport along a bijection $\sigma : U \longrightarrow V$ is carried out by setting, for any $(F \circ G)$-structure $s = (\pi, \varphi, (\gamma_p)_{p \in \pi})$ on U,

$$(F \circ G)[\sigma](s) = (\overline{\pi}, \overline{\varphi}, (\overline{\gamma}_{\overline{p}})_{\overline{p} \in \overline{\pi}}), \tag{3}$$

where
i) $\overline{\pi}$ is the partition of V obtained by transport of π along σ,
ii) for each $\overline{p} = \sigma(p) \in \overline{\pi}$, the structure $\overline{\gamma}_{\overline{p}}$ is obtained from the structure γ_p by G-transport along $\sigma|_p$,
iii) the structure $\overline{\varphi}$ is obtained from the structure φ by F-transport along the bijection $\overline{\sigma}$ induced on π by σ.

In a more visual fashion, we say that an $(F \circ G)$-structure is an *F-assembly* of (disjoint) G-structures. Figures 2 and 3 illustrate this concept graphically. The proof that $F \circ G$ as defined is a species of structures (i.e., that the transports satisfy the *functoriality* properties) is left to the reader. When F is the species of sets, an $(E \circ G)$-structure is more simply called an *assembly* of G-structures.

The passage from $F \circ G$ to its generating and cycle index series is more delicate to analyze than in the case of sum and product of species. In fact, although the formula $(F \circ G)(x) = F(G(x))$ is valid, the corresponding identity does not hold in general for unlabeled enumeration

$$(\widetilde{F \circ G})(x) \neq \widetilde{F}(\widetilde{G}(x)), \tag{4}$$

as shown in Exercise 3. This is one of the reasons for which the introduction of cycle index series is necessary, as shown by the following result. A complete proof will be given in Chapter 4.

Theorem 2. Let F and G be two species of structures and suppose that $G[\emptyset] = \emptyset$. Then the series associated to the species $F \circ G$ satisfy the equalities

a) $(F \circ G)(x) = F(G(x))$,

b) $(\widetilde{F \circ G})(x) = Z_F(\widetilde{G}(x), \widetilde{G}(x^2), \widetilde{G}(x^3), \ldots)$, (5)

c) $Z_{F \circ G}(x_1, x_2, x_3, \ldots) = Z_F(Z_G(x_1, x_2, \ldots), Z_G(x_2, x_4, \ldots), \ldots)$.

The index series given in the last formula is called the *plethystic substitution* of Z_G in Z_F, and is denoted by $Z_F \circ Z_G$ (or $Z_F(Z_G)$).

Definition 3. Let $f = f(x_1, x_2, x_3, \ldots)$ and $g = g(x_1, x_2, x_3, \ldots)$ be two formal power series. Then the *plethystic substitution* $f \circ g$ is defined by

$$(f \circ g)(x_1, x_2, x_3, \ldots) = f(g_1, g_2, g_3, \ldots), \tag{6}$$

where the following notational convention is used:

$$g_k = g(x_k, x_{2k}, x_{3k}, \ldots), \quad k = 1, 2, 3, \ldots, \tag{7}$$

i.e., the power series g_k is obtained by multiplying by k the index of each variable appearing in g.

Observe that $g_k = x_k \circ g = g \circ x_k$.

Example 4. From the combinatorial equation $\text{End} = \mathcal{S} \circ \mathcal{A}$, one immediately deduces the formulas

a) $\sum_{n \geq 0} n^n \dfrac{x^n}{n!} = \text{End}(x) = (\mathcal{S} \circ \mathcal{A})(x) = \mathcal{S}(\mathcal{A}(x)) = \dfrac{1}{1 - \mathcal{A}(x)}$,

b) $\widetilde{\text{End}}(x) = (\widetilde{\mathcal{S} \circ \mathcal{A}})(x) = Z_{\mathcal{S}}(\widetilde{\mathcal{A}}(x), \widetilde{\mathcal{A}}(x^2), \widetilde{\mathcal{A}}(x^3), \ldots)$

$$= \dfrac{1}{(1 - \widetilde{\mathcal{A}}(x))(1 - \widetilde{\mathcal{A}}(x^2))(1 - \widetilde{\mathcal{A}}(x^3)) \cdots}, \tag{8}$$

c) $Z_{\text{End}} = (Z_{\mathcal{S}} \circ Z_{\mathcal{A}})(x_1, x_2, x_3, \ldots)$

$$= (1 - x_1 \circ Z_{\mathcal{A}})^{-1}(1 - x_2 \circ Z_{\mathcal{A}})^{-1}(1 - x_3 \circ Z_{\mathcal{A}})^{-1} \cdots,$$

which relate the series $\text{End}(x)$, $\widetilde{\text{End}}(x)$, and Z_{End} to the series $\mathcal{A}(x)$, $\widetilde{\mathcal{A}}(x)$, and $Z_{\mathcal{A}}$. We will have the opportunity to study these series more deeply in Chapter 3.

Figure 4 shows that the species $\mathcal{A}$ of rooted trees satisfies the combinatorial equation

$$\mathcal{A} = X \cdot E(\mathcal{A}), \tag{9}$$

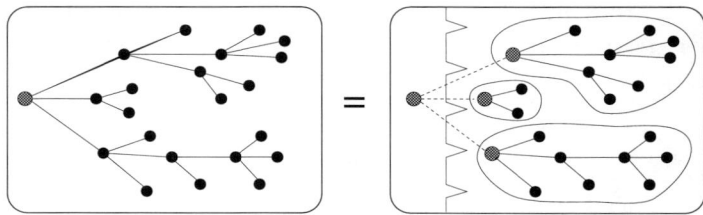

Fig. 4.

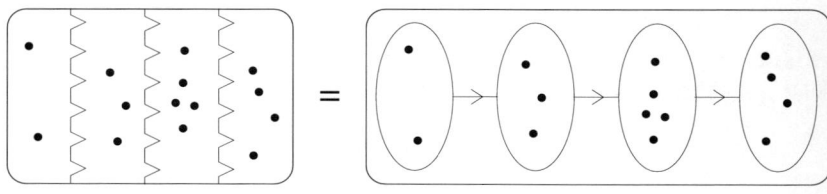

Fig. 5.

where E designates the species of sets. Passing to series gives the formulas

a) $\mathcal{A}(x) = x e^{\mathcal{A}(x)}$,

b) $\widetilde{\mathcal{A}}(x) = x \exp\left(\widetilde{\mathcal{A}}(x) + \frac{\widetilde{\mathcal{A}}(x^2)}{2} + \frac{\widetilde{\mathcal{A}}(x^3)}{3} + \cdots \right)$, (10)

c) $Z_\mathcal{A}(x_1, x_2, \ldots) = x_1 \exp\left(Z_\mathcal{A}(x_1, x_2, \ldots) + \frac{1}{2} Z_\mathcal{A}(x_2, x_4, \ldots) + \cdots \right)$.

We will see in Chapter 3 that these formulas allow recursive, and even explicit, calculation of the series $\mathcal{A}(x)$, $\widetilde{\mathcal{A}}(x)$, and $Z_\mathcal{A}(x_1, x_2, x_3, \ldots)$. Figure 5 shows that the species Bal of ballots satisfies the combinatorial equation

$$\text{Bal} = L \circ E_+,\tag{11}$$

where L is the species of linear orderings and E_+ that of nonempty sets. The identities (1.3.46) for the series associated to the species Bal can be deduced directly from this observation.

Consider Par, the species of *partitions*. Since every partition is naturally identified to a set of nonempty disjoint sets (see Figure 6), we obtain the combinatorial equation

$$\text{Par} = E(E_+).\tag{12}$$

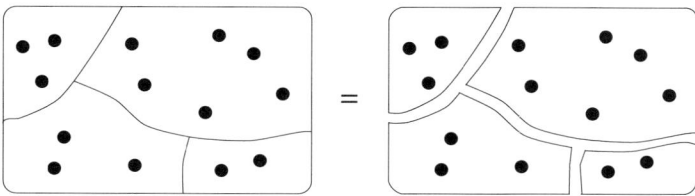

Fig. 6.

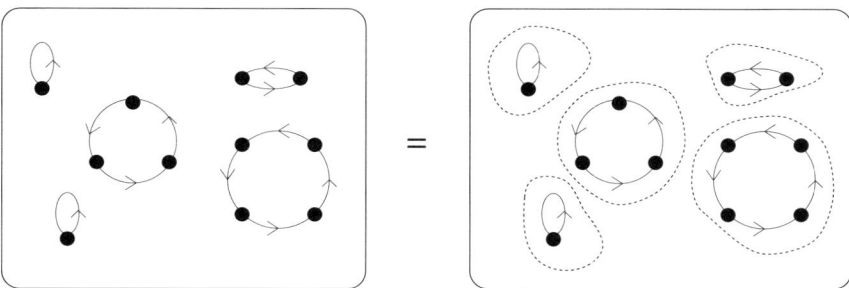

Fig. 7.

The following formulas are then immediately deduced:

a) $\mathrm{Par}(x) = e^{e^x - 1}$,

b) $\widetilde{\mathrm{Par}}(x) = \prod_{k \geq 1} \dfrac{1}{1 - x^k}$, (13)

c) $Z_{\mathrm{Par}}(x_1, x_2, x_3, \ldots) = \exp \sum_{k \geq 1} \dfrac{1}{k} \left(\exp\left(x_k + \dfrac{x_{2k}}{2} + \dfrac{x_{3k}}{3} + \cdots \right) - 1 \right)$.

In a similar fashion, since every permutation is a set of disjoint cycles (see Figure 7), we have the combinatorial equation

$$\mathcal{S} = E \circ \mathcal{C}, \qquad (14)$$

where $\mathcal{C}$ is the species of cycles (cyclic permutations).

It follows that

$$\mathcal{S}(x) = \dfrac{1}{1 - x} = e^{\mathcal{C}(x)} \qquad (15)$$

and we recover

$$\mathcal{C}(x) = \log \dfrac{1}{1 - x}. \qquad (16)$$

Moreover, we have the remarkable identities

a) $$\prod_{k\geq 1}\frac{1}{1-x^k} = \widetilde{S}(x) = Z_E(\widetilde{C}(x), \widetilde{C}(x^2), \ldots)$$

$$= \exp\sum_{n\geq 1}\frac{1}{k}\frac{x^k}{1-x^k}, \qquad (17)$$

b) $$\frac{1}{1-x_1}\frac{1}{1-x_2}\frac{1}{1-x_3}\cdots = Z_S(x_1, x_2, x_3, \ldots)$$

$$= \exp\sum_{k\geq 1}\frac{1}{k}Z_C(x_k, x_{2k}, x_{3k}, \ldots).$$

This last identity permits the explicit calculation of the index series Z_C of the species of cycles (see Exercise 8):

$$Z_C(x_1, x_2, x_3, \ldots) = \sum_{k=1}^{\infty}\frac{\phi(k)}{k}\log\frac{1}{1-x_k}, \qquad (18)$$

where ϕ denotes the arithmetic Euler ϕ-function.

The species $\mathcal{G}$ of graphs is related to the species $\mathcal{G}^c$ of connected graphs by the combinatorial equation $\mathcal{G} = E(\mathcal{G}^c)$, since every graph is an assembly of connected graphs. More generally, if two species F and F^c are related by a combinatorial equation of the form

$$F = E(F^c), \qquad (19)$$

we say that F^c is the species of *connected F-structures*. We then have

a) $F(x) = e^{F^c(x)}$,

b) $\widetilde{F}(x) = \exp\sum_{k\geq 1}\frac{1}{k}\widetilde{F}^c(x^k)$, $\qquad (20)$

c) $Z_F(x_1, x_2, x_3, \ldots) = \exp\sum_{k\geq 1}\frac{1}{k}Z_{F^c}(x_k, x_{2k}, x_{3k}, \ldots)$.

It is interesting to note that we can also express the series $F^c(x), \widetilde{F}^c(x)$, and Z_{F^c} as functions of the series $F(x), \widetilde{F}(x)$, and Z_F (see Exercise 9).

The species X of singletons is the *neutral element* for the substitution of species:

$$F = F(X) = F \circ X = X \circ F = X(F). \qquad (21)$$

Substitution is associative (up to isomorphism of species). For any species of structures G, the condition $G[\emptyset] = \emptyset$ is equivalent to $G(0) = 0$. If $G(0) = 0$, one recursively defines the successive iterates $G^{\langle n \rangle}$ of G by the recursive scheme

$$G^{\langle 0 \rangle} = X,$$

$$G^{\langle n+1 \rangle} = G \circ G^{\langle n \rangle} \quad (= G^{\langle n \rangle} \circ G). \qquad (22)$$

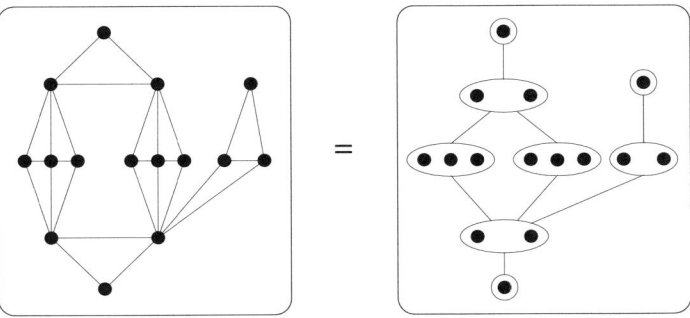

Fig. 8.

Consider the species Preo of all the preorders (i.e., reflexive and transitive relations) and the species Ord of all order relations (i.e., antisymmetric preorders). Since every preorder "$\prec$" induces, in a natural manner, an order on an appropriate quotient set (see Figure 8), we obtain the combinatorial equation

$$\text{Preo} = \text{Ord}(E_+). \tag{23}$$

The computation of the power series Preo(x), Ord(x), $\widetilde{\text{Preo}}(x)$, $\widetilde{\text{Ord}}(x)$, Z_{Preo}, and Z_{Ord} is an open problem. Nevertheless, Theorem 2 implies the following relations:

a) $\text{Preo}(x) = \text{Ord}(e^x - 1)$,

b) $\widetilde{\text{Preo}}(x) = Z_{\text{Ord}}\left(\dfrac{x}{1-x}, \dfrac{x^2}{1-x^2}, \dfrac{x^3}{1-x^3}, \ldots\right)$, $\tag{24}$

c) $Z_{\text{Preo}}(x_1, x_2, x_3, \ldots) = Z_{\text{Ord}}\left(e^{x_1 + \frac{x_2}{2} + \cdots} - 1, e^{x_2 + \frac{x_4}{2} + \cdots} - 1, \ldots\right)$.

The Derivative of a Species of Structures

Given an arbitrary species of structures F, we propose to construct another species G so that their respective generating series satisfy $G(x) = \frac{d}{dx}F(x)$. This is equivalent to requiring that $|G[n]| = |F[n+1]|$, $n = 0, 1, 2, \ldots$. Hence the number of G-structures on an arbitrary finite set U should be equal to the number of F-structures on the set U to which a "new" element has been added. This suggests the following definition:

Definition 5. Let F be a species of structures. The species F' (also denoted by $\frac{d}{dX}F(X)$), called the *derivative* of F, is defined as follows: An F'-structure on U is an F-structure on $U^+ = U \cup \{*\}$, where $* = *_U$ is a element chosen outside of U. In other words, for any finite set U, one sets

$$F'[U] = F[U^+], \quad \text{where} \quad U^+ = U + \{*\}. \tag{25}$$

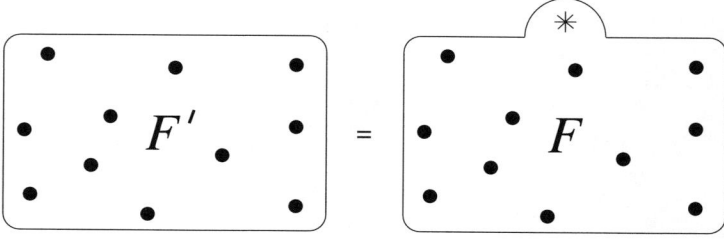

Fig. 9.

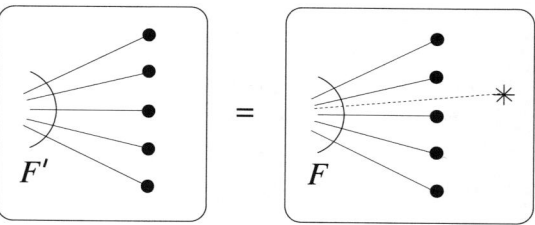

Fig. 10.

The transport along a bijection $\sigma : U \longrightarrow V$ is carried out by setting, for any F'-structure s on U,

$$F'[\sigma](s) = F[\sigma^+](s), \qquad (26)$$

where $\sigma^+ : U + \{*\} \longrightarrow V + \{*\}$ is the canonical extension of σ obtained by setting

$$\sigma^+(u) = \sigma(u), \quad \text{if } u \in U, \quad \text{and } \sigma^+(*) = *. \qquad (27)$$

Figures 9 and 10 illustrate graphically the concept of F'-structure.

Remark 6. Observe that the supplementary element $*$ is not a member of the underlying set of the F'-structure on U. Also note that the element $*$ has been placed in an arbitrary position in Figure 10 to emphasize that the set $U + \{*\}$ on which the F-structure is constructed is not otherwise structured. The careful reader may ask himself how does one systematically (and canonically) choose a element $* = *_U$ outside each given set U. Exercise 16 describes a classic solution to this problem.

Example 7. As a standard illustration, we analyze the derivative $\mathcal{C}'$ of the species $\mathcal{C}$ of cyclic permutations. By definition, a $\mathcal{C}'$-structure on the set $U = \{a, b, c, d, e\}$ is a $\mathcal{C}$-structure on $U + \{*\}$. It is identified in a natural manner (forgetting $*$) with

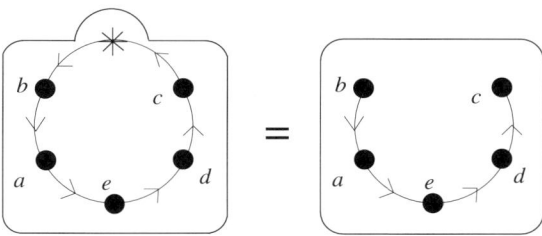

Fig. 11.

a linear ordering placed on U (see Figure 11). In other words, we have shown the combinatorial equation

$$\mathcal{C}' = \mathcal{L}. \tag{28}$$

Passing to generating series yields

$$\mathcal{C}'(x) = L(x) = \frac{1}{1-x}, \tag{29}$$

and, by integration,

$$C(x) = \int_0^x \frac{dx}{1-x} = \log\frac{1}{1-x}, \tag{30}$$

which gives a third way to obtain the series $C(x)$. The relation between the derivative of species and the corresponding series is summarized in the following proposition.

Proposition 8. *Let F be a species of structures. One has the equalities*

a) $F'(x) = \dfrac{d}{dx} F(x),$

b) $\widetilde{F'}(x) = \left(\dfrac{\partial}{\partial x_1} Z_F\right)(x, x^2, x^3, \ldots),$ \hfill (31)

c) $Z_{F'}(x_1, x_2, x_3, \ldots) = \left(\dfrac{\partial}{\partial x_1} Z_F\right)(x_1, x_2, x_3, \ldots).$

Example 9. Consider the species Par′, derivative of the species Par of partitions. Figure 12 shows that a Par′-structure on a set U can be identified in a natural way to a *partial partition* on U, that is to say, a partition on a part V of U: Simply take $V = U \backslash W$, where W is the class containing $*$. Let Par_P be the species of partial partitions. We then have the combinatorial equation

$$\text{Par}_P = \text{Par}'. \tag{32}$$

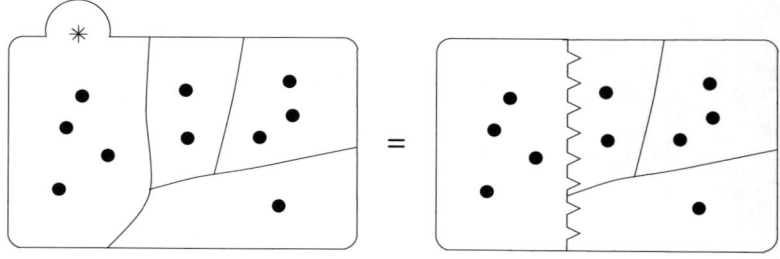

Fig. 12.

Applying the preceding proposition to the known series for the species Par yields

a) $\mathrm{Par}_P(x) = e^{x+e^x-1}$,

b) $Z_{\mathrm{Par}_P}(x_1, x_2, \ldots) = \exp \sum_{k\geq 1} \frac{1}{k}\left(x_k + \exp\left(x_k + \frac{x_{2k}}{2} + \cdots\right) - 1\right).$ (33)

In particular, letting $x_i := x^i$, we obtain

$$\widetilde{\mathrm{Par}}_P(x) = \left(\frac{1}{1-x}\right)^2 \prod_{k\geq 2} \frac{1}{1-x^k}. \tag{34}$$

Note that Par_P also satisfies the combinatorial equation

$$\mathrm{Par}_P = E \cdot \mathrm{Par}, \tag{35}$$

as is shown in the same Figure 12. This allows for a calculation of the series associated to Par_P in a different fashion. For example,

$$\mathrm{Par}_P(x) = (E \cdot \mathrm{Par})(x) = E(x)\mathrm{Par}(x) = e^x \mathrm{Par}(x), \tag{36}$$

agreeing with (33), a) obtained earlier.

The derivative E' of the species E of sets satisfies the combinatorial equation $E' = E$. This constitutes a combinatorial version of the classic equality $\frac{d}{dx}e^x = e^x$.

For the species L of linear orderings, Figure 13 shows that

$$L' = L^2 \quad (= L \cdot L), \tag{37}$$

reflecting combinatorially the identity

$$\frac{d}{dx}\left(\frac{1}{1-x}\right) = \left(\frac{1}{1-x}\right)^2. \tag{38}$$

The operation of differentiation can be iterated. For $F'' = (F')'$, we simply add successively two distinct elements, $*_1$ and $*_2$, to the underlying set. For example,

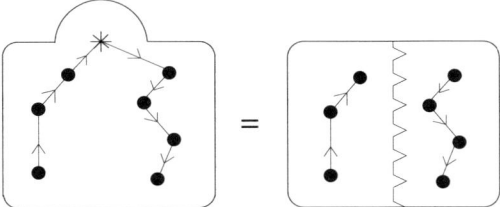

Fig. 13.

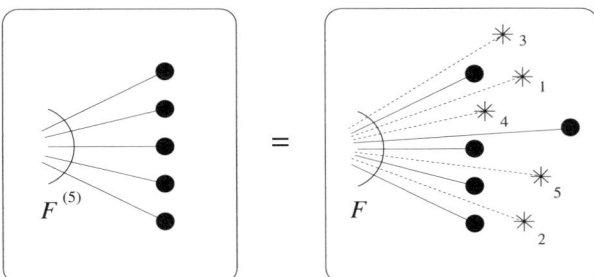

Fig. 14.

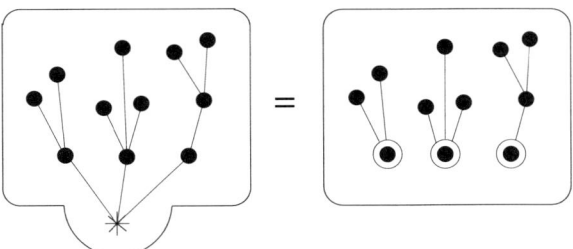

Fig. 15.

we have the combinatorial equation $\mathcal{C}'' = L^2$. More generally, we set

$$F^{(0)} = F \quad \text{and} \quad F^{(k)} = \left(F^{(k-1)}\right)', \quad k \geq 1. \tag{39}$$

An $F^{(k)}$-structure on U is then equivalent to an F-structure on $U \cup \{*_1, *_2, \ldots, *_k\}$, where $*_i$, $1 \leq i \leq k$, is an ordered sequence of k additional distinct elements (see Figure 14, where $k = 5$).

Consider the species $\mathfrak{a}$ of trees. Figure 15 immediately shows that

$$\mathfrak{a}' = \mathcal{F} = E(\mathcal{A}), \tag{40}$$

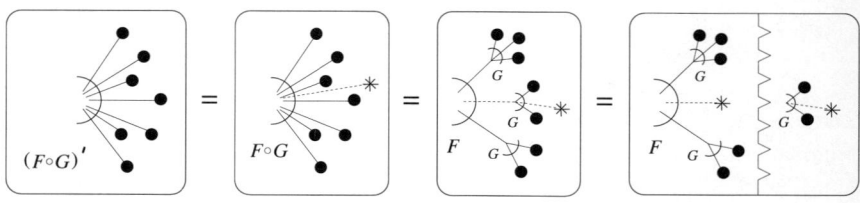

Fig. 16.

where $\mathcal{F}$ is the species of forests of rooted trees (i.e., disjoint sets of rooted trees). One can then assert, a priori, that the series for $\mathfrak{a}$ and $\mathcal{F}$ are related by the following equalities:

a) $\quad \mathcal{F}(x) = \dfrac{d}{dx} \mathfrak{a}(x),$

b) $\quad \widetilde{\mathcal{F}}(x) = \left(\dfrac{\partial}{\partial x_1} Z_\mathfrak{a} \right)(x, x^2, x^3, \ldots),$ (41)

c) $\quad Z_\mathcal{F}(x_1, x_2, x_3, \ldots) = \left(\dfrac{\partial}{\partial x_1} Z_\mathfrak{a} \right)(x_1, x_2, x_3, \ldots).$

Remark 10. To underline how the combinatorial differential calculus of species agrees with the classical differential calculus of formal power series, we mention that the chain rule admits the combinatorial equivalent

$$(F \circ G)' = (F' \circ G) \cdot G', \tag{42}$$

where G is a species such that $G(0) = 0$ (i.e., $G[\emptyset] = \emptyset$). Consideration of Figure 16 suffices to show the validity of this formula. It is easily verified that the other usual rules $(F + G)' = F' + G'$ and $(F \cdot G)' = F' \cdot G + F \cdot G'$ are also satisfied in the context of species.

Nevertheless, one must be prudent when establishing analogies with classical differential calculus. For instance, although the differential equation

$$y' = f(x), \qquad y(0) = 0, \tag{43}$$

always has a unique solution in the setting of formal power series, one can show that the equation

$$Y' = F(X), \qquad Y(0) = 0, \tag{44}$$

can have many nonisomorphic solutions in the algebra of species of structures (see Exercise 24). On the other hand, the equation

$$Y' = X E_3(X), \qquad Y(0) = 0, \tag{45}$$

1.4. Substitution and Differentiation

has no species of structures solution (see Exercise 2.5.12). However, it has an infinite number of solutions in the context of virtual species (see Section 2.5).

Remark 11. For other variants of the theory of the species of structures, for example that of $\mathbb{L}$-species introduced in Chapter 5, the existence and uniqueness of solutions of combinatorial differential equations coincides with the formal power series setting.

Exercises for Section 1.4

1. Let F, G, H, and K be species of structures, with $G[\emptyset] = \emptyset = H[\emptyset]$.
   a) Verify that the partitional composition $F \circ G$, defined by formulas (2) and (3), is a species of structures.
   b) Show, by an explicit describtion of the isomorphisms, that the partitional composition has the following properties:

   i) $(F \circ G) \circ H = F \circ (G \circ H)$, (associativity)

   ii) $F \circ X = X \circ F = F$, (neutral element)

   iii) $(F + K) \circ G = F \circ G + K \circ G$, (distributivity) (46)

   iv) $(F \cdot K) \circ G = (F \circ G) \cdot (K \circ G)$, (distributivity)

   v) $F_0 = F \circ 0$ and $F[\emptyset] = \emptyset \iff F(0) = 0$.

   c) Show, by enumerating the $F \circ G$-structures, that one indeed has

   $$(F \circ G)(x) = F(G(x)). \tag{47}$$

2. a) Let k be a fixed integer. Verify that the species $S^{[k]}$ of permutations having k cycles and the species $\mathrm{Par}^{[k]}$ of partitions having k blocks satisfy the combinatorial equations

   $$S^{[k]} = E_k \circ C, \qquad \mathrm{Par}^{[k]} = E_k \circ E_+, \tag{48}$$

   where E_k is the species of sets of cardinality k and C is that of oriented cycles (cyclic permutations).

   b) Let $c(n, k)$ be the number of permutations of a set of cardinality n having k cycles and $S(n, k)$ be the number of partitions of a set of cardinality n having k blocks. The numbers $s(n, k) := (-1)^{n-k} c(n, k)$ and $S(n, k)$ are called, respectively, *Stirling numbers of the first and second kind*. Deduce from a) the following identities:

   i) $\displaystyle\sum_{n \geq k} c(n, k) \frac{x^n}{n!} = \frac{(\log 1/(1 - x))^k}{k!},$ (49)

ii) $\displaystyle\sum_{n\geq k} S(n,k)\frac{x^n}{n!} = \frac{(e^x-1)^k}{k!}.$ (50)

c) Show that for $n \geq 0$, $k \geq 1$,

i) $S(n+1,k) = S(n,k-1) + k\, S(n,k),$ (51)

ii) $c(n+1,k) = c(n,k-1) + n\, c(n,k).$ (52)

d) Show that for $n \geq 0$,

i) $\displaystyle\sum_{k=0}^{n} s(n,k)\, x^k = x_{(n)} := x(x-1)\cdots(x-n+1),$ (53)

ii) $\displaystyle\sum_{k=0}^{n} S(n,k)\, x_{(k)} = x^n.$ (54)

3. Show that the generating series of unlabeled permutations is not the composite of the type generating series of the species of sets with that of cycles (i.e., $\widetilde{S}(x) \neq \widetilde{E}\,(\widetilde{C}(x))$).

4. Show that the formula (5), b), namely

$$(F\widetilde{\circ}G)(x) = Z_F\,(\widetilde{G}(x),\widetilde{G}(x^2),\widetilde{G}(x^3),\ldots),$$ (55)

is a consequence of formula (5), c).

5. a) Verify formulas (8) relating the series $\mathrm{End}(x)$, $\widetilde{\mathrm{End}}(x)$, and Z_{End} to the series $\mathcal{A}(x)$, $\widetilde{\mathcal{A}}(x)$, and $Z_\mathcal{A}$.

b) Show that the formulas (10) implicitly determine the series $\mathcal{A}(x)$, $\widetilde{\mathcal{A}}(x)$, and $Z_\mathcal{A}$.

6. a) Starting from combinatorial Equation (11), reprove the explicit formulas (1.3.46) for the series $\mathrm{Bal}(x)$, $\widetilde{\mathrm{Bal}}(x)$, and Z_{Bal}.

b) Starting from combinatorial Equation (12), establish formulas (13).

c) Establish identity (17), a), namely

$$\prod_{k\geq 1} \frac{1}{1-x^k} = \exp\sum_{n\geq 1} \frac{1}{n}\frac{x^n}{1-x^n},$$ (56)

starting from the combinatorial equation $\mathcal{S} = E(\mathcal{C})$.

7. For two formal power series $a = a(x_1, x_2, x_3, \ldots)$ and $b = b(x_1, x_2, x_3, \ldots)$ show that with notational convention (7),

$$b = \sum_{k\geq 1} \frac{1}{k} a_k \iff a = \sum_{k\geq 1} \frac{\mu(k)}{k} b_k.$$ (57)

1.4. Substitution and Differentiation

HINT: Use the following classic property of the Möbius function μ:

$$\sum_{d|n} \mu(d) = \begin{cases} 1, & \text{if } n = 1, \\ 0, & \text{otherwise.} \end{cases}$$

8. a) Taking the logarithm of the equality (17), b) and using Exercise 7, prove the explicit formula (18), namely

$$Z_C(x_1, x_2, x_3, \ldots) = \sum_{k \geq 1} \frac{\phi(k)}{k} \log \frac{1}{1 - x_k}. \tag{58}$$

HINT: Use the following formula for the Euler ϕ-function: $\phi(n) = \sum_{d|n} d\mu(n/d)$.

b) Deduce from a) the remarkable relation

$$\frac{x}{1-x} = \sum_{k \geq 1} \frac{\phi(k)}{k} \log \frac{1}{1-x^k}. \tag{59}$$

9. Show that if two species of structures F and F^c are related by the combinatorial equation $F = E(F^c)$, i.e., F^c is the species of connected F-structures, then

  i) $F^c(x) = \log F(x)$,

  ii) $\widetilde{F^c}(x) = \sum_{k \geq 1} \frac{\mu(k)}{k} \log \widetilde{F}(x^k)$, \hfill (60)

  iii) $Z_{F^c}(x_1, x_2, \ldots) = \sum_{k \geq 1} \frac{\mu(k)}{k} \log Z_F(x_k, x_{2k}, \ldots)$.

10. Consider the species Oct of octopuses, introduced in Exercise 1.1.7.

  a) Show that $\text{Oct} = C(L_+)$.

  b) Show, by calculus, that

$$\text{Oct}(x) = C(2x) - C(x). \tag{61}$$

  c) This formula suggests the combinatorial equation

$$\text{Oct}(X) + C(X) = C(2X). \tag{62}$$

  Prove the validity of this equation.

  d) Deduce the formulas

  i) $\text{Oct}(x) = \sum_{n \geq 1} (2^n - 1) x^n$,

  ii) $Z_{\text{Oct}}(x_1, x_2, x_3, \ldots) = \sum_{n \geq 1} \frac{\phi(n)}{n} \log \left(\frac{1 - x_n}{1 - 2x_n} \right)$. \hfill (63)

11. a) Show that the number of octopuses on n vertices ($n \geq 1$), where each tentacle is of odd length, is given by
$$(n-1)!\left(\mathcal{L}_n + (-1)^{n+1} - 1\right), \tag{64}$$
where
$$\mathcal{L}_n = \left(\frac{1+\sqrt{5}}{2}\right)^n + \left(\frac{1-\sqrt{5}}{2}\right)^n \tag{65}$$
designates the nth *Lucas number*. Note: The Lucas numbers satisfy the recurrence
$$\mathcal{L}_0 = 2, \quad \mathcal{L}_1 = 1, \quad \mathcal{L}_n = \mathcal{L}_{n-1} + \mathcal{L}_{n-2}, \quad \text{for } n \geq 2. \tag{66}$$
b) Compute the cycle index series of the species of octopuses with odd tentacles.

12. a) Let F and G be two species of structures, with $G(0) = 0$. Show that the family $(F(G_n))_{n \geq 1}$ is summable, where G_n is the species G restricted to the cardinality n. Also give a description of the $\sum_{n \geq 1} F(G_n)$-structures.

b) The combinatorial equation $\text{Oct}_{\text{reg}}(X) = \sum_{n \geq 1} C(X^n)$ defines the species Oct_{reg} of *regular octopuses*. Justify this terminology by giving a graphical example of a Oct_{reg}-structure and establish the following formulas for the series associated to the species Oct_{reg}:

i) $\text{Oct}_{\text{reg}}(x) = \sum_{n \geq 1} \left(\sum_{d|n} d \right) \frac{x^n}{n},$

ii) $\widetilde{\text{Oct}}_{\text{reg}}(x) = \sum_{n \geq 1} \tau(n) x^n,$ \hfill (67)

where $\tau(n) = $ number of divisors of n,

iii) $Z_{\text{Oct}_{\text{reg}}}(x_1, x_2, \ldots) = -\sum_{k,n \geq 1} \frac{\phi(k)}{k} \log(1 - x_k^n).$

c) Show that the species Cha of chains, introduced in Exercise 1.1.5, can be written in the form
$$\text{Cha} = (1+X)\left(1 + \sum_{n \geq 1} E_2(X^n)\right), \tag{68}$$
where E_2 designates the species of sets of cardinality two. Deduce the expressions for the various series associated to the species Cha.

13. Let F and G be two species of structures, with $F(0) = 1$. Show that the family $(F(G_n))_{n \geq 1}$ is multipliable. Also give a description of the $\prod_{n \geq 1} F(G_n)$-structures.

14. Let $G = L_+$ be the species of nonempty linear orderings. Show that the iterations $G^{\langle n \rangle}$ of G are given by the combinatorial equations $G^{\langle n \rangle} = XL(nX)$, $n = 0, 1, 2, \ldots$.

15. A partially ordered set is called *reduced* if it satisfies the following condition: For all x and y,
$$\left.\begin{array}{l}\{z \mid z \leq x\} \cup \{y\} = \{z \mid z \leq y\} \\ \{z \mid z \geq x\} = \{z \mid z \geq y\} \cup \{x\}\end{array}\right\} \Rightarrow x = y.$$
Denote by Red the species of reduced partial orderings.

a) Establish the combinatorial equation
$$\text{Ord} = \text{Red}(L_+). \tag{69}$$

b) Deduce the following relations:

i) $\text{Red}(x) = \text{Ord}\left(\frac{x}{1+x}\right)$,

ii) $\widetilde{\text{Red}}(x) = \widetilde{Z}_{\text{Ord}}\left(\frac{x}{1+x}, \frac{x^2}{1+x^2}, \ldots\right)$, (70)

iii) $Z_{\text{Red}}(x_1, x_2, x_3, \ldots) = Z_{\text{Ord}}\left(\frac{x_1}{1+x_1}, \frac{x_2}{1+x_2}, \ldots\right)$.

16. It is always possible to choose canonically a supplementary element $*_U$ outside of each finite set U by taking $*_U = U$ (i.e., the supplementary "element" is the set U itself!). In this case, we say that U^+ is the *set successor* of U and write $U^+ = U \cup \{U\}$. We have as well $*_U \notin U$, since according to classical set theory, one always has $U \notin U$ (Foundation Axiom). Show that with this choice for $*_U$, the operation of differentiation is well defined and that indeed $F'(x) = \frac{d}{dx}F(x)$.

17. a) Verify the equalities of Proposition 8 in the following cases: $F = C$, E, and L.

b) Prove formula (31), c).

18. Recall that the species Bal of ballots and the species Oct of octopuses are characterized by the combinatorial equations $\text{Bal} = L(E_+)$ and $\text{Oct} = C(L_+)$. Show, by graphical arguments, that the derivatives of these species satisfy

a) $\text{Bal}' = \text{Bal}^2 \cdot E$, b) $\text{Oct}' = L(X) \cdot L(2X)$. (71)

19. Establish rigorously, by giving the explicit isomorphisms, the following rules of combinatorial differential calculus.

a) $(F + G)' = F' + G'$, (additivity)

b) $(F \cdot G)' = F' \cdot G + F \cdot G'$, (product rule) (72)

c) $(F \circ G)' = (F' \circ G) \cdot G'$. (chain rule)

20. In a purely formal fashion (i.e., by using rules of combinatorial differential calculus, associativity, etc.), establish formulas a) and b) of Problem 18 starting from the combinatorial equations $\text{Bal} = L(E_+)$ and $\text{Oct} = \mathcal{C}(L_+)$.

21. Let $\mathcal{A} = X\,E(\mathcal{A})$ be the species of rooted trees, $\mathcal{F} = E(\mathcal{A})$ that of rooted forests, and $\mathfrak{a}$ that of trees. Establish, by combinatorial calculus as well as by a graphical argument, the following combinatorial equations:

$$\text{a)} \quad \mathcal{A}' = \mathcal{F} \cdot L(\mathcal{A}),$$
$$\text{b)} \quad \mathfrak{a}'' = \mathcal{F} \cdot \mathcal{A}', \tag{73}$$
$$\text{c)} \quad \mathcal{A}'' = (\mathcal{A}')^2 + (\mathcal{A}')^2 L(\mathcal{A}).$$

22. Calculate the successive derivatives of
    a) the species $\wp = E^2$ of subsets of a set,
    b) the species L of linear orderings.

23. Let $D = \frac{d}{dx}$ be the differentiation operator with respect to the variable x. In the classic theory of ordinary differential equations, the following identity is very useful: $(D-c)f(x) = e^{cx} D e^{-cx} f(x)$, where c is any constant and $f(x)$ any differentiable function or a formal power series. One also has the more general identity

$$(D + h'(x))f(x) = e^{-h(x)} D\, e^{h(x)} f(x), \tag{74}$$

which can be rewritten in the form

$$e^{h(x)}(f'(x) + h'(x)f(x)) = (e^{h(x)} f(x))'. \tag{75}$$

Establish, by a geometric argument, the following corresponding combinatorial identity, namely

$$E(H) \cdot (F' + H' \cdot F) = (E(H) \cdot F)', \tag{76}$$

where F and H are species of structures.

24. (See G. Labelle [183].) Verify that for any $m \geq 1$, the combinatorial differential equation

$$Y' = 3(m-1)X^2, \quad Y[\emptyset] = \emptyset, \tag{77}$$

possesses the m nonisomorphic solutions

$$Y = 3k\, \mathcal{C}_3 + (m-1-k)X^3, \quad k = 0, 1, \ldots, m-1, \tag{78}$$

where $\mathcal{C}_3$ denotes the species of oriented cycles on 3 elements sets.

2

Complements on Species of Structure

2.0. Introduction

This chapter contains important complements to the basic theory of species of structures. First, three new combinatorial operations are introduced in Sections 1 and 2, namely pointing (·), Cartesian product (×), and functorial composition (□). The pointing operation interprets combinatorially the operator $x\frac{d}{dx}$. The Cartesian product, $F \times G$, which consists of superimposing structures of species F and G, corresponds to a coefficient-wise product of exponential generating series known as the *Hadamard product*. The functorial composition, not to be confused with substitution, is a very natural operation if one recalls that a species of structures can be considered as a functor. Many varieties of graphs and multigraphs can be simply expressed with the help of this operation.

Subsequently, the theory is extended to weighted species (Section 3), where the structures are enumerated according to certain parameters, and to multisort species (Section 4), a combinatorial analogue of functions in many variables. These enrichments open the way to more refined enumeration related to complexity analysis of algorithms in computer science and as well as to classical analysis, notably orthogonal polynomials.

Finally, Sections 5 and 6 contain an extension to virtual species, making subtraction of species possible. This involves an analysis of molecular (i.e., indecomposable according to sum) and atomic (i.e., indecomposable according to sum and product) species of structures. These are useful tools which permit us to
- give a standard decomposition for each species, into irreducible components,
- give a rigorous combinatorial sense to the multiplicative inverse, $1/F$, of any species F such that $F(0) = 1$, and to the inverse under substitution, $G^{\langle -1 \rangle}$, of any species G such that $G(0) = 0$ and $G'(0) = 1$,
- solve certain differential and functional equations which do not have a solution in the context of species of structures.

2.1. Pointing and Cartesian Product

Pointing in a Species of Structures

Pointing corresponds at the combinatorial level to the differential operator $x\frac{d}{dx}$, whose effect on formal power series is:

$$x\frac{d}{dx}\sum_{n\geq 0} f_n \frac{x^n}{n!} = \sum_{n\geq 0} n f_n \frac{x^n}{n!}. \tag{1}$$

Definition 1. Let F be a species of structures. The species $F^\bullet$, called F dot, is defined as follows: An $F^\bullet$-structure on U is a pair $s = (f, u)$, where
i) f is an F-structure on U,
ii) $u \in U$ (a *distinguished* element).
The pair (f, u) is called a *pointed F-structure* (pointed at the distinguished element u). In other words, for any finite set U,

$$F^\bullet[U] = F[U] \times U \quad \text{(set-theoretic Cartesian product)}. \tag{2}$$

The transport along a bijection $\sigma : U \longrightarrow V$ is carried out by setting

$$F^\bullet[\sigma](s) = (F[\sigma](f), \sigma(u)), \tag{3}$$

for any $F^\bullet$-structure $s = (f, u)$ on U.

A typical $F^\bullet$-structure can be represented graphically by circling the pointed element (see Figure 1). The enumeration of the $F^\bullet$-structures satisfies

$$|F^\bullet[n]| = n|F[n]|, \quad n \geq 0. \tag{4}$$

As a first illustration, let us point the species $\mathfrak{a}$ of trees. We then obtain the species $\mathcal{A}$ of *rooted trees*:

$$\mathfrak{a}^\bullet = \mathcal{A}. \tag{5}$$

Indeed, a rooted tree is nothing more than a tree with a distinguished element, its *root* (see Figure 2). It is important to note that the distinguished element u of an $F^\bullet$-structure belongs to the underlying set U, whereas the element $*$ of an

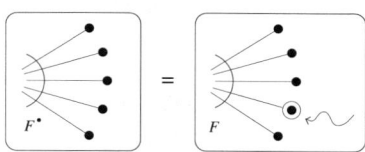

Fig. 1.

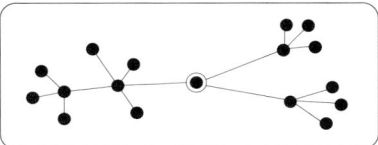

Fig. 2.

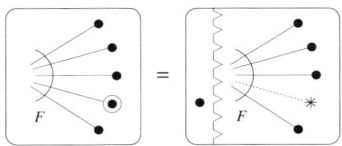

Fig. 3.

F'-structure is always outside of the underlying set U. The operations of pointing and derivation are related by the combinatorial equation

$$F^{\bullet} = X \cdot F', \qquad (6)$$

where X denotes the species of singletons. To see this, examine Figure 3. The distinguished element (the circled singleton) in the $F^{\bullet}$-structure on the left is taken aside and is replaced by a $*$. This gives, in a natural fashion an $X \cdot F'$-structure.

From the combinatorial equation $F^{\bullet} = X \cdot F'$, we can deduce the main properties of the operation of pointing (see Exercise 1), as well as the following proposition concerning the passage to the generating and index series.

Proposition 2. *Let F be a species of structures. One has the equalities*

a) $F^{\bullet}(x) = x \dfrac{d}{dx} F(x),$

b) $\widetilde{F^{\bullet}}(x) = x \left(\dfrac{\partial}{\partial x_1} Z_F \right)(x, x^2, x^3, \ldots),$ \qquad (7)

c) $Z_{F^{\bullet}}(x_1, x_2, x_3, \ldots) = x_1 \left(\dfrac{\partial}{\partial x_1} Z_F \right)(x_1, x_2, x_3, \ldots).$

Example 3. In [158], A. Joyal uses the operation of pointing to determine, in a simple and elegant fashion, the number α_n of trees on a set of n elements (see also Goulden–Jackson [133], 3.3.19). To this end, point twice the species $\mathfrak{a}$ of trees. This gives, by definition, the species $\mathcal{V}$ of *vertebrates*:

$$\mathcal{V} = \mathfrak{a}^{\bullet\bullet} \qquad (= \mathcal{A}^{\bullet}). \qquad (8)$$

Hence, a vertebrate is a *bipointed* tree (i.e., a pointed rooted tree). It possesses in a natural fashion a *vertebral column* (see Figure 4), that is, the unique elementary

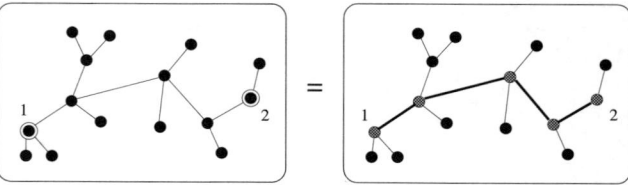

Fig. 4.

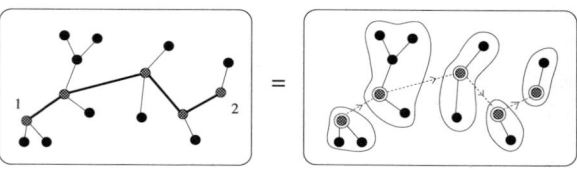

Fig. 5.

path (refer to the thick lines) from the first distinguished vertex (called *the tail* vertex, labeled by 1) to the second distinguished vertex (called *the head* vertex, labeled by 2). Note that the tail vertex can coincide with the head vertex. In this case, the vertebrate is called *degenerate*. Denote by v_n, the number of vertebrates on a set of n elements. We then have

$$v_n = n^2 \alpha_n, \qquad (9)$$

since there are n possible choices for the tail vertex and n other (independent) choices for the head vertex. We next calculate v_n in another fashion. The vertebral column determines in a natural manner a nonempty sequence of disjoint rooted trees (see Figure 5). Thus the species $\mathcal{V}$ of vertebrates satisfies the combinatorial equation

$$\mathcal{V} = L_+(\mathcal{A}). \qquad (10)$$

Replacing in this equality the species L_+ by the equipotent species $\mathcal{S}_+$ of nonempty permutations yields

$$\mathcal{V} \equiv \mathcal{S}_+(\mathcal{A}) \qquad \text{(equipotence)}. \qquad (11)$$

There are then, on a given set, as many vertebrates as (nonempty) permutations of rooted trees and then, as many as (nonempty) endofunctions, since $\mathcal{S}(\mathcal{A}) = \text{End}$. Thus, we have established the equipotence

$$\mathcal{V} \equiv \text{End}_+, \qquad (12)$$

from which we deduce $v_n = n^n$, $n \geq 1$, since $|\text{End}[n]| = n^n$. From $n^2 \alpha_n = v_n = n^n$, $n \geq 1$, we conclude that

$$\alpha_n = n^{n-2}, \qquad n \geq 1. \tag{13}$$

This is the classic formula of Cayley for the number of labeled trees. Incidentally, we have also shown the equality $a_n = n^{n-1}$, where a_n denotes the number of rooted trees on n elements, since $a_n = n\,\alpha_n$ (n choices for the root).

For the cycle index series, the combinatorial equation $\mathcal{V} = L_+(\mathcal{A})$ immediately gives

$$Z_{\mathcal{V}}(x_1, x_2, x_3, \ldots) = \frac{Z_{\mathcal{A}}(x_1, x_2, x_3, \ldots)}{1 - Z_{\mathcal{A}}(x_1, x_2, x_3, \ldots)}, \tag{14}$$

since $Z_{L_+} = x_1/(1-x_1)$. It is interesting to note that the equipotence $\mathcal{V} \equiv \text{End}_+$ is not an isomorphism. In fact,

$$\begin{aligned}Z_{\text{End}_+} &= Z_{\mathcal{S}_+} \circ Z_{\mathcal{A}} \\ &= \frac{1}{(1 - Z_{\mathcal{A}}(x_1, x_2, x_3, \ldots))(1 - Z_{\mathcal{A}}(x_2, x_4, x_6, \ldots)) \cdots} - 1,\end{aligned} \tag{15}$$

so that $Z_{\mathcal{V}} \neq Z_{\text{End}_+}$, which implies $\mathcal{V} \neq \text{End}_+$.

Consider now a species F possessing a notion of connected components, that is to say (as we have seen in Section 1.4) of the form

$$F = E(F^c), \tag{16}$$

where F^c is the species of *connected* F-structures. The operation of pointing permits a straightforward calculation of the average number, $\kappa_n(F)$, of connected components of a random F-structure on n elements, by the formula

$$\kappa_n(F) = \frac{|(F^c \cdot F)[n]|}{|F[n]|}, \qquad n \geq 0, \tag{17}$$

if $|F[n]| \neq 0$. Indeed, as an $E^\bullet(F^c)$-structure can be identified with an F-structure in which a connected component has been distinguished, it suffices to substitute the species F^c in the species $E^\bullet = X \cdot E$ to obtain (17). Applying formula (17) to the species Par, $\mathcal{S}$, and $\mathcal{G}$, gives:

– The average number of classes of a random partition on n elements is, in virtue of Equations (1.4.32) and (1.4.35),

$$\kappa_n(\text{Par}) = \frac{B_{n+1}}{B_n} - 1, \tag{18}$$

where B_n denotes the number of partitions of a set having n elements (Bell number).

– The average number of cycles of a random permutation on n elements is

$$\kappa_n(\mathcal{S}) = 1 + \frac{1}{2} + \frac{1}{3} + \cdots + \frac{1}{n} \sim \log(n). \tag{19}$$

– The average number of connected components of a random simple graph on n vertices is

$$\kappa_n(\mathcal{G}) = 2^{-\binom{n}{2}} \sum_{i=1}^{n} \binom{n}{i} 2^{\binom{n-i}{2}} |\mathcal{G}^c[i]|, \tag{20}$$

where $|\mathcal{G}^c[i]|$ is the number of connected graphs on i elements.

Cartesian Product of Species of Structures

The Cartesian product corresponds at a combinatorial level to the coefficient-wise product of exponential generating series, called the *Hadamard product* and denoted by $\times$:

$$\left(\sum_{n\geq 0} f_n \frac{x^n}{n!}\right) \times \left(\sum_{n\geq 0} g_n \frac{x^n}{n!}\right) = \sum_{n\geq 0} f_n g_n \frac{x^n}{n!}. \tag{21}$$

Definition 4. Let F and G be two species of structures. The species $F \times G$, called the *Cartesian product* of F and G, is defined as follows: An $(F \times G)$-structure on a finite set U is a pair $s = (f, g)$, where
i) f is an F-structure on U,
ii) g is a G-structure on U.
In other words, for all finite sets U, one has

$$(F \times G)[U] = F[U] \times G[U] \qquad \text{(Cartesian product)}. \tag{22}$$

The transport along a bijection $\sigma : U \longrightarrow V$ is carried out by setting

$$(F \times G)[\sigma](s) = (F[\sigma](f), G[\sigma](g)), \tag{23}$$

for any $(F \times G)$-structure $s = (f, g)$ on U.

An arbitrary $(F \times G)$-structure can be represented by a diagram of the type of Figure 6. The labeled enumeration of $(F \times G)$-structures satisfies

$$|(F \times G)[n]| = |F[n]| \cdot |G[n]|, \quad n \geq 0. \tag{24}$$

Remark 5. We underline that $F \times G$ is different from $F \cdot G$: Each of the structures f and g appearing in the formation of an $(F \times G)$-structure on U has underlying

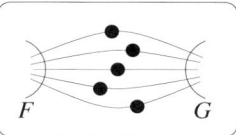

Fig. 6.

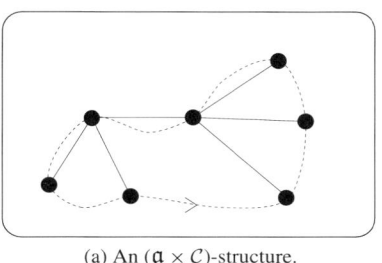

(a) An ($\mathfrak{a} \times \mathcal{C}$)-structure.

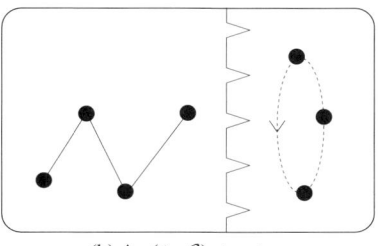

(b) An ($\mathfrak{a} \cdot \mathcal{C}$)-structure.

Fig. 7.

set U (in its entirety). However, for $(F \cdot G)$-structures (f, g) on U, the underlying sets U_1 and U_2 of f and g are disjoint (and $U_1 \cup U_2 = U$). The product $F \times G$ is sometimes called the *superposition* of F and G since an $(F \times G)$-structure on U is obtained by *superposing* an F-structure on U and a G-structure on U.

Example 6. Consider the species $\mathcal{C}$ of oriented cycles and the species $\mathfrak{a}$ of trees. Figure 7 illustrates the difference between an $(\mathfrak{a} \times \mathcal{C})$-structure and an $(\mathfrak{a} \cdot \mathcal{C})$-structure on a set of seven elements.

In order to describe the compatibility of the Cartesian product with passage to series, it is necessary to first define the *Hadamard product* $f \times g$ of two index series

$$f(\mathbf{x}) = \sum f_\mathbf{n} \frac{\mathbf{x}^\mathbf{n}}{\mathrm{aut}(\mathbf{n})}, \qquad g(\mathbf{x}) = \sum g_\mathbf{n} \frac{\mathbf{x}^\mathbf{n}}{\mathrm{aut}(\mathbf{n})},$$

where $\mathbf{x} = (x_1, x_2, x_3, \ldots)$, and $\mathbf{n} = (n_1, n_2, n_3, \ldots)$. These series are multiplied coefficient-wise:

$$(f \times g)(\mathbf{x}) = \sum f_\mathbf{n} g_\mathbf{n} \frac{\mathbf{x}^\mathbf{n}}{\mathrm{aut}(\mathbf{n})}. \tag{25}$$

We then have the following result, whose proof is left as an exercise.

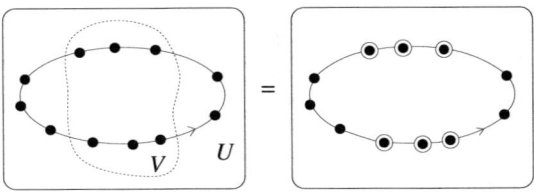

Fig. 8.

Proposition 7. *Let F and G be two species of structures. Then the series associated to the species $F \times G$ satisfy the equalities*

a) $(F \times G)(x) = F(x) \times G(x)$,

b) $(\widetilde{F \times G})(x) = (Z_F \times Z_G)(x, x^2, x^3, \ldots)$, (26)

c) $Z_{F \times G}(x_1, x_2, x_3, \ldots) = Z_F(x_1, x_2, x_3, \ldots) \times Z_G(x_1, x_2, x_3, \ldots)$.

Example 8. Consider the Cartesian product $\mathcal{C} \times \wp$ of the species $\mathcal{C}$ of oriented cycles by the species $\wp$ of subsets (of sets). A $(\mathcal{C} \times \wp)$-structure on a set U is then an oriented cycle on U on which one has *superimposed* a subset V of U (see Figure 8). In other words, a $(\mathcal{C} \times \wp)$-structure is an oriented cycle in which certain elements have been distinguished (the circled points in Figure 8). We now compute explicitly the series associated to this species. Since $|\mathcal{C}[n]| = (n-1)!$ if $n \geq 1$ and $|\wp[n]| = 2^n$, we obtain

$$|(\mathcal{C} \times \wp)[n]| = \begin{cases} (n-1)!2^n, & \text{if } n \geq 1, \\ 0, & \text{if } n = 0. \end{cases} \qquad (27)$$

We conclude that

$$(\mathcal{C} \times \wp)(x) = \sum_{n \geq 1} (n-1)! \frac{(2x)^n}{n!}$$
$$= \log\left(\frac{1}{1-2x}\right). \qquad (28)$$

A similar calculation, using the fact that fix $\wp[n_1, n_2, \ldots] = 2^{n_1+n_2+\cdots}$ and formula (1.4.18) for $Z_\mathcal{C}$, yields

$$Z_{\mathcal{C} \times \wp}(x_1, x_2, x_3, \ldots) = \sum_{k \geq 1} \frac{\phi(k)}{k} \log\left(\frac{1}{1-2x_k}\right). \qquad (29)$$

It follows that the type generating series of $(\mathcal{C} \times \wp)$-structures is of the form

$$(\widetilde{\mathcal{C} \times \wp})(x) = \sum_{k \geq 1} \frac{\phi(k)}{k} \log\left(\frac{1}{1-2x^k}\right). \qquad (30)$$

2.1. Pointing and Cartesian Product

We finally deduce that the number of unlabeled $(\mathcal{C} \times \wp)$-structures on a set having n elements is:

$$[x^n](\widetilde{\mathcal{C} \times \wp})(x) = \frac{1}{n} \sum_{d|n} \phi(d) 2^{\frac{n}{d}}. \tag{31}$$

The species E of sets is the *neutral element* for the Cartesian product, i.e., for any species F, one has

$$E \times F = F \times E = F. \tag{32}$$

Indeed, superimposing an F-structure on a set structure reduces to simply considering the F-structure. By restricting to the cardinality n, it is easy to verify that

$$E_n \times F = F \times E_n = F_n. \tag{33}$$

As the Cartesian product distributes over addition (see Exercise 7), we recover the canonical decomposition of a species

$$\begin{aligned} F &= F \times E \\ &= F \times (E_0 + E_1 + E_2 + \cdots + E_n + \cdots) \\ &= F_0 + F_1 + F_2 + \cdots + F_n + \cdots. \end{aligned} \tag{34}$$

For the Cartesian product, the operation of pointing can be distributed on one or the other factor:

$$(F \times G)^{\bullet} = F^{\bullet} \times G = F \times G^{\bullet}, \tag{35}$$

as can be seen from Figure 9. In particular, by taking $G = E$ (the species of sets), we obtain the equalities

$$F^{\bullet} = (F \times E)^{\bullet} = F \times E^{\bullet} = F \times (X \cdot E). \tag{36}$$

Hence, the operation of pointing can be expressed in terms of the Cartesian product and ordinary product.

It is interesting to note that the law of simplification (by a nonzero factor) is not valid for the Cartesian product of species. For example, $L \times L = S \times L$, but $L \neq S$.

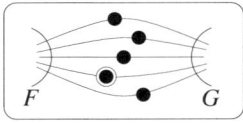

Fig. 9.

To end this section, let us now consider the series associated with the species $\mathfrak{a} \times \mathcal{C}$ mentioned in Example 6. As $|\mathfrak{a}[n]| = n^{n-2}$ and $|\mathcal{C}[n]| = (n-1)!$ if $n \geq 1$, we obtain the generating series

$$(\mathfrak{a} \times \mathcal{C})(x) = \sum_{n \geq 1} n^{n-3} x^n. \tag{37}$$

For the index series $Z_{\mathfrak{a} \times \mathcal{C}}$, the situation is more delicate since we do not yet know an expression for $Z_\mathfrak{a}$. Nevertheless, we can already affirm that many coefficients of $Z_{\mathfrak{a} \times \mathcal{C}}$ are zero. Indeed, we have seen the series $Z_\mathcal{C}$ in the form

$$Z_\mathcal{C} = \sum_{k \geq 1} \frac{\phi(k)}{k} \log\left(\frac{1}{1 - x_k}\right) = \sum_{k,m \geq 1} \frac{\phi(k) x_k^m}{km}. \tag{38}$$

Hence, only monomials of the form x_k^m enter into play in the expression of $Z_\mathcal{C}$. It then follows that $\operatorname{fix} \mathcal{C}[n_1, n_2, \ldots] = 0$, except if $(n_1, n_2, \ldots) = (0, 0, \ldots, i, 0, \ldots)$, $i = n_k$, $k \in \mathbb{N}$. We can thus assert (without knowing $Z_\mathfrak{a}$) that $Z_{\mathfrak{a} \times \mathcal{C}}$ is of the form

$$Z_{\mathfrak{a} \times \mathcal{C}} = w_1(x_1) + w_2(x_2) + \cdots + w_k(x_k) + \cdots, \tag{39}$$

for certain formal power series $w_k(x) \in \mathbb{Q}[\![x]\!]$. In fact, these series are identically zero, except $w_1(x)$ and $w_2(x)$ (see Exercise 4.1.11). Thus, $Z_{\mathfrak{a} \times \mathcal{C}}$ only depends upon x_1 and x_2:

$$Z_{\mathfrak{a} \times \mathcal{C}} = w_1(x_1) + w_2(x_2). \tag{40}$$

The series $Z_\mathfrak{a}$, $w_1(x_1)$, and $w_2(x_2)$ are explicitly calculated in Chapter 4.

Exercises for Section 2.1

1. Show that the pointing operation satisfies the following rules:

   a) $(F + G)^\bullet = F^\bullet + G^\bullet$, (additivity)

   b) $(F \cdot G)^\bullet = F^\bullet \cdot G + F \cdot G^\bullet$, (product rule) (41)

   c) $(F \circ G)^\bullet = (F' \circ G) \cdot G^\bullet$ (chain rule for pointing)

2. a) Show, by cutting off the head of nondegenerate vertebrates (see Example 3), that

$$\mathcal{V} = \mathcal{A} + \mathcal{V} \cdot \mathcal{A}. \tag{42}$$

   b) Deduce, for $n \geq 1$, the identity

$$n^n = \sum_{k=0}^{n-1} \binom{n}{k} k^k (n-k)^{n-k-1}. \tag{43}$$

3. For the species Bal, of ballots (see Example 1.3.10), and Oct, of octopuses (see Exercises 1.1.7 and 1.4.10), establish the isomorphisms

   a) $\text{Bal}^{\bullet} = \text{Bal}^2 \cdot E^{\bullet}$, (44)

   b) $\text{Oct}^{\bullet} + L(2X) = L(X) \cdot L(2X)$. (45)

4. The pointing operation of order n is defined by
$$F^{\bullet n} = (XD)^n F, \text{ where } D = d/dX. \quad (46)$$

   a) Show that
$$F^{\bullet n} = \sum_{k=0}^{n} S(n,k) X^k F^{(k)}, \quad (47)$$
   where the $S(n,k)$ are the Stirling numbers of the second kind (see Exercise 1.4.2).

   HINT: Use mathematical induction to show that $(XD)^n = \sum_{k=0}^{n} S(n,k) \times X^k D^k$.

   b) Express $X^n D^n$ with the help of the $(XD)^k$, $0 \le k \le n$, and the Stirling numbers of the first kind $s(n,k)$.

5. Establish formulas (18)–(20) giving the expected number of connected components of a random F-structure on n vertices in the cases $F = \text{Par}, S,$ and $\mathcal{G}$.

6. a) Let $F = G(H)$ and let n be an integer ≥ 0. Consider the random variable $\theta_n = $ the number of members of a G-assembly of random H-structures (i.e., of an F-structure) on $[n]$. Show that the expectation and the variance of θ_n are respectively given by

   i) $E(\theta_n) = \dfrac{|G^{\bullet}(H)[n]|}{|F[n]|}$,

   ii) $\text{Var}(\theta_n) = \dfrac{|G^{\bullet\bullet}(H)[n]|}{|F[n]|} - \left(\dfrac{|G^{\bullet}(H)[n]|}{|F[n]|}\right)^2$. (48)

   b) What happens to the preceding formulas in the case of cyclic assemblies (i.e., $G = \mathcal{C}$, the species of oriented cycles)?

7. a) Show that the Cartesian product of species of structures possesses the following properties: For all species F, G, and H,

   i) $(F \times G) \times H = F \times (G \times H)$, (associativity)

   ii) $F \times G = G \times F$, (commutativity)

   iii) $E \times F = F \times E = F$, (neutral element) (49)

   iv) $F \times (G + H) = F \times G + F \times H$, (distributivity)

   v) $(F \times G)^{\bullet} = F^{\bullet} \times G = F \times G^{\bullet}$.

b) Compare the following species
   i) $(F \times G)'$, $F' \times G$, and $F \times G'$,
   ii) $F \times (G \cdot H)$ and $(F \times G) \cdot (F \times H)$,
   iii) $(F \times G) \cdot H$ and $(F \cdot H) \times (G \cdot H)$,
   iv) $(F \times G) \circ H$ and $(F \circ H) \times (G \circ H)$.

c) Show that the passage to the generating series and the cycle index series is compatible with the Cartesian product (see Proposition 7).

8. Describe an isomorphism between the species $L \times L$ and $S \times L$. Deduce that the law of simplification is not valid for the Cartesian product.

9. Show the combinatorial equalities

$$
\begin{aligned}
&\text{a)} \quad C_3 \times C_3 = 2\,C_3, \\
&\text{b)} \quad X^3 \times C_3 = 2\,X^3, \\
&\text{c)} \quad (X \cdot E_2) \times (X \cdot E_2) = X \cdot E_2 + X^3, \\
&\text{d)} \quad \mathcal{C} \times \wp = \text{Oct} + \mathcal{C}.
\end{aligned}
\tag{50}
$$

10. a) For each $k \geq 0$, consider the species $\wp^{[k]}$ of subsets of cardinality k:

$$\wp^{[k]}[U] = \{V \mid V \subseteq U, |V| = k\}. \tag{51}$$

Establish the combinatorial equality

$$
\begin{aligned}
\wp^{[m]} \times \wp^{[n]} &= \sum_{k=0}^{\min(m,n)} E_{m-k} \cdot E_k \cdot E_{n-k} \cdot E \\
&= \sum_{k=0}^{\min(m,n)} E_{m-k} \cdot E_{n-k} \cdot \wp^{[k]}.
\end{aligned}
\tag{52}
$$

b) A lottery proceeds in the following manner: The "player" (resp., the "house") chooses a set V (resp., W) of k integers among the integers 1 to N. The player gains the pot i if $|V \cap W| = i$. In the case where $N = 49, k = 6$, calculate the probability that the player wins the pot i, for $i = 2, 3, 4, 5, 6$.

2.2. Functorial Composition

Definition 1. Let F and G be two species of structures. The species $F \,\square\, G$ (also denoted by $F[G]$), called the *functorial composite* of F and G, is defined as follows: An $(F \,\square\, G)$-structure on U is an F-structure placed on the set $G[U]$ of

2.2. Functorial Composition

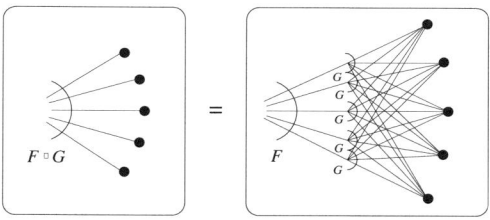

Fig. 1.

all the G-structures on U. In other words, for any finite set U,

$$(F \square G)[U] = F[G[U]]. \tag{1}$$

The transport along a bijection $\sigma : U \longrightarrow V$ is carried out by setting

$$(F \square G)[\sigma] = F[G[\sigma]] \tag{2}$$

(i.e., F-transport along the bijection $G[\sigma]$).

As a functor, the species $F \square G$ is the composite of the functors F and G, hence its name. The notation $\square$ is used in this text to avoid ambiguity with the substitution (partitional composition) of species of structures.

A generic $(F \square G)$-structure on U can be represented by a diagram of the type in Figure 1, where, on the right-hand side, it is understood that the totality of G-structures on U appear. The labeled enumeration of the $(F \square G)$-structures satisfies

$$|(F \square G)[n]| = |F[g_n]|, \tag{3}$$

with $g_n = |G[n]|$, which corresponds to an operation on the exponential formal power series, also denoted by $\square$:

$$\left(\sum_{n \geq 0} f_n \frac{x^n}{n!} \right) \square \left(\sum_{n \geq 0} g_n \frac{x^n}{n!} \right) = \sum_{n \geq 0} f_{g_n} \frac{x^n}{n!}, \tag{4}$$

under the hypothesis that $g_n \in \mathbb{N}$, for any $n \in \mathbb{N}$.

Example 2. Using functorial composition of species, we can express a variety of graph classes (simple, directed, with or without loops, etc.) in terms of simple species of structures. For example, the species $\mathcal{G}$ of all simple graphs (without loops) can be expressed as

$$\mathcal{G} = \wp \square \wp^{[2]}, \tag{5}$$

where $\wp$ denotes the species of subsets and $\wp^{[2]}$, that of subsets with two elements. Indeed a $\wp^{[2]}$-structure on a set amounts to considering a pair of elements, joined

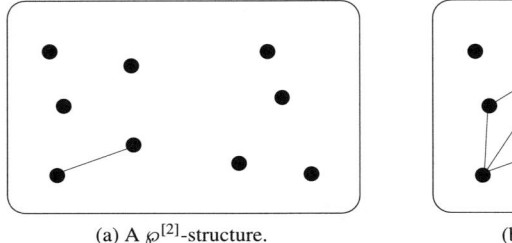
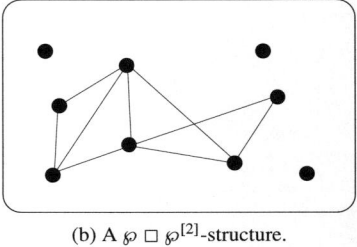

(a) A $\wp^{[2]}$-structure.   (b) A $\wp \,\square\, \wp^{[2]}$-structure.

Fig. 2.

by a segment in Figure 2(a). Such a pair of elements is called an *edge*. Moreover, a graph on a set U is nothing else but a selection among all possible edges. There is then a $\wp$-structure on the set of all $\wp^{[2]}$-structures on U. The reader can verify without difficulty that the transport of structures is as described in (2). Observe that, since $\wp = E \cdot E$ and $\wp^{[2]} = E_2 \cdot E$, the species of graphs can also be expressed in the form

$$\mathcal{G} = (E \cdot E) \,\square\, (E_2 \cdot E), \tag{6}$$

which only uses the species of sets, the product, the functorial composition, and the restriction to the cardinality 2.

The operation $\square$ is clearly associative but is not commutative (not even up to isomorphism). For example, the species $\wp^{[2]} \,\square\, \wp$ is identified with the species of *pairs of subsets*. It is not isomorphic to the species $\mathcal{G} = \wp \,\square\, \wp^{[2]}$ of graphs. The species $E^{\bullet}$ of pointed sets is the *neutral element* for the operation $\square$ (see Exercise 6), i.e., for any species F,

$$F \,\square\, E^{\bullet} = E^{\bullet} \,\square\, F = F. \tag{7}$$

In order to describe the behavior of the composition of species with respect to power series, it is convenient to first define a corresponding operation $Z_F \,\square\, Z_G$ in the context of cycle index series. It clearly follows from the definition of the transport of $(F \,\square\, G)$-structures that

$$\begin{aligned} \operatorname{fix}(F \,\square\, G)[\sigma] &= \operatorname{fix} F[G[\sigma]] \\ &= \operatorname{fix} F[(G[\sigma])_1, (G[\sigma])_2, \ldots], \end{aligned} \tag{8}$$

for any permutation σ where $[(G[\sigma])_1, (G[\sigma])_2, \ldots]$ denotes the cycle type of the permutation $G[\sigma]$, as defined in Section 1.2. The following proposition shows that the index series Z_G completely determines this cycle type.

2.2. Functorial Composition

Proposition 3. *Let G be a species of structures, $\sigma \in S_n$, and $k \geq 1$. Then the number of cycles of length k in $G[\sigma]$ is given by*

$$(G[\sigma])_k = \frac{1}{k} \sum_{d|k} \mu\left(\frac{k}{d}\right) \text{fix}\, G[\sigma^d], \tag{9}$$

where μ denotes the Möbius function for positive integers.

Proof. For any permutation β, and any $k \geq 1$,

$$\text{fix}\, \beta^k = \sum_{d|k} d\, \beta_d. \tag{10}$$

Indeed, an element is left fixed by the permutation β^k if and only if it is found in a cycle of β of length d, where d divides k. By applying Möbius inversion to (10), we deduce

$$\beta_k = \frac{1}{k} \sum_{d|k} \mu(k/d)\, \text{fix}\, \beta^d. \tag{11}$$

If $\beta = G[\sigma]$, then by functoriality $\beta^d = G[\sigma^d]$, giving the result. ∎

In other words, the coefficients of the cycle index series Z_G permits the calculation of all the $(G[\sigma])_k$. The following definition is thus legitimate.

Definition 4. The *composite* $Z_F \square Z_G$ is defined by the formula

$$Z_F \square Z_G = \sum_{n \geq 0} \frac{1}{n!} \sum_{\sigma \in S_n} \text{fix}\, F[(G[\sigma])_1, (G[\sigma])_2, \ldots] x_1^{\sigma_1} x_2^{\sigma_2} \cdots, \tag{12}$$

where $(\sigma_1, \sigma_2, \ldots)$ denotes the cycle type of a permutation $\sigma \in S_n, n = 0, 1, 2, \ldots$.
Immediately implied is the following result.

Proposition 5. *Let F and G be two species of structures. Then the series associated to the species $F \square G$ satisfy the equalities*

a) $(F \square G)(x) = F(x) \square G(x)$,

b) $(\widetilde{F \square G})(x) = (Z_F \square Z_G)(x, x^2, x^3, \ldots)$, (13)

c) $Z_{F \square G}(x_1, x_2, x_3, \ldots) = Z_F(x_1, x_2, x_3, \ldots) \square Z_G(x_1, x_2, x_3, \ldots)$.

A particularly interesting case is that where $F = \wp$, the species of subsets. Many varieties of graphs and relations can be described as the composite $\wp \square G$ of $\wp$ and of a given species G. Here are some examples:

- Simple graphs: $\mathcal{G} = \wp \,\square\, \wp^{[2]}$;
- Directed graphs: $\mathcal{D} = \wp \,\square\, (E^{\bullet} \times E^{\bullet})$;
- m-ary relations: $\mathrm{Rel}^{[m]} = \wp \,\square\, (E^{\bullet})^{\times m}$,

where

$$(E^{\bullet})^{\times m} = \underbrace{E^{\bullet} \times \cdots \times E^{\bullet}}_{(m \text{ factors})}. \tag{14}$$

These species have the generating series

$$\mathcal{G}(x) = \sum_{n \geq 0} 2^{\binom{n}{2}} \frac{x^n}{n!}, \tag{15}$$

$$\mathcal{D}(x) = \sum_{n \geq 0} 2^{n^2} \frac{x^n}{n!}, \tag{16}$$

$$\mathrm{Rel}^{[m]}(x) = \sum_{n \geq 0} 2^{n^m} \frac{x^n}{n!}. \tag{17}$$

We are now going to calculate the cycle index series of such species. Since for any permutation β,

$$\mathrm{fix}\,\wp[\beta] = 2^{\sum_{k \geq 1} \beta_k}, \tag{18}$$

we immediately obtain:

Proposition 6. *For any species of structures G and any permutation σ,*

$$\mathrm{fix}\,(\wp \,\square\, G)[\sigma] = 2^{\sum_{k \geq 1} (G[\sigma])_k}. \tag{19}$$

Thus, the index series of a species of the form $\wp \,\square\, G$ simply depends on the numbers $(G[\sigma])_k$. For the three preceding examples, these numbers are given by the following expressions:

Proposition 7. *For any permutation σ of cycle type $(\sigma_1, \sigma_2, \sigma_3, \ldots)$,*

a) $(\wp^{[2]}[\sigma])_k = \dfrac{1}{2} \displaystyle\sum_{[i,j]=k} (i,j)\sigma_i \sigma_j + \sigma_{2k} - \sigma_k + \dfrac{1}{2}(k \bmod 2)\sigma_k,$ (20)

b) $((E^{\bullet} \times E^{\bullet})[\sigma])_k = \displaystyle\sum_{[i,j]=k} (i,j)\sigma_i \sigma_j,$ (21)

c) $((E^{\bullet})^{\times m}[\sigma])_k = \displaystyle\sum_{[j_1,\ldots,j_m]=k} \dfrac{j_1 \cdots j_m}{k} \sigma_{j_1} \cdots \sigma_{j_m},$ (22)

where $[j_1, \ldots, j_m]$ denotes the least common multiple of $j_1, \ldots, j_m$ and $(j_1, \ldots, j_m)$, the greatest common divisor.

2.2. Functorial Composition

Proof. These formulas can be established directly by simple counting arguments. For example, for (21), recall that $(E^{\bullet} \times E^{\bullet})[U] = U \times U$. Then a pair of elements $(a, b) \in U \times U$, where a belongs to a cycle of length i of σ and b to a cycle of length j, generates a cycle $(a, b) \longrightarrow (\sigma(a), \sigma(b)) \longrightarrow (\sigma^2(a), \sigma^2(b)) \longrightarrow \cdots$ of length $[i, j]$. Each of the $\sigma_i \sigma_j$ cycles induces $(i, j) = ij/[i, j]$ such cycles, whence (21).

Formula (22) is proven in the same manner, by induction on m. A direct combinatorial proof of (20) is proposed in Exercise 1. We present in what follows a more algebraic proof, based on relation (9). Observe at first that for any permutation σ,

$$\text{fix } \wp^{[2]}[\sigma] = \binom{\sigma_1}{2} + \sigma_2 \qquad (23)$$

since a pair of elements left fixed by σ either consists of a pair of fixed points of σ or of a cycle of length 2 of σ. We obtain, in virtue of (9), (11), and (23) and the following lemma,

$$(\wp^{[2]}[\sigma])_k = \frac{1}{k} \sum_{d \mid k} \mu(k/d) \text{ fix } \wp^{[2]}[\sigma^d]$$

$$= \frac{1}{k} \sum_{d \mid k} \mu(k/d) \frac{1}{2} \left((\sigma^d)_1^2 - (\sigma^d)_1 + 2(\sigma^d)_2 \right)$$

$$= \frac{1}{2} \left(\sum_{[i,j]=k} (i, j)\sigma_i \sigma_j - \sigma_k + 2\sigma_{2k} - \sigma_k \chi(k \text{ is even}) \right),$$

whence formula (20). ∎

Lemma 8. *For any permutation σ and any $k \geq 1$,*

a) $\displaystyle \frac{1}{k} \sum_{d \mid k} \mu\left(\frac{k}{d}\right) (\sigma^d)_1^2 = \sum_{[i,j]=k} (i, j)\sigma_i \sigma_j,$ \hfill (24)

b) $\displaystyle \frac{1}{k} \sum_{d \mid k} \mu\left(\frac{k}{d}\right) (\sigma^d)_2 = \begin{cases} \sigma_{2k}, & \text{if } k \text{ is odd,} \\ \sigma_{2k} - \frac{1}{2}\sigma_k, & \text{if } k \text{ is even.} \end{cases}$ \hfill (25)

Proof. To prove (24), it suffices to apply Proposition 3 with $G = E^{\bullet} \times E^{\bullet}$, i.e., to join the identities (9) and (21), observing also that for any permutation τ, fix $(E^{\bullet} \times E^{\bullet})[\tau] = \tau_1^2$.

For (25), we have, by (11),

$$(\sigma^d)_2 = \frac{1}{2}\left((\sigma^{2d})_1 - (\sigma^d)_1\right) \qquad (26)$$

and

$$(\sigma^{2d})_1 = \sum_{i|2d} i\sigma_i$$
$$= \sum_{j|d} 2j\sigma_{2j} + \sum_{i|d} i\sigma_i \chi(i \text{ is odd}). \tag{27}$$

Using the fact that for any function f,

$$\sum_{d|k} \mu\left(\frac{k}{d}\right) \sum_{i|d} f(i) = f(k), \tag{28}$$

we obtain

$$\frac{1}{k} \sum_{d|k} \mu\left(\frac{k}{d}\right) (\sigma^d)_2 = \frac{1}{2k}(2k\sigma_{2k} + k\sigma_k \chi(k \text{ is odd}) - k\sigma_k),$$

whence (25). ∎

Combining Propositions 6 and 7, we find for the species $\mathcal{G}$ of simple graphs, $\mathcal{D}$ of directed graphs, and $\text{Rel}^{[m]}$ of m-ary relations, that for any permutation σ,

$$\text{fix } \mathcal{G}[\sigma] = 2^{\frac{1}{2} \sum_{i,j \geq 1}(i,j)\sigma_i \sigma_j - \frac{1}{2} \sum_{k \geq 1}(k \bmod 2)\sigma_k}, \tag{29}$$

$$\text{fix } \mathcal{D}[\sigma] = 2^{\sum_{i,j \geq 1}(i,j)\sigma_i \sigma_j}, \tag{30}$$

$$\text{fix } \text{Rel}^{[m]}[\sigma] = 2^{\sum_{i_1,\ldots,i_m \geq 1} i_1 \cdots i_m \, \sigma_{i_1} \cdots \sigma_{i_m}/[i_1,\ldots,i_m]}. \tag{31}$$

These expressions permit the calculation of the cycle index series and the type generating series of these species. For example,

$$\widetilde{\mathcal{G}}(x) = 1 + x + 2x^2 + 4x^3 + 11x^4 + 34x^5 + 156x^6$$
$$+ 1044x^7 + 12346x^8 + 274668x^9 + \cdots. \tag{32}$$

Exercises for Section 2.2

1. Establish directly formula (20) giving the number $(\wp^{[2]}[\sigma])_k$ for a given permutation σ of cycle type $(\sigma_1, \sigma_2, \sigma_3, \ldots)$, by considering separately the cycles of length k of $\wp^{[2]}[\sigma]$ originating from either two different cycles of σ or from the same cycle of σ.

2. (See [74].) Show that for any permutation σ of cycle type $(\sigma_1, \sigma_2, \ldots)$ and for all $k \geq 1$ and $m \geq 1$,

   a) $\displaystyle\sum_{d|k} \mu(k/d)(\sigma^d)_1^m = \sum_{[j_1,\cdots,j_m]=k} j_1 \cdots j_m \, \sigma_{j_1} \cdots \sigma_{j_m},$

   b) $\displaystyle(\sigma^k)_m = \sum_{d|k,(m,k/d)=1} d\sigma_{dm},$ (33)

   c) $\displaystyle\frac{1}{k}\sum_{d|k} \mu(k/d)(\sigma^d)_m = \sum_{d|(m,k)} \frac{\mu(d)}{d} \sigma_{mk/d}.$

3. Show that
   a) for any permutation β and any $\omega \geq 1$, with $\beta^\omega = Id$,

   $$\sum_{k \geq 1} \beta_k = \frac{1}{\omega} \sum_{d|\omega} \phi(\omega/d) \text{ fix } \beta^d, \tag{34}$$

   b) for any species of structures G and any permutation σ of order ω,

   $$\text{fix } (\wp \square G)[\sigma] = 2^{(1/\omega) \sum_{d|\omega} \phi(\omega/d) \text{ fix } G[\sigma^d]}, \tag{35}$$

   where ϕ denotes Euler's ϕ-function.

4. Denote by $\mathcal{G}_l$, the species of (undirected) graphs with at most one loop at each vertex and by $\mathcal{D}_w$, the species of directed graphs without loops.
   a) Verify the following relations:

   i) $\mathcal{G}_l = \wp \times \mathcal{G},$ ii) $\mathcal{D} = \wp \times \mathcal{D}_w.$ (36)

   b) Determine fix $\mathcal{G}_l[\sigma]$ and fix $\mathcal{D}_w[\sigma]$.

5. Write a program permitting the calculation of the first terms of the cycle index series and the type generating series of the following species:

   a) $\mathcal{G}$,   b) $\mathcal{G}_l$,   c) $\mathcal{D}$,   d) $\mathcal{D}_w$.

6. Starting from the definitions, show that
   a) the operation $\square$ is associative,
   b) the species $E^\bullet$ of pointed sets is the neutral element for the operation $\square$,
   c) the operation $\square$ is distributive on the right by the operation of Cartesian product $\times$:

   $$(F \times G) \square H = (F \square H) \times (G \square H), \tag{37}$$

   d) the operation $\square$ is not distributive on the left by $\times$ (give an example illustrating this fact),
   e) the Cartesian product squared $F \times F$ can be expressed with the help of the ordinary product and the operation $\square$:

   $$F \times F = ((X + X^2) \cdot E) \square F. \tag{38}$$

7. Calculate the following series $(\mathcal{C} \square \mathcal{C}_p)(x)$, $(\mathcal{C} \widetilde{\square} \mathcal{C}_p)(x)$, and $Z_{\mathcal{C} \square \mathcal{C}_p}$, where $\mathcal{C}_p$ denotes the species of oriented cycles restricted to the cardinality p, a prime number.
8. Let F and G be any species of structures.
    a) Compare the species $(F \cdot E) \square G$ and $F \square (E \cdot G)$.
    b) If $G(0) \neq 0$, show that $F \circ G$ is a subspecies (see Definition 2.5.1) of $(F \cdot E) \square (G \cdot E)$.
    c) What happens in b) in the case of the species of graphs (see formula (6))?
9. Let U be a finite set. A *covering* of U is a set of nonempty subsets of U whose union is U. Consider the species Cov (resp., Cov$^{[m]}$) of coverings of sets (resp., coverings of sets by exactly m nonempty subsets).
    a) Show that
    $$E \cdot \text{Cov} = \wp \square \wp^+, \quad E \cdot \text{Cov}^{[m]} = \wp^{[m]} \square \wp^+ \tag{39}$$
    where $\wp^+$ is the species of nonempty subsets.
    b) If $|U| = n$, show that
    $$\begin{aligned} \text{i)} \quad & |\text{Cov}[U]| = \sum_{i=0}^{n} (-1)^i \binom{n}{i} 2^{2^{n-i}-1}, \\ \text{ii)} \quad & |\text{Cov}^{[m]}[U]| = \sum_{i=0}^{n} (-1)^i \binom{n}{i} \binom{2^{n-i}-1}{m}. \end{aligned} \tag{40}$$

10. By definition, a *simplicial complex* on a finite set U is a collection C of subsets of U, called *simplices*, such that
    $$u \in U \Rightarrow \{u\} \in C \quad \text{and} \quad \emptyset \neq T \subseteq S \in C \Rightarrow T \in C. \tag{41}$$
    The *dimension* of a complex C is the maximum dimension of its simplices, the dimension of a simplex $S \in C$ being $|S| - 1$. Let $m \geq 0$; a complex C is called a pure m-complex if all of its maximal simplices are of dimension m. Show that the species $C^{[m]}$ of pure m-complexes satisfies
    $$\begin{aligned} \text{a)} \quad & C^{[m]} = \wp \square \wp^{[m+1]}, \\ \text{b)} \quad & \text{fix } C^{[m]}[\sigma] = 2^{P_m(\sigma_1, \sigma_2, \ldots)}, \end{aligned} \tag{42}$$
    where $P_m \in \mathbb{Z}[\sigma_1, \sigma_2, \ldots]$ is a polynomial of total degree $m+1$ in the variables $\sigma_1, \sigma_2, \ldots$.
11. (See [74].) Let F be any species and m, an integer ≥ 0. Define the species $F^{\langle m \rangle}$ of F-structured words of length m by setting, for any finite set V,
    $$F^{\langle m \rangle}[V] = (F[m] \times \Phi[m, V])/\mathcal{S}_m, \tag{43}$$

where $\Phi[m, V] = \{\varphi \mid \varphi : [m] \longrightarrow V\}$, $\mathcal{S}_m$ denotes the symmetric group of order m, and the quotient set (43) is interpreted as the set of orbits of the action

$$\mathcal{S}_m \times (F[m] \times \Phi[m, V]) \longrightarrow (F[m] \times \Phi[m, V]), \tag{44}$$

defined by $\beta \cdot (s, \varphi) = (F[\beta](s), \varphi \circ \beta^{-1})$. The elements of $F^{\langle m \rangle}[V]$ are called the *F-structured words of length m on the alphabet V* or also the *types of V-colored F-structures of order m*. In particular, taking $F = L, E, C$,

$L^{\langle m \rangle}$ is the species of words of length m,

$E^{\langle m \rangle}$ is the species of abelian words of length m,

$C^{\langle m \rangle}$ is the species of circular words of length m.

a) Prove that for any permutation $\sigma : V \longrightarrow V$,

$$\text{fix } F^{\langle m \rangle}[\sigma] = Z_{F_m}(\text{fix } \sigma, \text{fix } \sigma^2, \text{fix } \sigma^3 \cdots). \tag{45}$$

HINT: Refer to the case where F is a molecular species (see Section 2.6) of the form X^m/H. Show that then $F^{\langle m \rangle}[V] = \Phi[m, V]/H$ and use Exercise A.1.9.

b) Deduce expressions for the series $F^{\langle m \rangle}(x)$ and $\widetilde{F^{\langle m \rangle}}(x)$.

12. Every m-ary relation on U can be considered as a *set of words of length m on the alphabet U*, whence the combinatorial equation

$$\text{Rel}^{[m]} = \wp \,\square\, L^{\langle m \rangle}, \tag{46}$$

where $L^{\langle m \rangle}$ is the species of words of length m. More generally, given a species F, define the species $\text{Rel}_F^{[m]}$ of m-ary F-structured relations by

$$\text{Rel}_F^{[m]} = \wp \,\square\, F^{\langle m \rangle}, \tag{47}$$

where $F^{\langle m \rangle}$ is defined in the preceding exercise. Prove that for any permutation $\sigma : U \longrightarrow U$:

a) $\text{fix Rel}^{[m]}[\sigma] = 2^{\sum_{i_1,\ldots,i_m \geq 1} i_1 \cdots i_m \sigma_{i_1} \cdots \sigma_{i_m}/[i_1,\ldots,i_m]}$, (48)

b) $\text{fix Rel}_F^{[m]}[\sigma] = 2^{\frac{1}{\omega} \sum_{d \mid \omega} \phi(\omega/d) Z_{F_m}(f_d, f_{2d}, \ldots)}$, (49)

where $f_k = \text{fix } \sigma^k = \sum_{d \mid k} d\sigma_d$, ω is the order of σ, and ϕ denotes Euler's ϕ-function.

2.3. Weighted Species

In enumerative combinatorics, it is often required to consider some parameters related to the characteristics of the structures. For instance, in computer science, the complexity analysis of algorithms often involves the enumeration of structures according to certain descriptive parameters such as the number of leaves or the depth of binary rooted trees. It is this kind of enumeration problem that we are now going to address through the introduction of a variant of the concept of species of structures: *weighted species*. All constructions introduced in the preceding chapters and sections will be extended by taking into account this addition of weighting.

Example 1. Assign to each rooted tree $\alpha \in \mathcal{A}[U]$ a *weight* $w(\alpha)$ by setting

$$w(\alpha) = t^{f(\alpha)}, \tag{1}$$

where t is a formal variable and $f(\alpha)$ denotes the number of leaves of α (see Figure 1). This permits the regrouping of rooted trees according to the descriptive parameter "number of leaves." We say that the set $\mathcal{A}[U]$ of all rooted trees on U is *weighted* by (1) and also that the variable t acts as a leaf "counter." The "inventory" of rooted trees on U according to this weight w, denoted by $|\mathcal{A}[U]|_w$, is defined as the sum of the weights $w(\alpha)$, $\alpha \in \mathcal{A}[U]$:

$$|\mathcal{A}[U]|_w = \sum_{\alpha \in \mathcal{A}[U]} w(\alpha) = \sum_{\alpha \in \mathcal{A}[U]} t^{f(\alpha)}. \tag{2}$$

Regrouping terms according to the powers of t gives a polynomial in t. If $|U| = n$, we have

$$|\mathcal{A}[U]|_w = a_n(t) = \sum_{k=0}^{n} a_{n,k} t^k. \tag{3}$$

The coefficient $a_{n,k}$ gives the number of rooted trees on n elements of which k are leaves. This inventory is more refined than the simple enumeration of rooted trees; the substitution $t := 1$ has the effect of giving weight 1 to each rooted tree and we obtain

$$a_n(1) = |\mathcal{A}[U]| = n^{n-1}, \quad n \geq 1. \tag{4}$$

One can equally be interested in the enumeration of unlabeled rooted trees, and also of rooted trees left fixed by a given permutation, according to the same

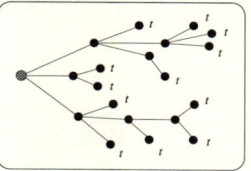

Fig. 1. A rooted tree α with $w(\alpha) = t^{12}$.

parameter "number of leaves." It is for this purpose that the concept of species of weighted structures is introduced.

Note that the weighting (1) is a function

$$w : \mathcal{A}[U] \longrightarrow \mathbb{A}, \quad (5)$$

where $\mathbb{A}$ is a polynomial ring[2] in the variable t. The pair $(\mathcal{A}[U], w)$ is said to be a weighted set in the ring $\mathbb{A}$ ($\mathbb{A}$-weighted, for short). Let us first study the general concept of $\mathbb{A}$-weighted sets as well as related constructions.

Let $\mathbb{K} \subseteq \mathbb{C}$ be an integral domain (for example $\mathbb{Z}$, $\mathbb{R}$, or $\mathbb{C}$) and $\mathbb{A}$, a ring of formal power series in an arbitrary number of variables, with coefficients in $\mathbb{K}$.

Definition 2. An $\mathbb{A}$-*weighted set* is a pair (A, w), where A is a (finite or infinite) set and

$$w : A \longrightarrow \mathbb{A} \quad (6)$$

is a function which associates a *weight* $w(\alpha) \in \mathbb{A}$ to each element $\alpha \in A$.

If the following sum, denoted by $|A|_w$, exists, the weighted set (A, w) is said to be *summable* (for a precise definition, see Exercise 1) and $|A|_w$ is called the *inventory* (or *total weight* or *cardinality*) of the weighted set (A, w):

$$|A|_w = \sum_{\alpha \in A} w(\alpha). \quad (7)$$

Many set-theoretic constructions can be extended to $\mathbb{A}$-weighted sets.

Definition 3. Let (A, w) and (B, v) be $\mathbb{A}$-weighted sets. A *morphism of $\mathbb{A}$-weighted sets*

$$f : (A, w) \longrightarrow (B, v) \quad (8)$$

is a function $f : A \longrightarrow B$ compatible with the weighting (one also says that the function f is *weight preserving*), that is to say, such that $w = v \circ f$. Moreover, if f is a bijection, f is called an *isomorphism of weighted sets* and we write $(A, w) \simeq (B, v)$.

Observe that

$$(A, w) \simeq (B, v) \implies |A|_w = |B|_v. \quad (9)$$

Definition 4. Let (A, w) and (B, v) be $\mathbb{A}$-weighted sets. Define

---

[2] Translator's note: $\mathbb{A}$ for the French "anneau."

i) The *sum* $(A, w) + (B, v)$ as the $\mathbb{A}$-weighted set $(A + B, \mu)$, where $A + B$ denotes the disjoint union of the sets A and B and μ is the weight function defined by

$$\mu(x) = \begin{cases} w(x), & \text{if } x \in A \\ v(x), & \text{if } x \in B; \end{cases} \quad (10)$$

ii) The *product* $(A, w) \times (B, v)$ as the $\mathbb{A}$-weighted set $(A \times B, \rho)$, where $A \times B$ denotes the Cartesian product of sets A and B and ρ is the weight function defined by

$$\rho(x, y) = w(x)v(y). \quad (11)$$

The proof of the following proposition is left as an exercise.

Proposition 5. *Let (A, w) and (B, v) be $\mathbb{A}$-weighted sets. With the preceding notation and conventions, we have*

a) $|A + B|_\mu = |A|_w + |B|_v$,

b) $|A \times B|_\rho = |A|_w |B|_v$. $\quad (12)$

To every finite set A, one can associate an $\mathbb{A}$-weighted set (A, w) by giving each element $\alpha \in A$ the weight $w(\alpha) = 1 \in \mathbb{A}$. This weighting is called *trivial*. Then $|A|_w = |A|$.

Weighted Species

We now introduce the notion of weighted species. It constitutes an important variant of the concept of species of structures, which allows a more refined enumeration of structures and their classification according to various *descriptive parameters*, by the addition of well-chosen *weights*.

Definition 6. Let $\mathbb{A}$ be a ring of formal power series or of polynomials over a ring $\mathbb{K} \subseteq \mathbb{C}$. An $\mathbb{A}$-*weighted species* is a rule F which
i) *produces*, for each finite set U, a finite or summable $\mathbb{A}$-weighted set

$$(F[U], w_U),$$

ii) *produces*, for each bijection $\sigma : U \longrightarrow V$, a function

$$F[\sigma] : (F[U], w_U) \longrightarrow (F[V], w_V)$$

preserving the weights (i.e., a weighted set morphism).

2.3. Weighted Species

Moreover, the functions $F[\sigma]$ must satisfy the following functoriality properties:
a) if $\sigma : U \longrightarrow V$ and $\tau : V \longrightarrow W$ are bijections, then

$$F[\tau \circ \sigma] = F[\tau] \circ F[\sigma], \tag{13}$$

b) for each set U, if Id_U denotes the identity bijection of U to U, then

$$F[\mathrm{Id}_U] = \mathrm{Id}_{F[U]}.$$

As before, an element $s \in F[U]$ is called an F-*structure* on U, and the function $F[\sigma]$, the *transport of F-structures* along σ.

It follows from the definition that the transport of structures $F[\sigma]$ along a bijection $\sigma : U \longrightarrow V$ is a weight preserving bijection and that

$$|F[U]|_{w_U} = |F[V]|_{w_V}. \tag{14}$$

Two F-structures s_1 and s_2 are called *isomorphic* (i.e., have the same *type*) if they are transportable one on the other along a bijection σ. As $F[\sigma]$ preserves weights, s_1 and s_2 are then forced to have the same weight. This permits the weighting of the set $F[U]/\sim$ of isomorphism types of F-structures (i.e., unlabeled F-structures), by defining the weight of a type as the weight of an arbitrary structure representing this type.

To be more precise, it is useful to write $F = F_w$ to denote the weighted species F together with the family of all the weight functions $w_U : F[U] \longrightarrow \mathbb{A}$ associated to F.

Example 7. Consider the species $\mathcal{S}_w$ of permutations with cycle counter α, i.e., the weight $w(\sigma)$ of a permutation σ is $w(\sigma) = \alpha^{\mathrm{cyc}(\sigma)}$, where α is a formal variable and $\mathrm{cyc}(\sigma)$ is the number of cycles of σ. This weighting has values in the ring of polynomials $\mathbb{Z}[\alpha]$, so $\mathcal{S}_w$ is a $\mathbb{Z}[\alpha]$-weighted species.

Observe that every species of structures F can be identified with an $\mathbb{A}$-weighted species by considering that the sets $F[U]$ are all provided with the trivial weighting. Moreover, this identification is compatible with all the operations and the passage to the diverse generating series introduced below.

Definition 8. Let $F = F_w$ be an $\mathbb{A}$-weighted species of structures. The *generating series* of F is the exponential formal power series $F_w(x)$ with coefficients in $\mathbb{A}$ defined by

$$F_w(x) = \sum_{n \geq 0} |F[n]|_w \frac{x^n}{n!}, \tag{15}$$

where $|F[n]|_w$ is the inventory of the set of F-structures on $[n]$. Its *cycle index series* is defined by

$$Z_{F_w}(x_1, x_2, x_3, \ldots) = \sum_{n \geq 0} \frac{1}{n!} \left(\sum_{\sigma \in S_n} |\text{Fix } F[\sigma]|_w x_1^{\sigma_1} x_2^{\sigma_2} x_3^{\sigma_3} \cdots \right), \quad (16)$$

where as before, σ_i is the number of cycles of length i in σ. Note that the set Fix $F[\sigma]$ inherits the weighting on $F[U]$ of which it is a subset, whence $|\text{Fix } F[\sigma]|_w$ is the inventory of all the F-structures on $[n]$ left fixed under the transport $F[\sigma]$. Since the weighting is preserved by transport of structures, we can define the *type generating series* of F_w:

$$\widetilde{F}_w(x) = \sum_{n \geq 0} |F[n]/\sim|_w x^n, \quad (17)$$

where $\sim$ is the isomorphism relation. Hence, $|F[n]/\sim|_w$ is the inventory of unlabeled F-structures on n points.

As in the nonweighted case, by suitably regrouping terms, the cycle index series can be written as

$$Z_{F_w}(x_1, x_2, \ldots) = \sum_{n_1 + 2n_2 + \cdots < \infty} |\text{Fix } F[n_1, n_2, \ldots]|_w \frac{x_1^{n_1} x_2^{n_2} \cdots}{1^{n_1} n_1! 2^{n_2} n_2! \cdots}, \quad (18)$$

where $|\text{Fix } F[n_1, n_2, \ldots]|_w$ denotes the inventory of the set F-structures left fixed under the action of a permutation σ of type $(n_1, n_2, n_3, \ldots)$. Moreover, the formulas

a) $\quad F_w(x) = Z_{F_w}(x, 0, 0, \ldots) \quad$ and $\quad$ b) $\quad \widetilde{F}_w(x) = Z_{F_w}(x, x^2, x^3, \ldots) \quad (19)$

remain valid.

The operations $+$, $\cdot$, $\circ$, $'$, $\overset{\cdot}{}$, $\times$, and $\square$ are defined in the same fashion on weighted species as in the nonweighted case, but the weights of the structures have to be carefully defined. See Table 2 for the precise definitions. Observe that for the product, the substitution, and the Cartesian product, a principle of multiplicativity, induced from (12), is used to define the weights. The goal, of course, is to reflect correctly the corresponding operations on the generating series. However, the plethystic substitution of the cycle index series must undergo an important modification in the weighted case.

Definition 9. Let F_w and G_v be two weighted species of structures, such that $G(0) = 0$ (i.e., there is no G-structure on the empty set). The *plethystic substitution* of Z_{G_v} in Z_{F_w}, denoted by $Z_{F_w} \circ Z_{G_v}$ (or $Z_{F_w}(Z_{G_v})$) is defined by

$$Z_{F_w} \circ Z_{G_v} = Z_{F_w}((Z_{G_v})_1, (Z_{G_v})_2, (Z_{G_v})_3, \ldots), \quad (20)$$

where, for $k = 1, 2, 3, \ldots$, $(Z_{G_v})_k(x_1, x_2, x_3, \ldots) = Z_{G_{v^k}}(x_k, x_{2k}, x_{3k}, \ldots)$.

2.3. Weighted Species

Table 2

Species	Structure	Weight
$F_w + G_v$	s	$w(s)$ if $s \in F[U]$ $v(s)$ if $s \in G[U]$
$F_w \cdot G_v$	$s = (f, g)$	$w(f)v(g)$
$F_w \circ G_v$	$s = (\pi, f, (\gamma_p)_{p \in \pi})$	$w(f) \prod_{p \in \pi} v(\gamma_p)$
F'_w	s	$w(s)$
$F_w^{\bullet}$	$s = (f, u)$	$w(f)$
$F_w \times G_v$	$s = (f, g)$	$w(f)v(g)$
$F_w \square G$ (G nonweighted)	s	$w(s)$

The reader should take note that in the series $Z_{G_{v^k}}(x_k, x_{2k}, x_{3k}, \ldots)$, the weighting v is raised to the power k (in the ring $\mathbb{A}$) and the indices of the variables x_i are multiplied by k.

Remark 10. It often happens that the weights of the structures of a species G_v are monomials in the variables $\alpha, \beta, \gamma, \ldots$. In this case, setting $g(\alpha, \beta, \gamma, \ldots; x_1, x_2, x_3, \ldots) = Z_{G_v}(x_1, x_2, x_3, \ldots)$,

$$g_k = (Z_{G_v})_k = g(\alpha^k, \beta^k, \gamma^k, \ldots; x_k, x_{2k}, x_{3k}, \ldots). \tag{21}$$

Proposition 11. *Let F_w and G_v be two $\mathbb{A}$-weighted species of structures. Then*

a) $(F_w + G_v)(x) = F_w(x) + G_v(x),$ a') $Z_{(F_w + G_v)} = Z_{F_w} + Z_{G_v};$

b) $(F_w G_v)(x) = F_w(x) G_v(x),$ b') $Z_{(F_w \cdot G_v)} = Z_{F_w} Z_{G_v};$

c) $F'_w(x) = \dfrac{d}{dx} F_w(x),$ c') $Z_{F'_w} = \dfrac{\partial}{\partial x_1} Z_{F_w};$ (22)

d) $(F_w \times G_v)(x) = F_w(x) \times G_v(x),$ d') $Z_{(F_w \times G_v)} = Z_{F_w} \times Z_{G_v};$

e) $(F_w \circ G_v)(x) = F_w(G_v(x)),$ e') $Z_{F_w \circ G_v} = Z_{F_w} \circ Z_{G_v};$

f) $(F_w \square G)(x) = F_w(x) \square G(x),$ f') $Z_{(F_w \square G)} = Z_{F_w} \square Z_G.$

It is necessary to assume, in the case e), that $G_v(0) = 0$ and, in case f), that G is an ordinary (nonweighted) species.

These equalities constitute powerful tools for calculation. Their proof is left as an exercise to the reader, except for formula (22), e') which is proved in Chapter 4.

Let α be a fixed element of the ring $\mathbb{A}$. To each species F, we can associate the $\mathbb{A}$-weighted species F_α by giving each structure of $F[U]$ the same weight α. Clearly then $F_\alpha(x) = \alpha F(x)$, and $Z_{F_\alpha} = \alpha Z_F$.

Example 12. In this manner, starting from $\mathcal{C}$, the species of oriented cycles, we can construct the species $\mathcal{C}_\alpha$ with weight α for each cycle. The species $\mathcal{S}_w$, of Example 7, is then isomorphic to the species $E(\mathcal{C}_\alpha)$, by the principle of multiplicativity. We deduce then that

a) $\quad \mathcal{S}_w(x) = \exp(-\alpha \log(1-x)) = \left(\dfrac{1}{1-x}\right)^\alpha,$

b) $\quad \widetilde{\mathcal{S}_w}(x) = \displaystyle\prod_{k\geq 1} \dfrac{1}{1-\alpha x^k} = \prod_{k\geq 1}\left(\dfrac{1}{1-x^k}\right)^{v_k(\alpha)},$ (23)

c) $\quad Z_{\mathcal{S}_w}(x_1, x_2, x_3, \ldots) = \displaystyle\prod_{k\geq 1}\left(\dfrac{1}{1-x_k}\right)^{v_k(\alpha)},$

where $v_n(\alpha) = \frac{1}{n}\sum_{d|n}\phi(d)\alpha^{n/d}$ and ϕ denotes Euler's ϕ-function.

Example 13. In a similar fashion, considering the species E_{+t} of nonempty sets of weight t, we can form the species $\mathrm{Par}_w = E(E_{+t})$ of partitions weighted by number of parts: $w(\pi) = t^{|\pi|}$, where $|\pi|$ is the number of blocks of a partition π. The following series are then obtained

a) $\quad \mathrm{Par}_w(x) = \exp t(e^x - 1),$

b) $\quad \widetilde{\mathrm{Par}_w}(x) = \displaystyle\prod_{k\geq 1}\dfrac{1}{1-tx^k} = \prod_{k\geq 1}\left(\dfrac{1}{1-x^k}\right)^{v_k(t)},$ (24)

c) $\quad Z_{\mathrm{Par}_w}(x_1, x_2, x_3, \ldots) = \exp\displaystyle\sum_{k\geq 1}\dfrac{1}{k}t^k\left(\exp\left(x_k + \dfrac{1}{2}x_{2k} + \dfrac{1}{3}x_{3k} + \cdots\right) - 1\right).$

Example 14. Consider now the species $\mathcal{A}_s$ of rooted trees each having weight s. By substitution in $\mathcal{S}_w$, we obtain the species $\mathrm{End}_v = \mathcal{S}_w \circ \mathcal{A}_s$ of endofunctions ψ, weighted by $v(\psi) = s^{\mathrm{rec}(\psi)}\alpha^{\mathrm{cyc}(\psi)}$, where $\mathrm{rec}(\psi)$ denotes the number of recurrent elements of ψ, and $\mathrm{cyc}(\psi)$ denotes the number of connected components of ψ. Since

$$\mathcal{A}_s(x) = s\mathcal{A}(x), \quad \widetilde{\mathcal{A}}_s(x) = s\widetilde{\mathcal{A}}(x), \quad \text{and} \quad Z_{\mathcal{A}_s} = sZ_{\mathcal{A}}, \quad (25)$$

2.3. Weighted Species

we obtain, after some calculation, the following series

a) $\mathrm{End}_v(x) = \left(\dfrac{1}{1-sA(x)}\right)^\alpha,$

b) $\widetilde{\mathrm{End}}_v(x) = \displaystyle\prod_{k\geq 1}\left(\dfrac{1}{1-s^k\widetilde{A}(x^k)}\right)^{v_k(\alpha)},$ (26)

c) $Z_{\mathrm{End}_v}(x_1, x_2, x_3, \ldots) = \displaystyle\prod_{k\geq 1}\left(\dfrac{1}{1-s^k Z_A(x_k, x_{2k}, x_{3k}, \ldots)}\right)^{v_k(\alpha)}.$

Expanding the series $\mathrm{End}_v(x)$ according to powers of x, the following expression is obtained for the total weight of all endofunctions on a set of $n \geq 1$ elements (see Exercise 5):

$$|\mathrm{End}\,[n]|_v = \sum_{k=1}^n \dfrac{k}{n}\binom{n}{k} n^{n-k} s^k \alpha(\alpha+1)\cdots(\alpha+k-1). \qquad (27)$$

In particular, setting $s = \alpha = 1$, we deduce the identity

$$n^n = \sum_{k=1}^n kn^{n-k}(n-1)(n-2)\ldots(n-k+1). \qquad (28)$$

Example 15. SYMMETRIC FUNCTIONS. Substituting the species X_t of singletons of weight t into an ordinary species F, we obtain the weighted species $F(X_t)$. Here, the weight of a structure of species $F(X_t)$ on a set U is given by $t^{|U|}$. Since an $F(X_t)$-structure is simply an F-structure on a set U together with a weight t on each of its elements, we have the identity

$$F(X_t) = F \times E(X_t). \qquad (29)$$

More generally, let τ be a finite sequence of distinct variables $t_1, t_2, \ldots, t_k$, and define the weighted species $X_\tau = X_{t_1} + X_{t_2} + \cdots + X_{t_k}$. Thus an X_τ-structure is a singleton of weight t_i, for some i between 1 and k. Now consider the composite species $F(X_\tau) = F(X_{t_1} + X_{t_2} + \cdots + X_{t_k})$; an $F(X_\tau)$-structure corresponds to placing an F-structure on a set of singletons of weights t_i, $1 \leq i \leq k$. Figure 3 shows that

$$F(X_{t_1} + X_{t_2} + \cdots + X_{t_k}) = F \times \left(E(X_{t_1})E(X_{t_2})\cdots E(X_{t_k})\right). \qquad (30)$$

Another way of visualizing $F(X_\tau)$-structures consists of assigning a "color" i, $1 \leq i \leq k$, to each element of the underlying set of an F-structure and to give to this F-structure the weight $t_1^{n_1} t_2^{n_2} \cdots t_k^{n_k}$, where n_i is the number of elements of color i in the underlying set. The general formula of weighted plethysm (22), e'), then gives

$$Z_{F\left(X_{t_1}+X_{t_2}+\cdots+X_{t_k}\right)} = Z_F(x_1(t_1+t_2+\cdots), x_2(t_1^2+t_2^2\cdots), \ldots) \qquad (31)$$

since

$$\left(Z_{X_{t_1}+X_{t_2}+\cdots+X_{t_k}}\right)_i = x_i(t_1^i + t_2^i + \cdots). \qquad (32)$$

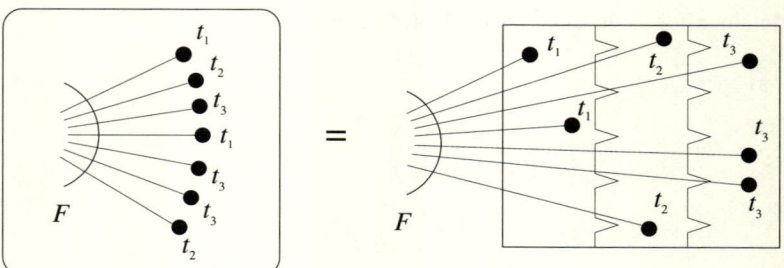

Fig. 3.

This cycle index series allows, in particular, the enumeration of unlabeled $F(X_\tau)$-structures. By restriction to n-element sets (i.e., by considering the species F_n), the enumeration of unlabeled F-structures on n points, colored with k colors, is given by the following polynomial in k variables $t_1, t_2, \ldots, t_k$

$$[x^n]F\widetilde{(X_\tau)}(x) = Z_{F_n}\big((t_1 + \cdots + t_k), (t_1^2 + \cdots + t_k^2), \ldots, (t_1^n + \cdots + t_k^n)\big), \quad (33)$$

of total degree n. As we will see in Chapter 4, this is essentially the classic enumeration theorem of Pólya. Moreover, the functions

$$p_n = t_1^n + t_2^n + t_3^n + \cdots \tag{34}$$

are the traditional *power sum* symmetric functions, so that for each n and each species F, we obtain a symmetric function

$$Z_{F_n}(p_1, p_2, p_3, \ldots)$$

by enumeration of unlabeled $F(X_\tau)$-structures on a set of cardinality n. Thus, for the species E_n of sets of cardinality n, we obtain the complete homogeneous symmetric functions

$$h_n(t_1, t_2, t_3, \ldots) = [x^n]E\widetilde{(X_\tau)}(x)$$
$$= \sum_{n_1+n_2+\cdots n_k = n} t_1^{n_1} t_2^{n_2} \cdots t_k^{n_k}, \tag{35}$$

and the following well known expression for complete homogeneous symmetric functions in terms of power sums

$$h_n = Z_{E_n}(p_1, p_2, p_3 \cdots)$$
$$= \sum_{\mathbf{d} \vdash n} \frac{p_1^{d_1} p_2^{d_2} p_3^{d_3} \cdots}{\text{aut}(\mathbf{d})}, \tag{36}$$

the sum being taken over all the sequences $\mathbf{d} = (d_1, d_2, \ldots)$ of integers ≥ 0 such that
$$n = d_1 + 2d_2 + 3d_3 + \cdots. \tag{37}$$
Equivalently
$$\sum_{n \geq 0} h_n = E(\widetilde{X_\tau})(x)\big|_{x=1}$$
$$= Z_E(p_1, p_2, \ldots) \tag{38}$$
$$= \exp\left(\sum_{k \geq 1} \frac{p_k}{k}\right).$$

Note that it is possible to take an infinite number of variables $t_1, t_2, \ldots$ in these formulas. In this case, the species $F(X_\tau)$ is no longer finite, but it is summable, that is to say, for any finite set U, $F(X_\tau)[U]$ is a summable weighted set in $\mathbb{A} = \mathbb{C}[\![t_1, t_2, \ldots]\!]$.

As the power sum functions are algebraically independent in the ring of symmetric functions, the identities (33) characterize the index series Z_F (see Section 4.3). We point out that according to Frobenius correspondence, $Z_{F_n}(p_1, p_2, p_3, \ldots)$ is the symmetric function corresponding to the representation of the symmetric group obtained by linearizing the group action $\mathcal{S}_n \times F[n] \longrightarrow F[n]$. In fact, for any permutation $\sigma \in \mathcal{S}_n$, fix $F[\sigma]$ is equal to the trace of the permutation matrix of $F[\sigma]$, that is, the character of the representation (see Serre [297]). With this point of view, we can deduce certain expressions already seen for the index series, for example (see Exercise 7)
$$Z_\mathcal{S}(p_1, p_2, \ldots) = \prod_{k \geq 1} \frac{1}{1 - p_k}. \tag{39}$$

To conclude the present section, we show how weighted species can lead to combinatorial models for certain classical families of orthogonal polynomials, such as, for example, Hermite polynomials. See also Exercise 8 as well as Exercise 9 where Laguerre polynomials are treated.

Example 16. HERMITE POLYNOMIALS. Consider the weighted species Inv_w of all involutions (i.e., permutations φ such that $\varphi \circ \varphi = \text{Id}$), weighted by $w(\varphi) = t^{\varphi_1}(-1)^{\varphi_2}$, where φ_1 is the number of fixed points of φ and φ_2 is the number of cycles of length 2 (the edges in Figure 4) in φ. We obtain the combinatorial equation
$$\text{Inv}_w = E(X_t + (C_2)_{-1}), \tag{40}$$
where $(C_2)_{-1}$ denotes the species of cycles of length 2 and weight -1, and it immediately follows that
$$\text{Inv}_w(x) = \exp\left(tx - \frac{1}{2}x^2\right) = \sum_{n \geq 0} H_n(t) \frac{x^n}{n!}, \tag{41}$$

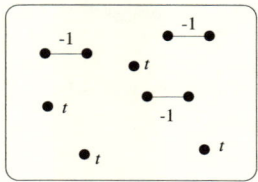

Fig. 4. An involution of weight $-t^4$.

where the coefficient $H_n(t)$ denotes, by definition, the (unitary) Hermite polynomial of degree n in t. Thus this classical polynomial appears as the inventory of all involutions on a set of n elements:

$$H_n(t) = \sum_{\varphi \in \text{Inv}[n]} t^{\varphi_1}(-1)^{\varphi_2}. \tag{42}$$

The other series associated with Inv_w are given by

$$\widetilde{\text{Inv}_w}(x) = \frac{1}{(1-tx)(1+x^2)},$$

$$Z_{\text{Inv}_w} = \exp \sum_{k \geq 1} \frac{1}{k}\left(t^k x_k + \frac{(-1)^k}{2}(x_k^2 + x_{2k})\right), \tag{43}$$

since $Z_X = x_1$ and $Z_{C_2} = Z_{E_2} = (x_1^2 + x_2)/2$.

Most classical properties of Hermite polynomials (recurrences, differential equations, Mehler's formula, coefficients of linearization) can be deduced from this combinatorial model (see Exercise 8).

Exercises for Section 2.3

1. Let $\mathbb{K} \subseteq \mathbb{C}$ be an integral domain and $\mathbb{A} = \mathbb{K}[\![t_1, t_2, \ldots]\!]$, a ring of formal power series in the variables $t_1, t_2, \ldots$. We say that the weighted set (A, w), where $w: A \longrightarrow \mathbb{A}$, is *summable* if, for any monomial $\mu = t_1^{n_1} t_2^{n_2} \cdots$ in the variables $t_1, t_2, \ldots$, the following set

$$W_\mu = \{a \in A \mid [\mu]w(a) \neq 0\}$$

is finite (recall that $[\mu]w(a) = [t_1^{n_1} t_2^{n_2} \cdots]w(a)$ denotes the coefficient of the monomial μ in the formal series $w(a)$). We define the *inventory* (or *total weight* or *cardinal*) of a summable weighted set (A, w) as being the unique element of $\mathbb{A}$ (i.e., the formal series), denoted by $|A|_w = \sum_{a \in A} w(a)$, satisfying

$$[\mu]|A|_w = \sum_{a \in W_\mu} [\mu]w(a) \tag{44}$$

for any monomial $\mu = t_1^{n_1} t_2^{n_2} \cdots$. Let (A, w) be a weighted summable set.

a) Show that if $S \subseteq A$ and $w_S : S \longrightarrow \mathbb{A}$ denotes the restriction of w to S, then the weighted set (S, w_S) is summable.

b) Show that if (B, v) is a weighted set such that $(A, w) \simeq (B, v)$ (see Definition 3), then (B, v) is summable and $|A|_w = |B|_v$.

c) Show that if (B, v) is a summable weighted set then the weighted sets sum, $(A, w) + (B, v)$, and product, $(A, w) \times (B, v)$ (see Definition 4), are summable and the formulas (12) are valid.

2. a) Let $A = \mathbb{N}_+ = \{1, 2, 3, \ldots\}$ and set $w(a) = t_1 t_2^2 \cdots t_a^a$ for all $a \in A$. Show that A is summable and calculate the inventory $|A|_w$.

b) Denote by $\tau(n)$, the number of divisors of an integer $n \geq 1$. Is the weighted set $(\mathbb{N}_+, v)$, where $v(n) = t^{\tau(n)}$, summable?

c) Let A be the (infinite) set of all the trees on the set of vertices $\{1, 2, \ldots, n\}$, where n runs over $\mathbb{N}_+$. Define the weight $w(a)$ of a tree a by $w(a) = t_1^{n_1} t_2^{n_2} \cdots$, where n_i is the number of vertices of degree i in a, $i = 1, 2, 3, \ldots$. Show that A is summable and interpret the inventory $|A|_w$ combinatorially.

3. Let A, B, and C be disjoint sets, with $|A| = a$, $|B| = b$, and $|C| = c$. Denote by $\mathrm{Inj}(A, A \cup B)$ the set of injective functions $f : A \longrightarrow A \cup B$ weighted by $w(f) = \lambda^{\mathrm{cyc}(f)}$.

a) Show that

$$|\mathrm{Inj}(A, A \cup B)|_w = (\lambda + b)^{(a)} = (\lambda + b)(\lambda + b + 1) \cdots (\lambda + b + a - 1).$$
(45)

b) Establish an isomorphism of weighted sets

$$\mathrm{Inj}(A \cup B, A \cup B \cup C) \overset{\sim}{\longrightarrow} \mathrm{Inj}(A, A \cup C) \times \mathrm{Inj}(B, A \cup B \cup C)$$

and write the identity which corresponds to it.

4. a) Establish Proposition 11, excluding (22), e').

b) Show that Definition 9 makes sense and verify the summability conditions which are implicit in formula (20).

5. a) Verify formulas (23) giving the generating and index series of the species S_w of permutations with cycle counter α.

b) Show moreover that

   i) $|S[n]|_w = \alpha^{\langle n \rangle} = \alpha(\alpha + 1) \cdots (\alpha + n - 1),$ (46)

   ii) for a permutation σ of type $(\sigma_1, \sigma_2, \ldots)$

$$|\mathrm{Fix}\, S[\sigma]|_\omega = \prod_{j \geq 1} \omega_j(\alpha) (\omega_j(\alpha) + j) \cdots (\omega_j(\alpha) + j(\sigma_j - 1)), \quad (47)$$

where $\omega_n(\alpha) = \sum_{d|n} \phi(d) \alpha^{n/d}$.

c) Verify formulas (24) giving the generating and index series of the species Par_w of weighted partitions by block count.

d) Show that

$$\sum_{n\geq 0}\sum_{k=0}^{n} S(n,k) t^k \frac{x^n}{n!} = \exp(t(e^x - 1)). \qquad (48)$$

6. a) Prove formulas (26) giving the generating and index series of the weighted species End_v of endofunctions, described in Example 14.
   b) Establish formula (27) for $|\mathrm{End}[n]|_v$.

   HINT: Take as known that the number of forests made up of k rooted trees on n vertices is given by formula (3.1.53), namely $\frac{k}{n}\binom{n}{k} n^{n-k}$, $n > 0$.

7. LYNDON WORDS AND TYPES OF COLORED PERMUTATIONS. Consider the alphabet

$$A = \mathbb{N}_+ = \{1, 2, 3, \ldots\},$$

weighted by the function $w(i) = t_i$. Then $|A|_w = p_1$, and more generally, for $k \geq 1$, $|A|_{w^k} = t_1^k + t_2^k + \cdots = p_k$. Denote by A^n the set of all words of length n ($n \geq 0$) and by A^* the set of all the words in the alphabet A. One extends the weighting w to A^* by associating to the word $ij \cdots k \in A^*$, the commutative word (or monomial) $t_i t_j \cdots t_k$.

A^* forms a monoid under concatenation. Two words m and m' are said to be *conjugates* if there exists a factorization $m = uv$ so that $m' = vu$. This constitutes an equivalence relation for which the classes are called the *circular words*. A *Lyndon word* is a *primitive* (i.e., not a positive power of another word) word which is smaller than all its conjugates with respect to the lexicographic order. Denote by $C(n)$ the set of circular words of length n and by $L(n)$, the set of Lyndon words of length n.

a) Show that for $n \geq 1$,

$$\begin{aligned}
\text{i)} \quad & |C(n)|_w = \sum_{d|n} |L(n/d)|_{w^d}, \\
\text{ii)} \quad & |A^n|_w = p_1^n = \sum_{d|n} \frac{n}{d} |L(n/d)|_{w^d}, \\
\text{iii)} \quad & |L(n)|_w = \frac{1}{n} \sum_{d|n} \mu(d) \, p_d^{n/d}, \\
\text{iv)} \quad & |C(n)|_w = \frac{1}{n} \sum_{d|n} \phi(d) \, p_d^{n/d}.
\end{aligned} \qquad (49)$$

HINT: Here μ denotes the Möbius function and ϕ Euler's ϕ-function. Use the fact that $\phi(n) = n \sum_{d|n} \frac{\mu(d)}{d}$.

b) Consider the weighted species $X_\tau = X_{t_1} + X_{t_2} + \cdots$, the species $\mathcal{C}$ of cyclic permutations, and the composite species $\mathcal{C}(X_\tau)$, whose structures are called *colored cycles*. Show that there exists an isomorphism of weighted sets between $(C(n), w)$ and unlabeled $C_n(X_\tau)$-structures and that

$$\widetilde{C(X_\tau)}(x) = \sum_{n\geq 1} \frac{x^n}{n} \sum_{d\mid n} \phi(d) \, p_d^{n/d}$$
$$= \sum_{m\geq 1} \frac{\phi(m)}{m} \log \frac{1}{1 - p_m x^m}. \tag{50}$$

Deduce formula (1.4.18) giving the index series Z_C.

c) A colored cycle is called *asymmetric* if its only automorphism is the identity. Denote by $\overline{C(X_\tau)}$, the species of colored asymmetric cycles. Show that there exists an isomorphism of weighted sets between Lyndon words and unlabeled colored asymmetric cycles. Deduce that

$$\overline{C(X_\tau)}(x) = \sum_{n\geq 1} \frac{x^n}{n} \sum_{d\mid n} \mu(d) \, p_d^{n/d}$$
$$= \sum\sum_{m\geq 1} \frac{\mu(m)}{m} \log \frac{1}{1 - p_m x^m}. \tag{51}$$

d) Show that every word $\mu \in A^*$ can be written in a unique manner as a product $\mu = \ell_1 \ell_2 \ldots \ell_k$, where the ℓ_i are Lyndon words and $\ell_1 \geq \ell_2 \geq \cdots \geq \ell_k$ with respect to the lexicographic order (see Chapter 5 of Lothaire [227] or Chapter 8 of Reutenauer [280]). Deduce that the series $|A^*|_w = 1/(1 - p_1)$ can be viewed as the inventory of unlabeled assemblies of asymmetric colored cycles.

e) Show that $|L(n)|_{w^k}$ is the inventory of words of length nk that are circular and k-*symmetric*, that is to say, of the form u^k, where u is primitive. Show that the series $1/(1 - p_k)$ can be considered as the inventory of unlabeled assemblies of k-symmetric colored cycles, that is to say, whose automorphism group is the cyclic group of order k.

f) An $S(X_\tau)$-structure is called a *colored permutation*. It is an assembly of colored cycles. By regrouping these colored cycles according to the order of their automorphism group, deduce from e) that

$$\widetilde{S(X_\tau)}(x) = \prod_{k\geq 1} \frac{1}{1 - p_k x^k}. \tag{52}$$

g) A colored permutation is called *asymmetric* if its only automorphism is the identity. It consists of an injective assembly (i.e., whose members are pairwise nonisomorphic) of colored asymmetric cycles. Denote by $\overline{S(X_\tau)}$, the species of asymmetric colored permutations. Show that the series $p_2/(1 - p_1)$ can be viewed as the inventory of unlabeled *noninjective* assemblies of asymmetric colored cycles (see Poulin [266]). Deduce that

$$\overline{S(X_\tau)}(x) = \frac{1 - p_2 x^2}{1 - p_1 x}. \tag{53}$$

8. HERMITE POLYNOMIALS. The (unitary) Hermite polynomials can be defined by formulas (41) or (42), that is to say from the combinatorial model of involutions φ weighted by $w(\varphi) = t^{\varphi_1}(-1)^{\varphi_2}$

a) Show that

$$H_n(t) = \sum_{k=0}^{\lfloor \frac{n}{2} \rfloor} (-2)^{-k} \frac{n!}{k!(n-2k)!} t^{n-2k}. \tag{54}$$

b) A variant $\overline{H}_n(t)$ of the Hermite polynomials is defined by the renormalization $\overline{H}_n(t) = 2^{n/2} H_n(t\sqrt{2})$. Show that

$$\sum_{n \geq 0} \overline{H}_n(t) \frac{x^n}{n!} = \exp(2tx - x^2). \tag{55}$$

c) (Chapter 5 of Constantineau [60].) Establish formulas (43) and show that for any permutation σ of type $(\sigma_1, \sigma_2, \ldots)$

$$|\text{Fix Inv}[\sigma]\|_w = \prod_{j \geq 1} ((-1)^{j-1} j)^{\frac{\sigma_j}{2}} H_{\sigma_j}(\xi_j), \tag{56}$$

where

$$\xi_j = \frac{t^j + \chi(j \text{ is even})(-1)^{j/2}}{((-1)^{j-1} j)^{1/2}}. \tag{57}$$

d) Show combinatorially that the polynomial $y = H_n(t)$ satisfies the differential equation

$$y'' - ty' + ny = 0. \tag{58}$$

HINT: Study the effect of the derivative on the weighted involutions.

e) Show combinatorially that

$$H_{n+1}(t) = t H_n(t) - n H_{n-1}(t). \tag{59}$$

f) MEHLER'S FORMULA. (SEE FOATA [107].) Show that

$$\sum_{n \geq 0} H_n(t_1) H_n(t_2) \frac{x^n}{n!} = \frac{1}{\sqrt{1 - x^2}} \exp \frac{t_1 t_2 x - (t_1^2 + t_2^2) x^2/2!}{1 - x^2}, \tag{60}$$

by considering an appropriate combinatorial model.

g) LINEARIZATION COEFFICIENTS. Let $\nu : \mathbb{R}[t] \longrightarrow \mathbb{R}$ be the linear functional defined by $\nu(1) = 1$ and

$$\begin{aligned} \nu(t^{2n}) &= (2n-1)(2n-3)\cdots 3 \cdot 1, & n \geq 1, \\ \nu(t^{2n+1}) &= 0, & n \geq 0. \end{aligned} \quad (61)$$

In fact, $\nu(t^n) = \frac{1}{\sqrt{2\pi}} \int_{-\infty}^{+\infty} t^n e^{-t^2/2} dt$. Denote by $I(n)$, the set of involutions without fixed points. Moreover, for a multicardinal $(n_1, n_2, \ldots, n_k)$, let $U_1, U_2, \ldots, U_k$ be disjoint sets such that $|U_j| = n_j$, $j = 1, 2, \ldots, k$, and denote by $I(n_1, n_2, \ldots, n_k)$, the set of involutions without any fixed color on $U = U_1 + U_2 + \cdots + U_k$, that is to say, the involutions φ on U such that $u \in U_j \Rightarrow \varphi(u) \notin U_j$. Show combinatorially that

i) $\nu(t^n) = |I(n)|, \quad n \geq 1,$

ii) $\nu(H_n(t)) = 0, \quad n \geq 1,$

iii) $\nu(H_n(t)H_m(t)) = \begin{cases} n!, & \text{if } n = m, \\ 0, & \text{if } n \neq m, \end{cases}$ (orthogonality) (62)

iv) $\nu\big(H_{n_1}(t)H_{n_2}(t)\cdots H_{n_k}(t)\big) = |I(n_1, n_2, \ldots, n_k)|.$

9. LAGUERRE POLYNOMIALS (SEE FOATA AND STREHL [113]). The Laguerre polynomials $\mathcal{L}_n^{(\alpha)}(x)$ can be defined by the generating series

$$\sum_{n\geq 0} \mathcal{L}_n^{(\alpha)}(x) \frac{u^n}{n!} = \left(\frac{1}{1-u}\right)^{\alpha+1} \exp\left(\frac{-xu}{1-u}\right). \quad (63)$$

a) Deduce a combinatorial model with the help of the weighted species

$$\text{Lag}^{(\alpha)} = E(\mathcal{C}_{\alpha+1}) \cdot E\big(L_{+(-x)}\big), \quad (64)$$

for which we have $\mathcal{L}_n^{(\alpha)}(x) = |\text{Lag}^{(\alpha)}[n]|_w$. The structures of this species are called *Laguerre configurations*.

b) Show that Laguerre configurations can be considered as injective partial endofunctions and that (see Exercise 3)

$$\mathcal{L}_n^{(\alpha)}(x) = \sum_{k=0}^{n} \binom{n}{k} (\alpha + 1 + n - k)^{\langle k \rangle} (-x)^{n-k}. \quad (65)$$

c) RECURRENCE FORMULA. Show, by a study of these configurations, that

$$\mathcal{L}_{n+1}^{(\alpha)}(x) = (\alpha + 2n - x + 1) \mathcal{L}_n^{(\alpha)}(x) - n(n+\alpha)\mathcal{L}_{n-1}^{(\alpha)}(x). \quad (66)$$

d) Give a combinatorial interpretation of the derivative $y' = dy/dx$ of the polynomial $y = \mathcal{L}_n^{(\alpha)}(x)$ and prove the following formulas using this interpretation:

i) $xy'' + (\alpha + 1 - x)y' + ny = 0,$

ii) $x\dfrac{d}{dx}\mathcal{L}_n^{(\alpha)}(x) = n\mathcal{L}_n^{(\alpha)}(x) - n(n+\alpha)\,\mathcal{L}_{n-1}^{(\alpha)}(x),$ (F.G. Tricomi)

iii) $\dfrac{d}{dx}\mathcal{L}_n^{(\alpha)}(x) = -n\mathcal{L}_{n-1}^{(\alpha+1)}(x).$

$$\tag{67}$$

e) Let Ψ be the linear functional defined on the polynomials by

$$\Psi(x^n) = (\alpha + 1)^{\langle n \rangle}, \quad n \geq 0. \tag{68}$$

For a real $\alpha > -1$, $\Psi(x^n) = \frac{1}{\Gamma(\alpha+1)}\int_0^\infty x^{n+\alpha}e^{-x}dx$. For $(n_1, n_2, \ldots, n_m)$ a multicardinal, let $U_1, U_2, \ldots, U_m$ be disjoint sets such that $|U_j| = n_j$, $j = 1, 2, \ldots, m$, and let $U = U_1 + U_2 + \cdots + U_m$. Denote by $L(n_1, n_2, \ldots, n_m)$, the set of *colored derangements* of U, that is to say, the permutations σ of U such that $u \in U_j \implies \sigma(u) \notin U_j$, weighted by $w(\sigma) = (-1)^{|U|}(\alpha + 1)^{\mathrm{cyc}(\sigma)}$. Show combinatorially that

i) $\Psi(x^n) = \displaystyle\sum_{\sigma \in S[n]} (\alpha + 1)^{\mathrm{cyc}(\sigma)},$

ii) $\Psi\bigl(\mathcal{L}_n^{(\alpha)}(x)\bigr) = 0, \quad n > 0,$

iii) $\Psi\bigl(\mathcal{L}_n^{(\alpha)}(x)\,\mathcal{L}_m^{(\alpha)}(x)\bigr) = \begin{cases} n!(\alpha+1)^{\langle n\rangle}, & \text{if } m = n, \\ 0, & \text{if } m \neq n, \end{cases}$ (orthogonality)

iv) $\Psi\bigl(\mathcal{L}_{n_1}^{(\alpha)}(x)\mathcal{L}_{n_2}^{(\alpha)}(x)\ldots\mathcal{L}_{n_m}^{(\alpha)}(x)\bigr) = |L(n_1, n_2, \ldots, n_m)|_w.$

$$\tag{69}$$

10. WEIGHTED EXPONENTIAL FORMULAS. Consider a weighted species $F = F_w$ and suppose that $F_w = E(F_w^c)$, where F_w^c is the species of connected F_w-structures.

a) Show that

i) $F_w(x) = \exp F_w^c(x),$

ii) $\widetilde{F_w}(x) = \exp \displaystyle\sum_{k \geq 1} \dfrac{1}{k}\widetilde{F_{w^k}^c}(x^k),$

iii) $Z_{F_w}(x_1, x_2, x_3, \ldots) = \exp \displaystyle\sum_{k \geq 1} \dfrac{1}{k} Z_{F_{w^k}^c}(x_k, x_{2k}, x_{3k}, \ldots).$

$$\tag{70}$$

b) Prove the inverse of the preceding formulas, analogues of (1.4.60).

   i) $F_w^c(x) = \log F_w(x)$,

   ii) $\widetilde{F_w^c}(x) = \sum_{k\geq 1} \frac{\mu(k)}{k} \log \widetilde{F_{w^k}}(x^k)$, \hfill (71)

   iii) $Z_{F_w^c}(x_1, x_2, x_3, \ldots) = \sum_{k\geq 1} \frac{\mu(k)}{k} \log Z_{F_{w^k}}(x_k, x_{2k}, x_{3k}, \ldots)$,

   where μ denotes the usual Möbius function.

c) Given a formal variable α, define the weighted species $F_{w^{(\alpha)}}$ by setting for any F_w-structure s
$$w^{(\alpha)}(s) = w(s) \cdot \alpha^{c(s)}, \tag{72}$$
where $c(s)$ is the number of connected components of s. Show that

   i) $F_{w^{(\alpha)}}(x) = (F_w(x))^\alpha$,

   ii) $\widetilde{F_{w^{(\alpha)}}}(x) = \prod_{k\geq 1} (\widetilde{F_{w^k}}(x^k))^{\lambda_k(\alpha)}$, \hfill (73)

   iii) $Z_{F_{w^{(\alpha)}}}(x_1, x_2, x_3, \ldots) = \prod_{k\geq 1} Z_{F_{w^k}}(x_k, x_{2k}, x_{3k}, \ldots)^{\lambda_k(\alpha)}$,

where
$$\lambda_n(\alpha) = \frac{1}{n} \sum_{d\mid n} \mu(n/d) \alpha^d. \tag{74}$$

d) Given two formal variables α and β, prove that
$$F_{w^{(\alpha+\beta)}}(x) = F_{w^{(\alpha)}}(x) F_{w^{(\beta)}}(x), \tag{75}$$
but that in general
$$\widetilde{F_{w^{(\alpha+\beta)}}}(x) \neq \widetilde{F_{w^{(\alpha)}}}(x) \widetilde{F_{w^{(\beta)}}}(x),$$
$$Z_{F_{w^{(\alpha+\beta)}}} \neq Z_{F_{w^{(\alpha)}}} Z_{F_{w^{(\beta)}}},$$
$$F_{w^{(\alpha+\beta)}} \neq F_{w^{(\alpha)}} F_{w^{(\beta)}}.$$

HINT: Use the fact that for $n \geq 2$, $\lambda_n(\alpha + \beta) \neq \lambda_n(\alpha) + \lambda_n(\beta)$.

e) Given two formal variables α and β, prove the formulas

   i) $F_{w^{(\alpha\beta)}}(x) = (F_{w^{(\alpha)}}(x))^\beta$,

   ii) $\widetilde{F_{w^{(\alpha\beta)}}}(x) = \prod_{k\geq 1} (\widetilde{F_{w^k(\alpha^k)}}(x^k))^{\lambda_k(\beta)}$, \hfill (76)

   iii) $Z_{F_{w^{(\alpha\beta)}}}(x_1, x_2, x_3, \ldots) = \prod_{k\geq 1} Z_{F_{w^k(\alpha^k)}}(x_k, x_{2k}, x_{3k}, \ldots)^{\lambda_k(\beta)}$.

HINT: Show at first that

$$F_{w^{(\alpha\beta)}} = F_{w^{(\alpha)(\beta)}}.\tag{77}$$

Deduce that for any $n \geq 1$,

$$\lambda_n(\alpha\beta) = \sum_{dk=n} \lambda_d(\alpha^k)\lambda_k(\beta).\tag{78}$$

f) Let $\alpha = |A|$, where A is a finite alphabet.
   i) Show that $\lambda_n(\alpha)$ is the number of Lyndon words of length n over A (see Exercise 7).
   ii) Interpret combinatorially formula (78) for $\lambda_n(\alpha\beta)$ in the context of Lyndon words.

11. a) Let $v_n(\alpha) = \frac{1}{n}\sum_{d|n} \phi(d)\alpha^{n/d}$ as in Example 12. Show that

$$v_n(\alpha) = \sum_{d|n} \lambda_d(\alpha)\tag{79}$$

(see (74)) and deduce from (73) formulas (23) and (24) giving the series associated with the species $\mathcal{S}_w$ and Par_w.

b) Deduce from (73) the identities (called *cyclotomic*; see Metropolis and Rota [249], Bergeron [17],)

i) $\dfrac{1}{1-\alpha x} = \prod_{k\geq 1}\left(\dfrac{1}{1-x^k}\right)^{\lambda_k(\alpha)},$

ii) $\exp\left(\alpha x_1 + \dfrac{1}{2}\alpha^2 x_2 + \cdots\right) = \prod_{k\geq 1}\exp\left(\lambda_k(\alpha)\left(x_k + \dfrac{1}{2}x_{2k} + \cdots\right)\right).$
(80)

HINT: Consider the species $E(X_\alpha)$, where X_α is the species of singletons of weight α.

12. Consider the weighted virtual species (see Section 2.5)

$$\Lambda^{(\alpha)} = E \circ X_\alpha \circ E_+^{\langle-1\rangle}.\tag{81}$$

a) Show that

   i) $\Lambda^{(\alpha)}(x) = (1+x)^\alpha,$

   ii) $\widetilde{\Lambda^{(\alpha)}}(x) = \prod_{k\geq 1}(1+x^k)^{\lambda_k(\alpha)},$

   iii) $Z_{\Lambda^{(\alpha)}}(x_1, x_2, x_3, \ldots) = \prod_{k\geq 1}(1+x_k)^{\lambda_k(\alpha)}.$
(82)

HINT: Identify the series associated to $\Lambda^{(\alpha)} \circ E_+$.

2.3. Weighted Species

b) Show that for a weighted species of the form $F_w = E(F_w^c)$,

$$F_{w^{(\alpha)}} = \Lambda^{(\alpha)} \circ (F_w)_+. \tag{83}$$

NOTE: $\Lambda^{(\alpha)}$ is called the *connected components weighting species*.

c) Use (83) to extend the exponential formulas of Exercise 10 to the case of a virtual weighted species Ψ_w, such that $(\Psi_w)_0 = 1$.

13. a) Returning to the notation of Exercise 10, show that

$$\lambda_n(-1) = \frac{1}{n} \sum_{d \mid n} \mu(n/d)(-1)^d = \begin{cases} -1, & \text{if } n = 1, \\ 1, & \text{if } n = 2, \\ 0, & \text{if } n > 2. \end{cases} \tag{84}$$

HINT: Show first that $-\lambda_n(-1)$ is a multiplicative function in n, using the fact that the convolution

$$(f * g)(n) = \sum_{d \mid n} f(n/d) g(d) \tag{85}$$

of two multiplicative functions f and g is a multiplicative function (see Apostol [3]).

b) Deduce the formulas

$$\text{i)} \quad F_{w^{(-1)}}(x) = \frac{1}{F_w(x)},$$

$$\text{ii)} \quad \widetilde{F}_{w^{(-1)}}(x) = \frac{\widetilde{F_{w^2}}(x^2)}{\widetilde{F_w}(x)}, \tag{86}$$

$$\text{iii)} \quad Z_{F_{w^{(-1)}}}(x_1, x_2, \ldots) = \frac{Z_{F_{w^2}}(x_2, x_4, \ldots)}{Z_{F_w}(x_1, x_2, \ldots)}.$$

c) In the case where $F_w = S$, the species of permutations, let $S_{(-1)} = F_{w^{(-1)}}$, that is, the species of permutations weighted by $(-1)^{\text{cyc}(\sigma)}$. Show that

i) $S_{(-1)}(x) = 1 - x$,

ii) $\widetilde{S_{(-1)}}(x) = (1-x)(1-x^3)(1-x^5)\cdots$, \hfill (87)

iii) $Z_{S_{(-1)}}(x_1, x_2, x_3, \ldots) = (1-x_1)(1-x_3)(1-x_5)\cdots$.

Interpret these formulas combinatorially.

d) Consider the case where $F_w = E(X_y)$.

14. Let $E_{(y)}$ be the species of sets, weighted by $w(U) = y^{|U|}$, and let $\wp_w$ be the species of subsets, weighted by $w(A) = y^{|A|}$, for $A \in \wp[U]$. In other words,

the variable y acts as an element counter. Show that

a) $E_{(y)} = E(X_y)$ and $\wp_w = E \cdot E_{(y)}$,

b) $\wp_w(x) = \sum_{n \geq 0} (1+y)^n \dfrac{x^n}{n!}$,

c) $\widetilde{\wp_w}(x) = \sum_{n \geq 0} \dfrac{1 - y^{n+1}}{1 - y} x^n$, (88)

d) $Z_{\wp_w}(x_1, x_2, \ldots) = \sum_{n \geq 0} \dfrac{1}{n!} \sum_{\sigma \in S_n} ((1+y) x_1)^{\sigma_1} ((1+y^2) x_2)^{\sigma_2} \cdots$,

e) $|\text{Fix}\,\wp_w[\sigma]| = (1+y)^{\sigma_1}(1+y^2)^{\sigma_2}\cdots$,   for any permutation σ.

2.4. Extension to the Multisort Context

Multisort Species

The theory of species can naturally be extended in yet another direction by considering structures constructed on sets containing several *sorts* of elements. It is an undertaking analogous to the introduction of functions in many variables.

Example 1. Consider tri-chromatic simple graphs, that is to say, simple graphs constructed on triplets of sets (U_1, U_2, U_3), corresponding to three distinct colors, in such a manner that adjacent vertices have different colors. Figure 1 represents one such graph, where $U_1 = \{a, b, c, d, e\}$, $U_2 = [7]$, and $U_3 = \{m, n, p, q\}$ correspond respectively to the colors white, gray, and black.

Another example is given by rooted trees constructed on a set having two sorts of elements: leaves and internal vertices. See Figure 2. In this example, the elements of leaf sort are not placed in an arbitrary manner; by convention, they are located at the end of the paths starting from the root. In other words, these are the vertices with empty "fibers."

For transport of structures, multisort species are distinguished by the fact that "the transports are carried out along bijections preserving the sort of the elements." We illustrate this with rooted trees. Figure 3 represents the transport of a rooted

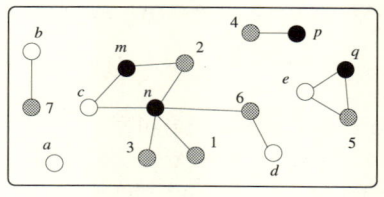

Fig. 1.

2.4. Extension to the Multisort Context

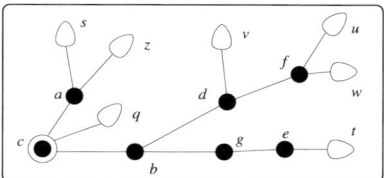

Fig. 2.

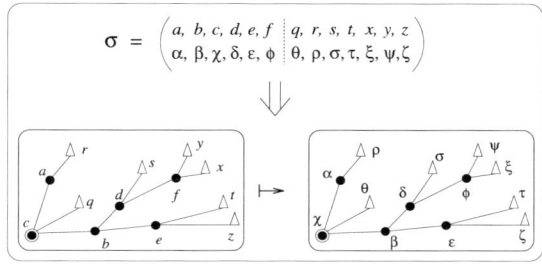

Fig. 3.

tree on two sorts (internal vertices and leaves). The bijection

$$\sigma : U_1 + U_2 \longrightarrow V_1 + V_2 \tag{1}$$

along which the transport is carried "must necessarily" send each internal vertex (ϵU_1) to an internal vertex (ϵV_1) and each leaf (ϵU_2) to a leaf (ϵV_2).

These preliminary considerations justify the definitions which follow.

Definition 2. Let $k \geq 1$ be an integer. A *multiset* (with k sorts of elements) is a k-tuple of sets

$$U = (U_1, \ldots, U_k). \tag{2}$$

For brevity we say that U is a k-*set*. An element $u \in U_i$ is called an element of U of *sort i*. The *multicardinality* of U is the k-tuple of cardinalities

$$|U| = (|U_1|, \ldots, |U_k|). \tag{3}$$

The *total cardinality* of U is the sum

$$\|U\| = |U_1| + \cdots + |U_k|. \tag{4}$$

Definition 3. A *multifunction* f from $(U_1, \ldots, U_k)$ to $(V_1, \ldots, V_k)$, denoted by

$$f : (U_1, \ldots, U_k) \longrightarrow (V_1, \ldots, V_k), \tag{5}$$

is a k-tuple of functions $f = (f_1, \ldots, f_k)$ such that $f_i : U_i \longrightarrow V_i$, for $i = 1, \ldots, k$.

The composition of two multifunctions is made componentwise. The multifunction f is said to be *bijective* if each function f_i is bijective.

Definition 4. Let $k \geq 1$ be an integer. A *species of k sorts* (or *k-sort species*) is a rule F which

i) *produces*, for each finite multiset $U = (U_1, \ldots, U_k)$, a finite set

$$F[U_1, \ldots, U_k],$$

ii) *produces*, for each bijective multifunction

$$\sigma = (\sigma_1, \ldots, \sigma_k) : (U_1, \ldots, U_k) \longrightarrow (V_1, \ldots, V_k),$$

a function

$$F[\sigma] = F[\sigma_1, \ldots, \sigma_k] : F[U_1, \ldots, U_k] \longrightarrow F[V_1, \ldots, V_k]. \quad (6)$$

Moreover, the functions $F[\sigma]$ must satisfy the functoriality properties, that is to say, for bijective multifunctions $\sigma = U \longrightarrow V$ and $\tau : V \longrightarrow W$, and for the multifunction identity $\mathrm{Id}_U : U \longrightarrow U$, it is required that

a) $F[\tau \circ \sigma] = F[\tau] \circ F[\sigma],$

b) $F[\mathrm{Id}_U] = \mathrm{Id}_{F[U]}.$ \quad (7)

An element $s \in F[U_1, \ldots, U_k]$ is called an *F-structure* on $(U_1, \ldots, U_k)$ (or a *structure of species F* on $(U_1, \ldots, U_k)$). The function $F[\sigma_1, \ldots, \sigma_k]$ is called *transport* of F-structures along $(\sigma_1, \ldots, \sigma_k)$. If $t = F[\sigma_1, \ldots, \sigma_k](s)$, then s and t are said to be *isomorphic F-structures*. This is an equivalence relation for which the classes are called *unlabeled F-structures*.

The usual graphical conventions to represent F-structures extend to the multisort context (see Figure 4, for $k = 3$) by associating to each element of the underlying set a number (or a shape, or a color, ...) identifying its sort.

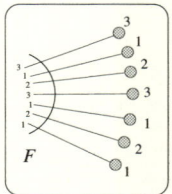 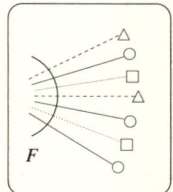

Fig. 4.

Of course, it is not necessary that there be at least one element of each sort in an F-structure. Thus, every multisort species can also be viewed as a multisort species on a *larger* set of sorts.

Remark 5. It is often useful, in practice, to represent the multiset $U = (U_1, \ldots, U_k)$ underlying a structure as being the set

$$U_1 + \cdots + U_k \quad \left(\text{disjoint union } \bigcup_{i=1}^{k} U_i \times \{i\}\right),$$

which is also denoted by U. There follows a function $\chi = \chi_U : U \longrightarrow [k]$, associating to each element its sort, that is to say, the fibers $\chi_U^{-1}(\{i\})$ are the U_i (more precisely, $\chi_U^{-1}(\{i\}) = U_i \times \{i\}$), $i = 1, \ldots, k$. In this case, the multifunctions are identified with *ordinary* functions $f : U \longrightarrow V$ preserving the sorts, that is to say, such that $\chi_V \circ f = \chi_U$.

We leave to the reader the task of formulating in functorial terms the definition of multisort species and of showing the equivalence of the different points of view presented in the preceding remark (see Exercise 6).

To each sort, one can associate a species of singletons.

Definition 6. For each i, $1 \leq i \leq k$, the (k-sort) species X_i of *singletons of sort i* is defined by

$$X_i[U] = \begin{cases} \{U\}, & \text{if } |U_i| = 1 \text{ and } U_j = \emptyset, \text{ for } j \neq i, \\ \emptyset, & \text{otherwise.} \end{cases} \quad (8)$$

In other words, there is an X_i-structure on $(U_1, \ldots, U_k)$ only if

$$U = \emptyset + \emptyset + \cdots + \{u\} + \emptyset + \cdots + \emptyset = \{u\}$$

is a singleton of sort i. In this case, $\{u\}$ is the unique X_i-structure.

Remark 7. Other variables can be used to designate the species of singletons. The variables currently utilized are the letters of the alphabet X, Y, Z, T, with or without indices. Note the abuse of language which consists in saying, for example, that a Y-structure is a singleton of *sort Y*. One often writes $F = F(X_1, \ldots, X_k)$ to indicate that F is a k-sort species. This notation is moreover compatible with the substitution of species defined later in this section.

Weighted Multisort Species

By combining the notions of multisort species and weighted species, the more general concept of weighted multisort species is obtained.

Definition 8. A *weighted multisort species* $F = F_w$ (of k sorts) is a rule which associates
i) to each multiset $U = (U_1, U_2, \ldots, U_k)$, a weighted set $(F[U], w_U)$,
ii) to each bijective multifunction $\sigma = (\sigma_1, \sigma_2, \ldots, \sigma_k)$ a function $F[\sigma_1, \sigma_2, \ldots, \sigma_k]$, which *preserves the weights*, in such a way that the functoriality conditions (7) are satisfied.

The reader will easily define for himself, in the weighted multisort context, the notions of isomorphic structures, isomorphism types, equipotent species, and isomorphic species (combinatorial equality). The operations of addition, multiplication, substitution, differentiation, Cartesian product, pointing, and functorial composition are defined essentially in the same manner as in the weighted unisort case, in analogy with functions in many variables.

We begin with some terminology for a k-set $U = (U_1, \ldots, U_k)$.
– A *dissection* of U is a pair of k-sets (V, W) such that for $i = 1, \ldots, k$, $U_i = V_i \cup W_i$ and $V_i \cap W_i = \emptyset$. We denote by $\Delta[U]$, the set of dissections of U.
– A *partition* π of U is a partition of the total set $U_1 + \cdots + U_k$. Each class $C \in \pi$ can be viewed as a multiset of k sorts, where $C_i = C \cap U_i$. We denote by $\text{Par}[U]$, the set of partitions of U.

For two weighted k-sort species F and G and a k-set $U = (U_1, U_2, \ldots, U_k)$, we set

$$(F + G)[U] = F[U] + G[U], \qquad (9)$$

and

$$(F \cdot G)[U] = \sum_{(V,W) \in \Delta[U]} F[V] \times G[W]. \qquad (10)$$

The sums and products on the right-hand sides of (9) and (10) are taken in the sense of weighted sets.

Let $F = F(Y_1, \ldots, Y_m)$ be a weighted m-sort species, and $(G_j)_{j=1,\ldots,m}$ be a family of weighted k-sort species. The *partitional composition* $F(G_1, \ldots, G_m)$ (*substitution* of the G_j in F) is a k-sort species defined by setting, for $U = (U_1, \ldots, U_k)$,

$$F(G_1, \ldots, G_m)[U] = \sum_{\substack{\pi \in \text{Par}[U] \\ \chi : \pi \to [m]}} F[\chi^{-1}] \times \prod_{\substack{j \in [m] \\ C \in \chi^{-1}(j)}} G_j[C], \qquad (11)$$

2.4. Extension to the Multisort Context

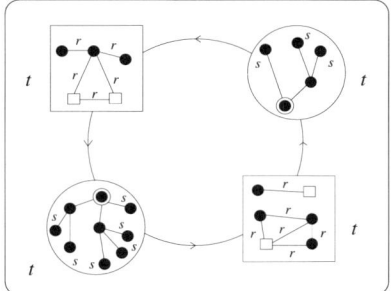

Fig. 5. An $F_{r,s,t}$-structure of weight $r^{11}s^9t^4$.

where, for each function $X : \pi \longrightarrow [m]$, X^{-1} denotes the m-set $(X^{-1}(i), \ldots, X^{-1}(m))$ associated to X. In descriptive terms, an $F(G_1, \ldots, G_m)$-structure is an F-structure in which each element of sort Y_j has been inflated into a cell which is a structure of species G_j. By definition, the weight of such a structure s is the product of the weights of the F-structure and the G_j-structures that form s.

Example 9. Let

$$F_{r,s,t}(X, Y) = C_{\text{alt}}(\mathcal{A}_w(X), \mathcal{G}_v(X + Y)),$$

where $C_{\text{alt}}(X, Y) = C(X \cdot Y)$ is the species of alternating oriented cycles of two sorts of elements, weighted by t^n (n being the length of the cycle), $\mathcal{G}_v$ is the species of simple graphs weighted by r^e (where e is the number of edges of the graph), and $\mathcal{A}_w$ is the species of rooted trees weighted by s^f (where f is the number of leaves of the rooted tree). Figure 5 represents an $F_{r,s,t}$-structure as well as its weight.

For $F = F_w(Y_1, \ldots, Y_m)$, the *functorial composition*

$$F \square (G_1, \ldots, G_m) = F[G_1, \ldots, G_m]$$

is only defined if each G_i is nonweighted. One then sets

$$F \square (G_1, \ldots, G_m)[U_1, \ldots, U_k] = F[G_1[U_1, \ldots, U_k], \ldots, G_m[U_1, \ldots, U_k]]. \quad (12)$$

The weight of an $F_w[G_1, \ldots, G_m]$-structure s on $(U_1, \ldots, U_k)$ is, by definition, its weight as an F_w-structure on $(G_1[U_1, \ldots, U_k], \ldots, G_m[U_1, \ldots, U_k])$.

There are as many notions of partial differentiations $\frac{\partial}{\partial X_i}$ as sorts. For a k-sort species $F = F_w(X_1, \ldots, X_k)$, one sets

$$\left(\frac{\partial}{\partial X_i} F\right)[U_1, \ldots, U_k] = F[U_1, \ldots, U_i + \{*_i\}, \ldots, U_k], \quad (13)$$

the weight of a $(\partial F/\partial X_i)$-structure s on $(U_1, \ldots, U_k)$ being equal to the weight of s as an F-structure of $(U_1, \ldots, U_i + \{*_i\}, \ldots, U_k)$.

The usual rules of differential calculus remain valid for multisort species. For example, the following partial differentiation chain rule holds: If F, G, and H are two-sort species (X and Y), then

$$\frac{\partial}{\partial X} F(G, H) = F_X(G, H) \frac{\partial G}{\partial X} + F_Y(G, H) \frac{\partial H}{\partial X}, \tag{14}$$

where $F_X = \partial F/\partial X$.

There are also k operations of pointing which are carried out by setting, for $i = 1, \ldots, k$,

$$F^{\bullet_i} = X_i \frac{\partial}{\partial X_i} F. \tag{15}$$

We leave to the reader the task of formulating the precise definitions of the diverse combinatorial operations introduced, with regard to the transport of structures.

The generating series of weighted multisort species are defined by introducing one formal variable $x, y, z, t, \ldots$ for each sort $X, Y, Z, T, \ldots$. For the case of index series, it is necessary to introduce an infinite number of formal variables

$$x_1, x_2, \ldots; y_1, y_2, \ldots; z_1, z_2, \ldots; t_1, t_2, \ldots; \ldots,$$

for each sort $X, Y, Z, T, \ldots$. Here is the definition of these series in the case of two sorts. It is straightforward to state the corresponding definitions in the general case.

Definition 10. Let $F = F_w(X, Y)$ be a weighted two-sort species. The *generating series* $F_w(x, y)$, the *type generating series* $\widetilde{F}_w(x, y)$, and the *cycle index series* Z_{F_w} are defined by

$$F_w(x, y) = \sum_{n,k \geq 0} |F[n, k]|_w \frac{x^n}{n!} \frac{y^k}{k!}, \tag{16}$$

where $|F[n, k]|_w$ is the total weight of F-structures on $([n], [k])$,

$$\widetilde{F}_w(x, y) = \sum_{n,k \geq 0} |F[n, k]/\sim|_w x^n y^k, \tag{17}$$

where $|F[n, k]/\sim|_w$ is the total weight of unlabeled F-structures on $([n], [k])$, and

$$Z_{F_w}(x_1, x_2, x_3, \ldots; y_1, y_2, y_3, \ldots) = \sum_{n,k \geq 0} \frac{1}{n!k!} \sum_{\substack{\sigma \in S_n \\ \tau \in S_k}} |\text{Fix } F[\sigma, \tau]|_w$$

$$\times x_1^{\sigma_1} x_2^{\sigma_2} \cdots y_1^{\tau_1} y_2^{\tau_2} \cdots, \tag{18}$$

where $|\text{Fix } F[\sigma, \tau]|_w$ is the total weight of F-structures on $([n], [k])$ that are left fixed under transport along (σ, τ).

One has the formulas

$$F_w(x, y) = Z_{F_w}(x, 0, 0, \ldots; y, 0, 0, \ldots), \tag{19}$$

$$\widetilde{F_w}(x, y) = Z_{F_w}(x, x^2, x^3, \ldots; y, y^2, y^3, \ldots), \tag{20}$$

and the passage to series is compatible with the combinatorial operations $+, \cdot, \circ,$ $', \times, \dot{}\,$, and $\square$. Let us describe explicitly the case of substitution. If $F = F_w(X, Y)$, $G = G_u(X, Y)$, and $H = H_v(X, Y)$ are three weighted two-sort species X and Y such that $G(0, 0) = 0 = H(0, 0)$, then

a) $F_w(G_u, H_v)(x, y) = F_w(G_u(x, y), H_v(x, y))$,

b) $\widetilde{F_w(G_u, H_v)}(x, y) = Z_{F_w}(\widetilde{G_u}(x, y), \widetilde{G_{u^2}}(x^2, y^2), \ldots;$
$$\widetilde{H_v}(x, y), \widetilde{H_{v^2}}(x^2, y^2), \ldots), \tag{21}$$

c) $Z_{F_w(G_u, H_v)} = Z_{F_w}(Z_{G_u}, Z_{H_v})$
$$= Z_{F_w}((Z_{G_u})_1, (Z_{G_u})_2, \ldots; (Z_{H_v})_1, (Z_{H_v})_2, \ldots),$$

where, for $k \geq 1$, $(Z_{G_u})_k(x_1, x_2, x_3, \ldots) = Z_{G_{u^k}}(x_k, x_{2k}, x_{3k}, \ldots)$.

This last substitution is the plethystic composition of weighted index series, in the variables x_i and y_i, $i = 1, 2, 3, \ldots$, defined in an fashion analogous to plethystic substitution for weighted unisort species (see Definition 2.3.9). These properties constitute, once more, powerful computational tools in concrete applications.

Example 11. Consider three sorts of elements X, Y, and Z. Their sum forms the species $X + Y + Z$. An $(X + Y + Z)$-structure is then a singleton of one of the three sorts X, Y, or Z. Substitution into a unisort species $F(X)$ gives the species $F(X + Y + Z)$. An $F(X + Y + Z)$-structure is an F-structure placed on a finite (multi-)set formed from three sorts of elements (possessing an arbitrary number, possibly zero, of elements of each sort). Thus, the species S^{tric} of tri-colored permutations can be written in the form

$$S^{\text{tric}}(X, Y, Z) = S(X + Y + Z), \tag{22}$$

$S = S(X)$ being the usual species of permutations on a single sort X. More generally, one can substitute any sum of sorts (weighted or not, with or without repetitions) for each variable in a weighted multisort species. For example, beginning with a weighted multisort species $F_w(X, Y, Z, T, \ldots)$, one can form the species

$$F_w(3X + T, X_s, X_s, 5X + 4Y_t, \ldots),$$

where X_s (respectively, Y_t) denotes the species of singletons of sort X and of weight s (respectively, of sort Y and of weight t). It is interesting to note the following combinatorial equations:

$$F_w(mX_s, nY_t, \ldots) = F_w(X, Y, \ldots) \times E(mX_s + nY_t + \cdots)$$
$$= F_w(X, Y, \ldots) \times \bigl(E^m(X_s) \cdot E^n(Y_t) \cdot \ldots\bigr). \quad (23)$$

The example of tri-colored permutations $\mathcal{S}^{\text{tric}}(X, Y, Z) = \mathcal{S}(X + Y + Z)$ could lead one to ask if for any multisort species $F = F(X, Y, Z, \ldots)$ there exists a species with one sort $H = H(X)$ such that $F(X, Y, Z, \ldots)$ is of the form $H(X + Y + Z + \cdots)$. This is false (just as for functions in many variables). For a combinatorial verification, it suffices, for example, to examine the species $\mathcal{C}_{\text{alt}}(X, Y)$ of alternating oriented cycles on two sorts. To obtain a $\mathcal{C}_{\text{alt}}$-structure, one must alternately place elements of sort X and of sort Y to form an oriented cycle (see Figure 6,(a)). We are going to show that for any unisort species $H = H(X)$,

$$\mathcal{C}_{\text{alt}}(X, Y) \neq H(X + Y). \quad (24)$$

Indeed, making the substitution $Y := 0$ in the equation gives $\mathcal{C}_{\text{alt}}(X, 0) = H(X)$. An H-structure must then be an alternating cycle not having any element of sort Y. But an alternating cycle must have as many elements of each sort, so we deduce that such a structure cannot exist (since a cycle must always contain at least one element). Thus $H = 0$ (the empty species) and we deduce $\mathcal{C}_{\text{alt}}(X, Y) = 0$, which is false.

Figure 6(b) shows, however, that one has the combinatorial equation

$$\mathcal{C}_{\text{alt}}(X, Y) = \mathcal{C}(X \cdot Y), \quad (25)$$

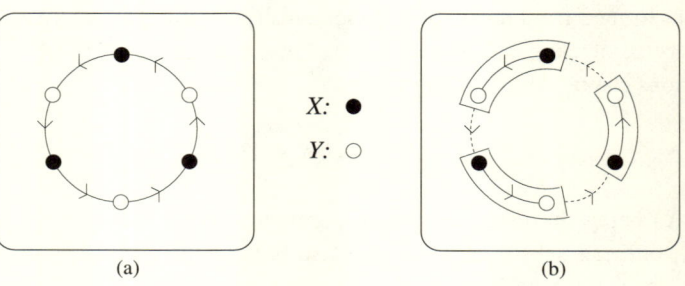

Fig. 6.

2.4. Extension to the Multisort Context

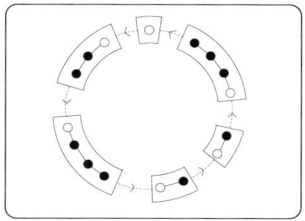

Fig. 7.

where $\mathcal{C} = \mathcal{C}(X)$ is the species of oriented cycles (on the sort X). The following series are immediately deduced:

a) $\mathcal{C}_{\text{alt}}(x, y) = \ln\left(\dfrac{1}{1-xy}\right),$

b) $\widetilde{\mathcal{C}_{\text{alt}}}(x, y) = \dfrac{xy}{1-xy},$ \hfill (26)

c) $Z_{\mathcal{C}_{\text{alt}}} = \displaystyle\sum_{k\geq 1} \dfrac{\varphi(k)}{k} \ln\left(\dfrac{1}{1-x_k y_k}\right).$

Another interesting example of a combinatorial equation comes from the following decomposition of the species $\mathcal{C}(X+Y)$ of oriented cycles on two sorts X, Y of elements:

$$C(X+Y) = C(X) + C(YL(X)). \qquad (27)$$

This formula expresses the fact that if a $\mathcal{C}(X+Y)$-structure has at least one element of sort Y, then it can be identified, in a natural manner, to an oriented cycle formed from disjoint chains of the form $yx_1x_2\cdots x_k$, where y is an element of sort Y and the x_i are distinct elements of sort X (see Figure 7). The reader is invited to analyze the precise form which this equation takes when passing to generating and index series (see Exercise 11).

Every weighted multisort species $F_w = F_w(X, Y, Z, \ldots)$ has a canonical decomposition of the form

$$F_w = \sum_{\mathbf{n}\geq 0} F_{w;\mathbf{n}},$$

where $F_{w;\mathbf{n}}$ denotes the multisort species F_w restricted to the multicardinality

$$\mathbf{n} = (m, n, p, \ldots),$$

defined by

$$F_{w;\mathbf{n}}[U, V, W, \ldots] = \begin{cases} F_w[U, V, W, \ldots], & \text{if } (|U|, |V|, |W|, \ldots) = \mathbf{n}, \\ \emptyset, & \text{otherwise.} \end{cases}$$

To end the present section, we examine an important operation that can be performed on multisort species: the passage to isomorphism types according to one of the sorts, illustrated here in the case of two-sort species.

Definition 12. Let $F = F_v(X, Y)$ be an $\mathbb{A}$-weighted two-sort species. Consider two F-structures $s \in F[U, V]$ and $t \in F[U', V']$. One says that s and t have the same *isomorphism type according to the sort* Y (and one writes $s \sim_Y t$) if
i) $U = U'$,
ii) t is obtained from s by transport of structures along a bijection of the form

$$\sigma = Id + \theta : U + V \longrightarrow U + V',$$

where $\theta : V \longrightarrow V'$ is a bijection.

The class of s according to the equivalence relation $\sim_Y$ is called the *type of s according to Y* and is denoted by $T_Y s$. One says that U is the underlying set of $T_Y s$. In more visual terms, one has $s \sim_Y t$ if and only if s and t become equal when the elements of sort Y are made indistinguishable in their underlying sets. The structure thus obtained represents the type $T_Y s$. For each finite set U, we can consider the set of types according to Y

$$T_Y F[U] = \{T_Y s \mid \exists V, s \in F[U, V]\}, \tag{28}$$

of which U is the underlying set. This set is infinite in general, but summable in $\mathbb{A}[\![y]\!]$ if we introduce a variable y as a counter for the points of sort Y, by analogy with the isomorphism types series of unisort species. In other words, we define the weight of a type $T_Y s$, for $s \in F_v[U, V]$, by setting

$$w(T_Y s) = v(s)\, y^{|V|}. \tag{29}$$

We then have

$$|T_Y F[U]|_w = F_v[\widetilde{U}, Y)(y), \tag{30}$$

where $F_v[U, Y)$ denotes the species of one sort Y derived from $F_v(X, Y)$ by keeping the first component fixed ($= U$). This gives a species of $\mathbb{A}[\![y]\!]$-weighted structures, denoted $\tau_{Y;y} F_v$, called *the species of types of $F(X, Y)$-structures according to the sort Y*. The definition of transport functions is left to the reader.

It is sometimes possible to set $y = 1$ in this process. A sufficient condition is that the sets $T_Y F[U]$ be finite, or that the species $F(X, Y)$ be polynomial in Y, in the following sense.

2.4. Extension to the Multisort Context

Definition 13. Let $F = F(X, Y)$ be a species on two sorts X, Y, whose canonical decomposition is

$$F = \sum_{n,k \geq 0} F_{n,k}.$$

We say that F is *polynomial in Y* if for any $n \geq 0$, there exists $N \geq 0$ such that

$$k \geq N \quad \text{implies} \quad F_{n,k} = 0.$$

When the species $F = F_v(X, Y)$ is polynomial in Y, then for each finite set U the set $T_Y F[U]$ of types according to Y is a finite union of sets that are finite or summable in $\mathbb{A}$, of the form $F[U, V]/\sim_Y$. It is then itself finite or summable in $\mathbb{A}$ and thus determines a species of structures, denoted by $T_Y F$, for which

$$T_Y F = \tau_{Y;y} F \big|_{y=1}, \tag{31}$$

and which could also be denoted by $F(X, 1)$ (see Joyal [163], Section 2.1).

More generally, one can define (see Exercise 15) the types of a species $F_w(X_1, X_2, \ldots, X_k)$ according to the sort X_i as well as the $(k-1)$-sort species

$$\tau_{X_i;x_i} F_w = \tau_{X_i;x_i} F_w(X_1, \ldots, X_{i-1}, X_{i+1}, \ldots, X_k). \tag{32}$$

Here is an example of a species of the form $T_Y F$.

Example 14. Consider the species $\Gamma = \Gamma(X, Y)$ of graphs constructed on vertices (of sort X) and edges (of sort Y). A Γ-structure on a pair (U, V) is then a graph in which the set of vertices is U and the edges (all distinguishable from one another) form the set V. We see also in this case that Γ is polynomial in Y, for we can select $N = \binom{n}{2} + 1$ in the preceding definition. Figure 8(a) illustrates a Γ-structure on the set $U = \{a, b, c, d, e, f\}$ of *vertices* and the set $V = \{m, n, p, q, r, s\}$ of *edges*, whereas Figure 8, (b) shows the $T_Y \Gamma$-structure to which it corresponds, of weight y^6. This $T_Y \Gamma$-structure is quite simply a graph in the usual sense since the edges (represented by line segments) have become indistinguishable. We have the equation

$$\mathcal{G}_v = T_{Y;y} \Gamma,$$

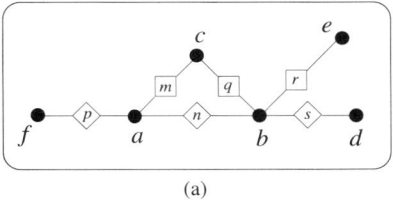 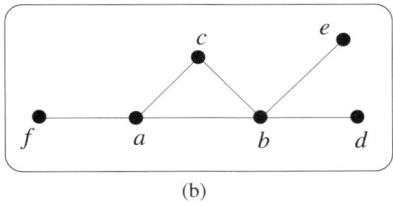

(a) (b)

Fig. 8.

where $\mathcal{G}_v = \mathcal{G}_v(X)$ denotes the species of simple graphs, with an edge counter y, that is with weight $v(g) = y^{e(g)}$, where $e(g)$ is the number of edges of the graph g. Moreover, by considering multigraphs, where multiple edges are allowed, we obtain a weighted species admitting a (summable) infinite set of structures (see Exercise 4).

Note the following formulas for the series associated to species of the form $T_{Y;y}F$:

a) $Z_{T_{Y;y}F}(x_1, x_2, x_3, \ldots) = Z_F(x_1, x_2, x_3, \ldots; y, y^2, y^3, \ldots)$,

b) $(T_{Y;y}F)\widetilde{}(x) = Z_F(x, x^2, x^3, \ldots; y, y^2, y^3, \ldots)$, (33)

c) $T_{Y;y}F(x) = Z_F(x, 0, 0, \ldots; y, y^2, y^3, \ldots)$,

where $Z_F(x_1, x_2, x_3, \ldots; y_1, y_2, y_3, \ldots)$ denotes the cycle index series of the weighted two-sort species $F = F_v(X, Y)$. (See Exercise 7, d).)

Exercises for Section 2.4

1. Let $\Phi = \Phi(X, Y)$ be the two-sort species of functions $f : U \longrightarrow V$, where X is the sort of elements of U and Y that of elements of V. We have

$$\Phi[U, V] = \{f \mid f : U \longrightarrow V\}. \tag{34}$$

The transport of the corresponding structures is obtained by the usual composition of functions:

$$\Phi[\sigma, \theta](f) = \theta \circ f \circ \sigma^{-1}.$$

Moreover, we introduce the three subspecies Inj, Sur, and Bij of Φ by setting

$$\text{Inj}[U, V] = \{f \mid f : U \hookrightarrow V, \text{ i.e., } f \text{ injective}\},$$

$$\text{Sur}[U, V] = \{f \mid f : U \twoheadrightarrow V, \text{ i.e., } f \text{ surjective}\}, \tag{35}$$

$$\text{Bij}[U, V] = \{f \mid f : U \xrightarrow{\sim} V, \text{ i.e., } f \text{ bijective}\}.$$

a) Show that

i) $\Phi(X, Y) = E(E(X) \cdot Y)$,

ii) $\text{Inj}(X, Y) = E((1 + X) \cdot Y)$,

iii) $\text{Sur}(X, Y) = E(E_+(X) \cdot Y)$, (36)

iv) $\text{Bij}(X, Y) = E(X \cdot Y)$.

2.4. Extension to the Multisort Context

b) Prove by combinatorial computations, as well as by geometric arguments, the following equalities (isomorphisms):

$$
\begin{aligned}
&\text{i)} \quad \frac{\partial}{\partial X}\Phi = Y\frac{\partial}{\partial Y}\Phi, \\
&\text{ii)} \quad (1+X)\frac{\partial}{\partial X}\text{Inj} = Y\frac{\partial}{\partial Y}\text{Inj}, \\
&\text{iii)} \quad \frac{\partial}{\partial X}\text{Sur} = Y\left(1+\frac{\partial}{\partial Y}\right)\text{Sur}, \quad\quad (37) \\
&\text{iv)} \quad \frac{\partial}{\partial X}\text{Bij} = Y\,\text{Bij}, \quad \frac{\partial}{\partial Y}\text{Bij} = X\,\text{Bij}, \\
&\text{v)} \quad \text{Bij}(X,Y) = \text{Bij}(Y,X).
\end{aligned}
$$

c) Show that the series $\widetilde{\Phi} = \widetilde{\Phi}(x, y)$ and $Z_\Phi = Z_\Phi(x_1, x_2, x_3, \ldots; y_1, y_2, y_3, \ldots)$ are given by the formulas

$$
\begin{aligned}
&\text{i)} \quad \widetilde{\Phi} = \frac{1}{(1-y)(1-xy)(1-x^2y)\cdots(1-x^k y)\cdots}, \\
&\text{ii)} \quad Z_\Phi = \exp\left(\sum_{n\geq 1}\frac{y_n}{n}\exp\sum_{m\geq 1}\frac{x_{mn}}{m}\right).
\end{aligned}
\quad\quad (38)
$$

d) Compute explicitly the six other series: Z_Inj, Z_Sur, Z_Bij, $\widetilde{\text{Inj}}$, $\widetilde{\text{Sur}}$, and $\widetilde{\text{Bij}}$.

e) More generally, for a species R on one sort, define the two-sort species

$$\Phi_R(X, Y) = E(R(X)\cdot Y) \quad\quad (39)$$

of *functions with R-enriched fibers*. Compute the series Z_{Φ_R} and $\widetilde{\Phi}_R$.

2. Denote by $\text{Oct}_\text{alt} = \text{Oct}_\text{alt}(X, Y)$, the species of *alternating octopuses* on two sorts of elements X, Y, i.e., the octopuses on X and Y in which the adjacent elements are of different sort.

a) Verify that

$$\text{Oct}_\text{alt}(X, Y) = \mathcal{C}_\text{alt}(X\cdot L(Y\cdot X)\cdot(1+Y),\, Y\cdot L(X\cdot Y)\cdot(1+X)). \quad (40)$$

b) Deduce the series

$$
\begin{aligned}
&\text{i)} \quad \text{Oct}_\text{alt}(x, y) = \log\frac{(1-xy)^2}{1-xy(3+x+y)}, \\
&\text{ii)} \quad \widetilde{\text{Oct}}_\text{alt}(x, y) = \sum_{n\geq 1}\frac{\phi(n)}{n}\log\frac{(1-x^n y^n)^2}{1-x^n y^n(3+x^n+y^n)}, \quad\quad (41) \\
&\text{iii)} \quad Z_{\text{Oct}_\text{alt}} = \sum_{n\geq 1}\frac{\phi(n)}{n}\log\frac{(1-x_n y_n)^2}{1-x_n y_n(3+x_n+y_n)}.
\end{aligned}
$$

3. Consider the species $S^{\text{mix}}(X, Y)$ of permutations on two sorts X and Y, where each cycle has *at least* one element of each sort.
   a) Show that
   $$S(X)S(Y)S^{\text{mix}}(X, Y) = S(X + Y). \tag{42}$$
   b) Deduce the formulas:
   i) $S^{\text{mix}}(x, y) = \dfrac{(1 - x)(1 - y)}{1 - x - y}$,

   ii) $\widetilde{S^{\text{mix}}}(x, y) = \displaystyle\prod_{n \geq 1} \dfrac{(1 - x^n)(1 - y^n)}{1 - x^n - y^n}$, $\tag{43}$

   iii) $Z_{S^{\text{mix}}} = \displaystyle\prod_{n \geq 1} \dfrac{(1 - x_n)(1 - y_n)}{1 - x_n - y_n}$.

4. Consider the species $\Gamma = \Gamma(X, Y)$ of graphs on vertices of sort X, and of edges of sort Y, introduced in Example 14, as well as the species of multigraphs $\Gamma_{\text{mult}} = \Gamma_{\text{mult}}(X, Y)$, where multiple edges are permitted.
   a) Verify the following equalities

   i) $\Gamma(x, y) = T_{Y;y}\Gamma(x) = \displaystyle\sum_{n \geq 0} (1 + y)^{\binom{n}{2}} \dfrac{x^n}{n!}$,

   ii) $\Gamma_{\text{mult}}(x, y) = \displaystyle\sum_{n \geq 0} (e^y)^{\binom{n}{2}} \dfrac{x^n}{n!}$, $\tag{44}$

   iii) $T_{Y;y}\Gamma_{\text{mult}}(x) = \displaystyle\sum_{n \geq 0} \left(\dfrac{1}{1 - y}\right)^{\binom{n}{2}} \dfrac{x^n}{n!}$.

   b) Show that

   i) $\Gamma(X, Y) = \text{Inj} \,\square\, (\varepsilon(Y), \wp^{[2]}(X))$,

   ii) $\Gamma_{\text{mult}}(X, Y) = \Phi \,\square\, (\varepsilon(Y), \wp^{[2]}(X))$, $\tag{45}$

   where ε denotes the species of elements (see Section 1.1), and $\wp^{[2]}$, that of subsets with two elements (see Section 2.2).
   c) More generally, for a unisort species R, define the two-sort species of multigraphs with an R-enrichment on the edges having the same endpoints, by $\Gamma_R(X, Y) = \Phi_R \,\square\, (\varepsilon(Y), \wp^{[2]}(X))$ (see Exercise 1). Verify that
   $$\Gamma_R(x, y) = \sum_{n \geq 0} R(y)^{\binom{n}{2}} \dfrac{x^n}{n!}. \tag{46}$$

5. Consider the two-sort species $\mathcal{B} = \mathcal{B}(X, T)$ of rooted trees on internal vertices of sort X and leaves of sort Y (see Figure 2). Let $\mathcal{A}_w = \mathcal{A}_w(X)$ be the weighted

species of rooted trees on vertices of sort X, the weight of each rooted tree α being given by $w(\alpha) = t^f$, where f is the number of leaves of α. Verify the combinatorial equality

$$\mathcal{A}_w(X) = \mathcal{B}(X, X_t), \tag{47}$$

where X_t is the species of singletons of weight t. Do not forget transports of structures.

6. FUNCTORIAL POINT OF VIEW. Let k be an integer ≥ 1. Denote by $\mathbb{B}$, the category of finite sets and bijections, and by $\mathbb{B}^k = \mathbb{B} \times \cdots \times \mathbb{B}$, the category, product of k copies of $\mathbb{B}$, of finite multisets (of k sorts) and of multibijections. Consider also the category, denoted by $\Phi(\mathbb{B}, [k])$, whose objects are the pairs (U, X), where U is a finite set and $X : U \longrightarrow [k]$ is a function, and whose morphisms $\sigma : (U, X) \longrightarrow (V, \psi)$ are the bijections $\sigma : U \longrightarrow V$ such that $\psi \circ \sigma = X$.

Let $\mathbb{A}$ be a ring of formal power series on a ring $\mathbb{K} \subseteq \mathbb{C}$. Denote by $\mathbb{E}_\mathbb{A}$, the category of summable $\mathbb{A}$-weighted sets and $\mathbb{A}$-weighted morphisms of sets (see Definitions 2.3.2 and 2.3.3).

a) Show that the categories $\mathbb{B}^k$ and $\Phi(\mathbb{B}, [k])$ are *equivalent* by explicitly describing functors $R : \mathbb{B}^k \longrightarrow \Phi(\mathbb{B}, [k])$ and $S : \Phi(\mathbb{B}, [k]) \longrightarrow \mathbb{B}^k$ as well as mutually inverse natural transformations $\alpha : S \circ R \longrightarrow \mathrm{Id}_{\mathbb{B}^k}$ and $\beta : R \circ S \longrightarrow \mathrm{Id}_{\Phi(\mathbb{B}, [k])}$.

b) Show that a weighted k-sort species $F = F_v$ can be considered as a functor $F : \mathbb{B}^k \longrightarrow \mathbb{E}_\mathbb{A}$ or equivalently, as a functor $F : \Phi(\mathbb{B}, [k]) \longrightarrow \mathbb{E}_\mathbb{A}$.

7. a) Complete the definitions of the combinatorial operations $(+, \cdot, \circ, \times, \square)$ on species in the weighted multisort context, not forgetting transports of structures in each case.

b) State and prove the properties of these combinatorial operations, in analogy with formulas (1.3.48), (1.3.55), (1.4.46), (1.4.47), (2.1.49), and (2.2.37).

c) Describe the behavior of generating and index series with respect to these operations. In particular, prove formulas (19), (20), and (21), a).

d) Prove formulas (33) for the series associated to the species of types according to sort.

HINT: Use Exercise A.1.9.

e) Extend these results to multisort species.

8. a) Let $F = F(X_1, X_2, \ldots, X_k)$ and $G = G(X_1, X_2, \ldots, X_k)$ be species on many sorts $X_1, X_2, \ldots, X_k$. Prove for $i = 1, \ldots, k$ the equalities

i) $\quad \dfrac{\partial}{\partial X_i}(F + G) = \dfrac{\partial}{\partial X_i} F + \dfrac{\partial}{\partial X_i} G,$

ii) $\quad \dfrac{\partial}{\partial X_i}(F \cdot G) = \left(\dfrac{\partial}{\partial X_i} F\right) \cdot G + F \cdot \left(\dfrac{\partial}{\partial X_i} G\right),$ \hfill (48)

iii) $\quad \dfrac{\partial^2}{\partial X_i \partial X_j} F = \dfrac{\partial^2}{\partial X_j \partial X_i} F.$

b) Consider the species $F = F(Y_1, Y_2, \ldots, Y_m)$ and $G_j = G_j(X_1, X_2, \ldots, X_k)$, $j = 1, \ldots, m$. Prove the chain rule

$$\frac{\partial}{\partial X_i} F(G_1, G_2, \ldots, G_m) = \sum_{j=1}^{m} \frac{\partial F}{\partial Y_j}(G_1, G_2, \ldots, G_m) \cdot \frac{\partial}{\partial X_i} G_j. \quad (49)$$

9. Consider a species on many sorts $F = F(X_1, X_2, \ldots, X_k)$. Interpret combinatorially the following species:

a) $\sum_{i=1}^{k} X_i \frac{\partial}{\partial X_i} F$, b) $X_2 \frac{\partial}{\partial X_1} F$, c) $\sum_{i \neq j} X_i \frac{\partial}{\partial X_j} F$.

10. Show that the species $\mathcal{G}_{\text{chro}} = \mathcal{G}_{\text{chro}}(X, Y, Z)$ of tri-chromatic graphs (see Example 1) is not of the form $F(X + Y + Z)$, where F is a unisort species.

11. Write explicitly the three identities between generating and cycle index series which correspond to the combinatorial equality $\mathcal{C}(X + Y) = \mathcal{C}(X) + \mathcal{C}(Y \cdot L(X))$.

12. a) Show that by taking the types according to Y of the species $\text{Sur}(X, Y)$ of surjections, one obtains the species $\text{Par}_v(X)$ of partitions weighted by $v(\pi) = y^{b(\pi)}$, where $b(\pi)$ is the number of blocks of π, i.e.,

$$\text{Par}_v(X) = T_{Y;y} \text{Sur}(X, Y). \quad (50)$$

b) Deduce that the Stirling numbers of the second kind $S(n, k)$ are given by

$$S(n, k) = \frac{|\text{Sur}[n, k]|}{k!}. \quad (51)$$

c) Establish the combinatorial identity

$$\Phi(X, Y) = \text{Sur}(X, Y) \cdot E(Y) \quad (52)$$

and deduce Touchard's formula,

$$\sum_{n,k \geq 0} k^n \frac{x^n}{n!} \frac{y^k}{k!} = \sum_{n,k \geq 0} S(n, k) \frac{x^n}{n!} y^k \cdot \sum_{j \geq 0} \frac{y^j}{j!}. \quad (53)$$

d) Deduce Dobinski's formula for Bell numbers,

$$B_n = \frac{1}{e} \sum_{k \geq 0} \frac{k^n}{k!}, \quad n \geq 0. \quad (54)$$

13. Consider the species $\mathcal{S}(X, Y) = \mathcal{S}(X + Y)$ of permutations of elements of sort X and elements of sort Y.
   a) Describe, using a geometric figure, a typical $T_Y \mathcal{S}$-structure on a finite set U.
   b) Compute the three series $(T_{Y;y} \mathcal{S})(x)$, $(T_{Y;y} \widetilde{\mathcal{S}})(x)$, and $Z_{T_{Y;y} \mathcal{S}}(x_1, x_2, x_3, \ldots)$.

14. a) Show that the species $\mathcal{B} = \mathcal{B}(X, T)$ of Exercise 5 is not polynomial in X nor in T.
    b) Let k be an integer ≥ 1, and consider the subspecies F of $\mathcal{B}$ formed of rooted trees in which each internal vertex is of degree $\leq k$. Show that $F = F(X, T)$ is polynomial in T but not in X.
    c) Let k be an integer ≥ 1, and consider the subspecies G of $\mathcal{B}$ formed of rooted trees in which each internal vertex is of degree between 2 and k. Show that $F = F(X, T)$ is polynomial both in T and in X.

15. Let $F_v = F_v(X_1, \ldots, X_k)$ be an $\mathbb{A}$-weighted multisort species. Let i be an integer such that $1 \leq i \leq k$.
    a) Define the type $T_{X_i} s$ of an F_v-structure s according to the sort X_i, as well as its weight $w(T_{X_i} s) \in \mathbb{A}[\![x_i]\!]$.
    b) Define the $\mathbb{A}[\![x_i]\!]$-weighted species $T_{X_i;x_i} F_v$ of types of F_v-structures according to the sort X_i.
    c) Give finiteness conditions ensuring existence of the species $T_{X_i} F_v = T_{X_i;x_i} \times F_v|_{x_i=1}$.

16. GENERALIZED POINTING. Consider two species of structures $F = F(X)$ and $G = G(X)$. Define the F-pointing of G as being the species, denoted by $F \langle\!\langle X \frac{d}{dX} \rangle\!\rangle G(X)$, given by

$$F \left\langle\!\!\left\langle X \frac{d}{dX} \right\rangle\!\!\right\rangle G(X) = (F(X) \cdot E(X)) \times G(X). \tag{55}$$

a) Give a combinatorial (geometrical) description of this definition.
b) Given a polynomial

$$p(u, v, w, \ldots) = \sum_{i,j,k,\ldots} p_{i,j,k,\ldots} u^i v^j w^k \cdots, \tag{56}$$

let us agree to write

$$p \langle\!\langle u, v, w, \ldots \rangle\!\rangle = \sum_{i,j,k,\ldots} p_{i,j,k,\ldots} u_{\langle i \rangle} v_{\langle j \rangle} w_{\langle k \rangle} \cdots, \tag{57}$$

where $x_{(n)} = x(x-1)(x-2)\cdots(x-n+1)$. Suppose that F is polynomial in X (see Exercise 19) and set $H(X) = F \langle\!\langle X \frac{d}{dX} \rangle\!\rangle G(X)$. Establish the equalities

i) $H(x) = F \left\langle\!\!\left\langle x \frac{d}{dx} \right\rangle\!\!\right\rangle G(x)$,

ii) $\widetilde{H}(x) = Z_F \left\langle\!\!\left\langle x_1 \frac{\partial}{\partial x_1}, 2x_2 \frac{\partial}{\partial x_2}, 3x_3 \frac{\partial}{\partial x_3}, \ldots \right\rangle\!\!\right\rangle Z_G \bigg|_{x_i := x^i,\ i \geq 1}$,

iii) $Z_H(x_1, x_2, x_3, \ldots) = Z_F \left\langle\!\!\left\langle x_1 \frac{\partial}{\partial x_1}, 2x_2 \frac{\partial}{\partial x_2}, 3x_3 \frac{\partial}{\partial x_3}, \ldots \right\rangle\!\!\right\rangle$
$\times Z_G(x_1, x_2, x_3, \ldots).$ \hfill (58)

17. a) Show that the species $\Gamma = \Gamma(X, Y)$, $\Gamma_{\text{mult}} = \Gamma_{\text{mult}}(X, Y)$, and $\Gamma_R = \Gamma_R(X, Y)$, of Exercise 4, satisfy the combinatorial equations (see Knuth [155])

i) $(1 + Y)\dfrac{\partial}{\partial Y}\Gamma(X, Y) = E_2\left\langle\!\left\langle X\dfrac{\partial}{\partial X}\right\rangle\!\right\rangle \Gamma(X, Y),$

ii) $\dfrac{\partial}{\partial Y}\Gamma_{\text{mult}}(X, Y) = E_2\left\langle\!\left\langle X\dfrac{\partial}{\partial X}\right\rangle\!\right\rangle \Gamma_{\text{mult}}(X, Y),$ \hfill (59)

iii) $R(Y)\dfrac{\partial}{\partial Y}\Gamma_R(X, Y) = R'(Y)E_2\left\langle\!\left\langle X\dfrac{\partial}{\partial X}\right\rangle\!\right\rangle \Gamma_R(X, Y).$

b) Write the implied relations for the generating and index series.

18. **GENERALIZED TAYLOR FORMULA.** Given two species of structures $F = F(X)$ and $G = G(X)$, we define the two-sort (X and T) species $F(T\frac{\partial}{\partial X})G(X)$ by setting

$$F\left(T\dfrac{\partial}{\partial X}\right)G(X) = (E(X) \cdot F(T)) \times G(X + T). \tag{60}$$

a) Interpret combinatorially definition (60) by describing geometrically a structure of species $F(T\partial/\partial X)G(X)$ on a pair of finite sets (U, V).

b) Set $H(X, T) = F(T\frac{\partial}{\partial X})G(X)$. Establish combinatorially the following relations:

i) $H(x, t) = F\left(t\dfrac{\partial}{\partial x}\right)G(x),$

ii) $\widetilde{H}(x, t) = \left.\left(Z_F\left(t_1\dfrac{\partial}{\partial x_1}, 2t_2\dfrac{\partial}{\partial x_2}, 3t_3\dfrac{\partial}{\partial x_3}, \ldots\right)Z_G\right)\right|_{x_i := x^i,\ t_i := t^i,\ i=1,2,\ldots},$

iii) $Z_H(x_1, x_2, \ldots; t_1, t_2, \ldots) = Z_F\left(t_1\dfrac{\partial}{\partial x_1}, 2t_2\dfrac{\partial}{\partial x_2}, \ldots\right)Z_G(x_1, x_2, \ldots).$ \hfill (61)

The *F*-Taylor expansion of $G(X)$ in T is defined by the formula

$$(E(X) \cdot F(T)) \times G(X + T) = \sum_{k \geq 0} F_k\left(T\dfrac{\partial}{\partial X}\right)G(X) \tag{62}$$

where $F = F_0 + F_1 + F_2 + \cdots$ is the canonical decomposition of F. By letting $X := 0$ in (62) we obtain, by definition, the *F*-Maclaurin expansion of $G(X)$ in T:

$$F(T) \times G(T) = \sum_{k \geq 0} F_k\left(T\dfrac{\partial}{\partial X}\right)G(X)\bigg|_{X:=0}. \tag{63}$$

c) Show that the series (62) and (63) are indeed summable and prove the combinatorial identities. In particular, when $F = E$, show that the series (62)

2.4. Extension to the Multisort Context

and (63) take the form of Taylor–Maclaurin expansions in T:

$$G(X+T) = G(X) + TG'(X) + E_2\left(T\frac{\partial}{\partial X}\right)G(X) + \cdots$$

$$+ E_k\left(T\frac{\partial}{\partial X}\right)G(X) + \cdots . \tag{64}$$

d) Verify that the following "classical" formulas are satisfied:

i) $G(x+t) = e^{t\frac{\partial}{\partial x}} G(x)$

$$= G(x) + tG'(x) + \frac{t^2}{2!}G''(x) + \cdots + \frac{t^k}{k!}G^{(k)}(x) + \cdots ,$$

ii) $G(t) = e^{t\frac{\partial}{\partial x}} G(x)\big|_{x:=0}$

$$= G(0) + tG'(0) + \frac{t^2}{2!}G''(0) + \cdots + \frac{t^k}{k!}G^{(k)}(0) + \cdots ,$$

iii) $Z_{G(X+T)}(x_1, x_2, \ldots; t_1, t_2, \ldots) = Z_G(x_1 + t_1, x_2 + t_2, \ldots)$

$$= e^{t_1\frac{\partial}{\partial x_1} + t_2\frac{\partial}{\partial x_2} + \cdots} Z_G(x_1, x_2, \ldots),$$

iv) $Z_G(t_1, t_2, \ldots) = e^{t_1\frac{\partial}{\partial x_1} + t_2\frac{\partial}{\partial x_2} + \cdots} Z_G(x_1, x_2, \ldots)\big|_{x_i:=0,\ i=1,2,\ldots}.$
$$\tag{65}$$

19. GENERALIZED DIFFERENTIATION. Let $F = F(X)$ be a polynomial species in X (i.e., there exists $N \geq 0$ such that $|U| > N \implies F[U] = \emptyset$). Given a species $G = G(X)$, define the species $F(\frac{d}{dX})G(X)$ by setting

$$F\left(\frac{d}{dX}\right)G(X) = T_Y[(E(X) \cdot F(Y)) \times G(X+Y)], \tag{66}$$

where Y is an auxiliary sort different from X.

a) Show that the species $(E(X) \cdot F(Y)) \times G(X+Y)$ is polynomial in Y and interpret combinatorially Definition (66) by describing geometrically an $F(\frac{d}{dX})G(X)$-structure on a finite set U.

b) Set $H(X) = F(\frac{d}{dX})G(X)$. Establish the relations

i) $H(x) = F\left(\dfrac{d}{dx}\right)G(x)$, if $F = X^n$,

but $H(x) \neq F\left(\dfrac{d}{dx}\right)G(x)$ in the general case,

ii) $\widetilde{H}(x) = \left\{Z_F\left(\dfrac{\partial}{\partial x_1}, 2\dfrac{\partial}{\partial x_2}, 3\dfrac{\partial}{\partial x_3}, \ldots\right) Z_G\right\}\bigg|_{x_i:=x^i,\ i\geq 1}$,

iii) $Z_H(x_1, x_2, x_3, \ldots) = Z_F\left(\dfrac{\partial}{\partial x_1}, 2\dfrac{\partial}{\partial x_2}, 3\dfrac{\partial}{\partial x_3}, \ldots\right) Z_G(x_1, x_2, x_3, \ldots).$
$$\tag{67}$$

2.5. Virtual Species

The *combinatorial calculus* that we have developed so far for species of structures has proven to be very useful, as illustrated by the various examples of preceding sections. However, for some purposes, it still suffers from one gap:

the absence of a convenient combinatorial operation of subtraction.

Establishing a combinatorial form of subtraction would have the advantage of widening the range of possibilities for our combinatorial calculus. This problem has been solved by Joyal [162, 161] and Yeh [333, 334] with the introduction of the concept of *virtual species*. We will see that virtual species allow us to
- give combinatorial meaning to the multiplicative inverse $1/E$ of the species E of sets (more generally: to $1/F$ for any species F such that $F(0) = 1$);
- solve certain combinatorial differential equations which do not have any solution in the context of species of structures;
- analyze the combinatorial nature of certain species in a better manner (for example, those of trees and rooted trees);
- give combinatorial meaning to the inverse for substitution $G^{\langle -1 \rangle}$, for suitable species G;
- extend the notion of connected components to a wider class of species.

The situation is analogous to the case of the semi-ring $\mathbb{N}$ of natural numbers for which subtraction is not everywhere defined. For instance, consider the species $\mathcal{G}$ (respectively $\mathcal{G}^c$, $\mathcal{G}^d$) of graphs (respectively connected graphs, disconnected graphs). The combinatorial equation

$$\mathcal{G} = \mathcal{G}^c + \mathcal{G}^d \tag{1}$$

allows us to define *subtraction* $\mathcal{G} - \mathcal{G}^c$ by setting

$$\mathcal{G} - \mathcal{G}^c = \mathcal{G}^d. \tag{2}$$

This combinatorial subtraction is possible because $\mathcal{G}^c$ is a subspecies of $\mathcal{G}$, in the sense of the following definition.

Definition 1. Let F be a (weighted, multisort) species. A species G is said to be a *subspecies* of F (written $G \subseteq F$) if it satisfies the following two conditions:
i) for any finite (multi-)set U, $G[U] \subseteq F[U]$ and, in the weighted case, the weighting on $G[U]$ is induced from that of $F[U]$;
ii) for any $\sigma : U \xrightarrow{\sim} V$, $G[\sigma] = F[\sigma]|_{G[U]}$.

But what significance can be given to a "difference" such as $\mathcal{C} - \mathcal{A}$, where $\mathcal{C}$ is the species of oriented cycles and $\mathcal{A}$ that of rooted trees, or more generally, to the difference $F - G$ whenever G is not a subspecies of F? At first glance, one might think that the notion of weighted species answers this question by defining

2.5. Virtual Species

$-G$, for a weighted species $G = G_w$, by $-G_w = G_{-w}$ (i.e., changing the sign of the weight of each G-structure). However, this *solution* is not acceptable since, for $G = G_w \neq 0$,

$$G_w + G_{-w} \neq 0, \tag{3}$$

where 0 denotes the empty species.

Not surprisingly our solution to this problem follows a course similar to the construction of the ring $\mathbb{Z}$ of integers from the semi-ring $\mathbb{N}$ of natural numbers. After all, the class of species of structures constitutes a (large) semi-ring with its addition (+) and combinatorial multiplication (·), where zero is the empty species 0 and one is the species 1 (with combinatorial equality playing the role of equality). However, we have to ensure that the other operations (∘, ′, •, ×, □) can be extended in a nice fashion to this new context. In particular, this extension should be compatible with the various series. For greater simplicity, we first treat the case of nonweighted species on a single sort X of points. The multisort weighted case follows easily.

Step 1. Completion

Consider the semi-ring Spe of (unisort, nonweighted) species of structures. By analogy with the construction of $\mathbb{Z}$ from $\mathbb{N}$, we make the following definition.

Definition 2. A *virtual species* is an element of the quotient set

$$\text{Virt} = (\text{Spe} \times \text{Spe})/\sim \tag{4}$$

where the equivalence relation $\sim$ is defined by

$$(F, G) \sim (H, K) \iff F + K \simeq G + H \tag{5}$$

(combinatorial equality in Spe; see Definition 1.2.12). One then writes

$$F - G = \text{class of } (F, G) \text{ according to } \sim, \tag{6}$$

to denote any virtual species. The pair (F, G) is called a *representative* of $F - G$.

The fact that $\sim$ is an equivalence relation relies on the cancelation law for the addition of species (see Exercise 3). We leave the proof of this property as well as that of the following proposition to the reader (see Exercise 4).

Proposition 3. *The set* Virt *of virtual species constitutes a commutative ring under the operations of addition and multiplication defined by*

a) $(F - G) + (H - K) = (F + H) - (G + K)$,

b) $(F - G) \cdot (H - K) = (F \cdot H + G \cdot K) - (F \cdot K + G \cdot H)$,

(7)

with zero $0 = (0 - 0)$, *and one* $1 = (1 - 0)$. *The additive inverse of* $(F - G)$ *is* $(G - F)$. *Moreover, the injection* Spe $\hookrightarrow$ Virt $(F \mapsto (F - 0))$ *is a semi-ring homomorphism.*

Definition 4. Let F and G be two species of structures and consider the virtual species $\Phi = F - G$. The *generating series*, *type generating series*, and *cycle index series* of Φ are respectively defined by

a) $\Phi(x) = F(x) - G(x)$,

b) $\widetilde{\Phi}(x) = \widetilde{F}(x) - \widetilde{G}(x)$, (8)

c) $Z_\Phi(x_1, x_2, x_3, \ldots) = Z_F(x_1, x_2, x_3, \ldots) - Z_G(x_1, x_2, x_3, \ldots)$.

Remark 5. It is easy to verify the correctness of the preceding definitions, that is to say, the independence with regard to the choice of representatives, and that for all virtual species $\Phi = F - G$ and $\Psi = H - K$, one has

a) $(\Phi \cdot \Psi)(x) = \Phi(x)\Psi(x)$,

b) $\widetilde{(\Phi \cdot \Psi)}(x) = \widetilde{\Phi}(x)\widetilde{\Psi}(x)$, (9)

c) $Z_{\Phi \cdot \Psi} = Z_\Phi Z_\Psi$.

Observe that for any species F, G, H, K,

$$H = K \implies (F + H) - (G + K) = (F - G). \tag{10}$$

This leads to the following definition.

Definition 6. Two species of structures F and G are said to be *unrelated* if the only subspecies of F which is isomorphic to a subspecies of G is the empty species. A virtual species Φ is said to be written in *reduced form* $\Phi = F - G$ if the species F and G are unrelated.

Proposition 7. *Every virtual species* Φ *can be written in reduced form*

$$\Phi = \Phi^+ - \Phi^-, \tag{11}$$

and furthermore, this can be done in a unique manner: If $\Phi = P - N$ is also in reduced form, then $P = \Phi^+$ and $N = \Phi^-$.

This important proposition will be proved in Section 2.6, using molecular decompositions.

Definition 8. Let Φ be a virtual species. The species Φ^+ (respectively Φ^-), in the reduced form (11) of Φ, is called the *positive* (respectively *negative*) part of Φ. If $\Phi^- \neq 0$, one says that Φ is *strictly virtual*. If $\Phi = \Phi^+$ (respectively $\Phi = -\Phi^-$) one says that Φ is *positive* (respectively, *negative*).

Evidently, the notion of positive virtual species is equivalent to that of species of structures.

Example 9. If G is a subspecies of a species of structures F ($G \subseteq F$), then $(F - G)^- = 0$ and $(F - G)^+ = F - G$. Consequently, $F - G$ is a species (see also Exercise 2). For a more concrete example, consider the virtual species $\Phi = C - A$ mentioned previously. One verifies that

$\Phi^+ = C - X =$ the species of oriented cycles on at least 2 points,

$\Phi^- = A - X =$ the species of rooted trees on at least 2 points.

The notion of summability is extended to the virtual case by simply saying that a family $(\Phi_i)_{i \in I}$ of virtual species is *summable* if each of the two families of species $(\Phi_i^+)_{i \in I}$ and $(\Phi_i^-)_{i \in I}$ is summable. When this condition is realized, the sum $\Phi = \sum_{i \in I} \Phi_i$ is defined by

$$\Phi = \sum_{i \in I} \Phi_i^+ - \sum_{i \in I} \Phi_i^-. \qquad (12)$$

In general, this representation is not in reduced form. A sufficient condition (but not necessary; see Exercise 6) for a family $\Phi_i = F_i - G_i$, $i \in I$, to be summable is that the families of species $(F_i)_i$ and $(G_i)_i$ be summable. In this case,

$$\Phi = \sum_{i \in I} \Phi_i = \sum_{i \in I} F_i - \sum_{i \in I} G_i. \qquad (13)$$

As an example of a summable family, we mention the *canonical decomposition* of a virtual species $\Phi = F - G$:

$$\Phi = \sum_{n \geq 0} \Phi_n = \sum_{n \geq 0} \left(\Phi_n^+ - \Phi_n^- \right), \qquad (14)$$

where

$$\Phi_n^+ = (F_n - G_n)^+ \quad \text{and} \quad \Phi_n^- = (F_n - G_n)^-. \tag{15}$$

Here F_n (respectively G_n) denotes the restriction of F (respectively G) to the cardinality n (see Example 1.3.6).

Example 10. THE MULTIPLICATIVE INVERSE OF THE SPECIES E. Consider the species E of sets. It is easy to verify that there does not exist a species of structures F satisfying the combinatorial equation $F \cdot E = 1$. However, the use of virtual species allows such an inversion. Indeed, the family of virtual species $(-1)^k (E_+)^k, k = 0, 1, \ldots$ is summable (see Exercise 7) and we set

$$E^{-1} = (1 + E_+)^{-1} = \sum_{k=0}^{\infty} (-1)^k (E_+)^k. \tag{16}$$

Distributing the product over the sum, we get

a) $E^{-1}(x) = e^{-x}$,

b) $\widetilde{E^{-1}}(x) = 1 - x$, $\qquad (17)$

b) $Z_{E^{-1}}(x_1, x_2, x_3, \ldots) = e^{-\left(x_1 + \frac{1}{2}x_2 + \frac{1}{3}x_3 + \cdots\right)}$.

It can be shown that

$$\begin{aligned}(E^{-1})^+ &= 1 + (E_+)^2 + (E_+)^4 + \cdots, \\ (E^{-1})^- &= E_+ + (E_+)^3 + (E_+)^5 + \cdots. \end{aligned} \tag{18}$$

Thus, an E^{-1}-structure is an ordered partition (ballot) either with an even number of classes (contributing to the positive part) or with an odd number of classes (contributing to the negative part).

We have already seen that $\mathbb{N} \subseteq \text{Spe}$ (via the identification of an integer n with the species $n = n\,1$). By extension, we obtain a corresponding embedding $\mathbb{Z} \subseteq \text{Virt}$. In fact, $\mathbb{Z}$ is a sub-ring of Virt.

All that has been said about virtual species can be extended to the context of weighted multisort species. This extension gives rise to the ring

$$\text{Virt}_\mathbb{A} \langle X, Y, Z, \ldots \rangle$$

of $\mathbb{A}$-weighted virtual species on the sorts $X, Y, Z, \ldots$ (see Exercise 8). We simply emphasize here that any $\mathbb{A}$-weighted virtual species

$$\Phi = \Phi_w(X, Y, Z, \ldots) = F_u(X, Y, Z, \ldots) - G_v(X, Y, Z, \ldots) \tag{19}$$

can be written in a unique reduced form

$$\Phi_w = \Phi_w^+(X, Y, Z, \ldots) - \Phi_w^-(X, Y, Z, \ldots). \tag{20}$$

Step 2. Extension of the Other Operations

To simplify the exposition, we again limit ourselves to the nonweighted context, with a single sort X of elements. We are going to extend the other operations $\circ$, $'$, $\bullet$, $\times$, $\square$ to the ring Virt of virtual species so that the *usual rules* of calculus be satisfied. The extension of the substitution $\circ$ is by far the most important and the most difficult to carry out. Thus we begin with the easiest operations: $'$, $\bullet$, $\times$, $\square$. For each of these, it suffices to use linearity and distributivity.

Definition 11. Let F, G, H, and K be species of structures. One sets, for the virtual species $\Phi = F - G$ and $\Psi = H - K$,

a) $\Phi' = F' - G'$,

b) $\Phi^\bullet = F^\bullet - G^\bullet$,

c) $\Phi \times \Psi = (F \times H + G \times K) - (F \times K + G \times H)$,

d) $\Phi \square H = F \square H - G \square H$.
$$\tag{21}$$

The reader should take note that the expressions on the right side of these equations are not, in general, written in reduced form even when $F - G$ and $H - K$ are. All usual properties of these operations remain valid in this new context (see Exercise 10).

Example 12. Consider, for fixed $n \geq 2$, the virtual species $\Phi = nC_n - X^n$, where C_n is the species of oriented cycles of cardinality n. One verifies that Φ is written in reduced form. Hence, a Φ-structure consists either of an oriented cycle on n points, together with a color i, $1 \leq i \leq n$ (these structures being considered as positive) or of a linear order on n points (these structures being considered as negative). Let us compute the derivative of Φ:

$$\Phi' = (nC_n)' - (X^n)'$$

$$= nX^{n-1} - nX^{n-1} = 0, \tag{22}$$

since $C_n' = L_{n-1} = X^{n-1}$. Thus, the differential equation $\Phi' = 0$ admits, in the context of virtual species, an infinite number of nonconstant solutions.

Virtual species permit the solution of differential equations which do not have a solution in the algebra of species. See for example Exercise 12. In fact, one has the following general result due to Joyal [162] (see Exercise 11).

Proposition 13. *For any virtual species Ψ, the combinatorial differential equation*

$$\Phi' = \Psi \tag{23}$$

admits the general solution

$$\Phi = \Omega + \int \Psi, \tag{24}$$

where Ω is a solution of $\Omega' = 0$ and

$$\int \Psi = E_1\Psi - E_2\Psi' + E_3\Psi'' + \cdots + (-1)^{n+1} E_n \Psi^{(n-1)} + \cdots. \tag{25}$$

We now have at our disposal all the tools needed in order to define the substitution

$$\Phi \circ \Psi = (F - G) \circ (H - K) \tag{26}$$

of a virtual species into another. We certainly want the property of right-distribution

$$(\Phi + \Theta) \circ \Psi = \Phi \circ \Psi + \Theta \circ \Psi \tag{27}$$

to be valid as it is in the case of ordinary species. This reduces the problem of virtual substitution to the case when the operand is an ordinary species since we then set

$$(F - G) \circ \Psi = F \circ \Psi - G \circ \Psi. \tag{28}$$

However, left-distribution is evidently not valid. We need to analyze the situation more deeply. Let $F = F(X)$, $H = H(X)$, and $K = K(X)$ and introduce a new sort Y of elements. Since the following equality should certainly hold,

$$F \circ (H - K) = F(X - Y)|_{X:=H, Y:=K}, \tag{29}$$

we see that it is sufficient to define the virtual two-sort species $F(X - Y)$. To do this, we first recall the formula

$$F(mX + nY) = F(X + Y) \times (E^m(X) E^n(Y)), \tag{30}$$

established in Section 2.4 and valid for all $m, n \in \mathbb{N}$. Since we have already given meaning to $E^m(X)$ and $E^n(Y)$ as virtual species, for all integers $m, n \in \mathbb{Z}$, it is possible to choose $m = 1$ and $n = -1$ in the preceding formula suggesting the following definition.

Definition 14. Let $F = F(X)$ be a species of structures. The virtual species $F(X - Y)$ on two sorts X, Y is defined by

$$F(X - Y) = F(X + Y) \times (E(X) E^{-1}(Y)), \tag{31}$$

where $E(X)$ (respectively, $E(Y)$) denotes the species of sets whose points are of sort X (respectively, Y).

More generally, for a virtual species $\Phi = \Phi(X) = F(X) - G(X)$, we extend this definition by linearity by setting

$$\Phi(X - Y) = \Phi(X + Y) \times (E(X)E^{-1}(Y)). \tag{32}$$

Whence the following definition of general substitution in the virtual case.

Definition 15. (Joyal [161], Yeh [333, 334].) Let $\Phi = \Phi(X)$ and $\Psi = \Psi(X)$ be two virtual species. Suppose that $\Psi = H - K$ and $\Psi_0 = 0$. The (partitional) composition (or substitution) of virtual species, $\Phi \circ \Psi$, is defined by the formula

$$\Phi \circ \Psi = \Phi(X + Y) \times (E(X)E^{-1}(Y))|_{X:=H, Y:=K}. \tag{33}$$

Using the geometric series expansion for E^{-1}, we obtain a more explicit form for $\Phi \circ \Psi$:

$$\Phi \circ \Psi = \sum_{k \geq 0} (-1)^k \Phi(X + Y) \times (E(X)(E_+(Y))^k)|_{X:=H, Y:=K}. \tag{34}$$

The above definition is independent of choice of representatives for Φ and Ψ and the main properties of substitution remain valid in this more general context (see Exercise 2.6.19):

$$\begin{array}{ll} \text{a)} & (\Phi \pm \Psi) \circ \Theta = \Phi \circ \Theta \pm \Psi \circ \Theta, \\ \text{b)} & (\Phi \cdot \Psi) \circ \Theta = (\Phi \circ \Theta) \cdot (\Psi \circ \Theta), \\ \text{c)} & (\Phi \circ \Psi) \circ \Theta = \Phi \circ (\Psi \circ \Theta), \\ \text{d)} & (\Phi \circ \Psi)' = (\Phi' \circ \Psi) \cdot \Psi', \end{array} \tag{35}$$

as well as the following formulas in the algebra of formal power series.

$$\begin{array}{ll} \text{a)} & (\Phi \circ \Psi)(x) = \Phi(\Psi(x)), \\ \text{b)} & (\widetilde{\Phi \circ \Psi})(x) = Z_\Phi(\widetilde{\Psi}(x), \widetilde{\Psi}(x^2), \ldots), \\ \text{c)} & Z_{\Phi \circ \Psi}(x_1, x_2, x_3, \ldots) = (Z_\Phi \circ Z_\Psi)(x_1, x_2, x_3, \ldots). \end{array} \tag{36}$$

Example 16. We show in Chapter 4 that the species $\mathfrak{a}$ of trees is related to the species $\mathcal{A}$ of rooted trees by the remarkable combinatorial equation

$$\mathfrak{a} + \mathcal{A}^2 = \mathcal{A} + E_2(\mathcal{A}), \tag{37}$$

where E_2 denotes the species of sets having cardinality 2. By subtraction, one obtains

$$\mathfrak{a} = \mathcal{A} + E_2(\mathcal{A}) - \mathcal{A}^2$$
$$= \Lambda \circ \mathcal{A}, \tag{38}$$

where $\Lambda = \Lambda(X)$ denotes the virtual species

$$\Lambda = X + E_2 - X^2. \tag{39}$$

This example shows that substitution of truly virtual species can very well give rise to an ordinary species of structures. The identity $\mathfrak{a} = \Lambda \circ \mathcal{A}$ yields not only a better understanding of the precise nature of the species of trees, but also an explicit calculation of its associated series.

Another interesting case of virtual substitution is that which consists of substituting the species $-X$ in an arbitrary species F. It suffices to set $m = -1$ and $n = 0$ in the general formula for $F(mX + nY)$, to obtain

$$F(-X) = F(X) \times E^{-1}(X)$$
$$= \sum_{k \geq 0} (-1)^k F(X) \times E_+^k(X). \tag{40}$$

After some calculations we get, for example, that

a) $E_2(-X) = X^2 - E_2(X),$

b) $L_n(-X) = X^n \circ (-X) = (-X)^n = (-1)^n X^n,$ \tag{41}

c) $C_4(-X) = E_2(X_2) - C_4(X).$

All of these operations can be extended to the ring $\text{Virt}_{\mathbb{A}}\langle X, Y, Z, \ldots \rangle$ of virtual $\mathbb{A}$-weighted species on many sorts of points $X, Y, Z, \ldots$.

Combinatorial Logarithm

The introduction of virtual species permit a better understanding of the notion of connected components in the theory of species. This important application is described in the following paragraphs. We first recall that a species of structures F, which can be written in the form

$$F = E \circ G, \qquad G(0) = 0, \tag{42}$$

where E is the species of sets, admits a notion of connected components, that is, the species G, hereafter denoted by $G = F^c$.

2.5. Virtual Species

Example 17. The species S (of permutations) and the species G (of graphs) are typical examples since we have the combinatorial equations

$$S = E(C) \quad \text{and} \quad G = E(G^c), \tag{43}$$

where C and G^c denote respectively the species of circular permutations (the connected permutations) and that of connected graphs.

If $F = E(G)$, then $F(0) = E(G(0)) = E(0) = 1$. Thus, the condition $F(0) = 1$ is necessary for F to admit a notion of connected component. However, this condition is not sufficient in the context of ordinary species. For example, consider the species L of linear orders. One has $L(0) = 1$ since there is only one linear order on the empty set. However, the equation $L = E(G)$ is false no matter what the species of structures G is. To see this, we proceed by contradiction. Observe first that each L-structure (i.e., each linear order) has a single automorphism: the identity. Suppose now that $L = E(G)$ for a certain species G. Evidently one has $G \neq 0$ (otherwise $L = E(0) = 1$). Consider an arbitrary G-structure g_1 and another G-structure g_2 isomorphic to g_1 but disjoint from g_1 (disjoint underlying sets). The assembly formed by g_1 and g_2 is clearly an $E(G)$-structure. This assembly has at least one nontrivial automorphism exchanging g_1 and g_2. Thus there exist $E(G)$-structures having at least two automorphisms. This contradicts the equation $L = E(G)$. Thus, the species L does not admit a notion of connected components when restricted to ordinary species.

The situation is quite different in the context of virtual species where it becomes possible to construct a virtual species Γ satisfying the equation $L = E(\Gamma)$. Indeed, the following general result is valid.

Proposition 18. *Let F be a species of structures satisfying the condition $F(0) = 1$. Then there exists a unique virtual species Γ satisfying the combinatorial equation*

$$F = E(\Gamma), \tag{44}$$

where E denotes the species of sets.

Proof. To prove this result, observe at first that $F = 1 + F_+$ (where $F_+ = F - 1$ denotes the species of F-structures on nonempty sets). By considering the species $E_+ = E - 1$ of nonempty sets, the proposed equation is equivalent to

$$1 + F_+ = (1 + E_+) \circ \Gamma$$
$$= 1 + E_+ \circ \Gamma. \tag{45}$$

That is, after simplification,

$$F_+ = E_+ \circ \Gamma. \tag{46}$$

The proposition will then follow if we succeed to construct a virtual species $E_+^{\langle -1 \rangle}$ which is the inverse of E_+ under substitution. Indeed, the desired virtual species Γ will necessarily be given by

$$\begin{aligned}
\Gamma &= \left(E_+^{\langle -1 \rangle} \circ E_+\right) \circ \Gamma \\
&= E_+^{\langle -1 \rangle} \circ (E_+ \circ \Gamma) \\
&= E_+^{\langle -1 \rangle} \circ F_+.
\end{aligned} \qquad (47)$$

The existence (and uniqueness) of $E_+^{\langle -1 \rangle}$ follows from the following proposition by taking $\Psi = E_+ = X + E_2 + E_3 + \cdots$. ∎

Proposition 19. *Let Ψ be a virtual species (in particular, a species of structures) whose canonical decomposition is of the form*

$$\Psi = X + \Psi_2 + \Psi_3 + \cdots \quad (\Psi_0 = 0, \ \Psi_1 = X). \qquad (48)$$

Then there exists a unique virtual species $\Psi^{\langle -1 \rangle}$ such that

$$\Psi^{\langle -1 \rangle} \circ \Psi = \Psi \circ \Psi^{\langle -1 \rangle} = X. \qquad (49)$$

Proof. Define the linear operator $\Delta_\Psi : \text{Virt} \longrightarrow \text{Virt}$ by the formula

$$\Delta_\Psi(\Phi) = \Phi \circ \Psi - \Phi \qquad (50)$$

and denote by $I : \text{Virt} \longrightarrow \text{Virt}$, the identity operator ($I(\Phi) = \Phi$). Clearly the equality $(I + \Delta_\Psi)\Phi = \Phi \circ \Psi$ holds for all virtual species Φ. Let $n \geq 0$ be any natural number. Iterating the operation $(I + \Delta_\Psi)$, n times, gives

$$(I + \Delta_\Psi)^n \Phi = \Phi \circ \Psi^{\langle n \rangle}, \qquad (51)$$

where $\Psi^{\langle n \rangle} = \Psi \circ \Psi \circ \ldots \circ \Psi$ denotes the nth iterate of Ψ for substitution. Now, it follows from Newton's binomial formula that

$$(I + \Delta_\Psi)^n = \sum_{k \geq 0} \binom{n}{2} \Delta_\Psi^k, \qquad (52)$$

which gives

$$\Phi \circ \Psi^{\langle n \rangle} = \sum_{k \geq 0} \binom{n}{2} \Delta_\Psi^k(\Phi). \qquad (53)$$

A remarkable phenomenon arises then (see Exercise 16): The family $\binom{n}{k}\Delta_\Psi^k(\Phi)$, $k \geq 0$, is summable even for negative values of n. This suggests that we define

$\Phi \circ \Psi^{\langle n \rangle}$ for all $n \in \mathbb{Z}$ by the same formula (53). This definition is correct since the *law of exponents* holds:

$$\Phi \circ \Psi^{\langle m+n \rangle} = \left(\Phi \circ \Psi^{\langle m \rangle}\right) \circ \Psi^{\langle n \rangle}, \quad m, n \in \mathbb{Z}. \tag{54}$$

In particular, taking $n = -1$ and $\Phi = X$, we obtain the desired virtual species:

$$\begin{aligned}\Psi^{\langle -1 \rangle} &= \sum_{k \geq 0} (-1)^k \Delta_\Psi^k(X) \\ &= X - (\Psi - X) + (\Psi \circ \Psi - 2\Psi + X) \\ &\quad - (\Psi \circ \Psi \circ \Psi - 3\Psi \circ \Psi + 3\Psi - X) + \cdots .\end{aligned} \tag{55}$$

We now return to the notion of connected component for any species F satisfying $F(0) = 1$. The equation $F = E(\Gamma)$ can be read "F is the combinatorial exponential of Γ." By solving for Γ, we find

$$\begin{aligned}\Gamma &= E_+^{\langle -1 \rangle} \circ F_+ \\ &= (X - (E_+ - X) + (E_+ \circ E_+ - 2E_+ + X) - \cdots) \circ F_+ \\ &= F_+ - (E_+ \circ F_+ - F_+) + (E_+ \circ E_+ \circ F_+ - 2E_+ \circ F_+ + F_+) - \cdots .\end{aligned} \tag{56}$$

This suggests the terminology which consists of saying that "Γ is the combinatorial logarithm of the species F." The following notation is appropriate:

$$\Gamma = F^c = \text{the virtual species of "connected } F\text{-structures."} \tag{57}$$

By inversion of formulas (1.4.20) we find, for the corresponding series,

a) $F^c(x) = \log F(x)$,

b) $\widetilde{F^c}(x) = \sum_{k \geq 1} \frac{\mu(k)}{k} \log \widetilde{F}(x^k)$, \hfill (58)

c) $Z_{F^c}(x_1, x_2, x_3, \ldots) = \sum_{k \geq 1} \frac{\mu(k)}{k} \log Z_F(x_k, x_{2k}, x_{3k}, \ldots)$,

where μ denotes the usual Möbius function. In the particular cases where $F = \mathcal{G}$ (graphs) and $F = \mathcal{S}$ (permutations), we find that the species $\mathcal{G}^c$ is the species of connected graphs and $\mathcal{S}^c = \mathcal{C}$ is the species of oriented cycles. For the species $F = L$ of linear orders, we obtain the strictly virtual species L^c of connected linear orders. We have

$$L^c = L_+ - (E_+ \circ L_+ - L_+) + (E_+ \circ E_+ \circ L_+ - 2E_+ \circ L_+ + L_+) - \cdots , \tag{59}$$

as well as

$$\text{a)} \quad L^c(x) = \log \frac{1}{1-x},$$

$$\text{b)} \quad \widetilde{L}^c(x) = x,$$

$$\text{c)} \quad L^c(x_1, x_2, x_3, \ldots) = \sum_{k \geq 1} \frac{\mu(k)}{k} \log \frac{1}{1-x_k} \qquad (60)$$

$$= \sum_{n \geq 1} \frac{1}{n} \sum_{d \mid n} \mu(d) \, x_d^{\frac{n}{d}}.$$

The existence of a substitutional inverse makes it possible to solve other combinatorial equations (for Ψ), for example,

$$F = \mathcal{S} \circ \Psi, \quad F = \mathcal{C} \circ \Psi, \quad F = \mathcal{A} \circ \Psi, \qquad (61)$$

where F is a given species and $\mathcal{S}$, $\mathcal{C}$, and $\mathcal{A}$ denote respectively the species of permutations, oriented cycles, and rooted trees.

The last of these equations merits special attention since the inverse $\mathcal{A}^{\langle -1 \rangle}$ has a very simple explicit description. In fact, since the species $\mathcal{A}$ of rooted trees is characterized by the combinatorial equation $\mathcal{A} = X \cdot (E \circ \mathcal{A})$, substitution of $\mathcal{A}^{\langle -1 \rangle}$ in this equality gives

$$X = \mathcal{A}^{\langle -1 \rangle} \cdot E. \qquad (62)$$

Thus,

$$\begin{aligned} \mathcal{A}^{\langle -1 \rangle} &= X \cdot E^{-1} \\ &= X \cdot \left(1 - E_+ + E_+^2 - \cdots \right). \end{aligned} \qquad (63)$$

The combinatorial equation $F = \mathcal{A} \circ \Psi$ (where $F(\emptyset) = \emptyset$) then has the explicit solution

$$\begin{aligned} \Psi &= \mathcal{A}^{\langle -1 \rangle} \circ F \\ &= F \cdot E^{-1}(F) \\ &= F \cdot \left(1 - E_+(F) + (E_+(F))^2 - \cdots \right). \end{aligned} \qquad (64)$$

See Example 4.2.5 for an application of this example. ∎

Example 20. In particular, one obtains the following result: There exists a virtual species Ψ such that

$$\mathcal{C} = \mathcal{A} \circ \Psi \qquad (65)$$

2.5. Virtual Species

where $\mathcal{C}$ denotes the species of oriented cycles, although the species $\mathcal{C}$ and $\mathcal{A}$ are of a very different combinatorial nature. Simple computations yield the following formulas in this particular case:

a) $\Psi = \mathcal{A}^{(-1)}(\mathcal{C}) = \mathcal{C} \cdot \mathcal{S}^{-1}$,

b) $\Psi(x) = -(1-x)\log(1-x)$,

c) $\widetilde{\Psi}(x) = x(1-x^2)(1-x^3)(1-x^4)\ldots$,

d) $Z_\Psi(x_1, x_2, x_3, \ldots) = (1-x_1)(1-x_2)(1-x_3)\ldots \cdot Z_\mathcal{C}(x_1, x_2, x_3, \ldots)$.

(66)

To end this section, here is an illustration of the concept of virtual species in the multisort and weighted case. We have seen in Section 2.4 that the combinatorial equation

$$\mathcal{C}(X+Y) = \mathcal{C}(X) + \mathcal{C}(Y \cdot L(X)) \tag{67}$$

is satisfied by the species $\mathcal{C}$ of oriented cycles. To make the equation more symmetric, we make the change of variable

$$T = YL(X) = Y/(1-X). \tag{68}$$

We deduce $Y = (1-X) \cdot T$, and by substitution in the initial equation,

$$\mathcal{C}(X) + \mathcal{C}(T) = \mathcal{C}(X + T - X \cdot T). \tag{69}$$

In particular, by setting $T := -X$, we find the remarkable relation

$$\mathcal{C}(-X) = \mathcal{C}(X^2) - \mathcal{C}(X). \tag{70}$$

Consider now the species $\mathcal{S}_w$ of permutations weighted by $w(\sigma) = \alpha^{c(\sigma)}$, where $c(\sigma)$ denotes the number of cycles of the permutation σ. We know that

$$\mathcal{S}_w = E_v \circ \mathcal{C}, \tag{71}$$

where E_v is the species of sets U weighted by $v(U) = \alpha^{|U|}$. We have

$$\begin{aligned}
\mathcal{S}_w(X) \cdot \mathcal{S}_w(T) &= E_v(\mathcal{C}(X)) \cdot E_v(\mathcal{C}(T)) \\
&= E_v(\mathcal{C}(X) + \mathcal{C}(T)) \\
&= E_v(\mathcal{C}(X + T - X \cdot T)) \\
&= \mathcal{S}_w(X + T - XT).
\end{aligned} \tag{72}$$

Setting $T := -X$, we deduce

$$S_w(-X) = \frac{S_w(X^2)}{S_w(X)}. \tag{73}$$

Exercises for Section 2.5

1. a) Returning to the definitions, verify that for any weighted species $G = G_w \neq 0$,

$$G_w + G_{-w} \neq 0,$$

where G_{-w} denotes the weighted species obtained from G_w by changing the sign of each G_w-structure.
  b) Show that the following formulas are nevertheless valid:

$$G_w(x) + G_{-w}(x) = 0, \quad \widetilde{G_w}(x) + \widetilde{G_{-w}}(x) = 0, \quad Z_{G_w} + Z_{G_{-w}} = 0. \tag{74}$$

2. COMPLEMENTARY SUBSPECIES.
  a) Define the concepts of intersection ($\cap$) and union ($\cup$) for subspecies of a species of structures F and give the main properties of these operations.
  b) Let F be a species of structures and $G \subseteq F$, a subspecies of F. Show that there exists a unique subspecies H of F such that $G \cap H = \emptyset$ and $F = G \cup H$. This subspecies H is called the *complementary subspecies* of G.

3. CANCELATION LAW FOR ADDITION. Given three species of structures F, G, and H, show that

$$F + H = G + H \implies F = G.$$

In other words, construct an isomorphism $\beta : F \longrightarrow G$ from an isomorphism $\alpha : F + H \longrightarrow G + H$ (see Definition 1.2.12).
HINT: Let U be a finite set. For all $s \in F[U]$, define $\beta_U(s)$ by

$$\beta_U(s) = \alpha_U^{\langle k \rangle}(s), \qquad \text{(the kth iterate of } \alpha_U) \tag{75}$$

where k is the smallest integer ≥ 0 such that $\alpha_U^{\langle k \rangle}(s) \in G[U]$. Verify that the family of β_U, where U ranges over finite sets, constitutes an isomorphism $\beta : F \longrightarrow G$ (do not forget naturality).

4. a) Show that the binary relation $\sim$ defined on Spe $\times$ Spe by relation (5) is indeed an equivalence relation (i.e., reflexive, symmetric, and transitive).
    HINT: To establish transitivity, refer to the preceding exercise.
  b) Prove Proposition 3 stating that Virt constitutes a commutative ring. Show, in particular, that operations (7) are well defined, that is to say, independent of the choice of representatives.

2.5. Virtual Species

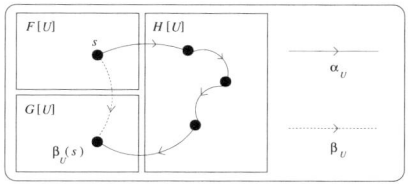

Fig. 1.

5. a) Show that definitions 4 a), b), c) for the series associated to a virtual species are correct (i.e., independent of the choice of representatives).

   b) Show that for any virtual species Φ and Ψ

   i) $(\Phi + \Psi)(x) = \Phi(x) + \Psi(x)$, ii) $(\widetilde{\Phi + \Psi})(x) = \widetilde{\Phi}(x) + \widetilde{\Psi}(x)$,

   iii) $Z_{\Phi+\Psi} = Z_\Phi + Z_\Psi$, iv) $(\Phi \cdot \Psi)(x) = \Phi(x) \cdot \Psi(x)$,

   v) $(\widetilde{\Phi \cdot \Psi})(x) = \widetilde{\Phi}(x) \cdot \widetilde{\Psi}(x)$, vi) $Z_{\Phi \cdot \Psi} = Z_\Phi \cdot Z_\Psi$. (76)

6. a) Show that for two species of structures F and G

$$G \subseteq F \implies (F - G)^- = 0 \quad \text{and} \quad (F - G)^+ = F - G. \quad (77)$$

   b) Let $\mathcal{C}$ be the species of oriented cycles and $\mathcal{A}$ be that of rooted trees and let $\Phi = \mathcal{C} - \mathcal{A}$. Show that $\Phi^+ = \mathcal{C} - X$, $\Phi^- = \mathcal{A} - X$.
   HINT: Every oriented cycle on at least two vertices has at least one automorphism without fixed points. This is not the case for a rooted tree.

   c) Show by an example that if $(\Phi_i)_{i \in I}$ is a summable family of virtual species then formula (12), which defines the sum of this family, is not always in reduced form.

   d) Show that if $(F_i)_{i \in I}$ and $(G_i)_{i \in I}$ are two summable families of species, then the family of virtual species $\Phi_i = F_i - G_i$, $i \in I$, is summable and that its sum is given by (13).

   e) Give an example of a summable family $\Phi_i = F_i - G_i$ of virtual species, where $(F_i)_{i \in I}$ and $(G_i)_{i \in I}$ are not summable.

7. Let E be the species of sets.

   a) Show that there are no species of structures F such that $F \cdot E = 1$.

   b) Show that the family of virtual species $(-1)^k(E_+)^k$, $k = 0, 1, \ldots$ is summable and that its sum, given by (16), also satisfies equalities (17) and (18).

   c) More generally, let G be a species of structures satisfying $G(0) = 1$ (i.e., there exists a unique G-structure on the empty set). Show that by defining G^{-1} as

$$G^{-1} = \frac{1}{G} = \sum_{k=0}^{\infty} (-1)^k (G_+)^k \quad (78)$$

   one obtains a virtual species satisfying $G^{-1} \cdot G = G \cdot G^{-1} = 1$.

d) Adapt the preceding result to the case where G is replaced by a virtual species Φ where $\Phi(0) = 1$. Moreover, with the help of the binomial formula, define Φ^n, for $n \in \mathbb{Z}$, and verify the law of exponents

$$\Phi^m \cdot \Phi^n = \Phi^{m+n}, \qquad m, n \in \mathbb{Z}. \tag{79}$$

8. Let $X, Y, Z, \ldots$ be a sequence of sorts of singletons, $\mathbb{K}$ a commutative unitary ring of characteristic zero, and $\mathbb{A} = \mathbb{K}[\![s_1, s_2, \ldots; t_1, t_2, \ldots]\!]$ the ring of formal power series in the variables $s_1, s_2, \ldots; t_1, t_2, \ldots$, with coefficients in $\mathbb{K}$.
   a) Define the notion of an $\mathbb{A}$-weighted virtual species on the sorts $X, Y, Z, \ldots$.
   b) Define a ring structure on the set $\mathrm{Virt}_{\mathbb{A}}(X, Y, Z, \ldots)$ of virtual species $\mathbb{A}$-weighted on the sorts $X, Y, Z, \ldots$.
   c) Under which conditions is an element of this ring invertible under the product?
9. a) Show that Definitions 21, a), b), c), d), for the operations $'$, $^{\bullet}$, $\times$, $\square$, in the context of virtual species, are correct (i.e., are independent of choice of representatives) and that they are coherent with passages to series.
   b) Show, by some explicit examples, that the expressions on the right sides of Equations (21), a), b), c), d) are in general not written in reduced form even if $F - G$ and $H - K$ are.

   HINT: For (21), a), b), see Example 12. For (21), c), take, for example, $F = H = E_3$ and $G = K = C_3$. For (21), d), take, for example, $F = X E_2$, $G = E_3$, $H = X^2 + E_2$, and verify that $F \square H = X^2 + E_2$, $G \square H = E_2$.
10. Let Φ, Ψ, Θ be virtual species and H be a species of structures.
    a) Show that the following formulas are valid:

$$\begin{aligned}
&\text{i)} \quad (\Phi \pm \Psi)' = \Phi' \pm \Psi', \\
&\text{ii)} \quad (\Phi \cdot \Psi)' = \Phi' \cdot \Psi + \Phi \cdot \Psi', \\
&\text{iii)} \quad (\Phi \pm \Psi)^{\bullet} = \Phi^{\bullet} \pm \Psi^{\bullet}, \\
&\text{iv)} \quad (\Phi \cdot \Psi)^{\bullet} = \Phi^{\bullet} \cdot \Psi + \Phi \cdot \Psi^{\bullet}, \\
&\text{v)} \quad (\Phi \pm \Psi) \times \Theta = \Phi \times \Theta \pm \Psi \times \Theta, \\
&\text{vi)} \quad (\Phi \pm \Psi) \square H = \Phi \square H \pm \Psi \square H.
\end{aligned} \tag{80}$$

   b) Show moreover that if Ψ^{-1} exists, then

$$\begin{aligned}
&\text{i)} \quad (\Psi^{-1})' = -\Psi^{-2}\Psi', \\
&\text{ii)} \quad (\Psi^n)' = n\Psi^{n-1} \cdot \Psi', \; n \in \mathbb{Z}, \\
&\text{iii)} \quad (\Phi/\Psi)' = \frac{\Phi' \cdot \Psi - \Phi \cdot \Psi'}{\Psi^2}.
\end{aligned} \tag{81}$$

2.5. Virtual Species 137

c) Show that the virtual species $F = E^{-1}$ and $G = S^{-1}$ satisfy the equalities

i) $F^{\bullet} = -X \cdot F$,   ii) $G^{\bullet} = -L \cdot G$.   (82)

11. Prove Proposition 13 concerning the general solution Φ of the equation $\Phi' = \Psi$ in the context of virtual species.
    HINT: First show that two solutions differ by an Ω such that $\Omega' = 0$. Then verify that $\int \Psi$, given by (25), is well defined and is a particular solution of $\Phi' = \Psi$.

12. a) Show that there does not exist a species of structures F satisfying $F' = E^{\bullet}$.
    HINT: Suppose that a natural bijection $\alpha : E^{\bullet} \longrightarrow F'$ exists and consider the set $U = \{0, 1, 2, 3\}$. There are four $E^{\bullet}$-structures on U, namely the pointed sets

    $s_0 = \{\underline{0}, 1, 2, 3\}, \ s_1 = \{0, \underline{1}, 2, 3\}, \ s_2 = \{0, 1, \underline{2}, 3\}, \ s_3 = \{0, 1, 2, \underline{3}\}.$

    There is then also four F'-structures $t_i = \alpha_U(s_i)$, $i = 0, 1, 2, 3$, on the set U. The F'-structures on U are identical to the F-structures on the set of five points $U^+ = \{0, 1, 2, 3, 4\}$ (4 being the supplementary element added to U). The t_i are transported in the same manner as the s_i along the transpositions $\tau_k = (0, k)$, for $k = 1, 2, 3$. Obtain a contradiction by analyzing the possibilities of transport of the t_i along the supplementary transposition $\tau_4 = (0, 4)$.
    b) Show that the virtual species $\Phi = E^{\bullet} - E$ satisfies the equation $\Phi' = E^{\bullet}$. Tie this in with the general solution given in Proposition 13.

13. Show that there exists an uncountable number of virtual species Ω satisfying $\Omega' = 0$.
    HINT: Let $W \subseteq \{2, 3, 4, \ldots\}$ and consider virtual species of the form

    $$\Omega = \sum_{n \in W} \left(nC_n - X^n \right).$$

14. Establish formulas (41), a), b), c) for the virtual species $E_2(-X)$, $L_n(-X)$, $C_4(-X)$.

15. Show how to define the operations $'$, $^{\bullet}$, $\times$, $\circ$, $\square$ in the context of multisort $\mathbb{A}$-weighted virtual species, on sorts $X, Y, Z, \ldots$ (see Exercise 8).
    NOTE: Regarding the definition of $\Phi \square (H_1, H_2, \ldots)$, it is only required to consider the case where the H_i are nonvirtual and nonweighted species on the sorts $X, Y, Z, \ldots$.

16. a) Let Φ and Ψ be two virtual species, with $\Psi = X + \Psi_2 + \Psi_3 + \cdots$, and let $n \in \mathbb{Z}$. Verify that the family $\binom{n}{k} \Delta^k_\Psi (\Phi)$, $k \geq 0$, appearing in (53), is summable.
    HINT: Use the fact that

    $$\Delta^{k+1}_\Psi (\Phi) = \Delta_\Psi \left(\Delta^k_\Psi (\Phi) \right) = \left(\Delta^k_\Psi (\Phi) \right) \circ \Psi - \Delta^k_\Psi (\Phi). \quad (83)$$

b) Starting from Definition (53) of the virtual species $\Phi \circ \Psi^{\langle n \rangle}$, verify that

$$\Phi \circ \Psi^{\langle m+n \rangle} = \left(\Phi \circ \Psi^{\langle m \rangle}\right) \circ \Psi^{\langle n \rangle}, \quad m, n \in \mathbb{Z}.$$

HINT: Use properties of the Newton binomial.

c) Calculate the first terms $(E_+^{\langle -1 \rangle})_1$, $(E_+^{\langle -1 \rangle})_2$, and $(E_+^{\langle -1 \rangle})_3$ of the canonical decomposition

$$E_+^{\langle -1 \rangle} = \left(E_+^{\langle -1 \rangle}\right)_1 + \left(E_+^{\langle -1 \rangle}\right)_2 + \left(E_+^{\langle -1 \rangle}\right)_3 + \cdots$$

of the inverse under the substitution of the species E_+ of nonempty sets.
HINT: Use formula (55) with $\Psi = E_+ = E_1 + E_2 + E_3 + \cdots$.

17. a) Show that the iterates $L_+^{\langle n \rangle}$ of the species of nonempty linear orders satisfy, for all $n \in \mathbb{Z}$,

$$L_+^{\langle n \rangle} = \frac{X}{1 - nX} = X + nX^2 + n^2 X^3 + \cdots. \tag{84}$$

b) Show that the species Ord and Red of partial orders and of reduced partial orders (see Exercise 1.4.15) are connected by the relation

$$\text{Red} = \text{Ord}\left(\frac{X}{1 + X}\right). \tag{85}$$

18. The operators $\mathcal{E}_0, \mathcal{E} : \text{Virt} \longrightarrow \text{Virt}$ are defined by

$$\Phi(X) \xmapsto{\mathcal{E}_0} \frac{1}{1-X} \Phi\left(\frac{-X}{1-X}\right), \quad \Phi(X) \xmapsto{\mathcal{E}} \frac{1}{1-X} \Phi\left(\frac{X}{1-X}\right). \tag{86}$$

a) Show that $\mathcal{E}_0$ is an involution, i.e., that $\mathcal{E}_0^2 = \text{Id}_{\text{Virt}}$, and that, for all $k \in \mathbb{Z}$, $\mathcal{E}^k$ is given by

$$\mathcal{E}^k \Phi(X) = \frac{1}{1-kX} \Phi\left(\frac{X}{1-kX}\right). \tag{87}$$

b) Interpret $\mathcal{E}\Phi$ combinatorially in the nonvirtual context.

c) Show that if $\Phi(x) = \sum_{n \geq 0} \varphi_n x^n$, then

$$\mathcal{E}_0 \Phi(x) = \sum_{n \geq 0} (-1)^n (\Delta^n \varphi_0) x^n, \quad \mathcal{E}\Phi(x) = \sum_{n \geq 0} (\Delta^n \varphi_0) x^n, \tag{88}$$

where Δ is the usual difference operator: $\Delta \varphi_k = \varphi_{k+1} - \varphi_k$.
NOTE: In the context of formal power series, the operators $\mathcal{E}_0$ and $\mathcal{E}$ are classically attributed to Euler.

19. By using Möbius inversion, prove formulas (58) giving the series associated to the virtual species F^c of connected F-structures. Verify the corresponding formulas (60) in the case where F is the species of linear orders.

20. Verify properties (66), a)–d) for the virtual species Ψ satisfying the combinatorial equation $\mathcal{C} = \mathcal{A} \circ \Psi$, where $\mathcal{C}$ is the species of oriented cycles and $\mathcal{A}$ is that of rooted trees.

21. Let $\mathcal{S}_w(X)$ be the species of permutations weighted by $w(\sigma) = \alpha^{c(\sigma)}$ where $c(\sigma)$ denotes the number of cycles of σ. Set

$$f_w(x_1, x_2, x_3, \ldots) = Z_{\mathcal{S}_w}(x_1, x_2, x_3, \ldots).$$

By passing to cycle index series, formula (73) gives rise to the equality

$$f_w(-x_1, -x_2, -x_3, \ldots) = \frac{f_w(x_1^2, x_2^2, x_3^2, \ldots)}{f_w(x_1, x_2, x_3, \ldots)}. \tag{89}$$

Verify this equality by explicitly calculating the expressions appearing on each side.

2.6. Molecular and Atomic Species

The purpose of this section is to establish two "standard" decompositions due to Yeh [333, 334]: the molecular decomposition and the atomic decomposition, for virtual species. The ring Virt of (nonweighted, single sort) virtual species then appears as a ring of formal power series

$$\text{Virt} = \mathbb{Z}[\![X, E_2, E_3, C_3, \ldots]\!] \tag{1}$$

in an infinite number of "variables" $X, E_2, E_3, C_3, \ldots$, corresponding to *atomic species*. The monomials $X^k E_2^h E_3^j C_3^m \cdots$ in these variables correspond to the molecular species. Similar results hold for the multisort case.

One of our goals is to establish rigorously the reduced form representation of virtual species. We have already seen that any species of structures $F = F(X)$ has a canonical decomposition of the form

$$F = F_0 + F_1 + \cdots + F_n + \cdots,$$

where F_n denotes the subspecies of F whose structures are restricted to cardinality n, $n = 0, 1, 2, \ldots$. Now let $G = G_0 + G_1 + \cdots + G_n + \cdots$ be another species of structures, and form the virtual species $\Phi = F - G$. Clearly, if $m \neq n$ then the species F_m and G_n are unrelated, that is to say, the only subspecies of F_m isomorphic to a subspecies of G_n is the empty species 0. We can write

$$\Phi = \sum_{n \geq 0}(F_n - G_n), \tag{2}$$

and conclude that determining the reduced form $\Phi^+ - \Phi^-$ of Φ requires a detailed study of the subspecies of F_n and G_n, for each $n \geq 0$.

Recall (see Appendix 1) that a group action $G \times Y \longrightarrow Y$ is called *transitive* if $Y \neq \emptyset$ and for any $x, y \in Y$, there exists a $g \in G$ for which $g \cdot x = y$. In the context of species of structures, the notion of transitive action translates into that of *molecular* species.

Definition 1. (Yeh [333, 334].) A species of structures M is said to be *molecular* if there is only one type of isomorphism of M-structures (i.e., two arbitrary M-structures are isomorphic).

In other words, the species M is molecular if and only if $M \neq 0$, $M = M_n$ for an integer n (M is then said to be *concentrated* on the cardinality n), and the action $S[n] \times M[n] \longrightarrow M[n]$, induced by transport of structures, is *transitive*. The classification of molecular species is then equivalent to the classification of transitive actions of the symmetric group. This classification is itself equivalent to the classification of conjugacy classes of subgroups of the symmetric group (see Proposition 7).

Example 2. The species C_n of oriented cycles of length n ($n \geq 1$) is molecular since two such cycles are necessarily isomorphic. The same is true for the species L_n and E_n ($n \geq 0$).

For $n \geq 3$, the species $\mathcal{A}_n$ of rooted trees on n points is not molecular. For example, Figure 1 illustrates two nonisomorphic $\mathcal{A}_4$-structures.

The product $M \cdot N$ of two molecular species M and N is always molecular. This is also the case for the substitution $M \circ N$, for $N \neq 1$. On the contrary, the sum $M + N$ of two molecular species M, N is never molecular. In fact, molecular species are precisely those which are (nontrivially) indecomposable under the addition of species:

Proposition 3. *A species of structures $M \neq 0$ is molecular if and only if for any species of structures P and Q,*

$$M = P + Q \quad \text{implies} \quad P = 0 \quad \text{or} \quad Q = 0.$$

Now consider any species $F = F_0 + F_1 + \cdots + F_n + \cdots$. By decomposing each action $\mathcal{S}_n \times F[n] \longrightarrow F[n]$, $n \geq 0$, in terms of transitive actions, we immediately obtain the following main result.

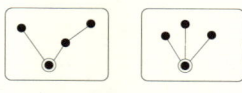

Fig. 1.

2.6. Molecular and Atomic Species

Proposition 4. *Every species of structures F is the sum of its molecular subspecies:*

$$F = \sum_{\substack{M \subseteq F \\ M \text{ molecular}}} M. \tag{3}$$

Moreover, every subspecies G of F is written in a unique manner (up to order of terms) as a sum of molecular subspecies of F.

The above sum is called the *molecular decomposition* of the species F. It constitutes a refinement of the canonical decomposition of F since

$$F = \sum_{n \geq 0} F_n = \sum_{n \geq 0} \sum_{\substack{M \subseteq F_n \\ M \text{ molecular}}} M. \tag{4}$$

It is clear that to obtain the molecular decomposition of a species F, it suffices to make a (countable) list of the isomorphism types of (i.e., unlabeled) F-structures and to form, for each type τ, the molecular subspecies M_τ which corresponds to it, and sum all the M_τ thus obtained.

Example 5. Consider the species $\mathcal{A}$ of all rooted trees. Figure 2 describes the unlabeled $\mathcal{A}$-structures on $n \leq 5$ points.

Working up to isomorphism of species, we can often identify the molecular species M_τ in terms of simpler ones, using the usual combinatorial operations. In this way, the molecular decomposition of $\mathcal{A}$, for example, takes the form

$$\mathcal{A} = X + X^2 + (X^3 + XE_2) + (2X^4 + X^2E_2 + XE_3)$$
$$+ (3X^5 + 3X^3E_2 + X^2E_3 + X(E_2 \circ X^2) + XE_4) + \cdots, \tag{5}$$

and a new phenomenon appears: Certain molecular species have a *coefficient* (or *multiplicity*). Thus, for $n = 4$, the two distinct types of $\mathcal{A}_4$-structures appearing in Figure 3 both represent the molecular species X^4, which explains the coefficient 2 in the monomial $2X^4$. The reader is invited to describe explicitly the isomorphism of species explaining the occurrence of the coefficients 3 in the above decomposition.

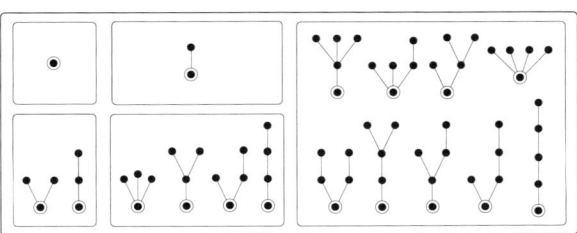

Fig. 2. The 17 unlabeled rooted trees on $n \leq 5$ points.

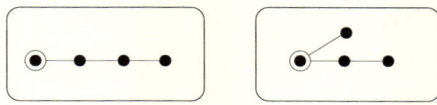

Fig. 3.

In certain cases, the molecular decomposition is obtained directly. For example, we have the following molecular decompositions:

a) $E = E_0 + E_1 + E_2 + E_3 + \cdots$,

b) $E^{\bullet} = X + X^2 + X \cdot E_2 + X \cdot E_3 + \cdots$,

c) $C = C_1 + C_2 + C_3 + \cdots$,

d) $L = 1 + X + X^2 + X^3 + \cdots$.

(6)

For the species S of permutations, one has successively

$$\begin{aligned} S = E(C) &= E(C_1 + C_2 + C_3 + \cdots) \\ &= E(C_1) \cdot E(C_2) \cdot E(C_3) \ldots \\ &= \prod_{k \geq 1}(1 + E_1(C_k) + E_2(C_k) + \cdots + E_i(C_k) + \cdots) \\ &= \sum_{n \geq 0} \sum_{i_1 + 2i_2 + \cdots = n} E_{i_1}(C_1) E_{i_2}(C_2) E_{i_3}(C_3) \cdots. \end{aligned}$$

(7)

This last decomposition reflects the fact that unlabeled permutations are encoded by partitions of an integer n: i_1 cycles of length 1, i_2 cycles of length 2, etc.

We are going to see that the totality $\mathfrak{M}$ of molecular species, up to isomorphism, constitutes a countable set,

$$\mathfrak{M} = \{1, X, X^2, E_2, X^3, X \cdot E_2, E_3, \ldots\}.$$

(8)

In fact, for each cardinality $n \in \mathbb{N}$, the set

$$\mathfrak{M}_n = \{M \in \mathfrak{M} \mid M \quad \text{is concentrated on the cardinality} \quad n\}$$

is finite. With this set $\mathfrak{M} = \mathfrak{M}_0 \cup \mathfrak{M}_1 \cup \ldots \cup \mathfrak{M}_n \cup \ldots$, it is now possible to give the molecular decomposition of an arbitrary species in "standard" form.

Proposition 6. *The molecular decomposition of every species F can be given in standard form*

$$F = \sum_{M \in \mathfrak{M}} f_M M \qquad \text{(finite or countable sum)},$$

(9)

where each $f_M \in \mathbb{N}$ denotes the number of molecular subspecies of F isomorphic to M. Moreover, the species F_n of F-structures concentrated on the cardinality n can be written

$$F_n = \sum_{M \in \mathfrak{M}_n} f_M M \qquad \text{(finite sum)}. \tag{10}$$

Finally, two species F and G are isomorphic if and only if for any $M \in \mathfrak{M}$, one has $f_M = g_M$.

Hence, with this molecular decomposition, we can determine whether two given species are isomorphic by simply comparing numerical coefficients. The explicit construction of the set $\mathfrak{M}$ uses the following result concerning transitive actions (see Appendix 1).

Proposition 7. *Let H be a subgroup of a group G, and consider the set*

$$G/H = \{gH \mid g \in G\} \tag{11}$$

of (left) cosets with respect to H. Then the action $G \times (G/H) \longrightarrow (G/H)$ defined by $g_1 \cdot (g_2 H) = (g_1 g_2) H$ is transitive. Moreover, two actions of this form, corresponding to two subgroups H and K, are isomorphic if and only if the subgroups H and K are conjugate (i.e., there exists $g \in G$ such that $K = gHg^{-1}$). Finally, any transitive action $G \times Y \longrightarrow Y$ is isomorphic to an action of the form $G \times (G/H) \longrightarrow (G/H)$, obtained by taking

$$H = \{g \in G \mid g \cdot y = y\}, \tag{12}$$

where $y \in Y$ is arbitrarily chosen.

By virtue of this result, the set $\mathfrak{M}$ of all molecular species (up to isomorphism) can be constructed as follows: For every $n \geq 0$,
i) choose a system of representatives $\text{Conj}(\mathcal{S}_n)$ of conjugacy classes of subgroups of $\mathcal{S}_n$;
ii) for each subgroup $H \in \text{Conj}(\mathcal{S}_n)$, construct a molecular species M_H corresponding to the transitive action

$$\mathcal{S}_n \times (\mathcal{S}_n/H) \longrightarrow (\mathcal{S}_n/H);$$

iii) set $\mathfrak{M}_n = \{M_H \mid H \in \text{Conj}(\mathcal{S}_n)\}$ and

$$\mathfrak{M} = \mathfrak{M}_0 \cup \mathfrak{M}_1 \cup \mathfrak{M}_2 \cup \ldots \cup \mathfrak{M}_n \cup \ldots.$$

Thus, each $\mathfrak{M}_n$ is finite since $|\mathfrak{M}_n| = |\text{Conj}(\mathcal{S}_n)|$ and $\mathfrak{M}$ is countable.

More explicitly, here is how the molecular species M_H is associated to a subgroup H of $\mathcal{S}_n$.

Definition 8. Let n be an integer ≥ 0, H a subgroup of $\mathcal{S}_n$, U a finite set, and $\beta : U \longrightarrow V$, a bijection. Define $M_H[U]$ to be the set of cosets with respect to H of bijections $[n] \longrightarrow U$, i.e.,

$$M_H[U] = \{\lambda H \mid \lambda : [n] \xrightarrow{\sim} U\}, \qquad (13)$$

where $\lambda H = \{\lambda \circ h \mid h \in H\}$. The transport $M_H[\beta] : M_H[U] \longrightarrow M_H[V]$ is defined by setting

$$M_H[\beta](\lambda H) = (\beta \circ \lambda) H. \qquad (14)$$

This species is often denoted X^n/H (see Exercise 27).

The reader will verify without difficulty that M_H as defined is indeed a molecular species of structures, concentrated on cardinality n, and that $M_H = M_K$ if and only if H and K are conjugate subgroups.

Example 9. a) For the trivial subgroup $H = \{1\}$ of $\mathcal{S}_n$, this construction gives the species $X^n/\{1\} = X^n$ of lists of length n. For the symmetric group itself, $H = \mathcal{S}_n$, the species $X^n/\mathcal{S}_n = E_n$ of sets of cardinality n is obtained. For the cyclic subgroup $H = \langle \rho \rangle$ generated by the cyclic permutation $\rho = (1, 2, 3, \ldots, n)$, the species $X^n/\langle \rho \rangle = C_n$ of oriented cycles of cardinality n is obtained. Finally, for the dihedral group $D_n = \langle \rho, \tau \rangle$, where $\tau = (1, n)(2, n-1)\cdots$, the species $X^n/D_n = P_n$ of polygons of order n is obtained.

b) THE SPECIES $E^{\pm}$ OF ORIENTED SETS. This example arises from the alternating subgroup $H = A_n$ formed of even permutations in $\mathcal{S}_n$ ($n \geq 0$). The species $X^n/A_n = E_n^{\pm}$ of *oriented sets* of cardinality n is obtained. A structure of oriented set on U is, by definition, a total order on U modulo an even permutation of its elements. It is equivalent to arranging the elements of U at the vertices of an oriented simplex (of dimension $|U| - 1$), since every even permutation of the vertices of the simplex preserves the orientation, whereas odd permutations reverse the orientation (if $|U| > 1$). There are exactly two oriented set structures (or orientations) on every set U such that $|U| > 1$, and a single structure if $|U| \leq 1$. We set $E^{\pm} = \sum_{n \geq 0} E_n^{\pm}$.

The behavior of molecular species under the basic operations is given by the following rules.

2.6. Molecular and Atomic Species

Table 3. *Atomic species for $n \leq 5$.*

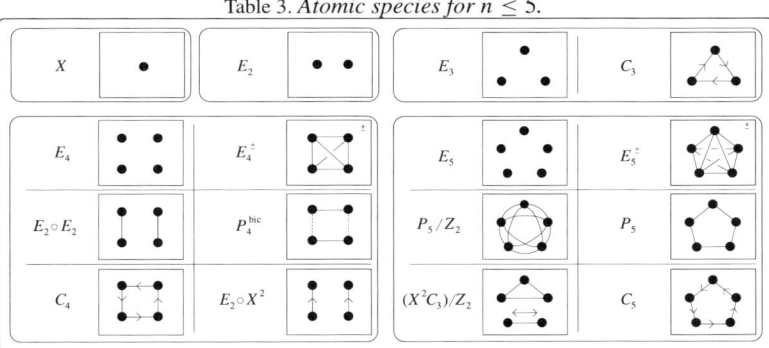

Proposition 10. (Yeh [333, 334].) *Given subgroups H of $\mathcal{S}_n$ and K of $\mathcal{S}_m$ $(m \geq 1)$, we have*

a) $\dfrac{X^n}{H} \cdot \dfrac{X^m}{K} = \dfrac{X^{n+m}}{H \times K}$,

b) $\dfrac{X^n}{H} \circ \dfrac{X^m}{K} = \dfrac{X^{nm}}{H \wr K}$,

c) $\dfrac{X^n}{H} \times \dfrac{X^m}{K} = \begin{cases} 0, & \text{if } m \neq n, \\ \sum_{\tau \in H \backslash \mathcal{S}_n / K} \dfrac{X^n}{H \cap (\tau K \tau^{-1})}, & \text{if } m = n, \end{cases}$ (15)

d) $\left(\dfrac{X^n}{H}\right)' = \sum_{i \in [n]/H} \dfrac{X^{n-1}}{(H \cap \mathcal{S}_{n \backslash \{i\}})^{\downarrow}}$.

Proof. a) See Exercise 10.
b) See Proposition 4.3.6. Here $H \wr K$ denotes the wreath product of the group actions $H : [n]$ and $K : [m]$.
c) See Exercise 11.
d) See Exercise 12. ∎

By making a systematic list of (conjugacy classes of the) subgroups of $\mathcal{S}_n$ for $n \leq N$, one obtains a complete classification of molecular species concentrated on $n \leq N$ points (see Exercise 4 and Table 3). For example, we have

$$\mathfrak{M}_0 = \{1\}, \quad \mathfrak{M}_1 = \{X\}, \quad \mathfrak{M}_2 = \{X^2, E_2\},$$
$$\mathfrak{M}_3 = \{X^3, X \cdot E_2, C_3, E_3\}. \tag{16}$$

Thus, every species F can be written in the form

$$F = a + bX + cX^2 + dE_2 + eX^3 + fX \cdot E_2 + gC_3 + hE_3 + \cdots, \qquad (17)$$

where $a, b, c, \ldots, h, \ldots \in \mathbb{N}$.

Example 11. An exhaustive analysis of unlabeled (simple) graphs on $n \leq 3$ points shows (see Exercise 5) that the species $\mathcal{G}$ of graphs has a molecular decomposition (see also Table A.2.6) beginning with

$$\mathcal{G} = 1 + X + 2E_2 + 2X \cdot E_2 + 2E_3 + \cdots.$$

Observe that certain molecular species, such as X^2, X^3, and $X \cdot E_2$, are products of simpler molecular species X and E_2. These last two species are called *atomic* since they are *indecomposable* under multiplication of species.

Definition 12. An *atomic species* is a molecular species of structures $A \neq 1$ which is not decomposable (except trivially) under product: For any species of structures P and Q,

$$A = P \cdot Q \quad \text{implies} \quad P = 1 \quad \text{or} \quad Q = 1. \qquad (18)$$

Note the following important result whose proof, due to Yeh [333], is relatively delicate (see Exercise 13).

Proposition 13. *Any molecular species M can be written as a finite product of atomic species*

$$M = A_1^{n_1} A_2^{n_2} \cdots A_k^{n_k}, \qquad A_i \text{ distinct atomic, } n_i \in \mathbb{N}, \ i = 1, \ldots, k, \qquad (19)$$

in a unique manner up to order of the factors and up to isomorphism.

Denote by

$$\mathfrak{A} = \mathfrak{A}_0 \cup \mathfrak{A}_1 \cup \mathfrak{A}_2 \cup \ldots \cup \mathfrak{A}_n \cup \ldots, \qquad (20)$$

the complete list of atomic species (up to isomorphism), where $\mathfrak{A}_n \subseteq \mathfrak{M}_n$. Through a detailed analysis of subgroups of $\mathcal{S}_n$ for $n \leq 5$, one can compile a list of all atomic species on $n \leq 5$ points. There are exactly 16 and they are described in Table 3, which gives in each case the name of the species as well as a typical structure. Note that there is no atomic species on 0 points. See also Table A.2.5 for a list of molecular species of order up to 5.

In this table, the atomic species denoted by $P_5/\mathbb{Z}_2$ and $(X^2 C_3)/\mathbb{Z}_2$ are examples of quotient species formed with the group $\mathbb{Z}_2$. By definition, a $P_5/\mathbb{Z}_2$-structure is a pair $\{p, q\}$, where p is a pentagon (as a nonoriented graph) and q is its complement. In a similar fashion, an $(X^2C_3)/\mathbb{Z}_2$-structure is a pair $\{s, t\}$, where s and t are two X^2C_3-structures which can be obtained, one from the other, by replacing simultaneously the X^2-structure orientation and the C_3-structure with opposite orientations (see Exercise 27 for a general definition of the notion of quotient species). Also note that the species $E_2 \circ E_2$ is isomorphic, by passing to the complementary graph, to the species P_4 of polygons of order 4 (squares). The species P_4^{bic} is, by definition, the molecular species X^n/H where $H = \langle (1, 2)(3, 4), (2, 3)(1, 4) \rangle$. A P_4^{bic}-structure can be identified with a P_4-structure whose edges are alternately colored with two colors (see Exercise 28).

Taking all the possible products of atomic species yields the list of molecular species. The numerical sequences $(\mu_n)_{n \geq 0}$ and $(\alpha_n)_{n \geq 0}$, defined by

$$\mu_n = |\mathfrak{M}_n|, \qquad \alpha_n = |\mathfrak{A}_n|, \tag{21}$$

start with

$$\mu_n : 1, 1, 2, 4, 11, 19, 56, 96, 296, 554, 1593, 3093 \ldots,$$

$$\alpha_n : 0, 1, 1, 2, 6, 6, 27, 20, 130, 124, 598, 640 \ldots.$$

It is interesting to note that they are related by the formula (see Exercise 14)

$$\sum_{n \geq 0} \mu_n x^n = \prod_{k \geq 1} \left(\frac{1}{1 - x^k} \right)^{\alpha_k}. \tag{22}$$

From the uniqueness of the atomic decomposition follows an algebraic characterization of the semi-ring Spe of species of structures and of the ring Virt of virtual species.

Proposition 14. (Yeh [329].) *The semi-ring* Spe *of species of structures is isomorphic to the semi-ring*

$$\mathbb{N}[\![\mathfrak{A}]\!] = \mathbb{N}[\![X, E_2, E_3, C_3, \ldots]\!] \tag{23}$$

of formal power series, with coefficients in $\mathbb{N}$, *on the set* $\mathfrak{A}$ *of atomic species.*

In view of this isomorphism, we can write a species $F \in$ Spe as a formal power series in $\mathbb{N}[\![X, E_2, E_3, C_3, \ldots]\!]$, i.e., as a sum of products of atomic species

$$F = a + bX + cX^2 + dE_2 + eX^3 + fXE_2 + gC_3 + hE_3 + \cdots. \tag{24}$$

This representation is called the *atomic decomposition* of F.

To characterize the ring Virt, we reason as follows: The molecular decomposition of a virtual species $F - G$ is obtained by taking the difference coefficientwise:

$$F - G = \sum_{M \in \mathfrak{M}} f_M M - \sum_{M \in \mathfrak{M}} g_M M$$

$$= \sum_{M \in \mathfrak{M}} (f_M - g_M) M. \tag{25}$$

Thus, every virtual species $\Phi = F - G$ has a unique molecular decomposition of the form

$$\Phi = \sum_{M \in \mathfrak{M}} \varphi_M M, \qquad \text{where} \quad \varphi_M \in \mathbb{Z}. \tag{26}$$

The reduced form $\Phi^+ - \Phi^-$ of the virtual species $\Phi = F - G$ (see Definition 2.5.6) is then given by

$$\Phi^+ = \sum_{\substack{M \in \mathfrak{M} \\ f_M > g_M}} (f_M - g_M) M, \quad \Phi^- = \sum_{\substack{M \in \mathfrak{M} \\ f_M < g_M}} (g_M - f_M) M. \tag{27}$$

This establishes Proposition 2.5.7 (see Exercise 7).

The atomic decomposition of Φ is obtained by factoring each $M \in \mathfrak{M}$ as a product of atomic species.

Proposition 15. *The ring* Virt *of virtual species is isomorphic to the ring*

$$\mathbb{Z}[[\mathfrak{A}]] = \mathbb{Z}[[X, E_2, E_3, C_3, \ldots]]$$

of formal power series, with coefficients in $\mathbb{Z}$, *over the set* $\mathfrak{A}$ *of atomic species.*

The other combinatorial operations, $\circ$, $'$, $\cdot$, $\times$, $\square$, can also be described within this ring of formal power series.

Remark 16. In principle, nothing prevents us from replacing the ring of coefficients $\mathbb{Z}$ in the molecular decomposition by any ring $\mathbb{K}$. We obtain in this manner the ring $\mathbb{K}[[\mathfrak{A}]]$ whose elements are called $\mathbb{K}$-species. Operations on species can be extended to this new ring $\mathbb{K}[[\mathfrak{A}]]$ whenever $\mathbb{K}$ has a nice algebraic structure such as a *binomial* ring structure or a lambda ring structure (see Knutson [174]). We will not undertake here the systematic study of $\mathbb{K}$-species (see Yeh [333] and G. Labelle [185]).

The molecular and atomic decompositions provide powerful algebraic tools for refined study of species of structures as well as for the construction of examples or counterexamples. For instance, consider a species of structures of the form

$$F = 1 + bX + cX^2 + dE_2 + \cdots + f_M M + \cdots, \tag{28}$$

2.6. Molecular and Atomic Species

where $F(0) = 1$ and $b, c, d, \ldots, f_M, \ldots$ are natural numbers. Then F has a notion of connected components (in the context of ordinary species of structures) if and only if the combinatorial logarithm of F, that is to say, the virtual species (see Exercise 24)

$$F^c = bX + \left(c - \frac{b(b-1)}{2}\right)X^2 + (d-b)E_2 + \cdots + f_M^c M + \cdots, \quad (29)$$

has all of its coefficients ≥ 0. In particular, F does not admit a notion of connected components in ordinary species if $c < \frac{b(b-1)}{2}$ or if $d < b$.

We know that two isomorphic species F and G have the same cycle index series $Z_F = Z_G$. The inverse is false in general, as shown by the example

$$F = X^3 + 2E_3, \quad G = C_3 + 2X \cdot E_2. \quad (30)$$

Indeed, it is readily verified that

$$Z_F = Z_G = \frac{4}{3}x_1^3 + x_1 x_2 + \frac{2}{3}x_3, \quad (31)$$

but $F \neq G$ since their molecular decompositions differ. This shows that the notion of molecular decomposition constitutes a strict generalization of the notion of cycle index series. Incidentally, the associated series can be calculated by the following formulas (see Exercise 8): For a virtual species Φ of the form

$$\Phi = \sum_{n \geq 0} \sum_{H \in \text{Conj}(S_n)} \varphi_H \frac{X^n}{H},$$

where $\varphi_H \in \mathbb{Z}$, we have

a) $\displaystyle \Phi(x) = \sum_{n \geq 0} \left(\sum_{H \in \text{Conj}(S_n)} \varphi_H \frac{n!}{|H|} \right) \frac{x^n}{n!},$

b) $\displaystyle \widetilde{\Phi}(x) = \sum_{n \geq 0} \sum_{H \in \text{Conj}(S_n)} \varphi_H x^n,$ \quad (32)

c) $\displaystyle Z_\Phi(x_1, x_2, x_3, \ldots) = \sum_{n \geq 0} \sum_{H \in \text{Conj}(S_n)} \varphi_H P_{H:[n]}(x_1, x_2, x_3, \ldots),$

where $P_{H:[n]}(x_1, x_2, x_3, \ldots)$ denotes the cycle index polynomial of the group action of $H \subseteq S_n$ on $[n]$. See Appendix 1.

The derivative of a molecular species is not molecular in general. Indeed, let M be a molecular species of the form $M = PQ$, where P and Q are molecular (different from the species 1). We have $M' = (PQ)' = P'Q + PQ'$, which gives a nontrivial decomposition of the species M' as a sum of at least two nonempty species and proves that M' is nonmolecular. This argument could lead us to believe that the derivative A' of an atomic species is necessarily molecular, since A cannot be written in the form $P \cdot Q$. This is false, as shown by the atomic species $A = E_2 \circ X^2$:

$$A' = (E_2 \circ X^2)' = 2X(E_2' \circ X^2) = 2X^3 = X^3 + X^3, \tag{33}$$

which is not molecular. This is the simplest example of this phenomenon.

Remark 17. The notions of molecular and atomic decompositions can be extended to the multisort context (see Exercises 15 and 22). It suffices to replace, in the study of transitive actions, the symmetric group $\mathcal{S}_n$ by the group

$$\mathcal{S}_{n_1} \times \mathcal{S}_{n_2} \times \cdots \times \mathcal{S}_{n_k}, \tag{34}$$

where $(n_1, n_2, \ldots, n_k)$ runs over all possible multicardinalities. Here k is the number of sorts and $\mathcal{S}_{n_1} \times \mathcal{S}_{n_2} \times \cdots \times \mathcal{S}_{n_k}$ forms the group of permutations, componentwise, of the product $[n_1] \times [n_2] \times \cdots \times [n_k]$. It is often useful (see Exercises 15 and 20) to replace this product of groups by the subgroup $\mathcal{S}_{n_1, n_2, \ldots, n_k}$ of $\mathcal{S}_n$ defined by

$$\mathcal{S}_{n_1, n_2, \ldots, n_k} = \{\sigma \in \mathcal{S}_n \mid \sigma(\{1, \ldots, n_1\}) = \{1, \ldots, n_1\},$$
$$\sigma(\{n_1 + 1, \ldots, n_1 + n_2\}) = \{n_1 + 1, \ldots, n_1 + n_2\}, \ldots\}, \tag{35}$$

where $n = n_1 + n_2 + \cdots + n_k$. This viewpoint suggests the identification of the multiset $[n_1] + [n_2] + \cdots + [n_k]$ with the set $[n]$, cut up into k successive intervals.

The extension to the weighted case is achieved by starting with the following observation. Let M_w be a weighted molecular species. Since by definition two arbitrary M_w-structures are necessarily isomorphic and transports preserve weight, every M_w-structure has the same weight $p \in \mathbb{A}$. Thus the weighted species M_w can be written in the *standard* form

$$M_w = M_p,$$

where M is nonweighted. The product of two molecular weighted species is given by the formula

$$M_p N_q = (MN)_{pq}$$

for all weights $p, q \in \mathbb{A}$. However, Proposition 13 is not valid in this context.

Exercises for Section 2.6

1. Consider two molecular species M and N. Show that
    a) $M \cdot N$ is molecular,

b) $M \circ N$ is molecular if $N \neq 1$,
c) $M + N$ is not molecular,
d) A species $F \neq 0$ is molecular if and only if

$$F = P + Q \implies P = 0 \text{ or } Q = 0.$$

2. a) Verify the accuracy of the molecular decomposition (5) for the species $\mathcal{A}$ of rooted trees, up to cardinality 5, by describing explicitly the required isomorphisms.
   b) Verify the molecular decompositions (6) a)–d) for the species E, $E^{\bullet}$, C, L.
3. For $n = 0, 1, 2, 3$, make a table of the subgroups of the symmetric group $\mathcal{S}_n$ and deduce that

$$\mathfrak{M}_0 = \{1\}, \qquad \mathfrak{M}_1 = \{X\},$$
$$\mathfrak{M}_2 = \{X^2, E_2\}, \qquad \mathfrak{M}_3 = \{X^3, X \cdot E_2, C_3, E_3\}.$$

4. a) Find two species of structures F and G such that

$$F \cdot G = 2 + 7X + 6X^2 + 2E_2 + 9X^3 + 6XE_2. \tag{36}$$

   b) Find all the species $F = F(X)$ satisfying the conditions $F(0) = 3$, $F'(X) = 4 + 3X + 6X^2 + 9E_2$, and $F_n(X) = 0$ if $n \geq 4$.
5. a) Make a list of all unlabeled simple graphs without loops on $n \leq 4$ vertices.
   b) Deduce that the molecular decomposition of the species $\mathcal{G}$ of simple graphs begins with the terms

$$\mathcal{G} = 1 + X + 2E_2 + 2E_3$$
$$+ 2X \cdot E_2 + 2E_4 + 2E_2 \cdot E_2 \tag{37}$$
$$+ 2E_2 \circ E_2 + 2X \cdot E_3 + 2X^2 \cdot E_2 + E_2 \circ X^2 + \cdots.$$

6. Starting from the table of molecular species concentrated on the cardinality $n \leq 5$ (see Table A.2.5), show (in a way different from Problem 2.5.12) that the combinatorial equation $Y' = E_4^{\bullet}$ does not have a solution in the context of species of structures.
   HINT: For a solution $Y = F$, one may assume, a priori, that $F_5 = \sum_{M \in \mathfrak{M}_5} f_M M$ where $f_M \geq 0$ are indeterminate coefficients.
7. REDUCED FORM OF VIRTUAL SPECIES.
   a) Let M be a molecular species. Show that the only subspecies of M are 0 and M.
   b) Let $k \in \mathbb{N}$ and M be a molecular species. One can interpret the species kM by setting, for any finite set U,

$$(kM)[U] = \{(i, \mu) \mid i \in [k], \mu \in M[U]\}. \tag{38}$$

Show that the subspecies G of kM are precisely those species defined by

$$G[U] = \{(i, \mu) \mid i \in I, \ \mu \in M[U]\}, \tag{39}$$

for some subset $I \subseteq [k]$. Moreover, the molecular decomposition of G is $k_1 M$ where $k_1 = |I|$.

c) If $k_1 \leq k$, show that $kM - k_1 M = (k - k_1)M$.

d) Let $F = \sum_{M \in \mathfrak{M}} f_M M$ and $G = \sum_{M \in \mathfrak{M}} g_M M$ be two species of structures. Show that F and G are unrelated if and only if

$$(\forall M)\,(f_M > 0 \implies g_M = 0) \quad \text{and} \quad (g_M > 0 \implies f_M = 0). \tag{40}$$

e) Let $\Phi = \sum_{M \in \mathfrak{M}} \varphi_M M$, $\varphi_M \in \mathbb{Z}$ be a virtual species. Show that the reduced form $\Phi = \Phi^+ - \Phi^-$ of Φ is given by

$$\Phi^+ = \sum_{\substack{M \in \mathfrak{M} \\ \varphi_M > 0}} \varphi_M M, \qquad \Phi^- = \sum_{\substack{M \in \mathfrak{M} \\ \varphi_M < 0}} (-\varphi_M) M. \tag{41}$$

8. a) Let $H \leq S_n$. Show that the cycle index series of the molecular species X^n/H is equal to the cycle index polynomial $P_{H:[n]}$ of the group H, in the sense of Pólya, given by

$$P_{H:[n]}(x_1, x_2, \ldots, x_n) = \frac{1}{|H|} \sum_{h \in H} x_1^{c_1(h)} x_2^{c_2(h)} \cdots x_n^{c_n(h)}, \tag{42}$$

where $c_i(h)$ denotes the number of cycles of length i in the permutation $h \in S_n$.

b) Establish formulas (32), a)–c) for the series $\Phi(x)$, $\widetilde{\Phi}(x)$, and $Z_\Phi(x_1, x_2, \ldots)$ associated to an arbitrary virtual species Φ.

9. Let $E^\pm = \sum_{n \geq 0} E_n^\pm$ be the species of oriented sets (see Example 9, b)).

a) Show that

i) $E_0^\pm = 1$, $\qquad E_1^\pm = X$, $\qquad E_2^\pm = X^2$, $\qquad E_3^\pm = C_3$,

ii) $E_n^\pm \neq 2E_n$, for any $n \geq 0$, $\hfill (43)$

iii) $\left(E_n^\pm\right)' = E_{n-1}^\pm$, for any $n \geq 3$.

b) Verify that

i) $E^\pm(x) = 2e^x - (1 + x)$,

ii) $\widetilde{E^\pm}(x) = \dfrac{1}{1-x}$, $\hfill (44)$

iii) $Z_{E^\pm}(x_1, x_2, x_3, \ldots) = \exp \sum_{k \geq 1} \dfrac{x_k}{k} + \exp \sum_{k \geq 1} (-1)^{k-1} \dfrac{x_k}{k} - 1 - x_1$.

10. Consider two molecular species X^n/H and X^m/K, where H is a subgroup of $\mathcal{S}_n$ and K is a subgroup of $\mathcal{S}_m$. Show that

$$\frac{X^n}{H} \cdot \frac{X^m}{K} = \frac{X^{n+m}}{H * K}, \tag{45}$$

where $H * K$ is the subgroup of $\mathcal{S}_{n+m}$, isomorphic to $H \times K$, defined by

$$H * K = \{h * k \mid h \in H, k \in K\}, \tag{46}$$

each $h * k \in \mathcal{S}_{n+m}$ being defined by

$$(h * k)(i) = \begin{cases} h(i), & \text{if } 1 \leq i \leq n, \\ n + k(i - n), & \text{if } n + 1 \leq i \leq n + m. \end{cases} \tag{47}$$

11. Let H and K be two subgroups of a group G. By definition, a *double coset* of H and K in G is a subset of G of the form

$$HgK = \{hgk \mid h \in H, k \in K\}, \text{ where } g \in G. \tag{48}$$

One sets

$$H \backslash G / K = \{HgK \mid g \in G\}. \tag{49}$$

a) Show that the set $H \backslash G / K$ of double cosets form a partition of G.
b) Show that if $H \leq \mathcal{S}_n$ and $K \leq \mathcal{S}_m$, then

$$\frac{X^n}{H} \times \frac{X^m}{K} = \begin{cases} 0, & \text{if } m \neq n, \\ \sum_{\tau \in H \backslash \mathcal{S}_n / K} \frac{X^n}{H \cap (\tau K \tau^{-1})}, & \text{if } m = n, \end{cases} \tag{50}$$

where $\tau \in H \backslash \mathcal{S}_n / K$ indicates τ ranges over a system of representatives of the double cosets.
HINT: Show that every $\frac{X^n}{H} \times \frac{X^n}{K}$-structure $(\alpha H, \beta K)$ is isomorphic to a structure of the form $(H, \tau K)$. Moreover, $(H, \tau K)$ and $(H, \tau' K)$ are isomorphic if and only if $H \tau K = H \tau' K$.

12. Show that for any subgroup H of $\mathcal{S}_n$, one has the equalities

$$\left(\frac{X^n}{H}\right)^{\bullet} = \sum_{i \in [n]/H} \frac{X^n}{H \cap \mathcal{S}_{n \setminus \{i\}}} \tag{51}$$

and

$$\left(\frac{X^n}{H}\right)' = \sum_{i \in [n]/H} \frac{X^{n-1}}{(H \cap \mathcal{S}_{n \setminus \{i\}})^{\downarrow}}, \tag{52}$$

where $i \in [n]/H$ indicates that i ranges over a system of orbit representatives of the natural action $H \times [n] \longrightarrow [n]$ (defined by $h \cdot x = h(x)$), $\mathcal{S}_{n \setminus \{i\}}$ denotes the subgroup of $\mathcal{S}_n$ composed of permutations σ such that $\sigma(i) = i$, and $(H \cap \mathcal{S}_{n\setminus\{i\}})^{\downarrow}$ denotes the subgroup of $\mathcal{S}_{n-1}$ obtained by restricting to $\mathcal{S}_{n-1}$ the group $\tau_i(H \cap \mathcal{S}_{n\setminus\{i\}})\tau_i^{-1}$, with τ_i as the transposition (i, n).

13. CHARACTERIZATION OF ATOMIC SPECIES. (Yeh [333].) Consider two sets U and V such that $U \subseteq V$. Let H be a subgroup of the group $S[V]$ of permutations of V and suppose that U is stable under H. For any $h \in H$, define the *restriction* $h_U : U \longrightarrow U$, of h to U, by letting $h_U(u) = h(u)$, for any $u \in U$. Inversely, if K is a subgroup of $S[U]$ and if $k \in K$, define the *extension* $k^V : V \longrightarrow V$, of k to V, by setting

$$k^V(v) = \begin{cases} k(v), & \text{if } v \in U, \\ v, & \text{if } v \in V\setminus U. \end{cases} \qquad (53)$$

Write H_U and K^V to denote the following two permutation groups:

$$H_U = \{h_U \mid h \in H\} \subseteq S[U], \quad K^V = \{k^V \mid k \in K\} \subseteq S[V]. \qquad (54)$$

One says that U *reduces* H if U is stable under H and

$$(H_U)^V \subseteq H. \qquad (55)$$

a) Show that
  i) if U reduces H then $V\setminus U$ reduces H,
  ii) if U_1, U_2 reduces H then $U_1 \cap U_2$ reduces H.
In other words, those $U \subseteq V$ which reduce H form a Boolean subalgebra of $\wp(V)$.

b) Let H be a subgroup of $\mathcal{S}_n$ and $M_H = X^n/H$, the molecular species associated to H. Show that if $U \subseteq [n]$ reduces H then

$$\frac{X^n}{H} = \frac{X^{|U|}}{H_U} \cdot \frac{X^{n-|U|}}{H_{[n]\setminus U}}. \qquad (56)$$

c) Show that X^n/H is atomic if and only if for any $U \subseteq [n]$, U reduces H implies $U = \emptyset$ or $U = [n]$.

d) Establish Proposition 13 which says that every molecular species M can be written as a product of atomic species, in a unique fashion up to order of the factors.

HINT: Show that the atoms (that is to say, the minimal nonempty elements) of the Boolean algebra of subsets $U \subseteq [n]$ which reduce H from a partition of $[n]$.

14. a) Let $\mathfrak{M}$ (respectively $\mathfrak{A}$) be the set of molecular (respectively atomic) species up to isomorphism. Show that

$$\sum_{M \in \mathfrak{M}} M = \prod_{A \in \mathfrak{A}} \frac{1}{1 - A}. \qquad (57)$$

b) By giving a weight x^n to every molecular species $M \in \mathfrak{M}_n$ concentrated on the cardinality n, deduce formula (22):

$$\sum_{n \geq 0} \mu_n x^n = \prod_{k \geq 1} \left(\frac{1}{1 - x^k} \right)^{\alpha_k}. \tag{58}$$

15. a) Define the concept of molecular k-sort species $M(X_1, X_2, \ldots, X_k)$.
    b) By considering the transitive action of the groups $\mathcal{S}_{n_1} \times \mathcal{S}_{n_2} \times \cdots \times \mathcal{S}_{n_k}$ and by adapting Definition 8 to the multisort context, define rigorously the species

$$M_H = M_H(X_1, \ldots, X_k) = \frac{X_1^{n_1} X_2^{n_2} \cdots X_k^{n_k}}{H}, \tag{59}$$

with H subgroup of $\mathcal{S}_{n_1} \times \mathcal{S}_{n_2} \times \cdots \times \mathcal{S}_{n_k}$. Show that every molecular species $M(X_1, \ldots, X_k)$ is of the form $M_H(X_1, \ldots, X_k)$ for a certain subgroup H of $\mathcal{S}_{n_1} \times \mathcal{S}_{n_2} \times \cdots \times \mathcal{S}_{n_k}$. Also show that $M_H(X_1, \ldots, X_k) = M_K(X_1, \ldots, X_k)$ implies that H and K are conjugates in $\mathcal{S}_{n_1} \times \mathcal{S}_{n_2} \times \cdots \times \mathcal{S}_{n_k}$.
    c) Deduce the existence and uniqueness of the molecular decomposition of an arbitrary multisort species $F = F(X_1, \ldots, X_k)$.
16. Let H be a subgroup of $\mathcal{S}_n$ and $M_H = X^n/H$ be the associated molecular species. Consider k sorts of points $X_1, \ldots, X_k$. Show that

$$M_H(X_1 + \cdots + X_k) = \sum_{n_1 + \cdots + n_k = n} \sum_{\tau \in \mathcal{S}_{n_1, \ldots, n_k} \backslash \mathcal{S}_n / H} \frac{X_1^{n_1} X_2^{n_2} \cdots X_k^{n_k}}{(\tau H \tau^{-1}) \cap \mathcal{S}_{n_1, \ldots, n_k}}, \tag{60}$$

where τ ranges over a system of (arbitrary) representatives of the set $\mathcal{S}_{n_1, \ldots, n_k} \backslash \mathcal{S}_n / H$ of double cosets (see Remark 17 and Exercise 11).
HINT: Let τH have an $M_H(X)$-structure on $[n]$. Show first that one can identify $\mathcal{S}_{n_1, \ldots, n_k} \backslash \mathcal{S}_n / H$ with the set of orbits of $M_H(X_1 + \cdots + X_k)$-structures on the multiset $[n_1] + \cdots + [n_k]$.
17. Show that the molecular decomposition of the species $C(X + Y)$ of oriented cycles on two sorts of points X and Y begins with the terms

$$C(X + Y) = X + Y + C_2(X) + X \cdot Y + C_2(Y) + C_3(X)$$
$$+ X^2 \cdot Y + X \cdot Y^2 + C_3(Y) + C_4(X) + X^3 \cdot Y$$
$$+ X^2 Y^2 + C_2(X \cdot Y) + X \cdot Y^3 + C_4(Y) + \cdots.$$

18. a) Consider the set $\mathfrak{M} = \bigcup_{k=0}^{\infty} \mathfrak{M}_k$ of molecular species and denote by

$$\mathfrak{M}_{\leq n} = \bigcup_{k=0}^{n} \mathfrak{M}_k \tag{61}$$

the set of molecular species of order $k \leq n$. Show that there exists a countable family of polynomials $(p_M)_{M \in \mathfrak{M}}$ such that, for $M \in \mathfrak{M}_n$,

$$p_M \in \mathbb{Q}[(x_N)_{N \in \mathfrak{M}_{\leq n}}, (y_N)_{N \in \mathfrak{M}_{\leq n}}], \tag{62}$$

satisfying

$$F \circ G = \sum_{n \geq 0} \sum_{M \in \mathfrak{M}_n} p_M((f_N)_{N \in \mathfrak{M}_{\leq n}}, (g_N)_{N \in \mathfrak{M}_{\leq n}}) M, \tag{63}$$

for any species of structures F and G, $G(0) = 0$, whose molecular decompositions are

$$F = \sum_{M \in \mathfrak{M}} f_M M, \quad G = \sum_{M \in \mathfrak{M}} g_M M, \quad g_1 = 0. \tag{64}$$

HINT: Consider the molecular decomposition

$$F \circ G = \sum_{M \in \mathfrak{M}} h_M M.$$

Fix $M \in \mathfrak{M}_n$ and use distributivity on the right,

$$F \circ G = \left(\sum_{P \in \mathfrak{M}} f_P P \right) \circ G = \sum_{P \in \mathfrak{M}} f_P P \circ G,$$

to show that h_M is a function of only f_N, $N \in \mathfrak{M}_{\leq n}$ and g_N, $N \in \mathfrak{M}_{\leq n}$, linear in the f_N. Conclude the polynomiality of h_M by introducing some auxiliary (sort) variables $(X_N)_{N \in \mathfrak{M}_{\leq n}}$ and by using the formulas

$$P \circ \sum_{N \in \mathfrak{M}_{\leq n}} g_N N = P\left(\sum_{N \in \mathfrak{M}_{\leq n}} g_N X_N \right)\Big|_{X_N := N, N \in \mathfrak{M}_{\leq n}}$$

$$= P\left(\sum_{N \in \mathfrak{M}_{\leq n}} X_N \right) \times \prod_{N \in \mathfrak{M}_{\leq n}} (E(X_N))^{g_N} \Big|_{X_N := N, N \in \mathfrak{M}_{\leq n}}$$

and the fact that

$$(E(X))^g = (1 + E_+(X))^g = \sum_{k \geq 0} \binom{g}{k} (E_+(X))^k,$$

where $\binom{g}{k} = g(g-1) \ldots (g-k+1)/k!$ is indeed a polynomial in g.

b) Show unicity of this family of polynomials $(p_M)_{M \in \mathfrak{M}}$.
HINT: Make use of the extension principle for algebraic identities (see Exercise 3.1.1)

19. a) Consider three species of structures

$$R = \sum_{M \in \mathfrak{M}} r_M M, \quad F = \sum_{M \in \mathfrak{M}} f_M M, \quad G = \sum_{M \in \mathfrak{M}} g_M M,$$

with $F(0) = G(0) = 0$, and a variable m taking on integer values ≥ 0.
Show that the polynomials $(p_M)_{M \in \mathfrak{M}}$ in the preceding exercise satisfy

$$R(X+Y) \times (E(X) \cdot E^m(Y))|_{X:=F,\ Y:=G}$$

$$= \sum_{n \geq 0} \sum_{M \in \mathfrak{M}_n} p_M\big((r_N)_{N \in \mathfrak{M}_{\leq n}}, (f_N + m g_N)_{N \in \mathfrak{M}_{\leq n}}\big) M. \qquad (65)$$

b) Conclude that Definition 2.5.15 for the composite $\Phi \circ \Psi$ of two virtual species is correct, that is to say, independent of the choice of representatives of Φ and of Ψ.

HINT: Observe that the preceding formula is a polynomial in m. Conclude, setting $m := -1$, that

$$\Phi \circ \Psi = \sum_{n \geq 0} \sum_{M \in \mathfrak{M}_n} p_M\big((\varphi_N)_{N \in \mathfrak{M}_{\leq n}}, (\psi_N)_{N \in \mathfrak{M}_{\leq n}}\big) M, \qquad (66)$$

where

$$\Phi = \sum_{M \in \mathfrak{M}} \varphi_M M, \quad \Psi = \sum_{M \in \mathfrak{M}} \psi_M M, \quad \Psi(0) = 0.$$

20. Let Φ, Ψ, and Θ be virtual species. Show that the following formulas are valid:

a) $(\Phi \pm \Psi) \circ \Theta = \Phi \circ \Theta \pm \Psi \circ \Theta$,

b) $(\Phi \cdot \Psi) \circ \Theta = (\Phi \circ \Theta) \cdot (\Psi \circ \Theta)$, $\qquad (67)$

c) $(\Phi \circ \Psi) \circ \Theta = \Phi \circ (\Psi \circ \Theta)$,

d) $(\Phi \circ \Psi)' = (\Phi' \circ \Psi) \cdot \Psi'$, si $\Psi_0 = 0$,

as well as formulas (2.5.36) concerning the series $(\Phi \circ \Psi)(x)$, $(\widetilde{\Phi \circ \Psi})(x)$, $Z_{\Phi \circ \Psi}$.

HINT: Treat at first the nonvirtual case and invoke the (unique) family $(p_M)_{M \in \mathfrak{M}}$ of polynomials appearing in the two preceding exercises.

21. Consider the molecular decomposition on k sorts

$$F(X_1, X_2, \ldots, X_k) = \sum_{n_1, n_2, \ldots, n_k \geq 0} \sum_{H \in \mathrm{Conj}(S_{n_1, \ldots, n_k})} f_H \frac{X_1^{n_1} \cdots X_k^{n_k}}{H}. \qquad (68)$$

Show that

$$F(X, X, \ldots, X) = \sum_{n \geq 0} \left(\sum_{n_1 + \cdots + n_k = n} \sum_{H \in \mathrm{Conj}(S_{n_1, \ldots, n_k})} f_H \frac{X^n}{H} \right). \qquad (69)$$

22. a) Define the concept of atomic k-sort species $X_1, \ldots, X_k$.
   b) Show that every molecular species $M = M(X_1, \ldots, X_k)$ can be written as a finite product of atomic species

   $$M = A_1^{n_1} A_2^{n_2} \cdots A_\nu^{n_\nu},$$

   with $A_i = A_i(X_1, \ldots, X_k)$ atomic, and $n_i \in \mathbb{N}$, $i = 1, \ldots, \nu$, in a unique fashion up to order of factors and up to isomorphism.
   HINT: Adapt Exercise 13 to the multisort context.
   c) For each multicardinality $\mathbf{n} = (n_1, n_2, \ldots, n_k)$, denote by
   – $\mathfrak{M}_\mathbf{n}$, the set of molecular species concentrated on the multicardinality $\mathbf{n}$,
   – $\mathfrak{A}_\mathbf{n}$, the set of atomic species concentrated on the multicardinality $\mathbf{n}$.
   Moreover, set $\mu_\mathbf{n} = |\mathfrak{M}_\mathbf{n}|$, $\alpha_\mathbf{n} = |\mathfrak{A}_\mathbf{n}|$. Establish the formal identity

   $$\sum_{n_1, \ldots, n_k \geq 0} \mu_{n_1, \ldots, n_k} t_1^{n_1} \cdots t_k^{n_k} = \prod_{m_1 + \cdots + m_k > 0} \left(\frac{1}{1 - t_1^{m_1} \cdots t_k^{m_k}} \right)^{\alpha_{m_1, \ldots, m_k}}. \tag{70}$$

23. Consider the species $\mathfrak{a}_w(X, Y)$ of trees having internal vertices of sort X and leaves of sort Y, weighted by $w(\alpha) = s^{n_2(\alpha)}$, where $n_2(\alpha)$ is the number of vertices of degree 2 in the tree α. Show that the first terms of the molecular decomposition of $\mathfrak{a}_w$ are given (up to total cardinality 5) by

$$\mathfrak{a}_w(X, Y) = Y + E_2(Y) + (XE_2(Y))_s$$
$$+ XE_3(Y) + (E_2(X \cdot Y))_{s^2} + XY^4$$
$$+ (X \cdot E_2(X \cdot Y))_{s^3} + \left(X^2 Y E_2(Y)\right)_s + \cdots. \tag{71}$$

24. Let $\mathcal{G}_w$ be the species of simple graphs weighted by the monomials

$$t_0^{n_0} t_1^{n_1} t_2^{n_2} t_3^{n_3} \cdots,$$

where n_i is the number of vertices of degree i in the considered graph, $i = 0, 1, 2, 3, \ldots$. Show that the weighted molecular decomposition of $\mathcal{G}_w$ begins with the terms

$$\mathcal{G}_w = 1 + X_{t_0} + (E_2)_{t_0^2} + (E_2)_{t_1^2}$$
$$+ (E_3)_{t_0^3} + (E_3)_{t_2^3} + (X \cdot E_2)_{t_0 t_1^2} + (X \cdot E_2)_{t_1^2 t_2}$$
$$+ (E_4)_{t_0^4} + (E_4)_{t_3^4} + (E_2 \cdot E_2)_{t_0^2 t_1^2} + (E_2 \cdot E_2)_{t_2^2 t_3^2}$$
$$+ (E_2 \circ E_2)_{t_1^4} + (E_2 \circ E_2)_{t_2^4} + (X \cdot E_3)_{t_0 t_2^3} + (X \cdot E_3)_{t_1^3 t_3}$$
$$+ \left(X^2 \cdot E_2\right)_{t_0 t_1^2 t_2} + \left(X^2 \cdot E_2\right)_{t_1 t_2^2 t_3} + \left(E_2 \circ X^2\right)_{t_1^2 t_2^2} + \cdots. \tag{72}$$

25. Let F be a species satisfying $F(0) = 1$, i.e., whose molecular decomposition is of the form

$$F = 1 + bX + cX^2 + dE_2 + eX^3 + fX \cdot E_2 + gC_3$$
$$+ hE_3 + \cdots + f_M M + \cdots,$$

and let F^c be the (possibly virtual) species of connected F-structures (see formula (2.5.56)):

$$F^c = \sum_{M \in \mathfrak{M}} f_M^c M.$$

a) Calculate the coefficients f_M^c for any molecular species M of order ≤ 3.
b) Show that for any molecular species $M \in \mathfrak{M}_n$, f_M^c is a polynomial in the f_N, where $N \in \mathfrak{M}_{\leq n}$.

26. Let $\Psi = X + \cdots$ be a virtual species satisfying $\Psi(0) = 0$, $\Psi'(0) = 1$. Show that there exists a unique family of polynomials $(\psi_M(x))_{M \in \mathfrak{M}}$ such that the iterates $\Psi^{\langle n \rangle}$ decompose as follows:

$$\Psi^{\langle n \rangle} = \sum_{M \in \mathfrak{M}} \psi_M(n) M, \ n \in \mathbb{Z}. \qquad (73)$$

In particular, the inverse $\Psi^{\langle -1 \rangle}$ of Ψ, under the (partitional) composition, is written

$$\Psi^{\langle -1 \rangle} = \sum_{M \in \mathfrak{M}} \psi_M(-1) M. \qquad (74)$$

HINT: For existence, use the operator $\Delta_\Psi : \text{Virt} \longrightarrow \text{Virt}$ defined by $\Delta_\Psi \Phi = \Phi \circ \Psi - \Phi$ (see Section 2.5). For uniqueness, invoke the extension principle for algebraic identities (see Exercise 3.1.1).

27. QUOTIENT SPECIES. (See J. Labelle [200].) A group G is said to *act naturally* on a species of structures F if, for any finite set U, there is a G-action (see Appendix 1)

$$\rho_U : G \times F[U] \longrightarrow F[U] \qquad (75)$$

so that for each bijection $\sigma : U \longrightarrow V$, the following diagram commutes:

$$\begin{array}{ccc} G \times F[U] & \xrightarrow{\rho_U} & F[U] \\ {\scriptstyle 1_G \times F[\sigma]} \downarrow & & \downarrow {\scriptstyle F[\sigma]} \\ G \times F[V] & \xrightarrow{\rho_V} & F[V] \end{array} \qquad (76)$$

2. Complements on Species of Structure

The quotient species of F by G, denoted F/G, is defined by setting, for each finite set U,

$$(F/G)[U] = F[U]/G, \tag{77}$$

that is, the set of orbits of the action of G on $F[U]$.

a) Define the transport functions $(F/G)[\sigma]$, and verify that F/G is indeed a species of structures.

b) Represent each of the atomic species described in Table 3 in the form X^n/H, where H is a subgroup of S_n, $n \leq 5$.

c) Let H be a subgroup of S_n ($n \geq 0$). Show that the notation X^n/H for the molecular species M_H (see Definition 8) is compatible with the concept of quotient species introduced here.

d) Show that the cyclic group with two elements, $\mathbb{Z}_2$, acts in a natural way on the species L of linear orders by replacing the order of an L-structure by the opposite order. Show that we have the species isomorphism

$$L/\mathbb{Z}_2 \simeq \text{Cha}, \tag{78}$$

where Cha is the species of chains (see Exercise 1.1.5). Show moreover that

i) $\text{Cha}(x) = \dfrac{2 - x^2}{2(1 - x)}$,

ii) $\widetilde{\text{Cha}}(x) = \dfrac{1}{1 - x}$, \hfill (79)

iii) $Z_{\text{Cha}}(x_1, x_2, x_3, \ldots) = \dfrac{2 - x_1^2}{2(1 - x_1)} + \dfrac{(1 + x_1)x_2}{2(1 - x_2)}$.

e) (See Joyal [163].) Consider a family of set actions

$$S_n \times F_n \longrightarrow F_n, \qquad n = 0, 1, 2, \ldots \tag{80}$$

of the symmetric group S_n. Recall that any finite set T can be considered as a species concentrated on the empty set by setting, for any finite set U, $T[U] = T$ if $U = \emptyset$ and $T[U] = \emptyset$ otherwise. Show that for any $n \geq 0$, S_n acts naturally on the species $F_n \cdot X^n$. We define a species F by setting

$$F = \sum_{n \geq 0} \dfrac{F_n \cdot X^n}{S_n}. \tag{81}$$

Show that the family of set actions associated to the species F (in the sense of Exercise 1.1.6, a)) is isomorphic, term by term, to the family of actions (80).

28. a) Describe the underlying isomorphisms in Examples 9 a).
   b) For n even, $n \geq 4$, define the species P_n^{bic} of polygons of order n whose edges are alternately colored with two colors. Show that

   i) $P_n^{\text{bic}} \simeq X^n/H$, where $H = \langle (1,n)(2, n-1)\cdots, (1,2)(3,n)\cdots \rangle$,

   ii) $P_n \simeq P_n^{\text{bic}}/\mathbb{Z}_2$.
   $$\tag{82}$$

   c) For n even, $n \geq 4$, show that P_n^{bic} is not isomorphic to $\mathcal{C}_n$, although

   $$\left(P_n^{\text{bic}}\right)' = \mathcal{C}_n' = X^{n-1}. \tag{83}$$

3

Combinatorial Functional Equations

3.0. Introduction

For the practitioner, the description of rooted tree structures often takes a recursive form. Hence, the computer scientist will probably describe the concept of (plane) binary rooted tree in the following manner:

i) the empty rooted tree, *Nil*, is a binary rooted tree,

ii) if r is a data item, and if ρ and λ are disjoint binary rooted trees not containing r, then $\beta = (\lambda, r, \rho)$ is a binary rooted tree. The data item r is called the *root* of β, and λ and ρ are respectively called the *left subtree* and the *right subtree* of β, written as $\lambda = \lambda(\beta), \rho = \rho(\beta)$.

In the language of species of structures, this description corresponds to the functional equation

$$\begin{aligned} \mathcal{B} &= 1 + \mathcal{B} \cdot X \cdot \mathcal{B} \\ &= 1 + X \cdot \mathcal{B}^2. \end{aligned} \tag{1}$$

A graphical representation (see Figure 1.1.7) is deduced from this, as well as the equations

$$\begin{aligned} &\text{a)} \quad B(x) = 1 + xB^2(x), \\ &\text{b)} \quad \widetilde{B}(x) = 1 + x\widetilde{B}^2(x), \\ &\text{c)} \quad Z_{\mathcal{B}}(x_1, x_2, x_3, \ldots) = 1 + x_1 Z_{\mathcal{B}}^2(x_1, x_2, x_3, \ldots), \end{aligned} \tag{2}$$

which imply that $\widetilde{B}(x) = B(x)$ and $Z_{\mathcal{B}}(x_1, x_2, x_3, \ldots) = B(x_1)$. These last relations are typical of asymmetric species of structures (see Section 4.4). Solving the quadratic equation, we find

$$B(x) = \frac{1 - \sqrt{1-4x}}{2x}, \tag{3}$$

and, expanding according to Newton's binomial formula,

$$\mathcal{B}(x) = 1 + x + 2x^2 + 5x^3 + 14x^4 + 42x^5 + 132x^6 + 429x^7$$
$$+ 1430 x^8 + 4862 x^9 + 16796 x^{10} + \cdots$$
$$= \sum_{n \geq 0} c_n x^n,$$

where $c_n = \frac{1}{n+1}\binom{2n}{n}$ is the nth *Catalan number*.

In fact, this procedure is very natural and can already be found in Euler's work. However, only in recent decades have general theories been proposed to justify rigorously such developments. Before this, the use of generating functions appeared to be only a mysterious artifice. The goal of the present chapter is to elaborate a general method for solving functional equations in the algebra of species of structures. There is a theorem analogous to the implicit function theorem in this context (see Theorem 3.2.1). It is the cornerstone of this chapter since from there follows explicit combinatorial constructions for a vast family of functional equations. Moreover, uniqueness, up to isomorphism, of the solutions permits an unambiguous description of these tree-like structures.

Section 1 is devoted to the combinatorial analysis of the functional equation

$$Y = X \cdot R(Y), \tag{4}$$

where R is a given species, whose solution consists of rooted trees that are enriched (or structured) by the species R. The explicit enumeration of these structures by a combinatorial procedure then furnishes a proof of classical Lagrange inversion formulas. In this section classes of R-enriched trees are also considered. Their enumeration, as well as that of R-enriched rooted trees, is related to the sequence of binomial type associated to the series $R(x)$. For a different proof of Lagrange inversion, also based on enriched rooted trees, which explains the cancelations that arise in the direct substitution, see Chen [48, 51].

Section 2 treats the more general functional equation

$$Y = H(X, Y), \qquad Y(0) = 0. \tag{5}$$

Conditions are given, ensuring the existence and uniqueness of combinatorial solutions to this equation and of systems of equations of this type. In particular, a multidimensional Lagrange inversion formula is established, following the approach of Gessel [127], and is applied to calculate the coefficients of the cycle index series of R-enriched rooted trees.

Section 3 approaches, from a combinatorial viewpoint, Newton–Raphson iterative methods to the solution of functional equations. There results a recursive process having quadratic convergence for the computation of species of structures Y defined by equations of the form $Y = X \cdot R(Y)$ or also

$$Y = X + G(Y), \tag{6}$$

where G is a given species of structures, or more generally $Y = H(X, Y)$, where H is a species of two sorts. Functional equations of the type

$$Y = H(X, Y \circ G), \qquad (7)$$

which we call of Read–Bajraktarević type, and which lend themselves to an iterative approach, are also studied in detail.

Section 4 is an introduction to methods used in the asymptotic enumeration of F-structures on $[n]$, whenever $n \to \infty$, by the analysis of dominating singularities of the associated generating series. Particular attention is given to tree-like structures.

3.1. Lagrange Inversion

Let $\mathbb{K}$ be a field of characteristic zero. It is well known that the set of formal power series $f(x) \in \mathbb{K}[\![x]\!]$ such that $f(0) = 0 \neq f'(0)$ constitutes a group under the operation of substitution. Using the Lagrange inversion formula one can write explicitly the coefficients of the inverse formal power series $f^{\langle -1 \rangle}(x)$ of $f(x)$ as follows:

$$f^{\langle -1 \rangle}(x) = \sum_{n \geq 1} \left(\frac{d}{dt}\right)^{n-1} \left(\frac{t}{f(t)}\right)^n \bigg|_{t=0} \frac{x^n}{n!}. \qquad (1)$$

Denote by $A(x)$ the inverse of $f(x)$ for substitution and set

$$R(x) = \frac{x}{f(x)} \in \mathbb{K}[\![x]\!]. \qquad (2)$$

The series $A(x)$ is determined by the functional equation

$$A(x) = x R(A(x)), \qquad (3)$$

obtained by the substitution of $A(x)$ in (2). With this setup, Lagrange inversion formula (1) takes the form

$$A(x) = \sum_{n \geq 1} a_n \frac{x^n}{n!}, \qquad (4)$$

where

$$a_n = \left(\frac{d}{dt}\right)^{n-1} (R(t))^n \bigg|_{t=0}. \qquad (5)$$

More generally, for any formal power series $F(x)$, if

$$F(A(x)) = \sum_{n \geq 0} b_n \frac{x^n}{n!}, \qquad (6)$$

3.1. Lagrange Inversion

then we have $b_0 = F(0)$ and, for any $n \geq 1$,

$$b_n = \left(\frac{d}{dt}\right)^{n-1} F'(t)(R(t))^n \bigg|_{t=0}. \tag{7}$$

Another formula of the same type, which is very useful (see, for example, Pólya–Szego [265]), gives the coefficients c_n of the formal power series

$$\frac{F(A(x))}{1 - xR'(A(x))} = \sum_{n \geq 0} c_n \frac{x^n}{n!} \tag{8}$$

in the form

$$c_n = \left(\frac{d}{dt}\right)^n F(t)(R(t))^n \bigg|_{t=0} = n! \, [t^n] F(t)(R(t))^n. \tag{9}$$

Note in passing the following equivalent forms of formulas (5), (7), and (9), in terms of coefficient extraction.

a) $[x^n]A(x) = \frac{1}{n}[t^{n-1}]R^n(t),$

b) $[x^n]F(A(x)) = \frac{1}{n}[t^{n-1}]F'(t)R^n(t),$ (10)

c) $[x^n]\dfrac{F(A(x))}{1 - xR'(A(x))} = [t^n]F(t)R^n(t).$

We will interpret these formulas and prove them combinatorially when R is a species of structures by explicitly constructing the species $Y = \mathcal{A}_R$ solution of the functional equation

$$Y = X \cdot R(Y). \tag{11}$$

As will be shown later in a more general context (see Section 3.2), this solution is unique up to isomorphism.

Let R be a fixed species of structures. The combinatorial equality $\mathcal{A}_R = X \cdot R$ ($\mathcal{A}_R$) suggests a recursive description of the species $\mathcal{A}_R$: An $\mathcal{A}_R$-structure on a set U is obtained by choosing an arbitrary element e of U and attaching to it an R-assembly of $\mathcal{A}_R$-structures on $U \setminus \{e\}$. See Figure 1. By reiterating this analysis for each member of the R-assembly, there appears a tree-like structure with root e (see Figure 2). It is an R-enriched rooted tree, in the following sense.

Definition 1. An R-enriched rooted tree on a finite set U is the data of
i) an arbitrary rooted tree α on U,
ii) an R-structure on the fiber of each vertex $u \in U$, in this rooted tree.
Here the *fiber* of a vertex u refers to the (possibly empty) set $\alpha^{-1}(u)$ of immediate predecessors of u, when all the edges of the rooted tree are oriented towards the

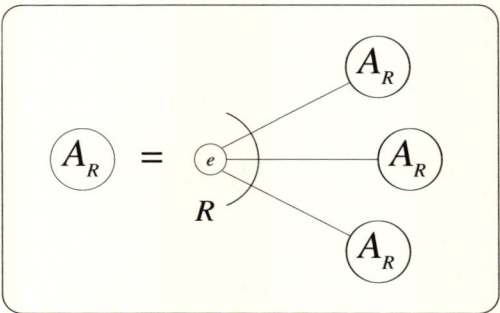

Fig. 1.

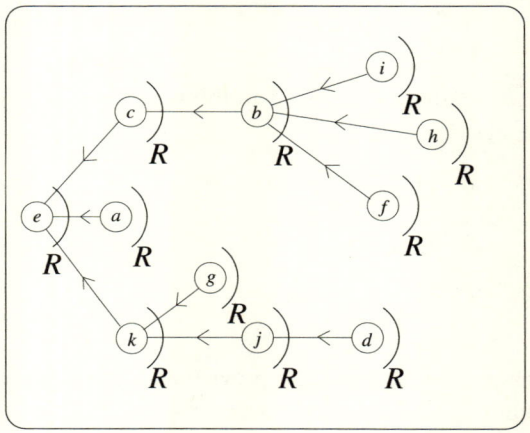

Fig. 2.

root. The *indegree* of u, denoted by $d^-(u)$, is, by definition, the cardinality of its fiber: $d^-(u) = |\alpha^{-1}(u)|$.

The *leaves* of a rooted tree are precisely those vertices with empty fiber. Observe that the empty fibers are also provided with an R-structure. It is customary to make the hypothesis that $R[0] \neq 0$. In the contrary case, where $R[0] = 0$, there cannot exist any $\mathcal{A}_R$-structure on a finite set, that is to say that $\mathcal{A}_R = 0$.

Theorem 2. *Let R be a species of structures. Then the species $\mathcal{A}_R$ of R-enriched rooted trees is uniquely determined, up to isomorphism, by the combinatorial equation*

$$\mathcal{A}_R = X \cdot R(\mathcal{A}_R). \tag{12}$$

Proof. It is clear, by construction, that $\mathcal{A}_R$ satisfies (12). Uniqueness is a corollary of the implicit species theorem (Theorem 3.2.1). ∎

Example 3. For $R = E$, the species of sets, we obtain the species $\mathcal{A}$ of ordinary rooted trees, which satisfies the equation

$$\mathcal{A} = X \cdot E(\mathcal{A}). \tag{13}$$

Its generating series $\mathcal{A}(x) = \sum_{n \geq 0} a_n x^n / n!$ satisfies $\mathcal{A}(x) = x\, e^{\mathcal{A}(x)}$ from which one can obtain, by the method of indeterminate coefficients,

$$\mathcal{A}(x) = x + 2\frac{x^2}{2!} + 9\frac{x^3}{3!} + 64\frac{x^4}{4!} + 625\frac{x^5}{5!} + 7776\frac{x^6}{6!} + \cdots . \tag{14}$$

In fact, an application of Lagrange inversion formula (5) gives the well-known result $a_n = n^{n-1}$, due to Cayley [46].

It also follows from (13) that the cycle index series of $\mathcal{A}$ satisfies

$$Z_{\mathcal{A}}(x_1, x_2, x_3, \ldots) = x_1 \exp\left(Z_{\mathcal{A}} + \frac{x_2}{2} \circ Z_{\mathcal{A}} + \frac{x_3}{3} \circ Z_{\mathcal{A}} + \cdots \right). \tag{15}$$

This equation can be solved by different methods, as will be seen later. One finds

$$Z_{\mathcal{A}}(x_1, x_2, x_3, \ldots) = x_1 + 2\frac{x_1^2}{2!} + 9\frac{x_1^3}{3!} + \frac{x_1 x_2}{2} + 64\frac{x_1^4}{4!} + 4\frac{x_1^2 x_2}{2!\,2}$$

$$+ \frac{x_1 x_3}{3} + \cdots . \tag{16}$$

Example 4. For $R = L$, the species of linear orders, we obtain the species $\mathcal{A}_L$ of *ordered rooted trees*, where the fiber of each vertex is given a linear order, for example from bottom to top or from left to right in a planar drawing (see Figure 3). These structures are sometimes called *planted plane* (or *planar*) rooted

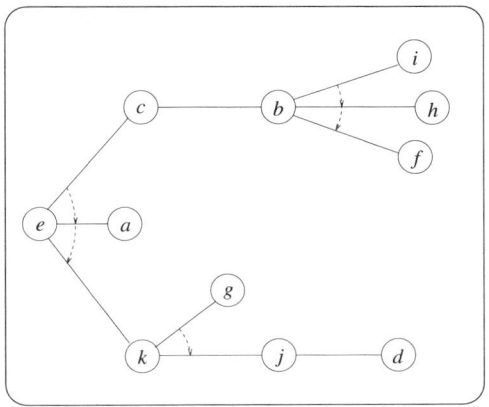

Fig. 3.

trees, the qualifiers "planted" and "rooted" being often implicit. This terminology is explained by the fact that the linear ordering of the fiber of each vertex can be represented by a planar drawing. One has

$$\mathcal{A}_L = X \cdot L(\mathcal{A}_L). \tag{17}$$

In this book, we call *plane rooted trees*, rooted trees that are embedded in an oriented plane (or more precisely the homeomorphism classes of such embeddings). Since the neighbors of the root of a rooted plane tree can rotate freely around the root as long as they preserve their respective positions, the species $\mathcal{P}$ of plane rooted trees is characterized by

$$\mathcal{P} = X + X \cdot C(\mathcal{A}_L), \tag{18}$$

where C denotes the species of circular (nonempty) permutations (see Proposition 15). The term X on the right-hand side corresponds to the rooted tree reduced to a single vertex.

From (17), we obtain

$$\mathcal{A}_L(x) = \frac{x}{1 - \mathcal{A}_L(x)} \tag{19}$$

and, solving the corresponding quadratic equation,

$$\mathcal{A}_L(x) = \frac{1 - \sqrt{1 - 4x}}{2}. \tag{20}$$

From (18), we deduce

$$\begin{aligned}\mathcal{P}(x) &= x + x \log \frac{1}{1 - \mathcal{A}_L(x)} \\ &= x + x \log \frac{1 - \sqrt{1 - 4x}}{2x}.\end{aligned} \tag{21}$$

Lagrange inversion formulas (5) and (7), with $F = 1 + C$, yield

$$\begin{aligned}|\mathcal{A}_L[n]| &= \frac{(2n-2)!}{(n-1)!}, \quad n \geq 1 \\ &= n! c_{n-1},\end{aligned} \tag{22}$$

where $c_n = \frac{1}{(n+1)}\binom{2n}{n}$ denotes the nth Catalan number, and

$$|\mathcal{P}[n]| = \frac{n(2n-3)!}{(n-1)!}, \quad n > 1. \tag{23}$$

3.1. Lagrange Inversion

In the general case, the computation of the various series associated to the species of R-enriched rooted trees involves functional equations of the form

$$\mathcal{A}_R(x) = x R(\mathcal{A}_R(x)) \tag{24}$$

$$Z_{\mathcal{A}_R} = x_1 Z_R(Z_{\mathcal{A}_R}). \tag{25}$$

From these equations, the coefficients of the series can be computed recursively. For example, if

$$R(x) = \sum_{n=0}^{\infty} r_n \frac{x^n}{n!} \quad \text{and} \quad \mathcal{A}_R(x) = \sum_{n=1}^{\infty} a_n \frac{x^n}{n!}, \tag{26}$$

we have $a_1 = r_0$ and

$$a_{n+1} = \sum_{\substack{k_1+k_2+\cdots+k_j=n \\ j \geq 0}} \frac{(n+1)!}{j! k_1! k_2! \ldots k_j!} r_j a_{k_1} a_{k_2} \ldots a_{k_j}. \tag{27}$$

It is possible to write analogous recurrences for the coefficients of the index series of $\mathcal{A}_R$.

We will develop a compact expression for a_n by considering the *powers* R^λ of the species R for positive integer values of λ. The generating series of R^λ is of the form

$$R^\lambda(x) = (R(x))^\lambda = \sum_{n \geq 0} r_n(\lambda) \frac{x^n}{n!},$$

where the coefficients

$$r_n(\lambda) = |R^\lambda[n]| \tag{28}$$

are functions of the parameter λ. The obvious isomorphism $R^{\lambda+\mu} = R^\lambda R^\mu$ shows that the functions $r_n(\lambda)$ satisfy the *binomial* identity

$$r_n(\lambda + \mu) = \sum_{k=0}^{n} \binom{n}{k} r_k(\lambda) r_{n-k}(\mu). \tag{29}$$

Definition 5. The sequence of functions $(r_n(\lambda))_{n \geq 0}$ defined by (28) is called the *binomial type sequence* associated to R, following Rota [286]. We write

$$R \vdash (r_n(\lambda))_{n \geq 0} \tag{30}$$

to express this fact.

Example 6. The following elementary examples illustrate the concept of binomial type sequence. If $R = E$, the species of sets, then $R^\lambda(x) = e^{\lambda x}$ and

$$r_n(\lambda) = \lambda^n. \tag{31}$$

For $R = L$, the species of linear orders, we have $R^\lambda(x) = (1-x)^{-\lambda}$ and we obtain

$$r_n(\lambda) = \lambda(\lambda + 1) \cdots (\lambda + n - 1), \tag{32}$$

the nth *rising factorial*, denoted by $\lambda^{\langle n \rangle}$ (or also $\lambda^{\overline{n}}$ or λ_n in the literature). Likewise, if $R = 1 + X$, the species of sets of cardinality ≤ 1, we get $R^\lambda(x) = (1+x)^\lambda$ and

$$r_n(\lambda) = \lambda(\lambda - 1) \cdots (\lambda - n + 1), \tag{33}$$

the nth *falling factorial*, denoted by $\lambda_{\langle n \rangle}$.

For an arbitrary species R and a positive integer λ, it is easy to give a combinatorial interpretation of $r_n(\lambda)$. Figure 4 illustrates a typical R^λ-structure on a set U of n elements ($\lambda = 24$, $n = 21$). We have a list of λ R-structures (some of which are empty), numbered 1 through λ, that can be identified in a natural manner to a function $f : U \longrightarrow \{1, 2, \ldots, \lambda\}$ of which each fiber $f^{-1}(k)$, $1 \leq k \leq \lambda$, is given an R-structure. The function f is said to be an *R-enriched function*. It follows that $r_n(\lambda)$ is the number of R-enriched functions whose domain and codomain contain respectively n and λ elements. The reader should reexamine formulas (31) to (33) in this combinatorial light. The principal interest of the functions $r_n(\lambda)$ is that they permit the enumeration of R-enriched rooted trees in a particularly compact fashion.

Proposition 7. *Let R be a species of structures and $(r_n(\lambda))_{n \geq 0}$ the associated binomial type sequence: $R \vdash (r_n(\lambda))_{n \geq 0}$. Then the number a_n of R-enriched rooted trees on a set of n elements is given by*

$$a_n = r_{n-1}(n). \tag{34}$$

Moreover, for any $k \geq 0$, the number $a_n^{\{k\}}$ of forests of k R-enriched rooted trees on a set of n elements is given by

$$a_n^{\{k\}} = \frac{k}{n} \binom{n}{k} r_{n-k}(n). \tag{35}$$

These identities are equivalent to Lagrange inversion formulas (5) and (7). To obtain combinatorial proofs, we introduce the concepts of R-enriched endofunctions and R-enriched partial endofunctions.

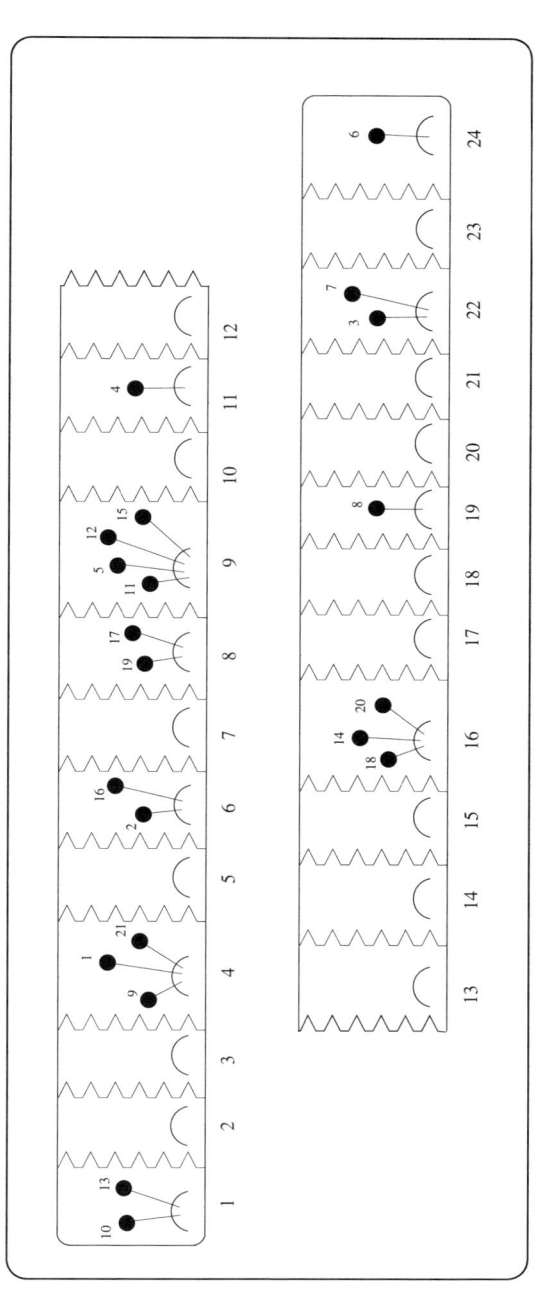

Fig. 4. An R^λ-structure on the set [21], $\lambda = 24$.

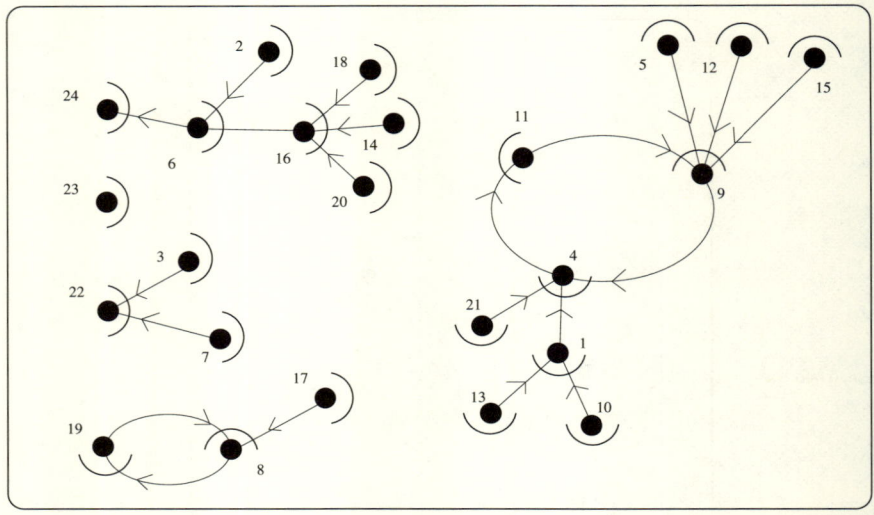

Fig. 5.

Definition 8. An *R-enriched partial endofunction* on a set U consists of a subset V of U and of a function $f : V \longrightarrow U$, of which each fiber $f^{-1}(u), u \in U$, is given an R-structure. When $V = U$, the function $f : U \longrightarrow U$ is called an *R-enriched endofunction*. Denote by End_R and $\text{End}_R^{\wp}$ the species of R-enriched endofunctions and partial endofunctions.

Figure 5 represents an R-enriched partial endofunction on the set $U = [24]$, with $V = [21]$. A close examination of this figure suggests the species isomorphism (see Lemma 11)

$$\text{End}_R^{\wp} = E(\mathcal{A}_R) \cdot \text{End}_R. \tag{36}$$

Example 9. For $R = E$, one obtains the species End and $\text{End}^{\wp}$ of (usual) endofunctions and partial endofunctions.

When $R = 1 + X$, the fibers are either empty or reduced to a single point. Hence we have $\text{End}_{1+X} = \mathcal{S}$, $\mathcal{A}_{1+X} = L_+$, and $\text{End}_{1+X}^{\wp} = E(L_+) \cdot \mathcal{S}$. The partial $(1 + X)$-enriched endofunctions have been introduced for the combinatorial study of Laguerre orthogonal polynomials (see Exercise 2.3.9).

Another interesting example is that of the Jacobi endofunctions defined as $(1 + X)^2$-enriched endofunctions. These structures serve as combinatorial models for Jacobi orthogonal polynomials (see Example 3.2.12 and Exercise 3.2.11).

Lemma 10. *Let R be a species of structures, $R \vdash (r_n(\lambda))_{n \geq 0}$, and let U be a finite set, $V \subseteq U$ a fixed subset, with $k = |V|$ and $n = |U|$. Then the number of R-enriched partial endofunctions with domain V is equal to $r_k(n) = |R^n[k]|$.*

Proof. Let $f : V \longrightarrow U$ be an R-enriched partial endofunction. Without loss of generality, we may assume that

$$V = \{1, 2, \ldots, k\} \subseteq \{1, 2, \ldots, n\} = U.$$

Such a function is determined by a family $\{\xi_u\}_{u \in U}$, where ξ_u is the R-structure on the fiber $f^{-1}(u)$ of the element $u \in U$. Using the usual order on the set $U = [n]$, this family may be identified with a list of n R-structures, that is to say, an R^n-structure on V. For example, the R^{24}-structure of Figure 4 on the set [21] corresponds exactly, by this transformation, to the R-enriched partial endofunction on [24] of Figure 5. The conclusion follows from the definition of $r_n(\lambda)$ (see (28)). ∎

Lemma 11. LEMMA OF R-ENRICHED ENDOFUNCTIONS. *Let R be a species of structures. Then we have the species isomorphisms*

$$\operatorname{End}_R^\wp = E(\mathcal{A}_R) \cdot \operatorname{End}_R, \tag{37}$$

$$\operatorname{End}_R = S(X \cdot R'(\mathcal{A}_R)) \tag{38}$$

and equipotence

$$\operatorname{End}_R \equiv L(X \cdot R'(\mathcal{A}_R)). \tag{39}$$

Proof. An examination of Figure 5 shows that an R-enriched partial endofunction $f : V \longrightarrow U$, $V \subseteq U$, naturally decomposes into two parts. The first consists of an assembly of R-enriched rooted trees whose roots are the elements of $U \setminus V$. The second part, complement of the first, is evidently an R-enriched endofunction. This establishes the isomorphism (37).

For (38), we introduce the auxiliary species

$$Q = X \cdot R'(\mathcal{A}_R)$$

represented schematically by Figure 6. An analysis similar to that of Example 2.1.3 connecting endofunctions to vertebrates reveals that an R-enriched endofunction may be canonically identified with a permutation of Q-structures (see Figure 7).

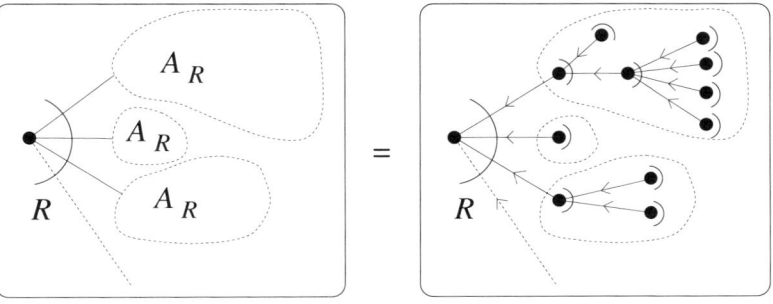

Fig. 6. The species $Q = X R'(\mathcal{A}_R)$.

174   3. Combinatorial Functional Equations

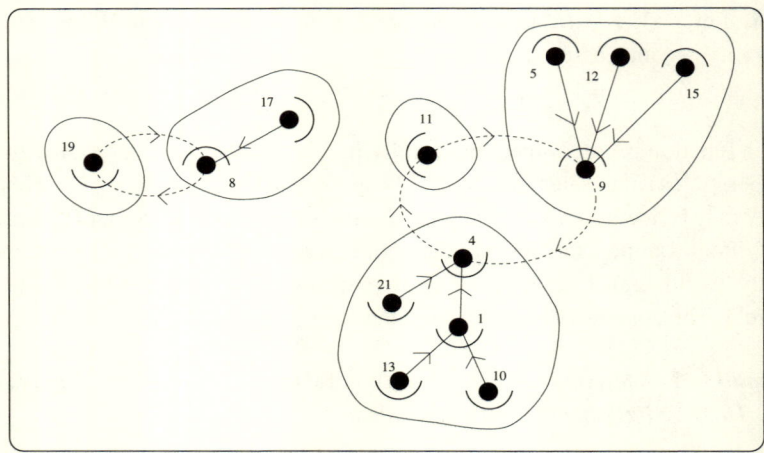

Fig. 7.

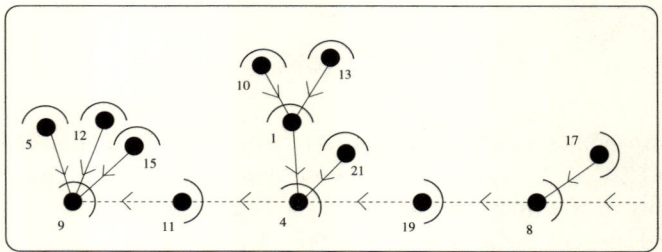

Fig. 8.

The equipotence in (39) follows immediately from (38) and the equipotence $\mathcal{S} \equiv \mathcal{L}$ (see Figure 8 where the "right to left minimum" fundamental transformation from $\mathcal{L}$ to $\mathcal{S}$ has been used). ∎

Theorem 12. LAGRANGE INVERSION, COMBINATORIAL VERSION. *Let R and F be two species of structures. Then, for any $n \geq 0$, there are bijections*

a) $\mathcal{A}_R^\bullet[n] \xrightarrow{\sim} (X \cdot R^n)[n]$,

b) $F(\mathcal{A}_R)^\bullet[n] \xrightarrow{\sim} (F^\bullet \cdot R^n)[n]$,   (40)

c) $(F(\mathcal{A}_R) \cdot \text{End}_R)[n] \xrightarrow{\sim} (F \cdot R^n)[n]$.

Proof. We are going to prove c), b), and a) in that order.

c) Consider an $(F(\mathcal{A}_R) \cdot dRR)$-structure on the set $U = [n]$ (see Figure 9 where $U = [24]$), and denote by W the set of roots of the R-enriched rooted trees

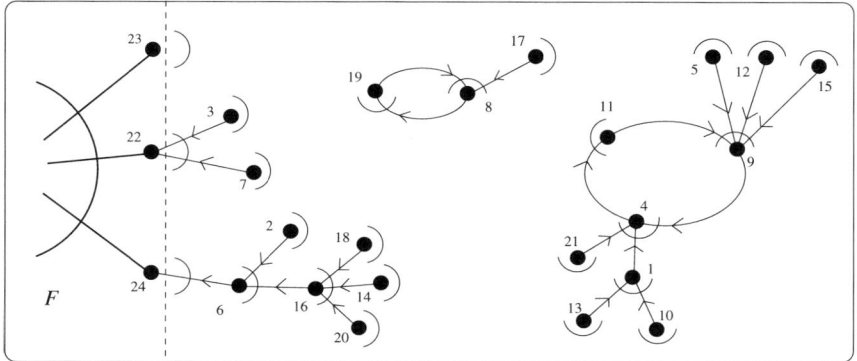

Fig. 9.

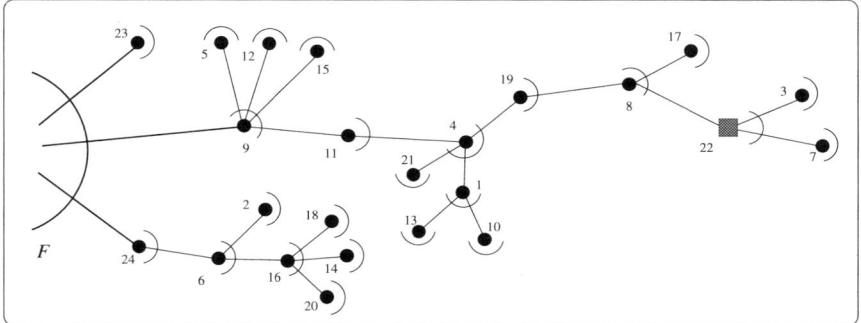

Fig. 10.

of the F-assembly. This structure can be decomposed in such a way as to make apparent an F-structure on W together with an R-enriched function, with domain $U \setminus W$ and codomain U. Since $|U| = n$, this last structure may be identified with an R^n-structure on $V = U \setminus W$ (see Lemma 10). Overall, this is an $(F \cdot R^n)$-structure on U.

b) Consider an $F(\mathcal{A}_R)^{\bullet}$-structure on U, that is to say, a pointed $F(\mathcal{A}_R)$-structure. We can carry out the following sequence of bijective transformations, illustrated by Figures 10, 11, and 12. First separate from the F-assembly the whole rooted tree α which contains the distinguished point (square). Then separate from α the subtree α' rooted at the square point and reattach it to the F-assembly in replacement of α. The remaining structure $\alpha \setminus \alpha'$, which is an $L(X \cdot R'(\mathcal{A}_R))$-structure, is then transformed bijectively, using the previous lemma, into an R-enriched endofunction. Globally, the resulting structure, represented by Figure 12 can be considered as an $(F^{\bullet} \cdot R^n)$-structure on U, in a manner analogous to the preceding proof.

a) This is a particular case of b), where $F = X$. ∎

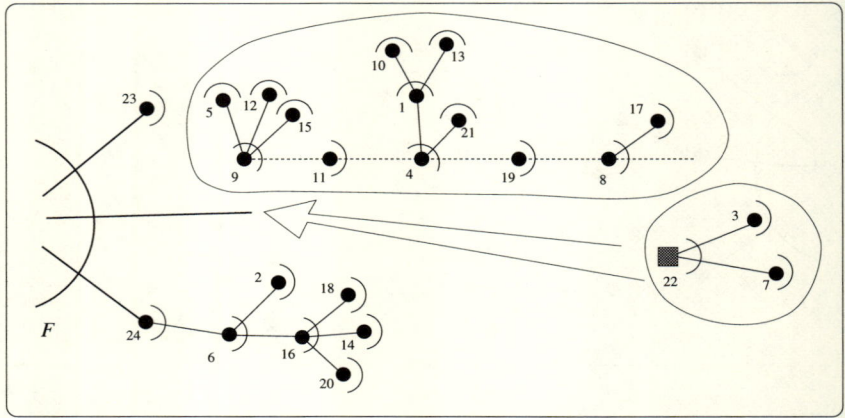

Fig. 11.

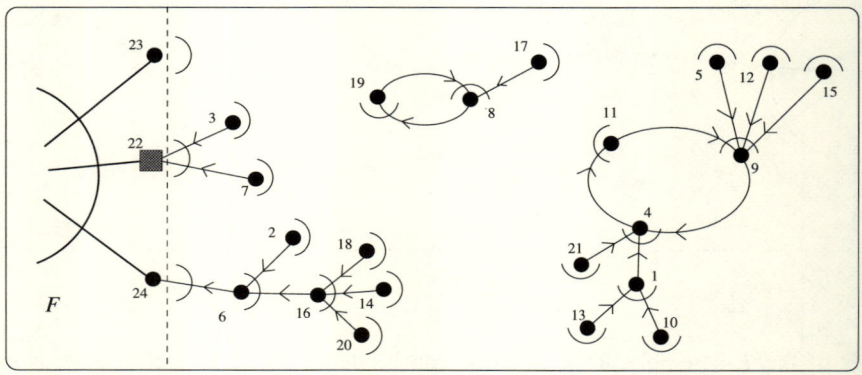

Fig. 12.

Proof of the Classical Lagrange Inversion Formulas

We can now prove Lagrange inversion formulas (5), (7), and (9) stated at the beginning of this section, as well as the compact expressions (34) and (35) of Proposition 7. First observe that in these formulas the expressions given for a_n, b_n, c_n, and $a_n^{[k]}$ are polynomials in $r_0, r_1, \ldots, r_n$ and $f_0, f_1, \ldots, f_n$, where

$$R(x) = \sum_{n \geq 0} r_n \frac{x^n}{n!} \quad \text{and} \quad F(x) = \sum_{n \geq 0} f_n \frac{x^n}{n!}. \tag{41}$$

Hence it suffices to prove these formulas in the particular case where the r_n and the $f_n, n \geq 0$, are arbitrary integers ≥ 0 and to invoke the extension principle of algebraic identities (see Exercise 1).

Choose then species of structures R and F with generating series given by (41). Let $\mathcal{A}_R = X \cdot R(\mathcal{A}_R)$ be the corresponding species of R-enriched rooted trees and take the cardinality of each of the sets appearing in the equipotences (40) of the combinatorial version of Lagrange inversion. This gives three equalities which may be written as follows:

$$na_n = |(X \cdot R^n)[n]| = n|R^n[n-1]|,$$
$$nb_n = |(F^{\bullet} \cdot R^n)[n]| = n|(F' \cdot R^n)[n-1]|, \quad (42)$$
$$c_n = |(F \cdot R^n)[n]|,$$

since $|\mathcal{A}_R^{\bullet}[n]| = n|\mathcal{A}_R^{\bullet}[n]| = n a_n$, $|F(\mathcal{A}_R)^{\bullet}[n]| = n|F(\mathcal{A}_R)[n]| = nb_n$, and, in virtue of Lemma 11 for R-enriched endofunctions,

$$(F(\mathcal{A}_R) \cdot \mathrm{End}_R)(x) = \frac{F(\mathcal{A}_R(x))}{1 - x R'(\mathcal{A}_R(x))} = \sum_{n \geq 0} c_n \frac{x^n}{n!}. \quad (43)$$

The classical Lagrange inversion formulas follow from equalities since, for any species of structures G, and any integer $k \geq 0$, one has

$$|G[k]| = k! \, [x^k] G(x) = \left(\frac{d}{dt}\right) G(t) \bigg|_{t=0}. \quad (44)$$

In order to obtain even more explicit expressions for a_n, b_n, and c_n, let $(r_k(\lambda))_{k \geq 0}$ be the binomial type sequence associated to the species R, that is to say,

$$R^n(x) = \sum_{k \geq 0} r_k(n) \frac{x^k}{k!}. \quad (45)$$

Formulas (42) give

$$a_n = r_{n-1}(n), \quad n > 0, \quad (46)$$

$$b_n = \sum_{k=1}^{n-1} \frac{k}{n} \binom{n}{k} f_k r_{n-k}(n), \quad n > 0, \quad (47)$$

$$c_n = \sum_{k=0}^{n} \binom{n}{k} f_k r_{n-k}(n), \quad n \geq 0. \quad (48)$$

Taking as a particular case $F = E_k$, the species of sets of cardinality k, formula (47) gives the number $a_n^{\{k\}}$ of forests consisting of k R-enriched rooted trees on a set of n elements:

$$a_n^{\{k\}} = \frac{k}{n} \binom{n}{k} r_{n-k}(n), \quad n > 0. \quad (49)$$

This completes the proof of Proposition 7.

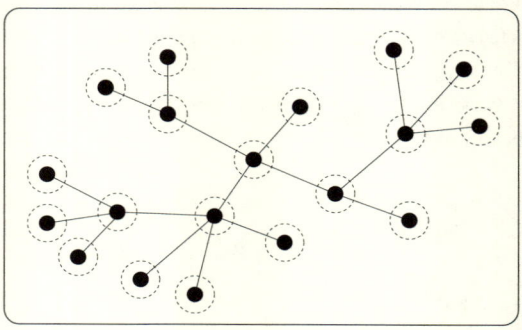

Fig. 13. An R-enriched tree.

R-Enriched Trees

Definition 13. An *R-enriched* tree on a finite set U is
i) an arbitrary tree a on U together with
ii) an R-structure on the set of neighbors of each of the vertices of the tree a. Here a vertex $v \in U$ is called a *neighbor* of a vertex $u \in U$ if it is joined to u by an edge of the tree a.

Denote by $\mathfrak{a}_R$ the species of R-enriched trees. Figure 13 presents a typical R-enriched tree. The dotted circle around each vertex represents the R-structure on the neighbors of that vertex.

Example 14. For $R = E$, the species of sets, we obtain the species $\mathfrak{a}$ of (ordinary) trees. For $R = 1 + C$ (the species of cycles, including the "empty cycle"), the species $\mathfrak{a}_R = \mathfrak{a}_{1+C}$ of $(1+C)$-enriched trees is identified with the species $\mathfrak{p}$ of *plane trees*. To see this, observe that the neighbors of each vertex u of a tree, embedded in an (oriented) plane, are arranged cyclically around u (i.e., the neighbors of u are given a natural C-structure). When the tree is reduced to a single vertex, the empty set of neighbors of this vertex is given its unique 1-structure. This explains the choice $R = 1 + C$ for the enriching species. Observe that the species $\mathcal{P}$ of plane rooted trees introduced in Example 4 satisfies $\mathcal{P} = \mathfrak{p}^\bullet$.

We have seen in Chapters 1 and 2 that the species $\mathfrak{a}$ of trees is related to the species $\mathcal{A}$ of rooted trees by the combinatorial equations

$$\mathfrak{a}' = E(\mathcal{A}), \qquad \mathfrak{a}^\bullet = \mathcal{A}. \tag{50}$$

In other words, the derivative of the species of trees is isomorphic to the species of forests of rooted trees and pointed trees are identified to rooted trees. These equations are false in the more general context of R-enriched context; that is to say that, in general,

$$\mathfrak{a}'_R \neq E(\mathcal{A}_R), \qquad \mathfrak{a}^{\bullet}_R \neq \mathcal{A}_R. \qquad (51)$$

However the following proposition holds, extending formula 18 for rooted plane trees.

Proposition 15. *Let R be a species of structures, and $\mathfrak{a}_R$ be the species of R-enriched trees. One has the combinatorial equations*

$$\mathfrak{a}'_R = R(\mathcal{A}_{R'}), \qquad \mathfrak{a}^{\bullet}_R = X R(\mathcal{A}_{R'}), \qquad (52)$$

where $\mathcal{A}_{R'} = X R'(\mathcal{A}_{R'})$ is the species of R'-enriched rooted trees.

Proof. Let U be a finite set. Consider an $\mathfrak{a}'_R$-structure on U, that is to say, an R-enriched tree on $U + \{*\}$ (see Figure 14(a)). Forgetting the vertex "$*$", we obtain, in a natural and bijective manner, an R-assembly of $\mathcal{A}_{R'}$-structures on U. See Figure 14(b), where the dotted circle denotes the R-structures on the neighbors of "$*$" and each circular arc denotes an R'-structure associated to the corresponding R-structure in Figure 14(a). This establishes the equation $\mathfrak{a}'_R = R(\mathcal{A}_{R'})$. Multiplying by X gives $\mathfrak{a}^{\bullet}_R = X \cdot R(\mathcal{A}_{R'})$. ∎

The enumeration of R-enriched trees can be carried out in an elegant and compact manner with the help of binomial type sequences.

Proposition 16. *Let R be a species of structures and denote by $(\rho_n(\lambda))_{n\geq 0}$ the binomial type sequence associated to the derived species R':*

$$R' \vdash (\rho_n(\lambda))_{n\geq 0}. \qquad (53)$$

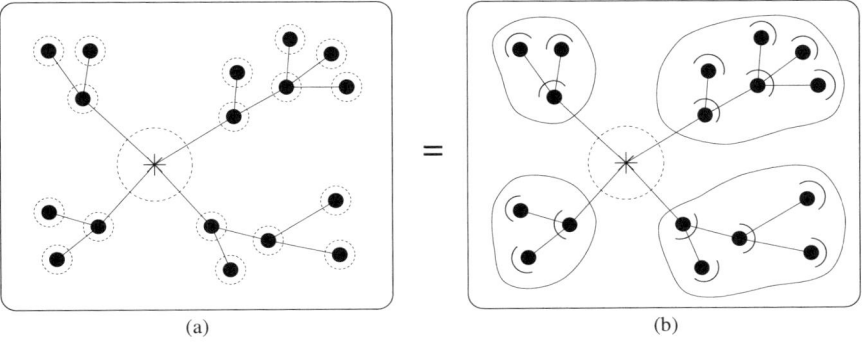

(a) = (b)

Fig. 14.

Then the number α_n of R-enriched trees on a set of n elements is given by

$$\alpha_n = \begin{cases} 0 & \text{if } n = 0, \\ r_0 & \text{if } n = 1, \\ \rho_{n-2}(n) & \text{if } n > 1, \end{cases} \tag{54}$$

where $r_0 = R(0)$.

Proof. Differentiate the first equation of (52). This gives

$$\mathfrak{a}_R'' = (\mathfrak{a}_R')' = (R(\mathcal{A}_{R'}))' = R'(\mathcal{A}_{R'}) \cdot \mathcal{A}_{R'}' \tag{55}$$

by the chain rule. Multiplying by X^2 yields

$$X^2 \cdot \mathfrak{a}_R'' = X \cdot R'(\mathcal{A}_{R'}) \cdot X \cdot \mathcal{A}_{R'}' = \mathcal{A}_{R'} \cdot \mathcal{A}_{R'}^{\bullet} \tag{56}$$

since $XR'(\mathcal{A}_{R'}) = \mathcal{A}_{R'}$. Consider now the species E_2 of sets of cardinality 2. The equality

$$(E_2(F))^{\bullet} = F \cdot F^{\bullet} \tag{57}$$

holds for any species F. From (56) and (57), we have

$$X^2 \mathfrak{a}_R'' = (E_2(\mathcal{A}_{R'}))^{\bullet}. \tag{58}$$

But the species $E_2(\mathcal{A}_{R'})$ is clearly that of forests of two R'-enriched rooted trees. It follows from Proposition 7 (formula (35)) that there are $\frac{2}{n}\binom{n}{2}\rho_{n-2}(n)$ such forests on a set of cardinality $n \geq 2$. Isomorphism (58) gives, for structures on $n \geq 2$ elements,

$$\begin{aligned} n(n-1)\alpha_n &= |(X^2 \cdot \mathfrak{a}_R'')[n]| \\ &= |(E_2(\mathcal{A}_{R'}))^{\bullet}[n]| \\ &= 2\binom{n}{2}\rho_{n-2}(n). \end{aligned} \tag{59}$$

We conclude that $\alpha_n = \rho_{n-2}(n)$, for $n \geq 2$. In the case $n = 1$, it is directly verified that $\alpha_1 = r_0$. ∎

Example 17. For $R = E$, the species of sets, we have $R' = E$, $\rho_n(\lambda) = \lambda^n$. Proposition 16 yields the classical formula of Cayley $\alpha_n = n^{n-2}$ for the number of trees on n elements. For $R = 1 + \mathcal{C}$, we have $R' = L$ and $\rho_n(\lambda) = \lambda(\lambda+1)\cdots(\lambda+n-1)$. Thus the number of plane trees on a set of n elements ($n \geq 2$) is given by

$$\alpha_n = \rho_{n-2}(n) = n^{\langle n-2 \rangle} = n(n+1)\cdots(2n-3). \tag{60}$$

Leaves and Binomial Type Sequences

We now turn to the concept of leaf. By definition, a *leaf* of an (enriched) tree or rooted tree is a vertex of degree one in the underlying graph (excluding the root in the case of a rooted tree). For technical reasons, when the tree or the rooted tree is reduced to a single vertex, this vertex is also considered to be a leaf.

Binomial type sequences are closely related to enumeration problems and to the computation of expectation and higher order moments involving leaves in random R-enriched trees and rooted trees. The proof of the following two propositions is left to the reader (see Exercises 15 and 16).

Proposition 18. *Let R be a species of structures and R_+ the species of nonempty R-structures and set*

$$R \vdash (r_n(\lambda))_{n \geq 0} \quad \text{and} \quad R_+ \vdash \left(r_n^+(\lambda)\right)_{n \geq 0}. \tag{61}$$

Then

a) The number $a_{n,k}$ of R-enriched rooted trees on a set of n elements, having k leaves, is given by

$$a_{n,k} = r_0^k \binom{n}{k} r_{n-1}^+(n-k). \tag{62}$$

b) The expectation m_n and the vth order factorial moment $m_n^{(v)}$ for the number of leaves of a random R-enriched rooted tree on a set of n elements are given by

$$m_n = r_0 \cdot n \frac{r_{n-1}(n-1)}{r_{n-1}(n)} \tag{63}$$

and, for $v > 1$,

$$m_n^{(v)} = r_0^v \cdot n(n-1) \cdots (n-v+1) \cdot \frac{r_{n-1}(n-v)}{r_{n-1}(n)}. \tag{64}$$

Proposition 19. *Let R be a species of structures and $(R')_+$ the species of nonempty R'-structures and set*

$$R' \vdash (\rho_n(\lambda))_{n \geq 0}, \qquad (R')_+ \vdash \left(\rho_n^+(\lambda)\right)_{n \geq 0}. \tag{65}$$

Then

a) The number $\alpha_{n,k}$ of R-enriched trees on a set of n elements, having exactly k leaves, is given by

$$\alpha_{n,k} = r_1^k \binom{n}{k} \rho_{n-2}^+(n-k), \qquad n > 1. \tag{66}$$

One also has $\alpha_{1,1} = r_0$ and $\alpha_{1,0} = 0$.

b) *The expectation μ_n and the vth order factorial moment $\mu_n^{(v)}$ for the number of leaves of a random R-enriched tree on a set of n elements are given by $\mu_1 = 1$, $\mu_1^{(v)} = 0$, for $v > 1$, and, for $n > 1$,*

$$\mu_n = r_1 \cdot n \frac{\rho_{n-2}(n-1)}{\rho_{n-2}(n)}, \tag{67}$$

and, more generally, for $v \geq 1$,

$$\mu_n^{(v)} = r_1^v \cdot n(n-1) \cdots (n-v+1) \frac{\rho_{n-2}(n-v)}{\rho_{n-2}(n)}. \tag{68}$$

Example 20. In the case of ordinary trees and rooted trees ($R = E$) we find the well-known expectations (see [251]):

a) $\quad m_n = n \cdot (1 - n^{-1})^{n-1} \sim \dfrac{n}{e}, \qquad n \to \infty, \tag{69}$

b) $\quad \mu_n = n \cdot (1 - n^{-1})^{n-2} \sim \dfrac{n}{e}, \qquad n \to \infty. \tag{70}$

For plane trees ($R = 1 + C$), we obtain the expectation and variance for the number of leaves in the form

a) $\quad \mu_n = \dfrac{n(n-1)}{2n-3} \sim \dfrac{n}{2}, \qquad 1 < n \to \infty, \tag{71}$

b) $\quad \sigma_n^2 = \mu_n^{(2)} + \mu_n - \mu_n^2$

$\qquad = \dfrac{n(n-1)(n-3)}{2(2n-3)^2} \sim \dfrac{n}{8}, \qquad 3 \leq n \to \infty. \tag{72}$

The reader will find in the exercises other applications of binomial type sequences, including R-enriched versions of Abel identities (see Exercise 22).

Exercises for Section 3.1

1. THE EXTENSION PRINCIPLE OF ALGEBRAIC IDENTITIES. Let $\mathbb{D}$ be an integral domain. Given a polynomial $P(x_1, x_2, \ldots, x_m) \in \mathbb{D}[x_1, x_2, \ldots, x_m]$, denote by $\deg_i P$ the degree of P relative to the variable x_i, $i = 1, \ldots, m$.
   a) Show that two polynomials $P(x_1, \ldots, x_m)$ and $Q(x_1, \ldots, x_m) \in \mathbb{D}[x_1, \ldots, x_m]$ are identical if they take the same value on each point $s = (s_1, s_2, \ldots, s_m)$ of a set $S \subseteq \mathbb{D}^m$ having the following property: For $i = 1, \ldots, m$, there exists $S_i \subseteq \mathbb{D}$ such that
   i) $S \supseteq \prod_{i=1}^m S_i$.
   ii) $|S_i| > \max(\deg_i P, \deg_i Q), \qquad i = 1, \ldots, m$.
   HINT: First prove the result in the case where $m = 1$ and then proceed by induction.

b) Generalize the preceding result using the concept of a *determining set* $S \subseteq \mathbb{D}^m$ for a polynomial $P(x_1, \ldots, x_m)$, defined inductively on m in the following manner:
- ($m = 1$) $S \subseteq \mathbb{D}$ is *determining for the polynomial* $P(x) \in \mathbb{D}[x]$ if $|S| > \deg P$.
- ($m > 1$) $S \subseteq \mathbb{D}^m$ is *determining for the polynomial* $P(x_1, \ldots, x_m)$ if there exist j, $1 \leq j \leq m$, and $T \subseteq \mathbb{D}$ such that:
  i) $|T| > \deg_j P$.
  ii) For any $t \in T$, the set $S_t \subseteq \mathbb{D}^{m-1}$ defined by

$$S_t = \{(s_1, \ldots, s_{j-1}, s_{j+1}, \ldots, s_m) \mid (s_1, \ldots, s_{j-1}, t,$$
$$s_{j+1}, \ldots, s_m) \in S\}$$

is determining for the polynomial

$$P_t = P(x_1, \ldots, x_{j-1}, t, x_{j+1}, \ldots, x_m)$$
$$\in \mathbb{D}[x_1, \ldots, x_{j-1}, x_{j+1}, \ldots, x_m].$$

2. Let $R(x)$ and $F(x) \in \mathbb{K}[\![x]\!]$, and let $A(x) \in \mathbb{K}[\![x]\!]$ be defined by $A(x) = xR(A(x))$.
a) Show that Lagrange inversion formulas (5), (7), and (9) are respectively equivalent to the formulas (10), a), b), and c).
b) Show that Lagrange inversion formulas (10), b) and c) are equivalent.
HINT: Calculate $F(A(x))'$ and use the fact that for $n \geq 1$, $[x^n]h(x) = \frac{1}{n}[x^{n-1}]h'(x)$, for any $h(x) \in \mathbb{K}[\![x]\!]$, as well as Exercise 1.2.11, a).
c) GOOD'S ONE-DIMENSIONAL FORMULA [132]. Suppose that $R(0) \neq 0$. Show that formula (10), c) is equivalent to the following formula:

$$[x^n]F(A(x)) = [t^n]F(t)\left(1 - \frac{tR'(t)}{R(t)}\right)R^n(t). \tag{73}$$

3. CENTRAL TRINOMIAL COEFFICIENTS. For $n \geq 1$, denote by c_n the constant term in the expansion of $(x^{-1} + 1 + x)^n$ and let

$$g(x) = \sum_{n \geq 0} c_n x^n = 1 + x + 3x^2 + 7x^3 + 19x^4 + \cdots. \tag{74}$$

Show that

$$g(x) = \frac{1}{\sqrt{1 - 2x - 3x^2}}. \tag{75}$$

HINT: Note that $c_n = [t^n](1 + t + t^2)^n$ and use (10), c).

4. Show that for any m,

a) $\displaystyle\sum_{k\geq 0}\binom{m+2k}{k}x^k = \frac{1}{\sqrt{1-4x}}\left(\frac{1-\sqrt{1-4x}}{2x}\right)^m$,

b) $\displaystyle\sum_{k\geq 0}\binom{m}{k}\binom{3m-2k}{n-k}(-3)^k = \binom{m}{n/3}$, $n \geq 0$. (76)

HINT: Note that if $y = x(1+y)^2$, then $1 - 3x = \frac{1+y^3}{(1+y)^3}$ and apply (10), c).

5. The classical orthogonal polynomials of Hermite, Laguerre, and Jacobi can be defined by Rodrigues' formulas:

a) $H_n(x) = (-1)^n e^{x^2/2}\dfrac{d^n}{dx^n}\left(e^{-x^2/2}\right)$,

b) $L_n^{(\alpha)}(x) = x^{-\alpha}e^x\dfrac{d^n}{dx^n}(x^{n+\alpha}e^{-x})$, (77)

c) $P_n^{(\alpha,\beta)}(x) = (-2)^{-n}(1-x)^{-\alpha}(1+x)^{-\beta}\dfrac{d^n}{dx^n}((1-x)^{n+\alpha}(1+x)^{n+\beta})$.

Use Lagrange inversion to deduce the generating series

a) $\displaystyle\sum_{n\geq 0} H_n(x)\frac{t^n}{n!} = \exp(xt - t^2/2)$,

b) $\displaystyle\sum_{n\geq 0} L_n^{(\alpha)}(x)\frac{t^n}{n!} = \left(\frac{1}{1-t}\right)^{\alpha+1}\exp\left(\frac{-xt}{1-t}\right)$, (78)

c) $\displaystyle\sum_{n\geq 0} P_n^{(\alpha,\beta)}(x)\frac{t^n}{n!} = 2^{\alpha+\beta}(1-t+R)^{-\alpha}(1+t+R)^{-\beta}R^{-1}$,

where $R = \sqrt{1 - 2xt + t^2}$.

HINT: Note that $(d^n/dx^n)f(x) = (d^n/d\theta^n)f(x+\theta)|_{\theta=0}$.

6. Apply Good's formula (see Exercise 2, c)) to the enumeration of $F(\mathcal{A}_R)$-structures on a set of n elements in the following cases:
   a) $R = E^2$ (oriented rooted trees), $F = 1/(1-2X) = L(2X)$.
   b) $R = E^k$, $F = S(kX)$.
   c) $R = L$, $F = L(L_+)$.
   d) $R = F$.

7. (See Strehl [311].) Let $R(x)$ and $F(x)$ be formal power series and $A(x)$ a formal power series satisfying $A(x) = xR(A(x))$.
   a) Show that the following inversion formula, due to Hurwitz [153], is a generalization of both Lagrange inversion formulas (10), b) and c):

   $$[x^n]\frac{F(A(x))}{(1-xR'(A(x)))^{1+\alpha}} = \sum_{r+s=n}\frac{\alpha^{\langle r\rangle}}{r!}[t^s](F(t)R''(t)R^s(t)). \quad (79)$$

   b) By introducing the counter $1+\alpha$ for the cycles of R-enriched endofunctions, prove this formula.

HINT: Introduce the auxiliary species $\text{End}_{R,(\alpha)}$ of R-enriched endofunctions provided with a cycle counter α, use formulas (2.3.73), i) and (2.3.75), and show that the total weight of $F(A_R) \cdot \text{End}_R \cdot \text{End}_{R,(\alpha)}$-structures on $[n]$ is given by

$$\sum_{r+s=n} \binom{n}{r} \alpha^{\langle r \rangle} |(F \cdot R'^r \cdot R^s)[s]|, \tag{80}$$

where r represents the number of recurrent elements in the $\text{End}_{R,(\alpha)}$-structure.

c) Deduce that Hurwitz's formula (79) is equivalent to the following generalization of Good's formula (73):

$$[x^n] \frac{F(A(x))}{(1 - xR'(A(x)))^{1+\alpha}} = [t^n]F(t)\left(1 - \frac{tR'(t)}{R(t)}\right)^{-\alpha} R^n(t). \tag{81}$$

8. Establish combinatorially the explicit formulas (31)–(33) for the polynomials $r_n(\lambda)$, using the notion of R-enriched function, in the cases $R = E$, $R = L$, $R = 1 + X$.

9. a) Let E_{even} be the species of sets with even cardinality. Show that the number of rooted E_{even}-enriched trees on a set with n elements is given by

$$\frac{1}{2^n} \sum_{k=0}^{n} \binom{n}{k}(2k - n)^{n-1}. \tag{82}$$

HINT: Expand $\left(\frac{e^x + e^{-x}}{2}\right)^\lambda$ with the help of the binomial formula and use Proposition 7.

b) Let c be an integer ≥ 1 and $R = c + E_+$. An R-enriched rooted tree can be viewed as a rooted tree where each leaf is colored by one of c colors given in advance. Show that

$$|\mathcal{A}_R[n]| = \sum_{k=0}^{n} \binom{n}{k}(c - 1)^k(n - k)^{n-1}. \tag{83}$$

HINT: Apply Newton's binomial formula to $R^\lambda(x) = ((c - 1) + e^x)^\lambda$.

c) For $R = \text{Inv} = E(X + E_2)$, the species of involutions, show that the number of R-enriched rooted trees on a set with n elements is given by

$$|\mathcal{A}_{\text{Inv}}[n]| = i^{n-1} n^{\frac{n-1}{2}} H_{n-1}(-i\sqrt{n}), \tag{84}$$

where $H_n(t)$ denotes the Hermite polynomial of degree n, in the variable t (see Exercise 2.3.8).

10. Let $\mathcal{A}_L = X \cdot L(\mathcal{A}_L)$ be the species of ordered rooted trees and $\mathcal{P} = X + X \cdot C(\mathcal{A}_L)$ that of rooted plane trees.

a) Establish formulas (22) and (23) for $|\mathcal{A}_L[n]|$ and $|\mathcal{P}[n]|$ by using Lagrange inversion formulas (5) and (7) with $F = 1 + C$, $R = L$.

b) Establish the same formulas (22) and (23) by using a Taylor expansion based on the analytic forms (20) and (21) for $\mathcal{A}_L(x)$ and $\mathcal{P}(x)$.
HINT: To expand $\mathcal{P}(x)$, first appeal to the fact that

$$\frac{d}{dx} \log\left(\frac{1 - \sqrt{1 - 4x}}{2x}\right) = \frac{(1 - 4x)^{-\frac{1}{2}} - 1}{2x}. \tag{85}$$

c) Show that $\widetilde{\mathcal{A}}_L(x) = \mathcal{A}_L(x)$ and

$$\widetilde{\mathcal{P}}(x) = x + x \sum_{m \geq 1} \frac{\phi(m)}{m} \sum_{k \geq 1} \frac{\mathcal{A}_L^k(x^m)}{k}. \tag{86}$$

Deduce the formula of Walkup [322] for the number of unlabeled plane rooted trees of order n:

$$[x^n]\widetilde{\mathcal{P}}(x) = \frac{1}{2(n-1)} \sum_{d \mid n-1} \phi((n-1)/d) \binom{2d}{d}. \tag{87}$$

11. Let $\mathcal{A}_R = X \cdot R(\mathcal{A}_R)$ be the species of R-enriched rooted trees and consider the cycle index series

$$Z_R(x_1, x_2, \ldots) = \sum_{n \geq 0} \sum_{i_1 + 2i_2 + \cdots = n} r_{i_1, i_2, \ldots} \frac{x_1^{i_1} x_2^{i_2} \cdots}{1^{i_1} i_1! \, 2^{i_2} i_2! \cdots}, \tag{88}$$

$$Z_{\mathcal{A}_R}(x_1, x_2, \ldots) = \sum_{n \geq 0} \sum_{i_1 + 2i_2 + \cdots = n} a_{i_1, i_2, \ldots} \frac{x_1^{i_1} x_2^{i_2} \cdots}{1^{i_1} i_1! \, 2^{i_2} i_2! \cdots}. \tag{89}$$

a) Starting from the equation $Z_{\mathcal{A}_R} = x_1 Z_R(\mathcal{A}_R)$, compute successively the coefficients $a_0, a_1, a_{2,0}, a_{0,1}, a_{3,0,0}, a_{1,1,0}, a_{0,0,1}$ as functions of the coefficients $r_0, r_1, r_{2,0}, r_{0,1}, r_{3,0,0}, r_{1,1,0}, r_{0,0,1}$.
NOTE: See Table A.2.12.

b) Show, by induction or otherwise, that for all $n \geq 0$ and all $i_1, i_2, i_3, \ldots$ such that $i_1 + 2i_2 + 3i_3 + \cdots = n$, the coefficient $a_{i_1, i_2, i_3, \ldots}$ can be expressed polynomially as a function of the $a_{j_1, j_2, j_3, \ldots}$ and of the $r_{k_1, k_2, k_3, \ldots}$ such that $j_1 + 2j_2 + 3j_3 + \cdots < n$ and $k_1 + 2k_2 + 3k_3 + \cdots \leq n$.

12. Let R be a species of structures. Recall that $R_+ = R_1 + R_2 + R_3 + \cdots$ denotes the species of nonempty R-structures. Prove that $(R')_+ \neq (R_+)'$, except in certain cases. Identify these exceptional cases.

13. Establish, by a combinatorial argument based on pictures, Equation (58), namely

$$X^2 a_R'' = (E_2(\mathcal{A}_{R'}))^{\bullet}.$$

3.1. Lagrange Inversion

14. a) Verify that the number of plane trees on a set of n elements is indeed given by (60), and that the expectation and variance for the number of leaves are given by (71) and (72).
    b) Let k be a fixed integer >1. A tree whose internal vertices all have degree k can be considered as an R-enriched tree, where $R = X + E_k$. For $n \geq 2$, determine the number of such trees on a set of n elements, as well as the expectation and the variance of their number of leaves.

15. Prove Proposition 18 concerning the leaves of R-enriched rooted trees.
    HINT: For (62), first develop the equality $(R(x))^\lambda = (r_0 + R_+(x))^\lambda$ by the binomial formula in order to show that

    $$r_n(\lambda) = \sum_{i=0}^{n} \binom{\lambda}{i} r_0^{\lambda - i} \, r_n^+(i). \tag{90}$$

    Observe then that the number $a_{n,k}$ of R-enriched rooted trees on a set of n elements, having k leaves, can be written in the form $a_{n,k} = c_{n,k} \, r_0^k$, where $c_{n,k}$ is independent of r_0 and that

    $$r_{n-1}(n) = \sum_{k=0}^{n} a_{n,k} = \sum_{k=0}^{n} c_{n,k} r_0^k. \tag{91}$$

    For (63), remark that $\sum k a_{n,k} = r_0 \frac{\partial}{\partial r_0} r_{n-1}(n)$ and that $r_0 \frac{\partial}{\partial r_0} R^\lambda(x) = r_0 \lambda R^{\lambda - 1}(x)$. For (64), make use of the differential operators $r_0^\nu (\frac{\partial}{\partial r_0})^\nu$, $\nu > 1$.

16. Prove Proposition 19 concerning the leaves of R-enriched trees.
    HINT: Proceed as in the previous problem, using this time R' and the differential operators $r_1^\nu (\frac{\partial}{\partial r_0})^\nu$.

17. a) Establish the exact and asymptotic formulas (69) and (70) for the expected number of leaves in rooted trees and ordinary trees ($R = E$).
    b) Establish analogous formulas in the case where $R = 1 + X$, $R = L$, $R = E^2$.

18. PERTURBATION OF A BINOMIAL TYPE SEQUENCE.
    a) Let $k \geq 0$ be an integer, t a real number, and $R(x) = \sum_{n \geq 0} r_n x^n / n!$ an arbitrary formal power series. Consider the following perturbation of the series $R(x)$,

    $$R(k, t, x) = R(x) + t \frac{x^k}{k!}, \tag{92}$$

    which consists in replacing the coefficient r_k in $R(x)$ by $r_k + t$, leaving all other coefficients of $R(x)$ unchanged. Show that the binomial type sequence $(r_n(k, t, \lambda))_{n \geq 0}$ associated to $R(k, t, x)$ is given by

    $$r_n(k, t, \lambda) = \sum_{\nu \geq 0} \binom{\lambda}{\nu} \frac{n! \, t^\nu}{k!^\nu (n - k\nu)!} \, r_{n-k\nu}(\lambda - \nu), \quad n \geq 0, \tag{93}$$

    where $(r_n(\lambda))_{n \geq 0}$ is the binomial type sequence associated to $R(x)$.

b) By definition, a *topological (or homeomorphically irreducible) tree* is a tree without any vertices of degree 2. In other words, it is an R-enriched tree, where $R = E - E_2$ is the species of sets of cardinality $\neq 2$. Show that the number of topological trees having n vertices is given by

$$\sum_{\nu=0}^{n-2} \binom{n}{\nu} \frac{(n-2)!}{(n-2-\nu)!} (-1)^\nu (n-\nu)^{n-\nu-2}, \qquad n \geq 2. \tag{94}$$

HINT: Use Proposition 16 and apply a) to $R'(x)$ with $k = 1, t = -1$.

19. Let N be an integer ≥ 0. If $R = E_0 + E_1 + \cdots + E_N = E_{\leq N}$, the species of sets of cardinality $\leq N$, an R-enriched rooted tree can then be interpreted as a genealogical tree (descended from a single individual) such that each individual gives birth to at most N children. Show that in this case, the binomial type sequence $(r_n(\lambda))_{n \geq 0}$ associated to R satisfies the recursion

$$\begin{cases} r_k(\lambda) = \lambda^k & \text{if } k \leq N, \\ r_{n+1}(\lambda) = \lambda \cdot (r_n(\lambda) - \binom{n}{N} r_{n-N}(\lambda - 1)) & \text{if } n \geq N. \end{cases} \tag{95}$$

HINT: First establish the identity

$$\frac{d}{dx} R^\lambda(x) = \lambda R^{\lambda-1}(x) \left(R(x) - \frac{x^N}{N!} \right). \tag{96}$$

20. GENERAL RECURSIVE SCHEMES FOR BINOMIAL TYPE SEQUENCES. Let R be a species of structures and $(r_n(\lambda))_{n \geq 0}$ the binomial type sequence associated to it.
a) Show that if $R(x) = \sum_{n \geq 0} r_n x^n / n!$, then

$$\begin{cases} r_0(\lambda) = r_0^\lambda, \\ r_{n+1}(\lambda) = \lambda \cdot \sum_{i+j=n} \binom{n}{i} r_{i+1} \cdot r_j(\lambda - 1), & \text{if } n \geq 0. \end{cases} \tag{97}$$

HINT: Use the identity $(d/dx) R^\lambda(x) = \lambda R^{\lambda-1}(x) R'(x)$.
b) Show that if $R(x) = \exp(l_1 \frac{x}{1!} + l_2 \frac{x^2}{2!} + l_3 \frac{x^3}{3!} + \cdots)$, then

$$\begin{cases} r_0(\lambda) = 1, \\ r_n(\lambda) = \sum_{i=1}^n \binom{n}{i} l_i \cdot \int_0^\lambda r_{n-i}(\lambda) d\lambda, & n \geq 1. \end{cases} \tag{98}$$

HINT: Use the identity $(d/d\lambda) R^\lambda(x) = R^\lambda(x) \log(R(x))$.
c) Apply the preceding schemes in the cases $R = 1 + X + \cdots + X^N$ and $R = E \circ (X + E_2)$ respectively.

21. Let $(r_n(\lambda))_{n \geq 0}$ be the binomial type sequence associated to $R(x)$. For each fixed $n \geq 0$, $r_n(\lambda)$ is a function of the coefficients $r_0, r_1, \ldots, r_n$ of the series $R(x) = \sum_{i \geq 0} r_i x^i / i!$.

a) Show that for any $k \geq 0$ and any v, $0 \leq v \leq n$,

$$\left(\frac{\partial}{\partial r_v}\right)^k r_n(\lambda) = \frac{\lambda(\lambda-1)\cdots(\lambda-k+1)n!}{(n-vk)!(v!)^k} r_{n-vk}(\lambda-k). \qquad (99)$$

b) Deduce an expression for the number of R-enriched rooted trees on a set of n elements, having exactly k vertices of degree v; likewise for the expected number of vertices of degree v in a random R-enriched rooted tree on n elements.

c) Show that for any sequence $(k_0, k_1, \ldots, k_n)$ of integers ≥ 0, one has

$$\left(\frac{\partial}{\partial r_0}\right)^{k_0}\left(\frac{\partial}{\partial r_1}\right)^{k_1}\cdots\left(\frac{\partial}{\partial r_n}\right)^{k_n} r_n(\lambda) = \frac{n!\,\lambda_{(k)}\,r_{n-m}(\lambda-k)}{0!^{k_0}1!^{k_1}\cdots n!^{k_n}(n-m)!}, \qquad (100)$$

where $k = k_0 + k_1 + \cdots + k_n$, $m = 0\,k_0 + 1\,k_1 + \cdots + nk_n$, and $\lambda_{(k)} = \lambda(\lambda-1)\cdots(\lambda-k+1)$.

22. GENERALIZATION OF ABEL IDENTITIES. The classical identities of Abel are stated as:

$$(u+v+n)^n = \sum_{i+j=n} \binom{n}{i} u \cdot (u+i)^{i-1}(v+j)^j, \qquad (101)$$

$$(u+v)(u+v+n)^{n-1} = \sum_{i+j=n} \binom{n}{i} u \cdot (u+i)^{i-1} \cdot v \cdot (v+j)^{j-1}. \qquad (102)$$

Their R-enriched versions are written as follows: If $R \vdash (r_n(\lambda))_{n \geq 0}$, then

$$r_n(u+v+n) = \sum_{i+j=n} \binom{n}{i} u \frac{r_i(u+i)}{u+i} r_j(v+j), \qquad (103)$$

$$(u+v)\frac{r_n(u+v+n)}{u+v+n} = \sum_{i+j=n} \binom{n}{i} u \frac{r_i(u+i)}{u+i} v \frac{r_j(v+j)}{v+j}. \qquad (104)$$

The classical identities correspond to the case $R = E$. Prove identities (103) and (104) by considering the species isomorphisms

$$\mathcal{A}_R^{[u+v]} = \mathcal{A}_R^u \cdot \mathcal{A}_R^{[v]} \quad \text{and} \quad \mathcal{A}_R^{u+v} = \mathcal{A}_R^u \cdot \mathcal{A}_R^v,$$

where $\mathcal{A}_R^{[k]}$ denotes the species of "hedges" formed from k R-enriched rooted trees where the last rooted tree is pointed.

NOTE: $v\mathcal{A}^{[v]}(x) = (\mathcal{A}^v)^{\bullet}(x)$.

23. WEIGHTED VERSION OF LAGRANGE INVERSION. Let $R = R_w(X)$ be a weighted species.

a) Show that the combinatorial equation

$$Y = X \cdot R_w(Y) \tag{105}$$

defines a unique weighted species $Y = \mathcal{A}_{R_w}(X)$, namely, the species of R_w-enriched rooted trees. Show that the weight of such a rooted tree is equal to the product of the weights of R_w-structures on the fibers of its vertices.

b) Show that the bijections (4), a), b), and c) described in the proof of the combinatorial version of Lagrange's inversion, Theorem 12, where $F = F_w$ is a weighted species, preserve weights. Show also that the bijections underlying (58) preserve weights. Deduce that Propositions 7 and 16, as well as formulas (10), are still valid in the weighted case.

c) The weighted species $\mathfrak{a}_{R_w}(X)$ of R_w-enriched trees is defined by agreeing that the weight of an R_w-enriched tree is equal to the product of the weights of R_w-structures placed on the neighbors of each of its vertices. Generalize Proposition 15 by showing that

$$\mathfrak{a}^{\bullet}_{R_w} = X R_w(\mathcal{A}_{(R_w)'}). \tag{106}$$

24. ENUMERATION OF PLANE ROOTED TREES ACCORDING TO DEGREE (also see Exercise 4.1.4). Let $\mathbf{r} = (r_0, r_1, r_2, \ldots)$ be a sequence of formal variables, utilized here as degree counters (in the sense of graph theory). Introduce the following species of weighted structures:

- $\mathfrak{p}_{\mathbf{r}}$, of plane trees weighted by degree, i.e., with $w(\alpha) = r_0^{n_0} r_1^{n_1} r_2^{n_2} \ldots$, where n_k is the number of vertices of degree k in the plane tree α,
- $\mathcal{P}_{\mathbf{r}}$, of plane rooted trees weighted by degree,
- $\mathcal{C}_{\mathbf{r}} = \sum_{k \geq 1} (\mathcal{C}_k)_{r_k}$, of oriented cycles weighted by r_ℓ, where ℓ is the cycle length,
- $L_{\mathbf{r}'} = (\mathcal{C}_{\mathbf{r}})' = \sum_{k \geq 0} (X^k)_{r_{k+1}}$, of lists weighted by $r_{\ell+1}$, where ℓ is the length of the list.

a) Show that the species of $L_{\mathbf{r}'}$-enriched rooted trees, $\mathcal{A}_{L_{\mathbf{r}'}}$, is the species of ordered rooted trees, weighted by degrees, agreeing that a supplementary half-edge contributes to the degree of the root, and that $X_{r_0} + (\mathcal{A}_{L_{\mathbf{r}'}})^2$ is isomorphic to the species $(\mathcal{A}_L)_w$ of ordered rooted trees, weighted by degrees (exactly, even for the root). Moreover, show that

$$\mathfrak{p}_{\mathbf{r}} = X_{r_0} + \mathfrak{a}_{R_w} \tag{107}$$

(see the preceding exercise), with $R_w = \mathcal{C}_{\mathbf{r}}$, and

$$\mathcal{P}_{\mathbf{r}} = X_{r_0} + X \cdot \mathcal{C}_{\mathbf{r}}(\mathcal{A}_{L_{\mathbf{r}'}}). \tag{108}$$

b) Let $n \geq 2$ and $\mathbf{i} = (i_0, i_1, i_2, \ldots)$ be a sequence of integers ≥ 0. Show that there exists a (plane) tree α on the set of vertices $[n]$ with degree distribution

i (i.e., of weight $w(\alpha) = \mathbf{r^i} = r_0^{i_0} r_1^{i_1} r_2^{i_2} \cdots$) if and only if

$$i_0 = 0, \quad n = i_1 + i_2 + \cdots, \quad \text{and} \quad 2(n-1) = i_1 + 2i_2 + 3i_3 + \cdots. \quad (109)$$

c) Let $n \geq 2$ and $\mathbf{i} = (i_0, i_1, i_2, \ldots)$, a sequence of integers ≥ 0 satisfying (109). Show that the number of plane trees, plane rooted trees, and ordered rooted trees on $[n]$, with distribution of degree $\mathbf{i}$, are given respectively by

$$\mathfrak{p}(\mathbf{i}) = (n-2)! \binom{n}{\mathbf{i}}, \qquad \mathcal{P}(\mathbf{i}) = n(n-2)! \binom{n}{\mathbf{i}},$$

$$\text{and} \quad (\mathcal{A}_L)_w(\mathbf{i}) = 2(n-1)! \binom{n}{\mathbf{i}}, \qquad (110)$$

where, to denote the multinomial coefficient, we set

$$\binom{n}{\mathbf{i}} = \binom{n}{i_0, i_1, i_2, \ldots}. \qquad (111)$$

d) Show that

$$\widetilde{\mathcal{P}}_\mathbf{r} = x\, r_0 + x \sum_{m \geq 1} \frac{\phi(m)}{m} \sum_{k \geq 1} \frac{r_{nk}}{k} \mathcal{A}_{L_{r'^m}}^k (x^m), \qquad (112)$$

where $\mathbf{r'^m} = (r_1^m, r_2^m, r_3^m, \ldots)$, and deduce that, for $n \geq 2$, the number of unlabeled plane rooted trees with degree distribution $\mathbf{i}$ (under the condition (109)) is given by

$$\widetilde{\mathcal{P}}(\mathbf{i}) = \frac{1}{n-1} \sum_{h,d} \phi(d) \binom{(n-1)/d}{(\mathbf{i} - \delta_h)/d}, \qquad (113)$$

the sum being taken over all pairs of integers $h \geq 1$, $d \geq 1$ such that $h \in \text{Supp}(\mathbf{i}) = \{k \mid i_k \neq 0\}$ and $d \in \text{Div}(h, \mathbf{i} - \delta_h)$, where we have set $\delta_h = (\delta_{h,k})_{k \geq 0}$ and $\text{Div}(h, \mathbf{i})$ is the set of common divisors to $h, i_0, i_1, i_2, \ldots$.

3.2. Implicit Species Theorem

We have seen in the preceding section that the functional equation

$$Y = X \cdot R(Y) \qquad (1)$$

characterizes the species $\mathcal{A}_R$ of R-enriched rooted trees. Other types of functional equations may be used to specify tree-like structures, for instance

$$Y = X + G(Y), \qquad (2)$$

$$Y = C(X) \cdot Y + D(X), \qquad (3)$$

or, more generally,

$$Y = H(X, Y), \qquad (4)$$

where G, C, and D are species of structures and $H = H(X, Y)$ is a two-sort species. It is important to make precise the conditions which ensure existence and uniqueness of a species solution of such an equation. This gives an analogue of the implicit function theorem, in the framework of the theory of species.

One form of the classical implicit function theorem is the following.

Theorem 1. IMPLICIT FUNCTION THEOREM. *Let $f(x, y)$ be a continuously differentiable function of two variables (defined in a neighborhood of $(0, 0)$) such that*

$$f(0, 0) = 0 \quad \text{and} \quad f_y(0, 0) = \frac{\partial f}{\partial y}(0, 0) \neq 0. \tag{5}$$

Then there exists a unique continuous function $a(x)$ (defined in a neighborhood of 0), such that for (x, y) sufficiently near $(0, 0)$, $f(x, y) = 0$ if and only if $y = a(x)$.

Consider the particular case where f takes the form $f(x, y) = y - h(x, y)$, with $h(0, 0) = 0$ and $h_y(0, 0) = 0$, for which conditions (5) are satisfied. Then the unique function $y = a(x)$, with $a(0) = 0$, whose existence is assured by Theorem 1, also satisfies the functional equation

$$a = h(x, a(x)). \tag{6}$$

In fact, this is but another form of the implicit function theorem, the "fixed point" version, which will be reformulated in the context of species of structures. So let $H(X, Y)$ be a two-sort species satisfying the conditions:

$$H(0, 0) = 0, \quad \text{and} \quad \frac{\partial H}{\partial Y}(0, 0) = 0. \tag{7}$$

We will show that there is one and only one (up to isomorphism) species of structures $A = A(X)$, with $A(0) = 0$, satisfying the relation

$$A(X) = H(X, A(X)). \tag{8}$$

The species A is constructed by expanding the recurrence implicit in this equation. Let us represent an H-structure as a diagram such as Figure 1. In order to better

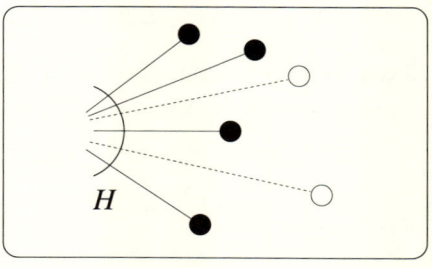

Fig. 1.

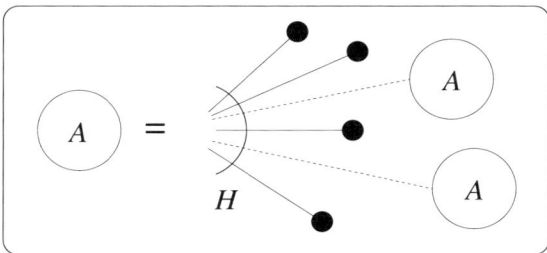

Fig. 2. $A = H(X, A)$.

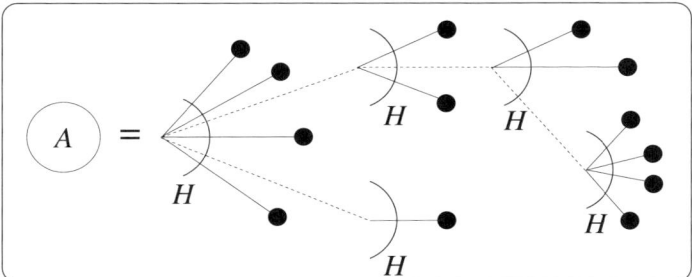

Fig. 3.

distinguish the two sorts of elements which occur in H-structures, the elements of sort X (black) are attached to solid lines, whereas the elements of the second sort (white) are attached to dotted lines. Condition (7) means that there is no H-structure on an empty set ($H(0, 0) = 0$) nor on a single white point ($\frac{\partial H}{\partial Y}(0, 0) = 0$).

In fact, Equation (8) should be interpreted as a species isomorphism, $\eta : A \longrightarrow H(X, A)$ describing the identification of each A-structure with an H-assembly of leaves (elements of sort X) and of A-structures. See Figure 2. Iterating this canonical decomposition gives enriched tree-like structures of a very particular type, described in Figure 3 and called H-*enriched rooted trees*. The hypotheses in (7) constitute a "stopping condition" which prevents an infinite spreading out of this decomposition. They imply that there can only be a finite number of H-enriched rooted trees on a finite set U (of sort X). We denote by $A = A_H$ this species of H-enriched rooted trees. It is clearly a solution of the functional equation (8), with $A(0) = 0$.

Suppose now that there exists two species A and B, solutions of Equation (4), with $A(0) = 0 = B(0)$, and hence two species isomorphisms

$$\eta : A \longrightarrow H(X, A), \qquad \gamma : B \longrightarrow H(X, B). \tag{9}$$

We construct a unique (canonical) isomorphism $\psi : A \longrightarrow B$ such that the following diagram commutes,

$$\begin{array}{ccc} A & \xrightarrow{\eta} & H(X,A) \\ \psi \downarrow & & \downarrow H(X,\psi) \\ B & \xrightarrow{\gamma} & H(X,B) \end{array} \qquad (10)$$

where the isomorphism $H(X, \psi)$ is obtained from ψ by functoriality, that is by using the transport of structures.

The construction proceeds by induction on the contact order between A and B. Recall (see Section 1.2) that two species A and B have a contact of order n (written $A =_n B$) if and only if there exists a species isomorphism $\varphi : A_{\leq n} \longrightarrow B_{\leq n}$, where $A_{\leq n}$ denotes the species A restricted to sets of cardinality less than or equal to n, i.e., $A_{\leq n} = \sum_{k \leq n} A_k$. Observe that any isomorphism $\alpha : F \longrightarrow G$ can also be restricted to an isomorphism $\alpha_{\leq n} : F_{\leq n} \longrightarrow G_{\leq n}$.

Evidently, the species A and B have a contact of order 0, since $A(0) = 0 = B(0)$ and we put $\psi_0 = Id_\emptyset$. Now suppose that they have a contact of order n, via an isomorphism $\psi_n : A_{\leq n} \longrightarrow B_{\leq n}$, for which the diagram (10) restricted to sets of cardinality $\leq n$ commutes. We verify that

$$\psi_{n+1} = (\gamma^{-1} \circ H(X, \psi_n) \circ \eta)_{\leq n+1} \qquad (11)$$

provides an isomorphism between $A_{\leq n+1}$ and $B_{\leq n+1}$. Indeed, in the canonical decomposition of an $A_{\leq n+1}$-structure as an $H(X, A)$-structure (see Figure 4), all A-structures are in fact $A_{\leq n}$-structures. This holds since, in virtue of conditions (7), there are at least two such A-structures, or else the unique A-structure that appears in the H-assembly comes along with other elements (at least one) of sort X. Moreover, ψ_0 clearly coincides with ψ_1 on $A_{\leq 0}$, thus recursively, ψ_n coincides with ψ_{n+1} on $A_{\leq n}$. By considering arbitrarily large n, we conclude that there exists a unique isomorphism ψ between A and B such that $\psi = \gamma^{-1} \circ H(X, \psi) \circ \eta$.

We have thus established the following theorem, due to Joyal [158].

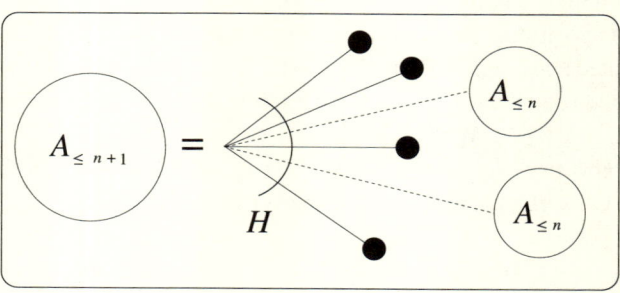

Fig. 4.

3.2. Implicit Species Theorem

Theorem 2. IMPLICIT SPECIES THEOREM. *Let $H = H(X, Y)$ be a two-sort species satisfying the conditions (7). Then the combinatorial equation $A = H(X, A)$ characterizes, up to canonical isomorphism, the unique species $A = A(X)$ for which $A(0) = 0$.*

Here is an example of a canonical isomorphism entirely deduced from the uniqueness of the solution of a functional equation.

Example 3. The species $Y = \mathcal{B}_+ = \mathcal{B} - 1$ of nonempty plane binary rooted trees satisfies the functional equation

$$Y = X(1+Y)^2 \tag{12}$$

equivalent to Equation (3.0.1) for the species $\mathcal{B}$. On the other hand, it is easily seen that Equation (12) is also satisfied by the species $Y = L_+(\mathcal{A}_L)$ of nonempty lists of ordered rooted trees (see Example 3.1.4). Indeed, for a nonempty list $\alpha = (\alpha_1, \alpha_2, \ldots, \alpha_k)$ of ordered rooted trees (see Figure 5(a)), we first obtain the factor X by extracting the root r from the rooted tree α_1 ($r = 6$ in the figure). There remains an ordered pair (β_1, β_2) of possibly empty lists of ordered rooted trees, i.e., a pair of $(1 + L_+(\mathcal{A}_L))$-structures: β_1 is the list of rooted trees attached to the root r and $\beta_2 = (\alpha_2, \ldots, \alpha_k)$. We conclude that there exists a canonical isomorphism

$$\tilde{\psi} : L_+(\mathcal{A}_L) \xrightarrow{\sim} \mathcal{B}_+ \tag{13}$$

compatible with Equation (12), obtained by iteration of the decomposition $\alpha \mapsto (r, \beta_1, \beta_2)$ described above. Figure 5(b) shows the binary rooted tree thus obtained from the $L_+(\mathcal{A}_L)$-structure of Figure 5(a).

Example 4. It sometimes happens that a functional equation can be solved explicitly. This is the case with the linear equation

$$Y = C(X) \cdot Y + D(X), \tag{14}$$

where C and D are given species of structures with $C(0) = 0$. The solution is obtained, either algebraically or combinatorially, by simple iteration, starting with $Y_0 = 0$ (see Figure 6):

$$Y = D + C \cdot D + C^2 \cdot D + \cdots$$
$$= L(C) \cdot D. \tag{15}$$

G-Rooted Trees

Another important family of species of structures defined by functional equations is that of G-rooted trees, denoted by $Y = {}_G A$, for given species G. They are

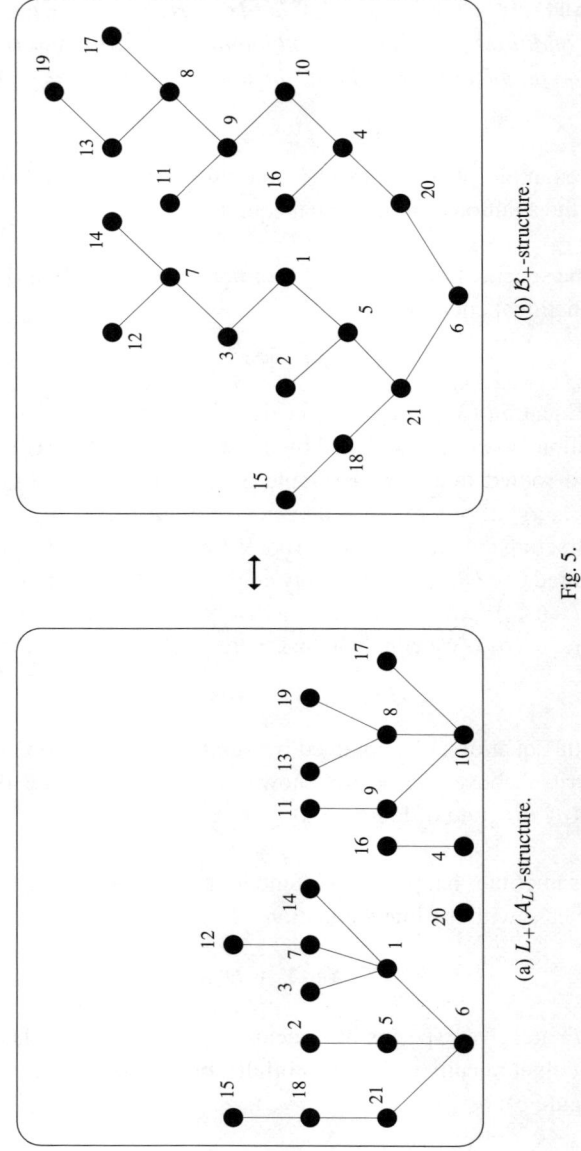

(a) $L_+(A_L)$-structure.

(b) B_+-structure.

Fig. 5.

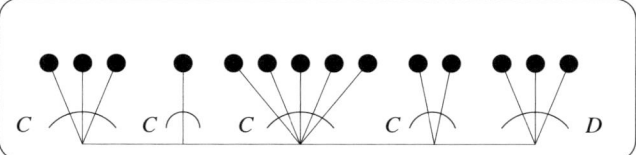

Fig. 6.

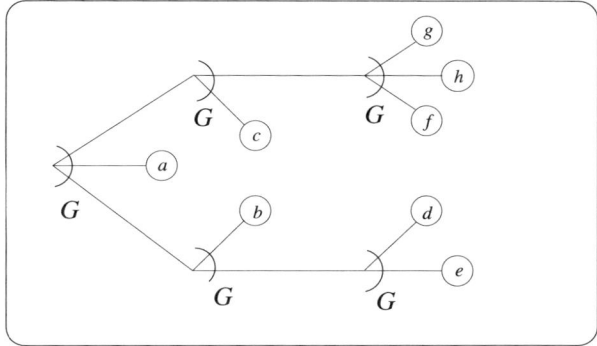

Fig. 7.

characterized by the functional equation

$$Y = X + G(Y). \tag{16}$$

Conditions (7) translate as

$$G(0) = 0 \quad \text{and} \quad G'(0) = 0. \tag{17}$$

Iterating (16), a rooted tree structure appears. These structures are variants of R-enriched rooted trees (with $R = G$) in the sense that internal vertices are not labeled, i.e., that the underlying set is found at the leaves of the rooted tree. See Figure 7.

Example 5. The species P of *parenthesizations* can be defined by the functional equation

$$P = X + P^2, \tag{18}$$

which is interpreted by saying that a P-structures is either an element (a symbol) or an ordered pair of already formed parenthesizations. Since this equation is of form (16), with $G = X^2$, the parenthesizations can be represented by G-rooted trees. See Figure 8 for such a representation of the parenthesization

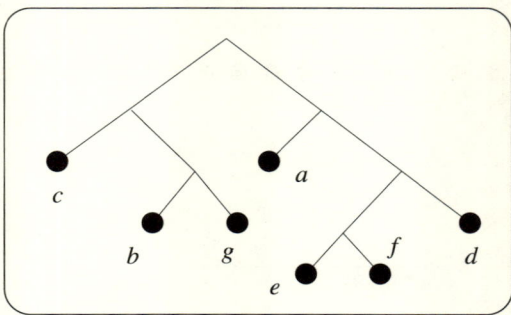

Fig. 8.

$((c, (b, g)), (a, ((e, f), d)))$. Other examples of G-rooted trees, appearing naturally in mathematics or in computer science, include:
- $G = E_2$; $_G A$ is the species $\mathcal{P}$, of *commutative parenthesizations*,
- $G = X^2 + X^3$; $_G A$ is the species of *(2, 3)-rooted trees* (see Exercise 5), not to be confused with the (2, 3)-trees considered in Section 3.3.
- $G = t_1 X + t_2 X^2 + t_3 X^3 + \cdots$, where $t_1, t_2, t_3, \ldots$ are formal variables representing unitary, binary, ternary, ... operations; in this case $_G A$ is the species of *well-formed expressions*. For example, the well-formed expression corresponding to the G-rooted tree of Figure 7, where the order is from top to bottom, is, in Polish notation, $t_3 t_2 t_3 g h f \, cat_2 b t_2 d e$.

Very few Lagrange-type inversion formulas are known for the solutions of functional equation (16). Note however the following series development, whose proof can be found in G. Labelle [117]: For any formal power series $F(x)$,

$$F(_G A(x)) = F(x) + \sum_{k \geq 1} \frac{1}{k!} \left(\frac{d}{dx} \right)^k (F'(x) G^k(x)). \tag{19}$$

Equivalently,

$$F(_G A(x)) = \sum_{k \geq 0} \frac{1}{k!} \left(\frac{d}{dx} \right)^k (F(x)(1 - G'(x)) G^k(x)). \tag{20}$$

Remark 6. When G factors as $G(Y) = Y \cdot D(Y)$, for some species of structures D, the equation $Y = X + G(Y)$ amounts to $Y = X \cdot R(Y)$, with $R = L(D)$, and the usual Lagrange inversion formulas (see Section 3.1) can be used. Note that for formal power series it is always possible, under conditions (17), to write $G(y) = y D(y)$, with $D(y) \in \mathbb{A}[\![y]\!]$, and to use Lagrange inversion with $R(y) = (1 - D(y))^{-1}$.

System of Functional Equations

The implicit species theorem can be generalized to multisort species. In fact, Joyal proved the following result (see [158], Section 5).

Theorem 7. *Every system of equations*

$$Y_1 = H_1(X_1, X_2, \ldots, X_n, Y_1, Y_2, \ldots, Y_k)$$
$$Y_2 = H_2(X_1, X_2, \ldots, X_n, Y_1, Y_2, \ldots, Y_k)$$
$$\vdots$$
$$Y_k = H_k(X_1, X_2, \ldots, X_n, Y_1, Y_2, \ldots, Y_k)$$
(21)

admits a solution $Y_1(X_1, \cdots, X_n), \ldots, Y_k(X_1, \ldots, X_n)$, unique up to isomorphism, whenever the following conditions for the species $H_1, H_2, \ldots, H_k$ are satisfied:

i) for any m, $1 \le m \le k$, $H_m(0, 0, \ldots, 0) = 0$,

ii) the Jacobian matrix : $\dfrac{\partial H_i}{\partial Y_j}(0, 0, \ldots, 0)$, $1 \le i, j \le k$, is nilpotent. (22)

An important special case is the system of functional equations

$$Y_1 = X_1 \cdot R_1(Y_1, Y_2, \ldots, Y_k)$$
$$Y_2 = X_2 \cdot R_2(Y_1, Y_2, \ldots, Y_k)$$
$$\vdots$$
$$Y_k = X_k \cdot R_k(Y_1, Y_2, \ldots, Y_k),$$
(23)

where the R_i are given k-sort species, $i = 1, \ldots, k$. Conditions (22) are then satisfied. This case is closely connected to multidimensional Lagrange inversion. Our goal here is to generalize the results of Section 3.1.

For series $R_i(x_1, \ldots, x_k)$, $i = 1, \ldots, k$, in the ring $\mathbb{A}[\![x_1, \ldots, x_k]\!]$ of formal power series in the (commutative) variables $x_1, \ldots, x_n$, there exists a unique formal power series $A_i(x_1, \ldots, x_k)$ satisfying the system of equations

$$A_i(x_1, \ldots, x_k) = x_i \, R_i(A_1, \ldots, A_k), \qquad i = 1, \ldots, k. \qquad (24)$$

In order to state the Good–Lagrange inversion formulas, we introduce the following matrices:

$$J(t_1, \ldots, t_k) = \left[\delta_{ij} - x_i \frac{\partial R_i}{\partial t_j}(t_1, \ldots, t_k) \right]_{1 \le i, j \le k} \qquad (25)$$

$$K(t_1, \ldots, t_k) = \left[\delta_{ij} - \frac{t_i}{R_i(t_1, \ldots, t_k)} \frac{\partial R_i}{\partial t_j}(t_1, \ldots, t_k) \right]_{1 \le i, j \le k}, \qquad (26)$$

where δ_{ij} is the Kronecker delta function. We also introduce the following vector notation: For given vectors (of variables, functions, species, etc.) $\mathbf{x} = (x_1, \ldots, x_k)$ and $\mathbf{y} = (y_1, \ldots, y_k)$, and $\mathbf{n} = (n_1, \ldots, n_k) \in \mathbb{N}^k$, we set

$$\mathbf{x} + \mathbf{y} := (x_1 + y_1, \ldots, x_k + y_k),$$

$$\mathbf{x}(\mathbf{y}) := (x_1(y_1, \ldots, y_k), \ldots, x_k(y_1, \ldots, y_k)),$$

$$\mathbf{x}^\mathbf{n} := x_1^{n_1} \cdots x_k^{n_k}, \tag{27}$$

$$\mathbf{0} := \underbrace{(0, \ldots, 0)}_{k \text{ terms}}.$$

Theorem 8. GOOD–LAGRANGE INVERSION FORMULAS [132]. *Let* $\mathbf{R}(\mathbf{x}) = (R_1(\mathbf{x}), \ldots, R_k(\mathbf{x}))$ *be a vector of formal power series in the variables* $\mathbf{x} = (x_1, \ldots, x_k)$ *and let* $\mathbf{A}(\mathbf{x})$ *be a vector of formal power series satisfying* (24). *Then for any formal power series* $F(\mathbf{x})$ *and all* $\mathbf{n} \in \mathbb{N}^k$,

a) $[\mathbf{x}^\mathbf{n}] \dfrac{F(\mathbf{A}(\mathbf{x}))}{\det J(\mathbf{A}(\mathbf{x}))} = [\mathbf{t}^\mathbf{n}] F(\mathbf{t}) \mathbf{R}^\mathbf{n}(\mathbf{t}),$

b) $[\mathbf{x}^\mathbf{n}] F(\mathbf{A}(\mathbf{x})) = [\mathbf{t}^\mathbf{n}] F(\mathbf{t}) \det(K(\mathbf{t})) \mathbf{R}^\mathbf{n}(\mathbf{t}).$ \hfill (28)

Proof. We first show that formulas (28), a) and b) are equivalent. To go from a) to b), we need only replace $F(\mathbf{t})$ by $F(\mathbf{t}) \det(K(\mathbf{t}))$ and observe that, by virtue of Equation (24), we have $K(\mathbf{A}(\mathbf{x})) = J(\mathbf{A}(\mathbf{x}))$. This procedure is clearly reversible since the formal power series $\det K(\mathbf{t})$ is invertible, having constant term equal to 1.

Observe that for $k = 1$, formula (24), a) coincides with (3.1.10), c). Following Gessel [127], we show that the concepts of R-enriched rooted trees and endofunctions, as well as the combinatorial proof that we have given of (3.1.10), c), extend naturally to the multidimensional case. Formula (28), a) is then a consequence of the results which follow. ∎

Suppose that we are given k-sort species $R_i(X_1, \ldots, X_k)$, $i = 1, \ldots, k$. We introduce the k-sort species $\text{End}_\mathbf{R}$ of **R**-*enriched endofunctions*. An $\text{End}_\mathbf{R}$-structure on a multiset $\mathbf{U} = (U_1, \ldots, U_k)$ is an endofunction f on the set $U = U_1 + \cdots + U_k$ together with, for each $i = 1, \ldots, k$ and each $u \in U_i$, an enrichment by an R_i-structure on the fiber $f^{-1}(u)$; more precisely, on the multiset $(f^{-1}(u) \cap U_1, \ldots, f^{-1}(u) \cap U_k)$. The concept of an **R**-*enriched partial endofunction on* **U**, *with domain* $\mathbf{V} \subseteq \mathbf{U}$, is defined in an analogous manner.

For $\mathbf{n} = (n_1, \ldots, n_k) \in \mathbb{N}^\mathbf{n}$, introduce the corresponding multiset

$$[\mathbf{n}] = ([n_1] \times \{1\}, \ldots, [n_k] \times \{k\}).$$

The following result is the multidimensional analogue of Lemma 3.1.10.

3.2. Implicit Species Theorem

Lemma 9. *Let* $\mathbf{n} \in \mathbb{N}^k$ *and let* $\mathbf{V}$ *be a multisubset of* $[\mathbf{n}]$. *Then there is a bijection between* $\mathbf{R}$*-enriched partial endofunctions with domain* $\mathbf{V}$ *on* $[\mathbf{n}]$ *and* $\mathbf{R}^{\mathbf{n}}$*-structures on* $\mathbf{V}$.

Proof. Indeed, an $\mathbf{R}$-enriched partial endofunction on $[\mathbf{n}]$ with domain $\mathbf{V}$ can be identified with the list of R_i-structures on the fibers of elements $u \in [n_i] \times \{i\}$, $i = 1, \ldots, k$, the union of these fibers being exactly $\mathbf{V}$. ∎

The concept of $\mathbf{R}$-enriched rooted trees on the multiset $\mathbf{U} = (U_1, \ldots, U_k)$ is defined in an analogous manner. It is a rooted tree on U together with an R_i-enrichment on the fiber of each $u \in U_i$, $i = 1, \ldots, k$. Denote by $\mathcal{A}_{\mathbf{R},i}$ the (k-sort) species of $\mathbf{R}$-enriched rooted trees whose root is of sort i (i.e., element of U_i; one also says "of color i"), and by $\mathcal{A}_{\mathbf{R}}$ the vector of species $(\mathcal{A}_{\mathbf{R},1}, \ldots, \mathcal{A}_{\mathbf{R},k})$. Evidently, the following functional equations hold:

$$\mathcal{A}_{\mathbf{R},i} = X_i \cdot R_i(\mathcal{A}_{\mathbf{R}}), \qquad i = 1, \ldots k, \tag{29}$$

Proposition 10. *Let* $\mathbf{R}(\mathbf{X})$ *be a vector of k-sort species,* $F(\mathbf{X})$ *a k-sort species, and* $\mathbf{n} \in \mathbb{N}^k$. *Then there exists a bijection*

$$(F(\mathcal{A}_{\mathbf{R}}) \cdot \text{End}_{\mathbf{R}})[\mathbf{n}] \xrightarrow{\sim} (F \cdot \mathbf{R}^{\mathbf{n}})[\mathbf{n}]. \tag{30}$$

Proof. The proof is essentially the same as that for Theorem 3.1.12, c): An $(F(\mathcal{A}_{\mathbf{R}}) \cdot \text{End}_{\mathbf{R}})$-structure on $[\mathbf{n}]$ can be identified with a pair consisting of an F-structure on a multisubset $\mathbf{W}$ of $[\mathbf{n}]$ and of an $\mathbf{R}$-enriched partial endofunction on $[\mathbf{n}]$, with domain $\mathbf{V} = [\mathbf{n}] \setminus \mathbf{W}$; there follows an $(F \cdot \mathbf{R}^{\mathbf{n}})$-structure on $[\mathbf{n}]$, in a bijective manner, in view of the preceding lemma. ∎

In order to enumerate $\mathbf{R}$-enriched endofunctions, our analysis in the one-dimensional case of relation (3.1.28), $\text{End}_R = S(X \cdot R'(\mathcal{A}_R))$, must be extended to the multidimensional case. To that end we need to take into account the sorts of the vertices that are located on the cycles of an $\text{End}_{\mathbf{R}}$-structure (the recurrent elements). Thus we introduce the species

$$Q_{i,j}(X_1, \ldots, X_k) = X_i \cdot R_i^{(j)}(\mathcal{A}_{\mathbf{R},1}(X_1, \ldots, X_k), \ldots, \mathcal{A}_{\mathbf{R},k}(X_1, \ldots, X_k)), \tag{31}$$

where we have set $R_i^{(j)}(X_1, \ldots, X_k) = \frac{\partial R_i}{\partial X_j}(X_1, \ldots, X_k)$ (see Figure 9). We also introduce the $k \times k$ matrix $Q = [Q_{i,j}]$, $1 \le i, j \le k$, of k-sort species. Observe that for generating series, we have

$$J(\mathcal{A}_{\mathbf{R}}(\mathbf{x})) = I - Q(\mathbf{x}), \tag{32}$$

where J is defined by (25) and $I = [\delta_{ij}]$ is the identity matrix.

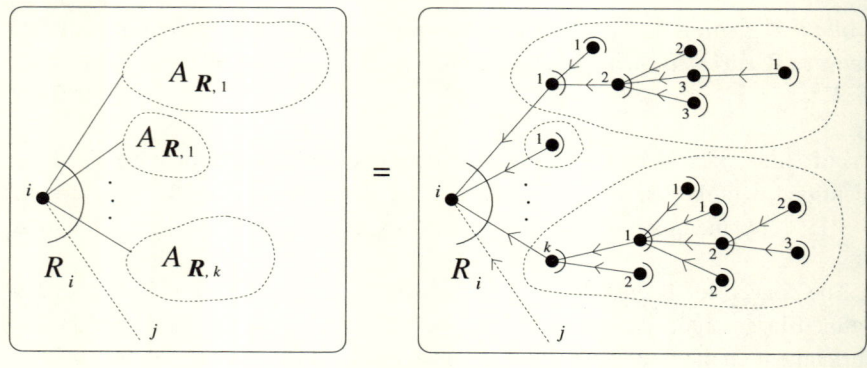

Fig. 9. The species $Q_{i,j} = X_i \cdot R_i^{(j)}(\mathcal{A}_\mathbf{R})$.

Denote by $\text{End}_\mathbf{R}^c$, the species of *connected* **R**-enriched endofunctions.

Proposition 11. *Let* $\mathbf{R} = (R_1, \ldots, R_k)$ *be a vector of k-sort species. Then the species* $\text{End}_\mathbf{R}$ *and* $\text{End}_\mathbf{R}^c$ *have the following generating series:*

$$\begin{aligned} \text{a)} \quad & \text{End}_\mathbf{R}^c(\mathbf{x}) = \text{trace log} \frac{1}{J(\mathcal{A}_\mathbf{R}(\mathbf{x}))}, \\ \text{b)} \quad & \text{End}_\mathbf{R}(\mathbf{x}) = \frac{1}{\det J(\mathcal{A}_\mathbf{R}(\mathbf{x}))}. \end{aligned} \qquad (33)$$

Proof. a) A connected **R**-enriched endofunction can be considered as an oriented cycle whose vertices have sorts in the set $\{1, 2, \ldots, k\}$ and whose arcs, labeled from j to i, are $Q_{i,j}$-structure. If the cycle is of length m, it can be pointed in m distinct ways on recurrent elements of the endofunction. Each of these pointings yields a circuit in the oriented graph whose (column to line) adjacency matrix is precisely $Q = [Q_{i,j}]$. Thus we have

$$\begin{aligned} \text{End}_\mathbf{R}^c(\mathbf{x}) &= \sum_{m \geq 1} \frac{1}{m} \text{trace } Q^m(\mathbf{x}) \\ &= \text{trace} \sum_{m \geq 1} \frac{Q^m(\mathbf{x})}{m} \\ &= \text{trace log} (I - Q(\mathbf{x}))^{-1} \\ &= \text{trace log} \frac{1}{J(\mathcal{A}_\mathbf{R}(\mathbf{x}))}. \end{aligned}$$

b) Since $\text{End}_\mathbf{R} = E(\text{End}_\mathbf{R}^c)$, one has $\text{End}_\mathbf{R}(\mathbf{x}) = \exp(\text{End}_\mathbf{R}^c(\mathbf{x}))$. Applying Jacobi's identity

$$\exp \operatorname{trace} M = \det \exp M, \tag{34}$$

with the matrix $M = -\log J(\mathcal{A}_{\mathbf{R}}(\mathbf{x}))$, we obtain the result. ∎

One may prove (33), b) directly by characterizing the species of **R**-enriched endofunctions in terms of "compatible permutations" of $Q_{j,i}$-structures (see Exercise 16). For a combinatorial interpretation and proof of Jacobi's identity (34) in the form

$$\exp \operatorname{trace} \log \frac{1}{\mathrm{I} - Z} = \frac{1}{\det(\mathrm{I} - Z)}, \tag{35}$$

as used here, the reader is referred to Foata [108], Viennot [320], or to Exercise 17. See also Zeng [339] and Zeilberger [336].

PROOF OF GOOD'S FORMULA (28), a).
For $\mathbf{n} = (n_1, \ldots, n_k)$, set

$$\mathbf{n}! = n_1! \cdots n_k!. \tag{36}$$

We can suppose that $R_i(\mathbf{x})$ and $F(\mathbf{x})$ are generating series of k-sort species, since the general case will follow from the extension principle of algebraic identities. Now, by virtue of (33), b), since Equations (29) imply (24), we have

$$[\mathbf{x}^{\mathbf{n}}] \frac{F(\mathbf{A}(\mathbf{x}))}{\det J(\mathbf{A}(\mathbf{x}))} = [\mathbf{x}^{\mathbf{n}}] \frac{F(\mathcal{A}_{\mathbf{R}}(\mathbf{x}))}{\det J(\mathcal{A}_{\mathbf{R}}(\mathbf{x}))}$$

$$= \frac{1}{\mathbf{n}!} |(F(\mathcal{A}_{\mathbf{R}}) \cdot \operatorname{End}_{\mathbf{R}})[\mathbf{n}]|$$

$$= \frac{1}{\mathbf{n}!} |(F \cdot \mathbf{R}^{\mathbf{n}})[\mathbf{n}]|$$

$$= [\mathbf{t}^{\mathbf{n}}] F(\mathbf{t}) \mathbf{R}^{\mathbf{n}}(\mathbf{t}).$$

Example 12. JACOBI ENDOFUNCTIONS. Jacobi endofunctions have been introduced in Foata and Leroux [111] as a combinatorial model for the classical Jacobi orthogonal polynomials. They constitute a two-sort species of **R**-enriched endofunctions, $\operatorname{Jac} = \operatorname{Jac}(X, Y)$, defined by

$$\operatorname{Jac}(X, Y) = \operatorname{End}_{\mathbf{R}}(X, Y), \tag{37}$$

with $\mathbf{R} = (R_1, R_2)$, where $R_1(X, Y) = (1 + X)(1 + Y) = R_2(X, Y)$. In other words, they are endofunctions on sets of black points (elements of sort X) or white points (elements of sort Y) having the property, called the *Jacobi property*, that (see Figure 10) "the fiber of any element (black or white) contains at most one black

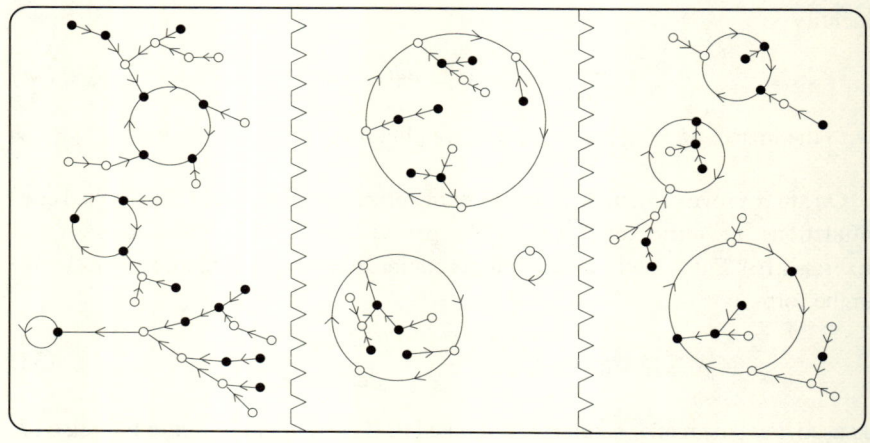

Fig. 10.

element and at most a white element." The associated **R**-enriched rooted trees form species $A = A(X, Y)$ and $B = B(X, Y)$ determined by the system of functional equations.

$$A = X(1 + A)(1 + B),$$
$$B = Y(1 + A)(1 + B). \tag{38}$$

These are (binary) rooted trees having the Jacobi property. In this case we can calculate directly the generating series $A(x, y)$ and $B(x, y)$. To this end, set

$$a(x, y) = A(x, y)/x = B(x, y)/y. \tag{39}$$

From (38) we obtain the quadratic equation $a(x, y) = (1+x\,a(x, y))(1+y\,a(x, y))$, whose solution is

$$a(x, y) = \frac{1}{2xy}\left(1 - x - y - \sqrt{(x + y - 1)^2 - 4xy}\right). \tag{40}$$

An application of formula (33), b), or direct reasoning, gives

$$\mathrm{Jac}(x, y) = \frac{1}{(1 - x(1 + B(x, y)) - y(1 + A(x, y)))}$$
$$= \frac{1}{\sqrt{(x + y - 1)^2 - 4xy}}. \tag{41}$$

We distinguish three types of Jacobi endofunctions, called *type a, b,* or *m*, according to whether all the cycles that they contain are composed uniquely of (recurrent) black points, uniquely of white points, or of points of both colors, respectively.

This determines three subspecies of Jac which are denoted by Jac_a, Jac_b, and Jac_m, respectively, and we have (see Figure 10).

$$\text{Jac} = \text{Jac}_a \cdot \text{Jac}_b \cdot \text{Jac}_m. \tag{42}$$

Note that

$$\text{Jac}_a(x,y) = \frac{1}{1 - x(1 + B(x,y))},$$

$$\text{Jac}_b(x,y) = \frac{1}{1 - y(1 + A(x,y))}, \tag{43}$$

$$\text{Jac}_m(x,y) = \frac{\text{Jac}(x,y)}{\text{Jac}_a(x,y)\text{Jac}_b(x,y)}.$$

The parameters α and β which occur traditionally in the Jacobi polynomials are introduced here by defining the weight $w(\varphi)$ of a Jacobi endofunction φ as

$$w(\varphi) = (\alpha + 1)^{\text{cyc}_a(\varphi)}(\beta + 1)^{\text{cyc}_b(\varphi)}, \tag{44}$$

where $\text{cyc}_a(\varphi)$ and $\text{cyc}_b(\varphi)$ denote respectively the number of purely black and purely white cycles of φ. Denote by $\text{Jac}^{(\alpha,\beta)}(X,Y)$ the weighted species defined in this way. We then have, in virtue of the weighted exponential formula (see Exercise 2.3.10, c) i)),

$$\begin{aligned}\text{Jac}^{(\alpha,\beta)}(x,y) &= (\text{Jac}_a(x,y))^{\alpha+1}(\text{Jac}_b(x,y))^{\beta+1}\text{Jac}_m(x,y) \\ &= (\text{Jac}_a(x,y))^{\alpha}(\text{Jac}_b(x,y))^{\beta}\text{Jac}(x,y) \\ &= 2^{\alpha+\beta}(1 - x + y + R)^{-\alpha}(1 + x - y + R)^{-\beta}R^{-1},\end{aligned} \tag{45}$$

where $R = R(x,y) = \sqrt{(x+y-1)^2 - 4xy}$. To recover the classical generating series (3.1.78), iii) for the Jacobi polynomials $P_n^{(\alpha,\beta)}(x)$ (see Exercise 3.1.5), we need only substitute for X and Y the species $T_{(x+1)/2}$ and $T_{(x-1)/2}$ of singletons of sort T and of weights $(x+1)/2$ and $(x-1)/2$ respectively, and remark (see Exercise 11, c)), setting

$$\text{Jac}^{(\alpha,\beta)}(T) = \text{Jac}^{(\alpha,\beta)}(T_{(x+1)/2}, T_{(x-1)/2}),\tag{46}$$

that

$$\text{Jac}^{(\alpha,\beta)}(t) = \sum_{n \geq 0} P_n^{(\alpha,\beta)}(x)\frac{t^n}{n!}. \tag{47}$$

See Exercise 11, as well as Leroux–Strehl [216] and Strehl [308–310] for a more thorough combinatorial study of Jacobi polynomials.

Example 13. BICOLORED TREES. Let us introduce the two-sort species $\mathfrak{a}(X, Y)$ of bicolored trees. If U denotes a set of black vertices and V a set of white vertices, then by definition $\mathfrak{a}[U, V]$ is the set of trees on $U + V$ with all edges having endpoints of different colors. A bicolored rooted tree is simply a pointed bicolored tree. Write

$$A(X, Y) = \mathfrak{a}^{\bullet X}(X, Y), \qquad B(X, Y) = \mathfrak{a}^{\bullet Y}(X, Y), \tag{48}$$

to denote, respectively, the species of bicolored rooted trees with black and white roots. We clearly have the functional equations

$$A = X \cdot E(B),$$
$$B = Y \cdot E(A). \tag{49}$$

Thus in this case $R_1(t_1, t_2) = e^{t_2}$, $R_2(t_1, t_2) = e^{t_1}$, and $\det K(t_1, t_2) = 1 - t_1 t_2$. Applying (28), b), the number of rooted trees with black roots is found to be

$$|A[n, m]| = n!m![x^n y^m] A(x, y)$$
$$= n!m![t_1^{n_1} t_2^{n_2}] t_1 (1 - t_1 t_2) e^{nt_2} e^{mt_1} \tag{50}$$
$$= n^m m^{n-1}.$$

Scoins' formula [294] for the number of bicolored trees follows immediately:

$$|\mathfrak{a}[n, m]| = n^{m-1} m^{n-1}. \tag{51}$$

R, S-Enriched Bicolored Trees.

In a slightly more general manner, consider the species $\mathfrak{a}_{R,S} = \mathfrak{a}_{R,S}(X, Y)$ of R, S-enriched bicolored trees, where R and S are given species of structures (of one sort). The enrichment is as follows. For each black vertex (respectively each white vertex), there is an R-structure (respectively S-structure) on the set of its neighbors (see Figure 11). If the species R and S are weighted, one defines the weight of an R, S-enriched bicolored tree to be the product of the weights of all the R-structures and S-structures attached to vertices of the tree. An example is given by bicolored plane trees weighted by degrees (see Exercise 14).

By analogy with (unisort) R-enriched trees (see Equations (3.1.52)), we have the functional equations

$$\mathfrak{a}_{R,S}^{\bullet X} = X \cdot R(\mathcal{B}_{R',S'}), \qquad \mathfrak{a}_{R,S}^{\bullet Y} = Y \cdot S(\mathcal{A}_{R',S'}), \tag{52}$$

where $\mathcal{A}_{R',S'}$ and $\mathcal{B}_{R',S'}$ denote the species of (R', S')-enriched (bicolored) rooted trees, defined by the equations

$$\mathcal{A}_{R',S'} = X \cdot R'(\mathcal{B}_{R',S'}),$$
$$\mathcal{B}_{R',S'} = Y \cdot S'(\mathcal{A}_{R',S'}). \tag{53}$$

3.2. Implicit Species Theorem

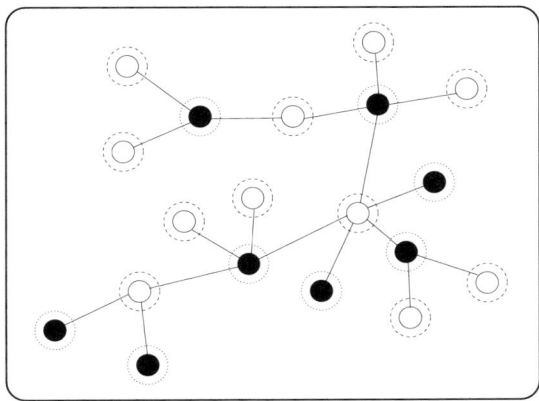

Fig. 11.

In this case, there exists a simpler (two-dimensional alternating) Lagrange inversion formula (see Exercise 13).

Let us mention finally the particular form which the dissymmetry theorem for trees (proved in Section 4.1) takes in the case of R, S-enriched bicolored trees (see Exercise 4.1.13):

$$\mathfrak{a}_{R,S}^{\bullet X} + \mathfrak{a}_{R,S}^{\bullet Y} = \mathfrak{a}_{R,S} + \mathcal{A}_{R',S'}\mathcal{B}_{R',S'}. \tag{54}$$

Computation of the Cycle Index Series of R-Enriched Rooted Trees

The Good–Lagrange inversion formula (28), b), conveniently adapted to the context of index series, provides a remarkable expression for the coefficients of the cycle index series $Z_{\mathcal{A}_R}(x_1, x_2, x_3, \ldots)$ of the (unisort) species $\mathcal{A}_R$ of R-enriched rooted trees.

Proposition 14. (G. Labelle [181].) *Let $R = R(X)$ be a species of structures and $\mathcal{A}_R = X \cdot R(\mathcal{A}_R)$ be the corresponding species of R-enriched rooted trees. Then, for any species of structures $F = F(X)$ and any finite sequence $\mathbf{n} = (n_1, n_2, n_3, \ldots)$ of nonnegative integers,*

$$[\mathbf{x}^{\mathbf{n}}]Z_{F(\mathcal{A}_R)}(\mathbf{x}) = [\mathbf{t}^{\mathbf{n}}]Z_F(\mathbf{t}) \prod_{i \geq 1} \rho(\mathbf{t}_{(i)}) Z_R(\mathbf{t}_{(i)})^{n_i}, \tag{55}$$

where $\mathbf{t}_{(i)} = (t_i, t_{2i}, t_{3i}, \ldots)$ and

$$\rho(t_1, t_2, t_3, \ldots) = 1 - t_1 \frac{\frac{\partial}{\partial t_1} Z_R(t_1, t_2, t_3, \ldots)}{Z_R(t_1, t_2, t_3, \ldots)}. \tag{56}$$

Proof. We have $Z_{F(A_R)} = Z_F(Z_{A_R})$ and $Z_{A_R} = x_1 Z_R(Z_{A_R})$. Let us simplify the notation by replacing Z_F, Z_{A_R}, and Z_R by f, a, and r respectively. We wish to compute $f(a) = f(a_1, a_2, a_3, \ldots)$ assuming the equality $a = x_1 r(a)$. This last equation is equivalent to the infinite system of equations

$$a_i = x_i r_i(a_1, a_2, a_3, \ldots), \qquad i = 1, 2, 3, \ldots, \tag{57}$$

where, according to our notational convention (1.4.7), $r_i(x_1, x_2, x_3, \ldots) = r(x_i, x_{2i}, x_{3i}, \ldots)$.

Consider a sequence of k nonnegative integers $\mathbf{n} = (n_1, n_2, \ldots, n_k)$. Then Good–Lagrange's inversion formula (28), b), applied to the "truncated" system of k equations

$$a_i = x_i r_i(a_1, a_2, \ldots, a_k, 0, 0, \ldots), \qquad i = 1, \ldots, k, \tag{58}$$

with $a_i = a_i(x_1, x_2, \ldots, x_k, 0, 0, \ldots)$, i.e., where $x_r = 0$ if $r > k$, takes the form

$$[\mathbf{x}^\mathbf{n}] f(a_1, a_2, \ldots, a_k, 0, 0, \ldots) = [\mathbf{t}^\mathbf{n}] f(\mathbf{t}) \det(K(\mathbf{t})) \prod_{i=1}^{k} r_i(\mathbf{t})^{n_i},$$

where we have set $t_r = 0$ for every index $r > k$ and identified

$$\mathbf{t} = (t_1, t_2, \ldots, t_k) = (t_1, t_2, \ldots, t_k, 0, 0, \ldots),$$

and where, as in (26),

$$\det(K(\mathbf{t})) = \det(\delta_{ij} - t_i (\partial r_i / \partial t_j) / r_i).$$

But

$$\frac{\partial r_i}{\partial t_j} = \frac{\partial}{\partial t_j} r(t_i, t_{2i}, \ldots) = \begin{cases} \left(\frac{\partial r}{\partial t_{j/i}}\right)_i & \text{if } i \text{ divides } j, \\ 0 & \text{otherwise.} \end{cases} \tag{59}$$

Hence, $\det(K(\mathbf{t}))$ is the determinant of an upper triangular matrix and we have

$$\det(K(\mathbf{t})) = \prod_{i=1}^{k} 1 - t_i \frac{(\partial r / \partial t_1)_i}{r_i}$$
$$= \prod_{i=1}^{k} \rho(t_i, t_{2i}, \ldots). \tag{60}$$

The conclusion follows by taking values of k to be arbitrarily large and by observing that the family $(\rho(\mathbf{t}_{(i)}) r(\mathbf{t}_{(i)})^{n_i})_{i=1,2,3\ldots}$ is multipliable, for any sequence $n_1, n_2, n_3, \ldots$ with a finite number of nonzero values. ∎

Corollary 15. Let $\mathcal{A} = XE(\mathcal{A})$ be the species of ordinary rooted trees. Then the cycle index series of $\mathcal{A}$ is given by

$$Z_{\mathcal{A}}(x_1, x_2, x_3, \ldots) = \sum_{n_1, n_2, n_3, \ldots} a_{n_1, n_2, n_3, \ldots} \frac{x_1^{n_1} x_2^{n_2} x_3^{n_3} \cdots}{1^{n_1} n_1! 2^{n_2} n_2! 3^{n_3} n_3! \cdots},$$

with $a_{n_1, n_2, n_3, \ldots} = 0$, if $n_1 = 0$, and, for $n_1 > 0$,

$$a_{n_1, n_2, n_3, \ldots} = n_1^{n_1 - 1} \prod_{\nu \geq 2} \phi_\nu^{n_\nu} - \nu n_\nu \phi_\nu^{n_\nu - 1}, \tag{61}$$

where

$$\phi_\nu = \phi_\nu(n_1, n_2, n_3, \ldots) = \sum_{d | \nu} d n_d. \tag{62}$$

Proof. Use Proposition 14 with $R = E$ and $F = X$ (see Exercise 19). ∎

Corollary 16. The cycle index series of the species End of endofunctions is given by

$$Z_{\text{End}}(x_1, x_2, x_3, \ldots) = \sum_{n_1, n_2, n_3, \ldots} \text{end}_{n_1, n_2, n_3, \ldots} \frac{x_1^{n_1} x_2^{n_2} x_3^{n_3} \cdots}{1^{n_1} n_1! 2^{n_2} n_2! 3^{n_3} n_3! \cdots},$$

with

$$\text{end}_{n_1, n_2, n_3, \ldots} = \phi_1^{n_1} \phi_2^{n_2} \phi_3^{n_3} \cdots, \tag{63}$$

where the ϕ_ν are defined by (62).

Proof. Use Proposition 14 with $R = E$ and $F = S$ (see Exercise 20). ∎

It is interesting to note that, as seen in Section 2.2,

$$\phi_\nu = \text{fix}(\sigma^\nu) \tag{64}$$

where σ is a permutation of cycle type $(n_1, n_2, n_3, \ldots)$ and that

$$a_{n_1, n_2, n_3, \ldots} = \text{fix } \mathcal{A}[\sigma] \quad \text{and} \quad \text{end}_{n_1, n_2, n_3, \ldots} = \text{fix End}[\sigma]. \tag{65}$$

Exercises for Section 3.2

1. a) Prove the implicit function theorem in the context of formal power series over a field of characteristic zero, i.e., show that, for any formal power series

$$f(x, y) = f_{00} + f_{10} x + f_{01} y + f_{20} x^2 + f_{11} xy + f_{02} y^2 + \cdots \tag{66}$$

satisfying $f(0, 0) = f_{00} = 0$ and $f_y(0, 0) = f_{01} \neq 0$, there exists a unique formal power series

$$a(x) = a_0 + a_1 x + a_2 x^2 + a_3 x^3 + \cdots$$

with $a(0) = 0$, satisfying $f(x, a(x)) = 0$.

HINT: Show that the conditions $f(x, a(x)) = 0$ and $a(0) = 0$ are equivalent to a recursive scheme (for the coefficients of $a(x)$) of the form $a_0 = 0$ and $a_{n+1} = \varphi_n(a+0, a_1, \ldots, a_n)$, where the $\varphi_n(x_0, x_1, \ldots, x_n)$ are polynomials whose coefficients only depend upon the $f_{i,j}$, with $i + f \leq n$.

b) Let H be a two-sort species satisfying conditions (7). Show that the series associated to the species A_H of H-enriched rooted trees: $a(x) = A_H(x)$, $\widetilde{a}(x) = \widetilde{A}_H(x)$, and

$$z(x_1, x_2, x_3, \ldots) = Z_{A_H}(x_1, x_2, x_3, \ldots),$$

are characterized by the equations

i) $a(x) = H(x, a(x)), \quad a(0) = 0,$

ii) $\widetilde{a}(x) = Z_H(x, x^2, x^3, \ldots; \widetilde{a}(x), \widetilde{a}(x^2), \widetilde{a}(x^3), \ldots), \quad \widetilde{a}(0) = 0,$

iii) $z(x_1, x_2, x_3, \ldots) = Z_H(x_1, x_2, x_3, \ldots; z_1, z_2, z_3, \ldots),$

$$z(0, 0, 0, \ldots) = 0. \quad (67)$$

2. Consider the species $A = A(X)$ defined recursively by the equation

$$A = X + C(X) \cdot A + E_{\geq 2}(X + A), \quad (68)$$

with initial condition $A(0) = 0$, where $C(X)$ is the species of cyclic permutations and $E_{\geq 2}(X)$ is that of sets of cardinality ≥ 2.

a) Give a few representative examples of A-structures on sets of cardinality ≥ 20.

b) Show that the species

$$X + 2X^2 + 2E_2 + 6X^3 + 9X E_2 + 2E_3$$

has a contact of order 3 with $A(X)$.

HINT: Adapt the proof of the implicit species theorem and use the fact that

$$E_{\geq 2}(P + Q + R + \cdots) = E(P + Q + R + \cdots) - 1$$
$$- (P + Q + R + \cdots)$$
$$= E(P) \cdot E(Q) \cdot E(R) \cdots - 1 \quad (69)$$
$$- (P + Q + R + \cdots).$$

c) Compute the coefficients of the series associated to A up to degree 4.

3. By adapting the proof of the implicit species theorem to the equation $\mathcal{A} = X \cdot E(\mathcal{A})$, show that the molecular decomposition of the species of rooted trees begins with the terms

$$\mathcal{A} = \mathcal{A}_0 + \mathcal{A}_1 + \mathcal{A}_2 + \mathcal{A}_3 + \mathcal{A}_4 + \mathcal{A}_5 + \mathcal{A}_6 + \cdots \tag{70}$$

where

$\mathcal{A}_0 = 0, \qquad \mathcal{A}_1 = X, \qquad \mathcal{A}_2 = X^2, \qquad \mathcal{A}_3 = XE_2 + X^3,$

$\mathcal{A}_4 = XE_3 + X^2 E_2 + 2X^4,$

$\mathcal{A}_5 = XE_4 + 3X^3 E_2 + XE_2(X^2) + X^2 E_3 + 3X^5,$

$\mathcal{A}_6 = XE_5 + X^2 E_2^2 + X^2 E_4 + 6X^4 E_2 + 3X^3 E_3 + 2X^2 E_2(X^2) + 6X^6.$

HINT: $\mathcal{A}_0 = 0$ and

$$\mathcal{A}_{n+1} = X \cdot E(\mathcal{A}_{n \leq 1})|_{n+1}$$
$$= X \cdot E(\mathcal{A}_0) \cdot E(\mathcal{A}_1) \cdots E(\mathcal{A}_n)|_{n+1}.$$

4. The species $Y = T$ of (plane) ternary rooted trees is characterized by the functional equation

$$Y = 1 + X \cdot Y^3. \tag{71}$$

a) Show that ternary rooted trees are asymmetric.

b) Show that

$$T(x) = \sum_{n \geq 0} \frac{1}{3n+1} \binom{3n+1}{n} x^n. \tag{72}$$

c) Generalize to m-ary rooted trees, $m \geq 1$.

5. Denote by $Y = {}_{2,3}A(X)$ the species of $(2, 3)$-rooted trees, which is the solution of the functional equation

$$Y = X + Y^2 + Y^3. \tag{73}$$

a) Show that these rooted trees are asymmetric.

b) Show that ${}_{2,3}A(X) = \sum_{n \geq 1} a_n x^n$, with

$$a_n = \frac{1}{n} \sum_{j \leq n-1} \binom{j}{n-1-j} \binom{n-i+j}{j}, \tag{74}$$

where the convention is that $\binom{n}{k} = 0$ if $k > n$.

6. a) Show analytically the equivalence of formulas (19) and (20).
   b) Use (19) or (20) to compute, up to degree 6, the coefficients of the series $F(_G A(x))$ in the following cases:

   $$
   \begin{array}{lll}
   \text{i)} & F = X, & G = E_2, \\
   \text{ii)} & F = L(G'), & G = X^2 + X^3, \\
   \text{iii)} & F = L(X^2), & G = C_3.
   \end{array}
   $$

7. Show that it is not always possible, in the context of species of structures, to reduce a combinatorial equation of the form $Y = X + G(Y)$, with $G(0) = G'(0) = 0$, to one of the form $Y = X \cdot R(Y)$, with $R(0) = 1$.
   HINT: Take $G = E_2$, consider a solution of $Y = X + G(Y) = X \cdot R(Y)$, and obtain a contradiction by passing to the cycle index series (or to the molecular decomposition).

8. Prove the following version of the implicit species theorem for virtual species. Let $H(X, Y)$ be a virtual species such that $H(0, 0) = 0$ and $\frac{\partial H}{\partial Y}(0, 0) = 0$. Then there exists a unique virtual species $A = A(X)$ satisfying the combinatorial equality $A = H(X, A)$, with $A(0) = 0$.

9. COMBINATORIAL ECLOSIONS (G. Labelle [180].) Let X and Y be species of singletons of two sorts, namely *leaves* and *buds*, respectively. Let also $G = G(Y)$ be a species of structures such that $G(0) = G'(0) = 0$. Define the two-sort species of *G-sheaves* by the formula

$$X \cdot L(G'(Y)). \tag{75}$$

Figure 12 describes a typical G-sheaf; the circular arcs represent G-structures. Let $\Psi = \Psi(X, Y)$ be a species of structures on the two sorts X and Y. One says that a Ψ-structure has undergone a *G-eclosion* whenever one of its buds has been replaced by a G-sheaf.

a) Considering the differential operator

$$\mathcal{D} = L(G'(Y)) \frac{\partial}{\partial Y}, \tag{76}$$

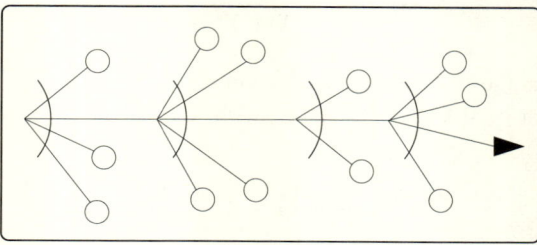

Fig. 12.

show that the species of Ψ-structures which have undergone a G-eclosion is given by the formula

$$XD\,\Psi(X,Y). \tag{77}$$

b) Interpret combinatorially the species

$$(XD)^n\Psi(X,Y) \quad \text{and} \quad \frac{(XD)^n}{n!}\Psi(X,Y), \qquad n\geq 0, \tag{78}$$

in terms of successive eclosions (forget the order of eclosion in the second case).

c) Show that for the species $A = {}_GA(X)$ of G-rooted trees (see (16)), one has

$$A(X) = e^{XD}Y\big|_{Y=0}, \tag{79}$$

where

$$e^{XD} = 1 + XD + \frac{(XD)^2}{2} + \cdots + \frac{(XD)^n}{n!} + \cdots. \tag{80}$$

d) Establish the more general formula

$$\Psi(X, A(X)) = e^{XD}\Psi(X,Y)\big|_{Y=0}. \tag{81}$$

10. Consider the species $K(X,Y) = e^{XD}Y$, with the notation of the preceding exercise, and introduce a new sort T (of leaves).

a) Show by direct computation and then by a combinatorial argument that the species $K(X,Y)$ satisfies the following addition theorem

$$K(X+T,Y) = K(X, K(T,Y)). \tag{82}$$

HINT: Use the fact that $e^{(X+T)D} = e^{XD}e^{TD}$.

b) Deduce that the species $A = {}_GA(X)$ of G-rooted trees satisfies the addition theorem

$$A(X+T) = K(X, A(T)). \tag{83}$$

c) Show, without appealing to differential operators, that the species $K(X,Y)$ is characterized by the combinatorial equation

$$K(X,Y) = X + Y + G(K(X,Y)) - G(Y). \tag{84}$$

11. JACOBI POLYNOMIALS. (See Example 12.)

a) One defines the weighted two-sort species $\text{Lag}^{(\alpha)}(X,Y)$ of Laguerre configurations by setting

$$\text{Lag}^{(\alpha)}[U,V] = \{\varphi : U \hookrightarrow U+V, \quad \text{injective function}\},$$

with $w(\varphi) = (\alpha + 1)^{\text{cyc}(\varphi)}$. Show that

$$\text{Lag}^{(\alpha)}(X, Y) = \mathcal{S}_{(\alpha+1)}(X) \, E(Y \cdot L(X)), \tag{85}$$

where $\mathcal{S}_{(\alpha+1)} = E \circ C_{\alpha+1}$, and that

$$\text{Lag}^{(\alpha)}(x, y) = \sum_{k,h \geq 0} (\alpha + 1 + h)^{\langle k \rangle} \frac{x^k y^h}{k! h!}. \tag{86}$$

Letting $\text{Lag}^{(\alpha)}(T) = \text{Lag}^{(\alpha)}(T, T_{-x})$, show that

$$\text{Lag}^{(\alpha)}(t) = \sum_{n \geq 0} L_n^{(\alpha)}(x) \frac{t^n}{n!},$$

where $L_n^{(\alpha)}(x)$ denotes the nth Laguerre polynomial (see Exercises 2.3.3 and 3.1.5).

b) Show that

$$\text{Jac}^{(\alpha,\beta)}(X, Y) = \text{Lag}^{(\alpha)}(X, Y) \times \text{Lag}^{(\beta)}(Y, X). \tag{87}$$

Conclude that

$$\text{Jac}^{(\alpha,\beta)}(x, y) = \sum_{k,h \geq 0} (\alpha + 1 + h)^{\langle k \rangle} (\beta + 1 + k)^{\langle h \rangle} \frac{x^k y^h}{k! \, h!}. \tag{88}$$

c) Jacobi polynomials $P_n^{(\alpha,\beta)}(x)$ are defined by formula (3.1.77), c). Computing the n–th derivative by Leibniz' product rule, show that

$$P_n^{(\alpha,\beta)}(x) = \sum_{k+h=n} \binom{n}{k} (\alpha + 1 + h)^{\langle k \rangle} (\beta + 1 + k)^{\langle h \rangle} \left(\frac{x+1}{2}\right)^k \left(\frac{x-1}{2}\right)^h. \tag{89}$$

Show moreover that the formulas giving the generating series (45) and (3.1.78), c) are equivalent.

d) Show that the cycle index series

$$Z^{(\alpha,\beta)} = Z^{(\alpha,\beta)}(x_1, x_2, x_3, \ldots; y_1, y_2, y_3, \ldots)$$

of the species $\text{Jac}^{(\alpha,\beta)}(X, Y)$ is given by

$$\prod_{k \geq 1} 2^{\alpha_k + \beta_k} (1 - x_k + y_k + R_k)^{-\alpha_k} (1 + x_k - y_k + R_k)^{-\beta_k} R_k^{-1}, \tag{90}$$

where $\alpha_k = v_k(\alpha + 1) - 1$ (see Example 2.3.12 for the definition of v_k), $\beta_k = v_k(\beta + 1) - 1$, and $R_k = R(x_k, y_k) = ((x_k + y_k - 1)^2 - 4x_k y_k)^{1/2}$.
HINT: Use the fact that the species of binary rooted trees $A(X, Y)$ and $B(X, Y)$ of Example 12 are asymmetric.

e) Deduce an explicit formula for the cycle index series of $\text{Jac}^{(\alpha,\beta)}(T)$, the one-sort weighted species defined by (46), as well as for $|\text{Jac}^{(\alpha,\beta)}[\sigma]|$ for any permutation σ (see Décoste [70] and Constantineau [60]).

12. Prove directly Scoins' formula (51) by introducing the species $\mathcal{V}(X, Y)$ of bicolored vertebrates and showing bijectively that for any $n, m \geq 1$,

$$|\mathcal{V}[n, m]| = n^m m^n. \tag{91}$$

13. ALTERNATING BIDIMENSIONAL LAGRANGE INVERSION (See [57]). Let $\varphi(x)$ and $\psi(x)$ be two formal power series in $\mathbb{A}[\![x]\!]$.

   a) Show that there exists unique formal power series $A(x, y)$ and $B(x, y)$ in $\mathbb{A}[\![x, y]\!]$ such that

   $$A(x, y) = x\varphi(A(x, y))$$
   $$B(x, y) = y\psi(B(x, y)). \tag{92}$$

   b) Show moreover, using Good–Lagrange's inversion formula (28), b), that for any integer $\alpha, \beta \geq 0$, one has

   $$[x^n y^k] A^\alpha B^\beta = \left(1 - \frac{(n-\alpha)(k-\beta)}{nk}\right)[x^{n-\alpha} y^{k-\beta}]\varphi^n(y)\psi^k(x). \tag{93}$$

14. ENUMERATION OF BICOLORED PLANE TREES ACCORDING TO DEGREE DISTRIBUTIONS. Let $\mathbf{r} = (r_0, r_1, r_2, \ldots)$ and $\mathbf{s} = (s_0, s_1, s_2, \ldots)$ be two sequences of formal variables, used here as degree counters of black points and white points, respectively. Introduce the following weighted two-sort species:
   - $\mathcal{P}_{\mathbf{r},\mathbf{s}} = \mathcal{P}_{\mathbf{r},\mathbf{s}}(X, Y)$, of plane (i.e., embedded in the plane) trees which are bicolored and weighted by degree, by putting

   $$w(\alpha) = r_0^{n_0} r_1^{n_1} r_2^{n_2} \cdots s_0^{m_0} s_1^{m_1} s_2^{m_2} \cdots,$$

   – where n_k and m_k denote respectively the number of black (of sort X) vertices and white (of sort Y) vertices, of degree k in the tree α. For example, the bicolored (plane) tree of Figure 11 has weight $r_1^4 r_3^2 r_4^2 s_1^8 s_2 s_3 s_5$.
   - $\mathcal{P}_{\mathbf{r},\mathbf{s}}^{\cdot X}$, rooted bicolored plane trees with black root, weighted by degree.
   - $\mathcal{G}_{\mathbf{r},\mathbf{s}}$, bicolored plane trees with distinguished edge, weighted by degree.
   Moreover, recall the following weighted (one-sort) species:
   - $\mathcal{C}_{\mathbf{r}}$ and $\mathcal{C}_{\mathbf{s}}$, oriented cycles weighted according to length (see Exercise 3.1.24) and $L_{\mathbf{r}'} = (\mathcal{C}_{\mathbf{r}})'$, $L_{\mathbf{s}'} = (\mathcal{C}_{\mathbf{s}})'$.

   a) Show that the species $A_{L_{\mathbf{r}'}, L_{\mathbf{s}'}}$ and $B_{L_{\mathbf{r}'}, L_{\mathbf{s}'}}$ of R, S-enriched bicolored rooted trees, with $R = L_{\mathbf{r}'}$ and $S = L_{\mathbf{s}'}$, are indeed the species of ordered bicolored rooted trees, weighted by degree, where we agree that a supplementary half-edge contributes to the degree of the root (black for $A_{L_{\mathbf{r}'}, L_{\mathbf{s}'}}$ and white for

$B_{L_{r'},L_{s'}}$). Moreover, show that

$$G_{r,s} = A_{L_{r'},L_{s'}} B_{L_{r'},L_{s'}}, \tag{94}$$

$$\mathfrak{p}_{r,s} = X_{r_0} + Y_{s_0} + \mathfrak{a}_{C_r,C_s}, \tag{95}$$

where $\mathfrak{a}_{C_r,C_s}$ denotes the species of R, S-enriched bicolored trees, with $R = C_r$ and $S = C_s$. Also show that

$$\mathfrak{p}_{r,s}^{\bullet x} = X_{r_0} + X \cdot C_r(B_{L_{r'},L_{s'}}), \qquad \mathfrak{p}_{r,s}^{\bullet y} = Y_{s_0} + Y \cdot C_s(A_{L_{r'},L_{s'}}), \tag{96}$$

$$\mathfrak{p}_{r,s} = \mathfrak{p}_{r,s}^{\bullet x} + \mathfrak{p}_{r,s}^{\bullet y} - G_{r,s}. \tag{97}$$

b) Let n and k be nonnegative integers with $n+k \geq 2$, and let $\mathbf{i} = (i_0, i_1, i_2, \ldots)$ and $\mathbf{j} = (j_0, j_1, j_2, \ldots)$ be two sequences of nonnegative integers. Show that there exists a bicolored (plane) tree α on the 2-set of vertices $([n], [k])$ with degree distribution $(\mathbf{i}, \mathbf{j})$ (i.e., of weight $w(\alpha) = \mathbf{r}^{\mathbf{i}}\mathbf{s}^{\mathbf{j}}$) if and only if

$$\begin{gathered} i_0 = 0 = j_0, \qquad n = i_1 + i_2 + i_3 + \cdots, \qquad k = j_1 + j_2 + j_3 + \cdots, \\ i_1 + 2i_2 + 3i_3 + \cdots = j_1 + 2j_2 + 3j_3 + \cdots = n + k - 1. \end{gathered} \tag{98}$$

c) Show that

$$A_{L_{r'},L_{s'}}(x, y) = \widetilde{A}_{L_{r'},L_{s'}}(x, y), \quad \text{and} \quad B_{L_{r'},L_{s'}}(x, y) = \widetilde{B}_{L_{r'},L_{s'}}(x, y), \tag{99}$$

and that these two series, more simply denoted by $A_{\mathbf{r}',\mathbf{s}'}(x, y)$ and $B_{\mathbf{r}',\mathbf{s}'}(x, y)$ respectively, satisfy the system of equations

$$\begin{aligned} A_{\mathbf{r}',\mathbf{s}'}(x, y) &= x\psi_\mathbf{r}(B_{\mathbf{r}',\mathbf{s}'}(x, y)), \\ B_{\mathbf{r}',\mathbf{s}'}(x, y) &= y\psi_\mathbf{s}(A_{\mathbf{r}',\mathbf{s}'}(x, y)), \end{aligned} \tag{100}$$

with

$$\psi_\mathbf{r} = r_1 + r_2 x + r_3 x^2 + \cdots, \quad \text{and} \quad \psi_\mathbf{s} = s_1 + s_2 x + s_3 x^2 + \cdots. \tag{101}$$

d) Let n, k, with $n + k \geq 2$, and $\mathbf{i}, \mathbf{j}$ be as in part b), satisfying conditions (98). Show that the number $\mathfrak{p}(\mathbf{i}, \mathbf{j})$ of plane trees on $([n], [k])$ with degree distribution $(\mathbf{i}, \mathbf{j})$ is given by

$$\mathfrak{p}(\mathbf{i}, \mathbf{j}) = (n-1)!(k-1)!\binom{n}{\mathbf{i}}\binom{k}{\mathbf{j}}, \tag{102}$$

where $\binom{n}{\mathbf{i}}$ denotes the multinomial coefficient $(n \, atopi_1, i_2, i_3, \ldots)$. Also show that the total number of plane trees on $([n], [k])$ is given by

$$|\mathfrak{p}[n, k]| = \frac{((n+k-2)!)^2}{(n-1)!(k-1)!} = n^{\langle k-1 \rangle} k^{\langle n-1 \rangle}. \tag{103}$$

HINT: Use one of the formulas in (96), pass to the generating series, and use the alternating two-dimensional Lagrange inversion formula (see Exercise 13).

e) Under the same hypotheses as in d), show that the numbers $\widetilde{g}(\mathbf{i}, \mathbf{j})$ and $\widetilde{p}(\mathbf{i}, \mathbf{j})$ of unlabeled bicolored plane trees with a distinguished edge and of unlabeled bicolored plane trees, respectively, whose degree distribution is $(\mathbf{i}, \mathbf{j})$, are given by

$$\widetilde{g}(\mathbf{i}, \mathbf{j}) = \frac{n+k-1}{nk}\binom{n}{\mathbf{i}}\binom{k}{\mathbf{j}} \tag{104}$$

and

$$\widetilde{p}(\mathbf{i}, \mathbf{j}) = \widetilde{f}(\mathbf{i}, \mathbf{j}) + \widetilde{f}(\mathbf{j}, \mathbf{i}) - \widetilde{g}(\mathbf{i}, \mathbf{j}), \tag{105}$$

where

$$\widetilde{f}(\mathbf{i}, \mathbf{j}) = \frac{1}{n}\sum_{h,d}\phi(d)\binom{n/d}{\mathbf{i}/d}\binom{(k-1)/d}{(\mathbf{j}-\delta_h)/d}, \tag{106}$$

the sum being taken over all pairs of integers $h, d \geq 1$ such that $h \in \text{Supp}(\mathbf{j}) = \{\ell \mid j_\ell \neq 0\}$ and $d \in \text{Div}(h, \mathbf{i}, \mathbf{j} - \delta_h)$, where $\delta_h = (\delta_{h,\ell})_{\ell \geq 0}$ and $\text{Div}(h, \mathbf{i}, \mathbf{j})$ denotes the set of common divisors of $h, i_0, j_0, i_1, j_1, i_2, j_2, \ldots$.
HINT: Use Equations (96) and (97), set $F_{\mathbf{r},\mathbf{s}} = \mathbf{p}_{\mathbf{r},\mathbf{s}}^{\bullet y}$, pass to type generating series, and use alternating two-dimensional Lagrange inversion to show that

$$\widetilde{f}_{n,k}(\mathbf{r}, \mathbf{s}) = [x^n y^k]\widetilde{F}_{\mathbf{r},\mathbf{s}}(x, y)$$

$$= \frac{1}{n}\sum_{\substack{d\mid(n,k-1)\\1\leq\ell\leq n/d}}\phi(d)s_{\ell d}\left[x^{\frac{n}{d}-\ell}y^{\frac{k-1}{d}}\right]\psi_{\mathbf{r}^d}^{n/d}(y)\psi_{\mathbf{s}^d}^{(k-1)/d}(x), \tag{107}$$

where $\mathbf{r}^d = (r_0^d, r_1^d, r_2^d, \ldots)$, and observe that

$$\widetilde{f}(\mathbf{i}, \mathbf{j}) = [\mathbf{r}^\mathbf{i}\mathbf{s}^\mathbf{j}]\widetilde{f}_{n,k}(\mathbf{r}, \mathbf{s}). \tag{108}$$

NOTE: Unlabeled bicolored plane trees are closely related to Shabat's polynomials; see Bétréma–Svonkin [31].

f) (See Goulden–Jackson [137] and Bédard–Goupil [7].) Under the same hypotheses as in d), show that $\widetilde{g}(\mathbf{i}, \mathbf{j})$ is equal to the number of ways to decompose a given cyclic permutation σ in the symmetric group into a product of two permutations α and β with cycle types $(1^{i_1}2^{i_2}3^{i_3}\cdots)$ and $(1^{j_1}2^{j_2}3^{j_3}\cdots)$, respectively.

15. MACMAHON'S MASTER THEOREM [231].
    a) Let $\mathbf{R} = (R_1, R_2, \ldots, R_k)$ be a vector of k-sort species such that $R_i(0) = 0$, $i = 1, \ldots, k$. Show that then $\mathcal{A}_{\mathbf{R},i} = 0, i = 1, \ldots, k$ (see (29)), and that the

species $\text{End}_{\mathbf{R}}$ of **R**-enriched endofunctions coincides with the subspecies $\mathcal{S}_{\mathbf{R}}$ of **R**-enriched permutations. Conclude that only the linear part $\mathbf{R}_{\leq 1}$ of **R** occurs in $\text{End}_{\mathbf{R}}$.

HINT: No fiber can be empty.

b) Let $R_i(X) = a_{i1}X_1, a_{i2}X_2 + \cdots + a_{ik}X_k, i = 1, \ldots, k$, where the a_{ij} are formal variables. Show that

i) $\mathcal{S}_{\mathbf{R}}(\mathbf{x}) = (\det(\delta_{ij} - x_i a_{ij}))^{-1}$,

ii) for any $\mathbf{n} = (n_1, n_2, \ldots, n_k) \in \mathbb{N}^k$, one has a bijection

$$\mathcal{S}_{\mathbf{R}}[\mathbf{n}] \xrightarrow{\sim} \mathbf{R}^{\mathbf{n}}[\mathbf{n}].$$

HINT: Adapt the combinatorial proof of Good–Lagrange's formulas (with $F = 1$) to the present case.

c) Deduce the following formula of Macmahon:

$$[\mathbf{x}^{\mathbf{n}}](\det(\delta_{ij} - x_i a_{ij}))^{-1} = [\mathbf{t}^{\mathbf{n}}]\prod_{i=1}^{k}(a_{i1}t_1 + a_{i2}t_2 + \cdots + a_{ik}t_k)^{n_i}. \quad (109)$$

16. COMPATIBLE PERMUTATIONS (Strehl [311]). Let k be an integer ≥ 1. Introduce the weighted species $^{k \times k}\mathcal{S}_{(\alpha)}(\mathbf{Z})$ of k^2 sorts Z_{ij}, $i \leq i, j \leq k$, of "compatible permutations," weighted by a cycle counter α. On a matrix of sets $U = (U_{ij})_{1 \leq i, j \leq k}$, a *compatible permutation* is, by definition, a permutation σ of the set $U = \sum_{i,j} U_{ij}$ satisfying the following compatibility condition: For $j = 1, \ldots, k$,

$$\sigma\left(\sum_{i} U_{ij}\right) \subseteq \sum_{h} U_{jh}. \quad (110)$$

The weight of the compatible permutation $\sigma \in {}^{k \times k}\mathcal{S}_{(\alpha)}[U]$ is given by $w(\sigma) = \alpha^{\text{cyc}(\sigma)}$. The goal of this exercise is to prove that the generating series of the species $^{k \times k}\mathcal{S}_{(\alpha)}$ is given by

$$^{k \times k}\mathcal{S}_{(\alpha)}(\mathbf{z}) = \det(I - \mathbf{z})^{-\alpha}, \quad (111)$$

where **z** denotes the matrix of variables $(z_{ij})_{1 \leq i, j \leq k}$ associated to the sorts Z_{ij}. As this species admits a concept of connected structures $^{k \times k}\mathcal{S}^c_{(\alpha)}$ (the circular permutations), one can appeal to the exponential formula (2.3.75) and it suffices to prove (111) for one particular value of α. This will be done with $\alpha = -1$.

a) Let $U = (U_{ij})$ be a matrix of sets for which there exists a j such that $\sum_i U_{ij} \geq 2$. Show that then

$$\left|{}^{k \times k}\mathcal{S}_{(-1)}[U]\right| = \sum_{\sigma \in {}^{k \times k}\mathcal{S}[U]} (-1)^{\text{cyc}(\sigma)} = 0. \quad (112)$$

HINT: Let a and b be two distinct elements of $\sum_i U_{ij}$. Show that the transposition (a, b) induces (by composition) an involution without fixed points $\sigma \mapsto \sigma \circ (a, b)$ on $^{k \times k}\mathcal{S}[U]$ which reverses the sign of the permutations.

b) Show that
$$^{k \times k}\mathcal{S}_{(-1)}(\mathbf{z}) = \det(I - \mathbf{z}). \tag{113}$$

HINT: Let $U = (U_{ij})_{1 \le i, j \le k}$ be a matrix of sets such that for any j, $|\sum_i U_{ij}| \le 1$, and admitting a compatible permutation σ. Set $m_{ij} = |U_{ij}|$ and $B = \{j \mid \sum_i m_{ij} = 1\} \subseteq [k]$. Then show that $B = \{j \mid \sum_h m_{jh} = 1\}$, that σ is uniquely determined by the matrix $\mathbf{m} = (m_{ij})$, and that indeed $\mathbf{m}$ is the matrix representation of a partial permutation $\hat{\sigma}$ on $[k]$, with domain and codomain B, and with $\mathrm{cyc}(\sigma) = \mathrm{cyc}(\hat{\sigma})$. Conclude that

$$^{k \times k}\mathcal{S}_{(-1)}(\mathbf{z}) = \sum_{\substack{\mathbf{m} \in \mathbb{N}^{k \times k} \\ \sum_i m_{ij} \le 1}} \sum_{\sigma \in {}^{k \times k}\mathcal{S}[\mathbf{m}]} (-1)^{\mathrm{cyc}(\sigma)} \frac{\mathbf{z}^{\mathbf{m}}}{\mathbf{m}!}$$

$$= \sum_{B \subseteq [k]} \sum_{\sigma \in \mathcal{S}[B]} (-1)^{\mathrm{cyc}(\sigma)} \prod_{i \in b} z_{i \sigma(i)}$$

$$= \det(I - \mathbf{z}).$$

c) Establish formula (111) as well as, for $\alpha = 1$,
$$^{k \times k}\mathcal{S}(\mathbf{z}) = \det(I - \mathbf{z})^{-1}. \tag{114}$$

d) Establish directly identity (33), b), by substituting in $^{k \times k}\mathcal{S}(Z)$ the transpose of the species matrix $Q_{ij} = X_i \cdot R_i^{(j)}(A_{\mathbf{R}})$, $1 \le i, j \le k$.

17. JACOBI'S IDENTITY.

a) Let $^{k \times k}\mathcal{S}^c(\mathbf{Z})$ be the species of connected compatible permutations (see the preceding exercise). Show that
$$^{k \times k}\mathcal{S}^c(\mathbf{z}) = \mathrm{trace}\, \log(I - \mathbf{z})^{-1}. \tag{115}$$

HINT: A cyclic compatible permutation can be considered as a circuit in the complete oriented graph K_k, where an arc of the circuit, say from i to j, is labeled by elements of sort Z_{ij}. If the circuit is of length $m \ge 1$, it can be pointed in m distinct ways, each time giving a closed walk in the oriented graph whose adjacency matrix is $(Z_{ij})_{1 \le i, j \le k}$. One has then

$$^{k \times k}\mathcal{S}^c(\mathbf{z}) = \sum_{m \ge 1} \frac{1}{m} \mathrm{trace}\, \mathbf{z}^m$$

$$= \mathrm{trace} \sum_{m \ge 1} \frac{\mathbf{z}^m}{m}$$

$$= \mathrm{trace}\, \log(I - \mathbf{z})^{-1}.$$

b) Deduce Jacobi's identity

$$\exp \operatorname{trace} \log(I - \mathbf{z})^{-1} = \det(I - \mathbf{z})^{-1} \quad (116)$$

for any matrix of variables $\mathbf{z} = (z_{ij})_{1 \le i,j \le k}$.

18. By introducing a cycle counter $\alpha + 1$ for the **R**-enriched endofunctions, prove the $\alpha + 1$-extension of Good–Lagrange's inversion formula (see (28)), due to Strehl [311] (see also Zeng [339]):

$$[\mathbf{x}^\mathbf{n}] \frac{F(\mathbf{A}(\mathbf{x}))}{\det(J(\mathbf{A}(\mathbf{x})))^{\alpha+1}} = [\mathbf{t}^\mathbf{n}] F(\mathbf{t}) \det(K(\mathbf{t}))^{-\alpha} \mathbf{R}^\mathbf{n}(\mathbf{t}). \quad (117)$$

19. Prove Corollary 15 which gives explicitly the individual coefficients of the cycle index series $Z_\mathcal{A}$ of the species of rooted trees.
    HINT: Use Proposition 14 with $R = E$ and $F = X$ to show that

$$[\mathbf{x}^\mathbf{n}] Z_\mathcal{A} = [\mathbf{t}^\mathbf{n}] t_1 \prod_{\nu \ge 1} (1 - t_\nu) \exp(\phi_\nu t_\nu / \nu) \quad (118)$$

and develop each factor of the infinite product into a series, before extracting the desired coefficient.

20. Prove Corollary 16 which gives explicitly the individual coefficients of the cycle index series Z_{End} of the species of endofunctions.
    HINT: Use Proposition 14 with $R = E$ and $F = S$ to first show that

$$[\mathbf{x}^\mathbf{n}] Z_\mathcal{A} = [\mathbf{t}^\mathbf{n}] t_1 \prod_{\nu \ge 1} \exp(\phi_\nu t_\nu / \nu). \quad (119)$$

21. a) Give direct combinatorial proofs of formulas (61) and (63), based on observations (65) (see Constantineau–Labelle [60] and [63]).
    b) Let $\sigma = (\sigma_1, \sigma_2, \ldots)$ be a sequence of integers ≥ 0 such that $n = \sum_k k\sigma_k < \infty$. For each of the following species F, prove the given formulas for fix $F[\sigma]$.
       i) $F = S$, the species of permutations,

$$\operatorname{fix} S[\sigma] = \prod_{k \ge 1} k^{\sigma_k} \sigma_k!. \quad (120)$$

       ii) $F = \mathcal{C}$, the species of cyclic permutations,

$$\operatorname{fix} \mathcal{C}[\sigma] = \phi(d) d^{(n/d)-1} ((n/d) - 1)!, \quad (121)$$

       if $\sigma_d = n/d$, and fix $\mathcal{C}[\sigma] = 0$, otherwise.
       iii) $F = \operatorname{Der}$, the species of derangements,

$$\operatorname{fix} \operatorname{Der}[\sigma] = \prod_{k \ge 1} \sum_{j=0}^{\sigma_k} \binom{\sigma_k}{j} |\operatorname{Der}[j]| k^j (k-1)^{\sigma_k - j}. \quad (122)$$

iv) $F = \mathrm{I}$, the species of involutions without fixed points,

$$\mathrm{fix}\,\mathrm{I}[\sigma] = \prod_{k \geq 1} J_k, \tag{123}$$

where

$$J_k = \begin{cases} \sum_{j \geq 0} \binom{\sigma_k}{2j} |\mathrm{I}[2j]|\, k^j, & \text{if } k \text{ is even,} \\ k^{\sigma_k/2}\, |\mathrm{I}[\sigma_k]| \chi(\sigma_k \text{ is even }), & \text{if } k \text{ is odd.} \end{cases} \tag{124}$$

NOTE: Recall that $|\mathrm{I}[2j]| = 1 \cdot 3 \cdot 5 \cdots (2j-1)$.

v) $F = \mathrm{Oct}$, the species of octopuses,

$$\mathrm{fix}\,\mathrm{Oct}[\sigma] = \phi(d) d^{(n/d)-1} (2^{n/d} - 1)((n/d) - 1)!, \tag{125}$$

if $\sigma_d = n/d$, and $\mathrm{fix}\,\mathrm{Oct}[\sigma] = 0$, otherwise.

vi) $F = E \circ \mathrm{Oct}$, the species of assemblies of octopuses,

$$\mathrm{fix}\,E \circ \mathrm{Oct}[\sigma] = \prod_{k \geq 1 : \sigma_k \geq 1} 2^{\sigma_k - 1} \sigma_k! k^{\sigma_k}. \tag{126}$$

vii) F is the species of idempotent endofunctions,

$$\mathrm{fix}\,F[\sigma] = \sum_{0 \leq \lambda \leq \sigma} \prod_{k=1}^{n} \binom{\sigma_k}{\lambda_k} \left(\sum_{d \mid k} d \lambda_d \right)^{\sigma_k - \lambda_k}. \tag{127}$$

viii) $F = \mathrm{Par}$, the species of set partitions,

$$\mathrm{fix}\,\mathrm{Par}[\sigma] = \sum_{\pi \in \mathrm{Par}[C_\sigma]} \prod_{b \in \pi} \sum_{d \mid \gcd(b)} d^{|b|-1}, \tag{128}$$

where C_σ denotes the set of cycles of a given permutation on $[n]$ with cycle type σ and $\gcd(b)$ denotes the greatest common divisor of the cycle lengths appearing in any block b of π.

HINT: Extract the coefficients of the cycles index series or else proceed by direct combinatorial constructions (see [64]).

22. Joni's multidimensional inversion theorem (see [156]) is an extension of formula (20) and is expressed as follows. Given m formal power series $g_i = g_i(x_1, \ldots, x_m)$, $1 \leq i \leq m$, such that $0 = g_i(0, \ldots, 0) = (\partial g_i/\partial x_j)(0, \ldots, 0)$, $1 \leq i, j \leq m$, define m formal power series $b_i(x_1, \ldots, x_m)$, by the system of equations

$$b_i = x_i + g_i(b_1, \ldots, b_m), \qquad 1 \leq i \leq m. \tag{129}$$

Then for any formal power series $f(\mathbf{x}) = f(x_1, \ldots, x_m)$, one has

$$f(b_1, \ldots, b_m) = \sum_{k_1, \ldots, k_m \geq 0} D_1^{k_1} \cdots D_m^{k_m} f(\mathbf{x}) |M(\mathbf{x})| \prod_{i=1}^{m} \frac{g_i(\mathbf{x})^{k_i}}{k_i!}, \qquad (130)$$

where $D_j = \partial/\partial x_j$, and

$$|M(\mathbf{x})| = |M(x_1, \ldots, x_m)| = \det(\delta_{ij} - \partial g_i/\partial x_j). \qquad (131)$$

a) Taking Joni's theorem for granted, show that for any species of structures F and G such that $G(0) = G'(0) = 0$, the index series $Z_{F(_GA)}$ (see (16)) is given by

$$Z_{F(_GA)} = \sum_{k_1, \ldots, k_m \geq 0} D_1^{k_1} \cdots D_m^{k_m} Z_F(\mathbf{x}) \prod_{i \geq 1} \frac{\gamma(\mathbf{x_i}) Z_G(\mathbf{x_i})^{k_i}}{k_i!}, \qquad (132)$$

where $\gamma(\mathbf{x}) = 1 - \frac{\partial}{\partial x_1} Z_G(\mathbf{x})$, and $\mathbf{x}_i = (x_i, x_{2i}, x_{3i}, \ldots)$.

b) Show that the cycle index series of the species $\mathcal{P} = {}_{E_2}\mathcal{A}$ of commutative parenthesizations is given by the formula

$$Z_\mathcal{P} = \sum_{k_1, k_2, \ldots} D_1^{k_1} D_2^{k_2} \cdots x_1 \prod_{i \geq 1} \frac{(1 - x_i)[(x_i^2 + x_{2i})/2]^{k_i}}{k_i!}. \qquad (133)$$

23. Let $\mathcal{A}_S$ be the species of S-enriched rooted trees (i.e., enriched by permutations). Show that for any species of structures F and any finite sequence $\mathbf{n} = (n_1, n_2, \ldots)$ of nonnegative integers, one has

$$[\mathbf{x^n}] Z_{F(\mathcal{A}_S)}(\mathbf{x}) = [\mathbf{t^n}] Z_F(\mathbf{t}) \prod_{k \geq 1} \frac{(1 - 2x_k)}{(1 - x_k)^{1 + \sum_{d|k} n_d}} \qquad (134)$$

and deduce an explicit formula for fix $\mathcal{A}_S[n_1, n_2, \ldots]$.
HINT: $(1 - 2x)/(1 - x)^{1+\alpha} = 2/(1-x)^\alpha - 1/(1-x)^{1+\alpha} = 1 + \sum_{\nu \geq 1} (\alpha - \nu)(\alpha + 1)(\alpha + 2) \cdots (\alpha + \nu - 1) x^\nu / \nu!$.

3.3. Quadratic Iterative Methods

In this section we present some acceleration methods for the iterative computation of species of structures defined by functional equations, as well as for their associated series. These methods are based, for the most part, on Newton–Raphson iteration. In addition to functional equations of the type $Y = H(X, Y)$, already encountered in the preceding sections, equations of the form

$$Y(X) = X + Y(G(X)), \qquad (1)$$

3.3. Quadratic Iterative Methods

for a given species G, or, more generally, of the form

$$Y = H(X, Y \circ G), \qquad (2)$$

which lend themselves to an iterative approach, will also be treated here.

Newton–Raphson's iterative method, also called the tangent method, is well known in analysis. It allows for successive approximations α_n, $n = 0, 1, 2, \ldots$, of a root a of an equation of form $f(t) = 0$, by starting with a (well-chosen) initial approximation α_0 and defining subsequent ones as $\alpha_{n+1} = \alpha_n^+$, where

$$\alpha^+ = \alpha - f'(\alpha)^{-1} f(\alpha) \qquad (3)$$

is the point of intersection of the tangent to the graph of f at the point $(\alpha, f(\alpha))$ with the t-axis. The strength of this method comes from the fact that fairly weak hypotheses ensure quadratic convergence of the sequence α_n toward the root a, that is,

$$a - \alpha_{n+1} = O((a - \alpha_n)^2) \to 0, \qquad \text{if } n \to \infty. \qquad (4)$$

For instance, if f is in the class C^2 in a neighborhood of a and if $f'(a) \neq 0$, then a Taylor expansion of (3) in a neighborhood of a shows that

$$a - \alpha^+ = -\frac{f''(a)}{2 f'(a)} \cdot (a - \alpha)^2 + \cdots . \qquad (5)$$

This phenomenon of rapid convergence is informally described by saying that the number of exact decimal places for the root a is "doubled" with each iteration step.

It is natural, in the context of formal power series, to adapt Newton–Raphson iteration to the *rapid* inversion of series for the operation of substitution. Brent and Kung [39] have studied the efficiency of this approach from a computational complexity point of view.

Let $\mathbb{K}$ be a commutative field of characteristic zero and let $c(x) \in \mathbb{K}[\![x]\!]$, with $c(0) = 0$ and $c'(0) \neq 0$. We may write $c(x) = x/R(x)$, where $R(x) \in \mathbb{K}[\![x]\!]$, with $R(0) \neq 0$. By taking

$$f(t) = x R(t) - t, \qquad (6)$$

where $t = t(x)$, we immediately see that the solution $t = A$ of the equation $f(t) = 0$ is the formal power series

$$A = A(x) = c^{\langle -1 \rangle}(x) \in \mathbb{K}[\![x]\!], \qquad (7)$$

since the equation $F(A) = 0$ is equivalent to $A = x R(A)$ and thus $c(A) = x$. Using (3), a sequence of approximations $\alpha_i = \alpha_i(x)$, $i \geq 0$, of this solution A is obtained by setting $\alpha = 0$ and $\alpha_{i+1} = \alpha_i^+$, $i \geq 0$, with

$$\alpha^+ = \alpha + \frac{xR(\alpha) - \alpha}{1 - xR'(\alpha)}. \tag{8}$$

Since (5) is still valid, there is again a phenomenon of *quadratic* convergence, which is expressed as follows: *If $\alpha(x)$ has a contact of order n (see (1.2.27)) with the solution $A(x)$, then $\alpha^+(x)$ has a contact of order $2n + 2$ with $A(x)$.* That is to say, the number of exact terms is at least doubled at each iteration step.

We propose to deduce combinatorially version (8) of Newton–Raphson iteration by looking for approximations of the root T of the combinatorial equation $f(T) = XR(T) - T = 0$, that is,

$$T = X \cdot R(T), \tag{9}$$

where R is a fixed species of structures. As we know, the solution to this equation is the species $T = \mathcal{A}_R$ of R-enriched rooted trees. Thus for a fixed integer $n \geq 0$, we consider the following combinatorial approximation problem: Starting from a species α having a contact of order n with the species $\mathcal{A}_R$ (i.e., $\alpha =_n \mathcal{A}_R$; see Definition 1.2.14), construct a new approximation α^+ of $\mathcal{A}_R$ where this time

$$\alpha^+ =_{2n+2} \mathcal{A}_R. \tag{10}$$

Let us say that an (R-enriched) rooted tree on a finite set U is *light* if $|U| \leq n$ and *heavy* if

$$n + 1 \leq |U| \leq 2n + 2. \tag{11}$$

Given the nature of the problem, we may suppose that α is itself the species of light rooted trees, that is, $\alpha = (\mathcal{A}_R)_{\leq n}$. Denote by β that of heavy rooted trees. One evidently has

$$\alpha + \beta =_{2n+2} \mathcal{A}_R, \tag{12}$$

and thus $\alpha^+ =_{2n+2} \alpha + \beta$.

Hence, our approximation problem comes down to the construction of the species β of heavy rooted trees starting from the species α of light rooted trees. Let us then analyze the structure of a generic rooted tree on a set of cardinality $\leq 2n + 2$.

A simple counting argument shows that the root of such a heavy rooted tree is attached to an R-assembly of rooted trees with *at most one* heavy one (see Figure 1). Indeed, if two of these rooted trees were heavy, the cardinality of the underlying set would be $> 2n + 2$ (counting the root), contrary to hypothesis.

This implies that $\beta =_{2n+2} \gamma$, where γ is the species of structures defined by the linear functional equation

$$\gamma = D(X) + C(X) \cdot \gamma, \tag{13}$$

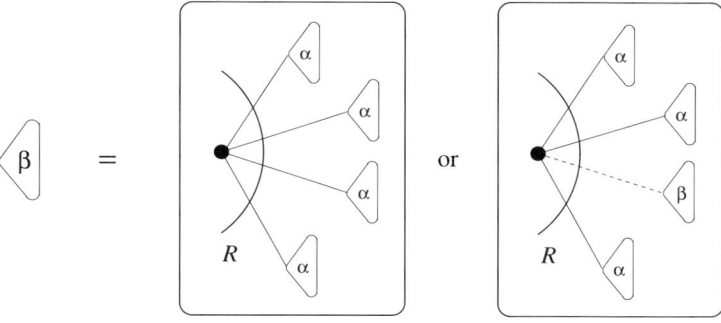

Fig. 1.

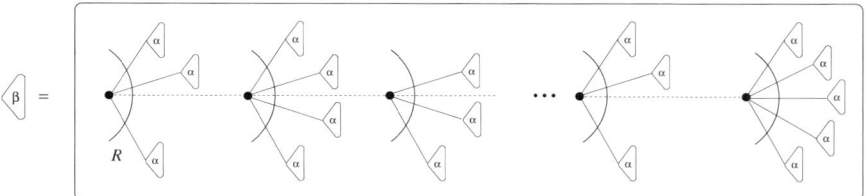

Fig. 2.

with $D(X) = X \cdot R(\alpha) - \alpha$ and $C(X) = X \cdot R'(\alpha)$. The term $D(X)$ corresponds to the first case of Figure 1: that of heavy rooted trees whose branches are all light; the subtraction of α comes from the fact that light rooted trees, although of the form $X \cdot R(\alpha)$, should be excluded here. The term $C(X) = X \cdot R'(\alpha)$ denotes the species of R-enriched rooted trees having a "free" branch and whose other branches are light; this factor is obtained by removing the unique heavy rooted tree β corresponding to the second case in Figure 1. The general solution (see Example 3.2.4) to Equation (13) is given by

$$\gamma = L(XR'(\alpha)) \cdot (XR(\alpha) - \alpha). \tag{14}$$

This expresses the fact that any heavy rooted tree decomposes canonically into a finite sequence of $XR'(\alpha)$-structures followed by a last heavy rooted tree with all branches light (see Figure 2).

Taking Equation (12) into account, we have thus combinatorially obtained the following version of Newton–Raphson iteration (compare with (8)).

Proposition 1. *Let α be a species of R-enriched light (i.e., supported by cardinalities $\leq n$) rooted trees. Then the species of structures defined by*

$$\alpha^+ = \alpha + L(XR'(\alpha)) \cdot (XR(\alpha) - \alpha) \tag{15}$$

has a contact of order $2n + 2$ with the species $\mathcal{A}_R$ of R-enriched rooted trees.

Passing to generating series, the validity of Newton–Raphson's method for formal power series is deduced from this proposition. However, the real interest of Proposition 1 is that it allows the iterative computation of the species $\mathcal{A}_R$ itself, with a quadratic rate of convergence, as well as its molecular decomposition, cycle index series, or any other associated series. Hence we obtain successive approximations $\alpha^{(n)}$ of the species $\mathcal{A}_R$ by letting $\alpha^{(0)} = 0$ and, for $n \geq 0$,

$$\alpha^{(n+1)} = (\alpha^{(n)^+})_{\leq 2n+2}. \tag{16}$$

Note that α^+ has been replaced by $\alpha^+_{\leq 2n+2}$ to keep only the exact terms of the approximation at each step. We then have $\alpha^{(n)} =_{2^{n+1}-2} \mathcal{A}_R(X)$. Here is an efficient algorithm to generate the sequence $\alpha^{(n)}$ without having to compute the preliminary auxiliary species:

$$L(XR'(\alpha)) = 1 + XR'(\alpha) + (XR'(\alpha))^2 + \cdots. \tag{17}$$

Set $A = \mathcal{A}_R(X)$, and recall that A_k denotes the species A restricted to the cardinality k.

Algorithm 2. INPUT: $\alpha = A_{\leq k} = A_0 + A_1 + \cdots + A_k$; OUTPUT; $A_{\leq 2k+2}$.

For i from 1 to $k+2$, do
    $A_{k+i} := (XR(\alpha))_{k+i}$
    For j from 1 to $i-1$, do
        $A_{k+i} := A_{k+i} + (XR'(\alpha))_j \cdot A_{k+i-j}$
End.

Proof. To simplify notation, set $P = XR(\alpha)$ and $Q = XR'(\alpha)$. For any i with $1 \leq i \leq k+2$, an A_{k+i}-structure (i.e., a β-structure on the cardinality $k+i$) is, according to Figure 1, either a P_{k+i}-structure or a $Q_j \cdot A_{k+i-j}$-structure, with $1 \leq j \leq i-1$. Hence

$$A_{k+i} = P_{k+i} + Q_1 A_{k+i-1} + \cdots + Q_{i-1} A_{k+1}. \tag{18}$$

∎

Example 3. Let us apply this algorithm to the species $\mathcal{A}$ of rooted trees, characterized by the equation $\mathcal{A} = X \cdot E(\mathcal{A})$, to obtain the molecular decomposition of $\mathcal{A}_{\leq 6}$, starting from $\alpha = \mathcal{A}_{\leq 2} = X + X^2$, i.e., with $k = 2$. We have $P = Q = XE(\alpha) = XE(X)E(X^2)$ so that

$$Q_1 = X, \quad Q_2 = X^2, \quad Q_3 = P_3 = XE_2 + X^3,$$
$$P_4 = XE_3 + X^4, \quad P_5 = XE_4 + XE_2(X^2) + X^3 E_2, \tag{19}$$
$$P_6 = XE_5 + X^3 E_3 + X^2 E_2(X^2).$$

3.3. Quadratic Iterative Methods

Consequently, the algorithm gives

$$\mathcal{A}_3 = P_3 = XE_2 + X^3, \tag{20}$$

$$\mathcal{A}_4 = P_4 + Q_1\mathcal{A}_3 = XE_3 + X^2E_2 + 2X^4, \tag{21}$$

$$\mathcal{A}_5 = P_5 + Q_1\mathcal{A}_4 + Q_2\mathcal{A}_3$$
$$= XE_4 + XE_2(X^2) + X^2E_3 + 3X^3E_2 + 3X^5. \tag{22}$$

Figure 2 in Section 2.6 gives a graphical representation of the 17 isomorphism types of rooted trees obtained up to now, that is, for $\mathcal{A}_{\leq 5}$. Finally,

$$\mathcal{A}_6 = P_6 + Q_1\mathcal{A}_5 + Q_2\mathcal{A}_4 + Q_3\mathcal{A}_3$$
$$= XE_5 + 2X^2E_2(X^2) + X^2E_2^2 + X^2E_4$$
$$+ 3X^3E_3 + 6X^4E_2 + 6X^6, \tag{23}$$

with 20 isomorphism types. The reader may compare the efficiency of this method with that of simple iteration as suggested in Exercise 3.2.3.

Remark 4. In Proposition 1, we have assumed that α was the approximation of order n of $\mathcal{A}_R$ obtained by truncation. The result is nevertheless valid more generally, no matter which approximation of order n of $\mathcal{A}_R$ we start from. Indeed, let α be a species of structures such that $\alpha =_n \mathcal{A}_R$ and set $\alpha = \alpha^- + \gamma$, where $\alpha^- = \alpha_{\leq n} = (\mathcal{A}_R)_{\leq n}$ and $\gamma = \alpha - \alpha^-$, the latter being supported only on cardinalities $\geq n+1$. Denote as before by β the species of heavy rooted trees, i.e., supported by sets of cardinality ranging from $n+1$ to $2n+2$. A counting argument shows that

$$R(\alpha) = R(\alpha^- + \gamma) = R(\alpha^-) + \gamma R'(\alpha^-) + \cdots$$
$$=_{2n+1} R(\alpha^-) + \gamma R'(\alpha^-) + \cdots \tag{24}$$
$$=_{2n+1} R(\alpha^-) + \gamma R'(\alpha) + \cdots$$

and

$$\beta R'(\alpha) =_{2n+1} \beta R'(\alpha^-). \tag{25}$$

Also, resuming the previous reasoning for (13), we find:

$$\beta =_{2n+2} XR(\alpha^-) - \alpha^- + X\beta R'(\alpha^-)$$
$$=_{2n+2} XR(\alpha) - X\gamma R'(\alpha) - (\alpha - \gamma) + X\beta R'(\alpha)$$
$$= XR(\alpha) - \alpha + \gamma(1 - XR'(\alpha)) + XR'(\alpha)\beta.$$

Using solution (3.2.15) of this linear functional equation and the fact that for any species F, $(1 - F)L(F) = 1$, we deduce that

$$\beta =_{2n+2} L(XR'(\alpha))(XR(\alpha) - \alpha) + \gamma. \tag{26}$$

Setting

$$\alpha^+ = \alpha + L(XR'(\alpha))(XR(\alpha) - \alpha), \tag{27}$$

we find

$$\alpha^+ =_{2n+2} \alpha^- + \beta$$
$$=_{2n+2} \mathcal{A}_R$$

as before.

In other words, in the expression (27), with $\alpha = \alpha^- + \gamma$, all the structures containing a γ-structure are eliminated by telescoping, at least for cardinalities $\leq 2n + 2$, and only the α^--structures really count. This argument is equally valid for the generating and index series.

Hence, if $\alpha(x)$ is an approximation having contact of order n with the generating series $\mathcal{A}_R(x)$, then the series

$$\alpha^+(x) = \alpha(x) + \frac{xR(\alpha(x)) - \alpha(x)}{1 - xR'(\alpha(x))} \tag{28}$$

will have contact of order $2n + 2$ with $\mathcal{A}_R(x)$. Moreover, for k between $n + 1$ and $2n + 2$, Equation (28) shows that the coefficients $[x^k]\alpha^+(x)$ are fixed polynomials of coefficients of $\alpha(x)$ and $R(x)$. This implies that Newton–Raphson's iterative process is valid for any formal power series.

Corollary 5. *Let $R(x)$ be a formal power series with coefficients in an integral domain $\mathbb{K}$ containing $\mathbb{Q}$ and let $A = A(X)$ be the unique formal power series satisfying the equation $A = x R(A)$. If a series $\alpha(x)$ is an approximation of $A(X)$ having contact of order n, then the series $\alpha^+(x)$ defined by formula (28) has a contact of order $2n + 2$ with $A(x)$.* ∎

Finally, Newton–Raphson iterative schemes for types generating series $\widetilde{\mathcal{A}}_R(x)$ and cycle index series $Z_{\mathcal{A}_R}(x_1, x_2, x_3, \ldots)$ of the species $\mathcal{A}_R$ are deduced in the same manner.

Corollary 6. *Let R be a species of structures and $\mathcal{A} = \mathcal{A}_R$ be the associated species of R-enriched rooted trees. Set $r(x_1, x_2, x_3, \ldots) := Z_R(x_1, x_2, x_3, \ldots)$.*

a) If $\alpha(x) =_n \widetilde{A}_R(x)$, then $\alpha^+(x) =_{2n+2} \widetilde{A}_R(x)$, where

$$\alpha^+(x) = \alpha(x) + \frac{xr(\alpha(x), \alpha(x^2), \alpha(x^3), \ldots) - \alpha(x)}{1 - x\frac{\partial r}{\partial x_1}(\alpha(x), \alpha(x^2), \alpha(x^3), \ldots)}. \tag{29}$$

b) If $\alpha(x_1, x_2, x_3, \ldots) =_n Z_A(x_1, x_2, x_3, \ldots)$ (see Section 1.2 for the definition of contact of order n in the case of index series), then $\alpha^+ =_{2n+2} Z_A$, where

$$\alpha^+ = \alpha + \frac{x_1 (r \circ \alpha) - \alpha}{1 - x_1 \frac{\partial r}{\partial x_1} \circ \alpha}. \tag{30}$$

Similar counting arguments can be used to obtain recursive schemes in the case of equations of the form $Y = X + G(Y)$ or more generally $Y = H(X, Y)$. The two main results are simply stated here, leaving to the reader the task of giving proofs and applications to the associated series (see Exercises 4 and 7).

Proposition 7. *Let G be a species of structures such that $G(0) = 0 = G'(0)$, and let $Y = {}_G A$ be the species of G-rooted trees (see Section 3.2), characterized by the functional equation $Y = X + G(Y)$. Let α be a species of structures having contact of order n with ${}_G A$, for an integer $n \geq 1$. Then the species α^+ defined by*

$$\alpha^+ = \alpha + (X + G(\alpha) - \alpha) \cdot L(G'(\alpha)) \tag{31}$$

has contact of order $2n + 1$ with ${}_G A$.

Proposition 8. *Let $H(X, Y)$ be a two-sort species satisfying the conditions $H(0, 0) = 0 = \frac{\partial H}{\partial Y}(0, 0)$ and let $Y = A_H$ be the species of H-enriched rooted trees (see Section 3.2), characterized by the functional equation $Y = H(X, Y)$, with $Y(0) = 0$. Let α be a species of structures having contact of order n with A_H, for an integer $n \geq 1$. Then the species α^+ defined by*

$$\alpha^+ = \alpha + (H(X, \alpha) - \alpha) \cdot L\left(\frac{\partial H}{\partial Y}(X, \alpha)\right) \tag{32}$$

has contact of order $2n + 1$ with A_H.

The Equation $Y = H(X, Y \circ G)$

As we have seen, it is possible to define many classes of rooted trees using functional equations of the form $Z = H(X, Z)$. However, some of the fundamental tree–like data structures used in computer science cannot be described in this way. Let us mention, for instance, AVL rooted trees and other classes of rooted trees with some equilibrium condition. We will see that it is possible to specify these structures with functional equations of the form

$$Y(X) = H(X, Y(G(X))), \tag{33}$$

where $H(X, Y)$ and $G(X)$ are given species. Equations of this form have already appeared in theoretical computer science in the work of Flajolet, Gonnet, Odlyzko, and others (see [90], [98], [96], and [5]), in functional analysis (see Fatou [89] and Kuczma [176]), as well as in the study of fractal curves (see Dubuc [81] and [82]).

The first study of functional equations of the form (where $\varphi(z)$ is the unknown)

$$\varphi(z) = h(z, \varphi(g(z))), \tag{34}$$

appears to be due to A.H. Read [273] who gave conditions ensuring the existence of a solution. M. Bajraktarević [6] generalized this study to the equation

$$\varphi(x) = f(x, \varphi(g_1(x)), \varphi(g_2(x)), \ldots, \varphi(g_n(x))), \tag{35}$$

where $x \in \mathbb{R}^m$. Examples of combinatorial equations of the form (35) are given (with $m = 2$ and $n = 1$) by AVL trees (see Example 14) as well as (with $m = 1$) by the isomorphism types of R-enriched rooted trees: If $\mathcal{A}_R = XR(\mathcal{A}_R)$ and if $\alpha(x) = \mathcal{A}_R(x)$, then

$$\alpha(x) = xZ_R(\alpha(x), \alpha(x^2), \alpha(x^3), \ldots). \tag{36}$$

See, for example, Pólya [263], Section 40.

Note that the case of a linear functional equation of the form

$$\varphi(x) = q(x) + p(x)\varphi(g(x)) \tag{37}$$

has also been considered by Ghermanescu (and others, see [176]) who gave the solution

$$\varphi(x) = \sum_{k \geq 0} \prod_{j=0}^{k-1} p(g^{(j)}) q(g^{(k)}). \tag{38}$$

Let us agree to call the functional equations in (33)–(37) equations of *Read-Bajraktarević type*.

Example 9. Consider the species T of (2–3)-*trees* composed of (2–3)-rooted trees with all leaves at the same depth (see Figure 3). Recall that (2–3)-rooted trees are a particular case of G-rooted trees, introduced in Section 3.2 and characterized by the functional equation $Y = X + G(Y)$, with $G = X^2 + X^3$ (see Figure 3.2.7). The internal vertices of these rooted trees are not labeled and possess two or three (linearly ordered) children; the underlying set is "at" the leaves of the rooted tree. The species T then satisfies the equation

$$T(X) = X + T(X^2 + X^3). \tag{39}$$

Indeed, by virtue of the uniform depth condition for the leaves, a T-structure that is not reduced to a point can be obtained from a smaller T-structure by replacing all of its leaves by an ordered list of two or three leaves.

3.3. Quadratic Iterative Methods

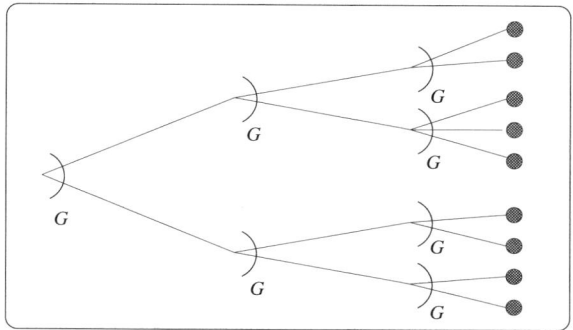

Fig. 3.

This example is a particular case ($G = X^2 + X^3$) of a more general situation treated in the following proposition.

Proposition 10. *Let G be a species of structures such that*

$$G(0) = 0 = G'(0). \tag{40}$$

Then there exists a unique species of structures $\mathcal{A} = Y(X)$, up to isomorphism, satisfying the condition $Y(0) = 0$ and the functional equation

$$Y(X) = X + Y(G(X)). \tag{41}$$

The species $\mathcal{A}$ is given as the sum of iterates of G:

$$\mathcal{A} = \sum_{k=0}^{\infty} G^{\langle k \rangle}. \tag{42}$$

Proof. First observe that the family of iterates $G^{\langle k \rangle}$ of G, $k \geq 0$, is summable, under conditions (40) since the species $G^{\langle k \rangle}$ is supported by sets of cardinality $\geq 2^k$. Since $G^{\langle 0 \rangle} = X$ and $G^{\langle k+1 \rangle} = G^{\langle k \rangle}(G)$, we have

$$\mathcal{A} = \sum_{k \geq 0} G^{\langle k \rangle} = X + \sum_{k \geq 1} G^{\langle k \rangle} = X + \sum_{k \geq 0} G^{\langle k \rangle}(G) = X + \mathcal{A}(G), \tag{43}$$

so that Equation (41) is satisfied. Uniqueness is shown by induction on the cardinality $n \geq 0$ of the underlying sets. Indeed, if $Y = \mathcal{B} = \mathcal{B}(X)$, with $\mathcal{B}(0) = 0$, is another solution of (41) and if $\mathcal{B}_{\leq n} = \mathcal{A}_{\leq n}$, then

$$\mathcal{B} = X + \mathcal{B}(G)$$
$$=_{2n+1} X + \mathcal{B}_{\leq n}(G)$$
$$= X + \mathcal{A}_{\leq n}(G)$$
$$=_{2n+1} \mathcal{A}. \qquad \blacksquare$$

Figure 3 represents a typical $\mathcal{A}$-structure (in fact a $G^{(3)}$-structure). Observe that the elements of the underlying set are found at the leaves and that all leaves have the same depth. This depth corresponds to the *recursive* (or *iterative*) order of the rooted tree, a $G^{(k)}$-structure being, by definition, of order k. The careful reader will also observe that this procedure has a quadratic convergence in the sense that after k iterations, one has a contact of order 2^k with the solution.

In a more general fashion, let $H(X, Y)$ be a two-sort species, and G be a species of structures with $G(0) = 0$, satisfying the conditions a) or b) given below:

$$\begin{aligned} &\text{a)} \quad H(0, 0) = 0, \qquad G'(0) = 0, \\ &\text{b)} \quad H(0, 0) = 0, \qquad \frac{\partial H}{\partial Y}(0, 0) = 0. \end{aligned} \qquad (44)$$

Then Read–Bajraktarević's functional equation

$$Y(X) = H(X, Y(G(X))), \qquad (45)$$

has a unique solution (up to isomorphism) $Y = \mathcal{A}$, with $Y(0) = 0$. The proof of this result is left as an exercise (see Exercise 11).

A typical $\mathcal{A}$-structure is represented in Figure 4. Such a structure has the following characteristic: For any path from the root to a leaf, if $w = w_1 w_2 \cdots w_\ell$ is the word in the letters H and G defined by setting $w_i = H$ (respectively, $w_i = G$) and if the set of children of the ith node of this path is given an H-structure (respectively, a G-structure), then w is of the form $H^{k+1} G^k$, where $k \geq 0$. Note that

$$H(X, 0) \subseteq H(X, H(G(X), 0)) \subseteq H(X, H(G(X), H(G^{(2)}(X), 0))) \subseteq \cdots$$

and it is natural to write

$$\mathcal{A} = H(X, H(G(X), H(G^{(2)}(X), \ldots))) \qquad (46)$$

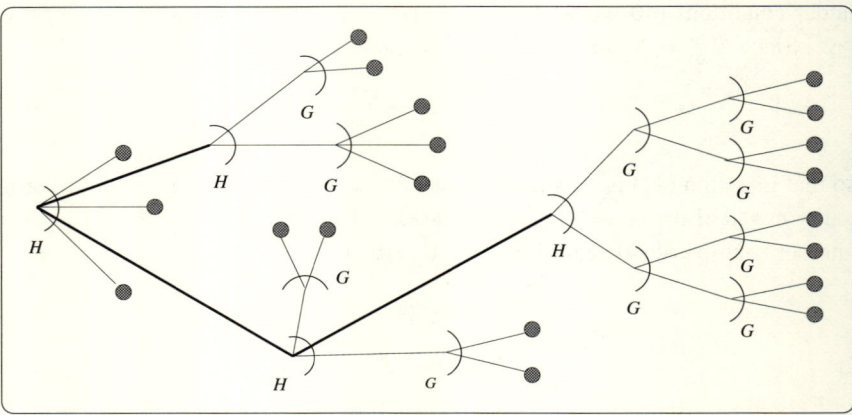

Fig. 4.

3.3. Quadratic Iterative Methods

for this class of rooted trees. On a finite set U of cardinality n, any $\mathcal{A}$-structure α has an *iterative order* k, that is to say, a minimum integer k (≥ 0) such that

$$\alpha \in H(X, H(G(X), H(G^{(2)}(X), \ldots, H(G^{(k)}(X), 0)\cdots)))[U] \qquad (47)$$

and, under the hypotheses (44), a), one has $k \leq \log_2(n)$.

Example 11. Consider the functional equation

$$\mathcal{A}(X) = P(X)\,\mathcal{A}(G(X)), \qquad (48)$$

where P and G are species of structures with $P(0) = 1$ and $G(0) = 0 = G'(0)$. We have then $H(X, Y) = P(X)\,Y$ and condition 44), a) is satisfied. Observe that in this case the arrangement of H-structures is necessarily linear and that the solution of (48) is given by Fatou's formula (see [89])

$$\mathcal{A}(X) = \prod_{k \geq 0} P(G^{(k)}(X)). \qquad (49)$$

An important family of Read–Bajraktarević functional equations is given by the *linear equations*

$$\mathcal{A}(X) = Q(X) + P(X)\,\mathcal{A}(G(X)), \qquad (50)$$

where Q, P, and G are given species, with $Q \neq 0$ and $G(0) = 0$, satisfying either conditions a) or b) below:

$$\begin{array}{lll} \text{a)} & Q(0) = 0, & G'(0) = 0; \\ \text{b)} & Q(0) = \alpha, & P(0) = 0. \end{array} \qquad (51)$$

In this case, we have $H(X, Y) = Q(X) + P(X)\,Y$ and conditions a) or b), with $\alpha = 0$, assure that the corresponding conditions of (44), a) or b), are satisfied. The more general case $\alpha \neq 0$ is also admissible. Under these hypotheses, the following explicit form (see Ghermanescu, (38)) can be given for the unique solution $\mathcal{A}(X)$ of (50):

$$\mathcal{A}(X) = \sum_{k \geq 0} P_k(X)\,Q\bigl(G^{(k)}(X)\bigr), \qquad (52)$$

where the species $P_k = P_k(X)$ is defined by $P_0(X) = 1$ and, for $k > 1$,

$$P_k(X) = P(X)P(G(X))\cdots P\bigl(G^{(k-1)}(X)\bigr). \qquad (53)$$

The proof of this result is similar to that of Proposition 10 (see Exercise 12; also see Kuczma [176] and Flajolet–Prodinger [98]).

If $\mathcal{A} = \mathcal{A}(X)$ is a solution of (50), the (iterative) order of an $\mathcal{A}$-structure α on a set U is the integer $k \geq 0$ such that $\alpha \in (P_k \cdot Q(G^{\langle k \rangle}))[U]$. Denote by $\mathcal{A}^{[k]}$ the species of $\mathcal{A}$-structures of order k, and set $\mathcal{A}^{[\leq k]} = \sum_{j \leq k} \mathcal{A}^{[j]}$. In an iterative manner, one has $\mathcal{A}^{[0]}(X) = Q(X)$, and

$$\mathcal{A}^{[\leq k]} = Q(X) + P(X)\mathcal{A}^{[\leq k-1]}(G(X)). \tag{54}$$

Observe that if α is an $\mathcal{A}$-structure of order k on U with $|U| = n$, then $n \geq 2^k$ under the hypothesis (51), a) and $n \geq k$, under (51), b). In the first case one has quadratic convergence, while in the second case, the kth iteration (54) adds one exact term to the approximation of $\mathcal{A}$.

For $k = 2^r - 1$, here is an algorithm accelerating the process of constructing all $\mathcal{A}$-structures of order $\leq k$ in r iterative steps. The method is inspired from the classical identity

$$1 + X + X^2 + X^3 + X^4 + \cdots = (1 + X)(1 + X^2)(1 + X^4)(1 + X^8) \cdots. \tag{55}$$

Algorithm 12. QUADRATIC ITERATION.
Let $\mathcal{A} = \mathcal{A}(X)$ be the solution of the linear Read–Bajraktarević equation (50), and let $k = 2^r - 1$. To obtain $S_r := \mathcal{A}^{[\leq k]}$, the species of $\mathcal{A}$-structures of order $\leq k$, set

$$S_0(X) = \mathcal{A}^{[0]}(X) = Q(X), \qquad G^{\langle 1 \rangle}(X) = G(X), \qquad P_1(X) = P(X), \tag{56}$$

and for $j = 0, 1, \ldots, r - 1$, compute successively

a) $S_{j+1}(X) = S_j(X) + P_{2^j}(X) S_j(G^{\langle 2^j \rangle}(X))$,

b) $P_{2^{j+1}}(X) = P_{2^j}(X) P_{2^j}(G^{\langle 2^j \rangle}(X))$, \hfill (57)

c) $G^{\langle 2^j + 1 \rangle}(X) = G^{\langle 2^j \rangle}(G^{\langle 2^j \rangle}(X))$.

One particular important case of the linear equation, with $P(X) = 1$, is the equation

$$Y(X) = Q(X) + Y(G(X)), \tag{58}$$

whose solution, under the hypothesis that

$$Q(0) = 0 = G(0) = G'(0), \tag{59}$$

is given by

$$Y(X) = \sum_{k \geq 0} Q(G^{\langle k \rangle}). \tag{60}$$

For instance, Odlyzko [258] has shown that if $Q(z)$ and $G(z)$ are polynomials or formal power series with real positive coefficients satisfying (59) and some additional condition, and if $a(z) = \sum_{n \geq 1} a_n z^n$ is the series solution of

$$a(z) = Q(z) + a(G(z)), \tag{61}$$

then

$$a_n \sim \frac{\alpha^n}{n} u(\log n), \tag{62}$$

where α is the unique positive solution of $G(z) = z$ and $u(z)$ is a periodic function of period $\log(G(\alpha))$.

Note also that the resolution of the linear Read–Bajraktarević equation (50) can be reduced, when $P(0) = 1$, to the solution of two linear equations of the form (58). Indeed, if $B(X)$ and $F(X)$ are solutions of the equations

$$B(X) = P(X) B(G(X)), \tag{63}$$

$$F(X) = \frac{Q(X)}{B(X)} + F(G(X)), \tag{64}$$

then $\mathcal{A}(X) = F(X)B(X)$ satisfies Equation (50). Using the combinatorial logarithm (see Section 2.5), which amounts to the ordinary logarithm for generating series, Equation (63) is equivalent to

$$\log B(X) = \log P(X) + \log B(G(X)), \tag{65}$$

which is of the form (58).

It is possible to take into consideration the iterative order of the structures in (52) by replacing the species P by P_t, where t is a uniform weight given to every P-structure. Here is an example.

Example 13. STRAHLER ORDER. (See Françon [119].) The function $s(\beta)$ is defined recursively on binary rooted trees β (see the introduction to this chapter) by

$$s(\beta) = \begin{cases} 0, & \text{if } \beta \text{ is the empty rooted tree,} \\ \max(s(\lambda), s(\rho)), & \text{if } s(\lambda) \neq s(\rho), \\ 1 + \max(s(\lambda), s(\rho)), & \text{if } s(\lambda) = s(\rho), \end{cases}$$

where λ and ρ are respectively the left and right subtrees of β. This parameter $s(\beta)$ is equal to the minimum number of registers necessary to evaluate the arithmetic expression represented by β (see Exercise 14). It is called *the Strahler order* of the binary rooted tree β. We will obtain a functional equation of form (50) for the species $\mathcal{B}_w$ of binary rooted trees β weighted by the Strahler order, i.e., by $w(\beta) = t^{s(\beta)}$. Observe first that the species $\mathcal{B}$ is related to the species $\mathcal{C}$ of *complete binary rooted trees* (solution of $C = X(1 + C^2)$), by the combinatorial equation

$$C = X\mathcal{B}(X^2), \tag{66}$$

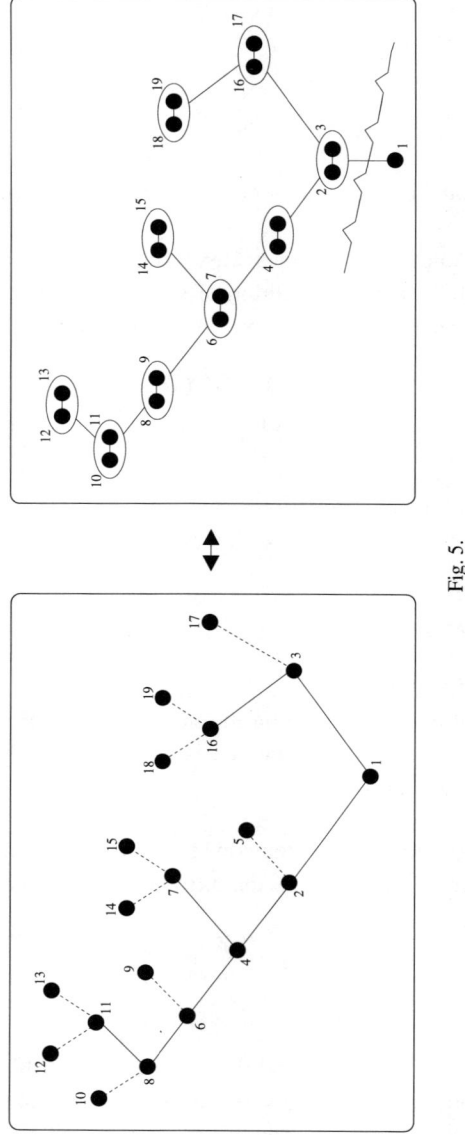

Fig. 5.

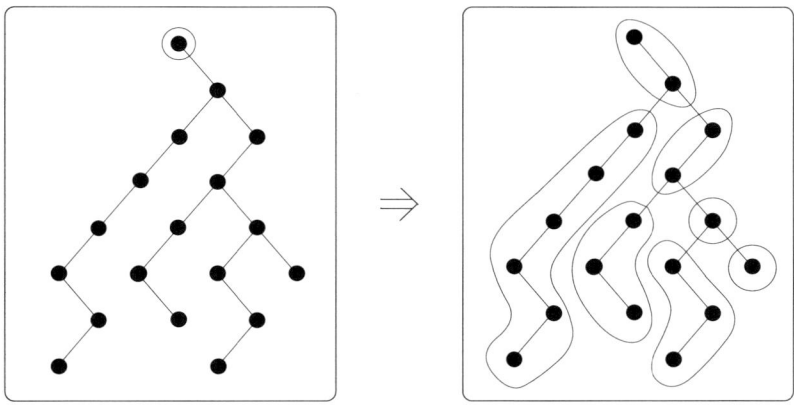

Fig. 6. $\mathcal{B}_w = 1 + \mathcal{C}_w(XL(X + X))$.

as shown by the "billiard stroke" effect (Figure 5). This equation admits the weighted form

$$\mathcal{C}_w = X_t \mathcal{B}_w(X^2), \tag{67}$$

since in the bijection of Figure 5, the Strahler order of the left binary rooted tree is clearly one more than that of the right one. Also, Figure 6 shows that there is a second (weight preserving) relation between the species $\mathcal{B}_w$ and $\mathcal{C}_w$, namely

$$\mathcal{B}_w = 1 + \mathcal{C}_w(XL(2X)), \tag{68}$$

where the species $XL(2X)$ represents the zigzag (left-right) linear parts of a binary tree. Combining Equations (67) and (68) gives the functional equation

$$\mathcal{B}_w = 1 + (XL(2X))_t \mathcal{B}_w((XL(2X))^2), \tag{69}$$

for the species $\mathcal{B}_w$ of binary rooted trees, weighted by the Strahler order. Denoting $\mathcal{B}_w(x)$ by $\mathcal{B}(x, t)$, the following equation, due to Françon, is deduced:

$$\mathcal{B}(x, t) = 1 + \frac{tx}{1 - 2x} \mathcal{B}\left(\left(\frac{x}{1 - 2x}\right)^2, t\right). \tag{70}$$

These are linear equations of type (50), with $G(X) = (XL(2X))^2$, $Q(X) = 1$, and $P(X) = (XL(2X))_t$, and we see that the Strahler order coincides with the iterative order k defined by formula (52).

In order to obtain information on the average Strahler order of binary rooted trees, we can differentiate each side of (70) with respect to t, and substitute $t = 1$

in the result. A new linear Read–Bajraktarević equation results, namely

$$\mathcal{B}_S(x) = \frac{1 - 2x - \sqrt{1 - 4x}}{2x} + \frac{x}{1 - 2x} \mathcal{B}_S\left(\left(\frac{x}{1 - 2x}\right)^2\right), \tag{71}$$

where we have set

$$\mathcal{B}_S(x) := \left.\frac{\partial \mathcal{B}_w(x, t)}{\partial t}\right|_{t=1} \tag{72}$$

and used the fact that

$$\mathcal{B}(x) = \frac{1 - \sqrt{1 - 4x}}{2x}. \tag{73}$$

Another direction of generalization is Read–Bajraktarević's equation in two (or more) variables, for example of the form

$$V(X, Y) = F(X, Y) + V(G(X, Y), H(X, Y)), \tag{74}$$

where F, G, and H are fixed two-sort species (see Exercise 16). Here is an example drawn from the theory of balanced rooted trees which illustrates the interest of such a generalization. See also Exercise 18.

Example 14: AVL TREES. (See Aho–Hopcroft–Ullman [1], Knuth [172], vol. 3, Ottmann–Wood [260], and Mehlhorn [244].) The *height* $h(\beta)$ (also called the *depth*) of a binary rooted tree β is defined recursively by

$$h(\beta) = \begin{cases} 0, & \text{if } \beta \text{ is empty,} \\ 1 + \max(h(\lambda), h(\rho)), & \text{otherwise,} \end{cases} \tag{75}$$

where $\lambda = \lambda(\beta)$ and $\rho = \rho(\beta)$ are respectively the left and right subtrees of β. An AVL *tree* is defined as a complete binary rooted tree β, labeled at the leaves, such that for any vertex s of β one has

$$|h(\lambda(s)) - h(\rho(s))| \leq 1, \tag{76}$$

where $\lambda(s)$ and $\rho(s)$ denote $\lambda(\beta(s))$ and $\rho(\beta(s))$, and $\beta(s)$ denotes the maximal rooted subtree of β with root s.

Denote by $A = A(X)$ the species of AVL trees. Take note again that the underlying set of an AVL tree β is the set of labels of the leaves of β. It can be seen that $A(X) = V(X, 0)$, where $V(X, Y)$ is the species characterized by the functional equation

$$V(X, Y) = X + V(X^2 + 2XY, X). \tag{77}$$

Indeed, when the heights of the left and right branches of an internal vertex of β are not equal, a virtual vertex of sort Y, of degree 1, can be added in order to

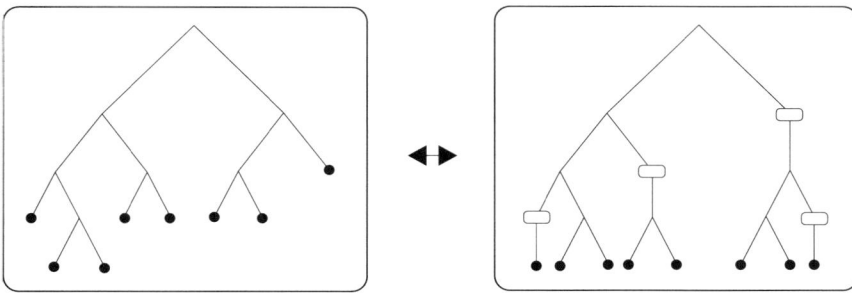

Fig. 7.

balance the two branches. See Figure 7. Equation (77) can be solved by simple iteration. Setting

$$V_0(X, Y) = X, \qquad V_{k+1} = X + V_k(X^2 + 2XY, X), \tag{78}$$

and $A_k(X) = V_k(X, 0)$, one finds

$$\begin{aligned}
A_0 &= X, \qquad A_1 = X + X^2, \qquad A_2 = X + X^2 + 2X^3 + X^4, \\
A_3 &= X + X^2 + 2X^3 + X^4 + 4X^5 + 6X^6 + 4X^7 + X^8, \\
A_4 &= X + X^2 + 2X^3 + X^4 + 4X^5 + 6X^6 + 4X^7 + 17X^8 + 32X^9 + 44X^{10} \\
&\quad + 60X^{11} + 70X^{12} + 56X^{13} + 28X^{14} + 8X^{15} + X^{16}, \\
A_5 &= X + X^2 + 2X^3 + X^4 + 4X^5 + 6X^6 + 4X^7 + 17X^8 + 32X^9 + 44X^{10} \\
&\quad + 60X^{11} + 70X^{12} + 184X^{13} + 476X^{14} + 872X^{15} + 1553X^{16} \\
&\quad + 2720X^{17} + 4288X^{18} + 6312X^{19} + 9004X^{20} + 11992X^{21} \\
&\quad + 14372X^{22} + 15400X^{23} + 14630X^{24} + 11968X^{25} + 8104X^{26} \\
&\quad + 4376X^{27} + 1820X^{28} + 560X^{29} + 120X^{30} + 16X^{31} + X^{32},
\end{aligned}$$

etc.

$$\tag{79}$$

Exercises for Section 3.3

1. R-ENRICHED ROOTED TREES: GENERIC SERIES. Consider a set of formal variables $(r_0, r_1, r_2, \ldots)$ and set $\mathbb{K} = \mathbb{Q}[r_0, r_1, r_2, \ldots]$, the ring of polynomials in the r_i with rational coefficients. Let $R(x) = \sum_{i \geq 0} r_i x^i / i! \in \mathbb{K}[\![x]\!]$.
   a) Show that the equation $A(x) = xR(A(x))$ has a unique solution $A(x) \in \mathbb{K}[\![x]\!]$ and that Lagrange's inversion formula (3.1.10), a) is valid.
   NOTE: The series $R(x)$ can be viewed as the "generic" generating series of an arbitrary species of structures R. In this case, the series $A(x)$ is the "generic" generating series of the species $\mathcal{A}_R$ of R-enriched rooted trees. The first few coefficients of this generic series are given in Table 9 of Appendix 2.

b) Establish the validity of (28) for the recursive computation of the series $A(x)$ (see Corollary 5), including in the generic case.

2. R-ENRICHED ROOTED TREES: GENERIC SERIES (continued). Associate a formal variable $r_{\mathbf{n}}$ to each type of permutation $\mathbf{n} = (n_1, n_2, n_3, \ldots)$, where $|\mathbf{n}| = n_1 + 2n_2 + 3n_3 + \cdots < \infty$. Let

$$\mathbb{K} = \mathbb{Q}[(r_{\mathbf{n}})_{|\mathbf{n}|<\infty}] \qquad (80)$$

be the polynomial ring in the variables $r_{\mathbf{n}}$ with rational coefficients.

a) Show that for any "formal" index series $\in \mathbb{K}[\![x_1, x_2, x_3, \ldots]\!]$

$$r(x_1, x_2, x_3, \ldots) = \sum_{|\mathbf{n}|<\infty} r_{\mathbf{n}} \frac{x_1^{n_1} x_2^{n_2} \cdots}{1^{n_1} n_1! 2^{n_2} n_2! \cdots}, \qquad (81)$$

i) the equation $a(x_1, x_2, \ldots) = x_1 r(a_1, a_2, \ldots)$, where $a_k = a(x_k, x_{2k}, x_{3k}, \ldots)$, has a unique solution $a(x_1, x_2, \ldots) \in \mathbb{K}[\![x_1, x_2, \ldots]\!]$,

ii) the equation $a(x) = xr(a(x), a(x^2), a(x^3), \ldots)$ has a unique solution $a(x) \in \mathbb{K}[\![x]\!]$.

HINT: Proceed by induction or invoke combinatorial uniqueness (i.e., in the case where $r_{\mathbf{n}} \in \mathbb{N}$) together with the extension principle of algebraic identities.

NOTE: Series (81) can be viewed as the "generic" cycle index series of an arbitrary species of structures R. In this case, the series $a(x_1, x_2, \ldots)$ and $a(x)$ can be viewed respectively as the generic cycle index series Z_{A_R} and the generic types generating series $\widetilde{A}_R(x)$ of the corresponding species of R-enriched rooted trees. Tables 10 and 12 in Appendix 2 contain the first few terms of these series.

b) Establish the validity of (29) and (30) for the iterative computation of the series $\widetilde{A}_R(x)$ and $Z_{A_R}(\mathbf{x})$ (see Corollary 6), including in the generic case.

3. MOBILES. Let $\mathcal{A} = \mathcal{A}_{1+C}$ be the species of rooted $(1 + C)$-enriched trees characterized by the combinatorial equation $\mathcal{A} = X \cdot (1 + C(\mathcal{A}))$. An $\mathcal{A}$-structure is called a *mobile* (suspended by the root).

a) Show that the molecular decomposition of $\mathcal{A}_{\leq 6}$ is given by

$$\begin{aligned}\mathcal{A}_{\leq 6} = & X + X^2 + X^3 + XE_2(X) + 2X^4 + XC_3(X) + X^2 E_2(X) \\ & + 4X^5 + XE_2(X^2) + XC_4(X) + 2X^3 E_2(X) + X^2 C_3(X) \\ & + 10X^6 + XC_5(X) + 5X^4 E_2(X) + X^2 E_2(X^2) \\ & + 2X^3 C_3(X) + X^2 C_4(X)\end{aligned} \qquad (82)$$

by applying Algorithm 2 of Newton–Raphson starting with the initial approximation $\mathcal{A}_0 = 0$.

HINT: Use the molecular decomposition of the species $C_k(X+Y)$, $k = 1, 2, \ldots, 5$ (see Table 14 in Appendix 2).

b) Using (28) as well as Corollary 6, show that the formal power series $\mathcal{A}(x)$ and $\widetilde{\mathcal{A}}(x)$ are given, up to order 6, by

$$\mathcal{A}(x) = x + 2\frac{x^2}{2!} + 9\frac{x^3}{3!} + 68\frac{x^4}{4!} + 730\frac{x^5}{5!} + 10164\frac{x^6}{6!} + \cdots, \quad (83)$$

$$\widetilde{\mathcal{A}}(x) = x + x^2 + 2x^3 + 4x^4 + 9x^5 + 20x^6 + \cdots. \quad (84)$$

Check that this is coherent with the result found in a).

c) Proceed in a similar fashion to compute the cycle index series Z_A truncated to order 6.

4. Let G be a species of structures such that $G(0) = 0 = G'(0)$, and $Y = {}_G A$, the species of G-rooted trees, characterized by the functional equation $Y = X + G(Y)$. For a given integer $n \geq 1$, let $\alpha = {}_G A_{\leq n}$ be the species of *light* G-rooted trees and $\beta = {}_G A_{\leq 2n+1} - \alpha$ that of *heavy* G-rooted trees.

a) Show that

$$\beta =_{2n+1} X + G(\alpha) - \alpha + G'(\alpha) \cdot \beta. \quad (85)$$

Deduce Proposition 7.

b) Adapt Algorithm 2 and show how to construct a sequence of approximations $\alpha^{(n)}$ of the species ${}_G A$, such that $\alpha^{(n)} =_{2^{n+1}-1} {}_G A$.

c) Show that the associated series $y(x) = {}_G A(x)$, $\tilde{y}(x) = \widetilde{{}_G A}(x)$, and $z = Z_{{}_G A}$ are characterized by the functional equations

   i) $y(x) = x + G(y(x))$,

   ii) $\tilde{y}(x) = x + Z_G(\tilde{y}(x), \tilde{y}(x^2), \ldots)$, \quad (86)

   iii) $z = x_1 + Z_G(z_1, z_2, \ldots)$.

d) Give recursive schemes with quadratic convergence for the associated series ${}_G A(x)$, $\widetilde{{}_G A}(x)$, $Z_{{}_G A}(\mathbf{x})$.

5. COMMUTATIVE PARENTHESIZATIONS. Let $\mathcal{P}$ be the species of commutative parenthesizations, defined by the functional equation

$$\mathcal{P} = X + E_2(\mathcal{P}). \quad (87)$$

a) For $n \geq 1$, let $p_n = |\mathcal{P}[n]|$. Show that

   i) $p_{n+1} = (2n-1)p_n$,

   ii) $P(x) = 1 - \sqrt{1-2x}$, \quad (88)

   iii) $p_1 = 1$, and for $n \geq 2$, $p_n = 1 \cdot 3 \cdots (2n-3)$.

b) Use Newton–Raphson's scheme of the preceding exercise to calculate the molecular decomposition of $\mathcal{P}$ and the associated series $\widetilde{\mathcal{P}}(x)$ and $Z_{\mathcal{P}}(x_1, x_2, \ldots)$ up to order 15.

c) (Joyal [158, Example 26].) Show that the index series $z = Z_{\mathcal{P}}$ only depends on the variables x_{2^i}, $i \geq 0$, and that moreover, if $y = \psi(x)$ is the solution of the equation $y^2 - 2y + 2x = 0$ (i.e., $\psi(x) = \mathcal{P}(x)$), one has the functional equations

$$\text{i) } \widetilde{\mathcal{P}}(x) = \psi\left(x + \frac{1}{2}\widetilde{\mathcal{P}}(x^2)\right),$$

$$\text{ii) } z(x_1, x_2, \ldots) = \psi\left(x_1 + \frac{1}{2}z(x_2, x_4, \ldots)\right). \tag{89}$$

Conclude that, for $N = 2^{n+1} - 1$,

a) $\widetilde{\mathcal{P}}(x) =_N 1 - \sqrt{-2x + \sqrt{-2x^2 + \cdots + \sqrt{1 - 2x^{2^n}}}}$,

b) $Z_{\mathcal{P}}(x_1, x_2, \ldots) =_N 1 - \sqrt{-2x_1 + \sqrt{-2x_2 + \cdots + \sqrt{1 - 2x_{2^n}}}}$. $\tag{90}$

6. **PQ-TREES.** PQ-trees are data structures used in certain algorithms on graphs, notably for planarity testing (see Booth and Lueker [34], as well as Chiba, Nishizeki, Abe, and Ozawa [52] and Karabeg [165]). Formally the species $Y = {}_{PQ}A$ of PQ-trees is defined by the functional equation

$$Y = X + P(Y) + Q(Y), \tag{91}$$

where $P = \text{Cha}_{\geq 3}$ is the species of (nonoriented) chains of order (number of elements) ≥ 3 and $Q = E_{\geq 2}$, with $Y(0) = 0$.

a) Describe PQ-trees and analyze the series associated to this species.

b) Apply the methods of Exercise 4 to the recursive computation of the species ${}_{PQ}A$ and its associated series.

7. a) Prove Proposition 8 and deduce the corresponding results for the generating and index series.

b) Apply these results to the species of structures A characterized by functional equation (3.2.68) (see Exercise 3.2.2). The first terms of the generating series $A(x)$ are

$$x + 6\frac{x^2}{2!} + 65\frac{x^3}{3!} + 1092\frac{x^4}{4!} + 25272\frac{x^5}{5!} + 749034\frac{x^6}{6!} + 27108440\frac{x^7}{7!} + \cdots. \tag{92}$$

8. **HIGHER ORDER NEWTON–RAPHSON ITERATION.** Let R be a species of structures and $n \geq 0$, $k \geq 1$ be fixed integers. An (R-enriched) rooted tree on a set U is called k-heavy if

$$n + 1 \leq |U| \leq (k+1)(n+1). \tag{93}$$

3.3. Quadratic Iterative Methods

Denote by $\mathcal{B}_k$ the species of k-heavy rooted trees and by α, the species of light rooted trees, that is, with $|U| \leq n$. Setting $r = (k+1)(n+1)$, one has

$$\mathcal{A}_R =_r \alpha + \mathcal{B}_k. \tag{94}$$

a) Show by a counting argument that $\mathcal{B}_k$ has contact of order r with the species of structures γ implicitly defined by the combinatorial equation (polynomial in γ)

$$\gamma = XR(\alpha) - \alpha + X \sum_{i=1}^{k} E_i \left(Y \frac{d}{dx} \right) R(X) \bigg|_{X:=\alpha, Y:=\gamma}, \tag{95}$$

where E_i denotes the species of sets of cardinality i (see Exercise 2.4.18), and that, consequently, the species of structures α^+ defined by $\alpha^+ = \alpha + \gamma$ has contact of order $(k+1)(n+1)$ with $\mathcal{A}_R$.

b) Let $f(t)$ be of the form $f(t) = xR(t) - t$, where $R(t)$ is a formal power series (or a C^{k+1} function), and let $a = a(x)$ be a root of the equation $f(t) = 0$. Show that if α is an approximation of a, then a new approximation α^+ is obtained by finding a root of the Taylor polynomial of degree k

$$f(\alpha) + \frac{f'(\alpha)}{1!}(t-\alpha) + \cdots + \frac{f^{(k)}(\alpha)}{k!}(t-\alpha)^k \tag{96}$$

of the function $f(t)$, expanded around α. Show moreover that the convergence is of order $k+1$ in the sense that, for $\alpha_{n+1} = \alpha_n^+$, $n = 0, 1, 2, \ldots$, one has

$$a - \alpha_{n+1} = O((a - \alpha_n)^{k+1}). \tag{97}$$

9. (a, b)-TREES. (See Mehlhorn [244] and Gonnet [5].) Let a and b be two integers, with $a \geq 2$ and $b \geq 2a - 1$. An (a, b)-tree is a G-rooted tree α (see Section 3.2), with $G = L$, having the additional following properties:
   i) the leaves of α are all of the same height,
   ii) the internal vertices all have indegree (number of children) $\leq b$,
   iii) the internal vertices, except the root, have indegree $\geq a$,
   iv) the root is of degree ≥ 2, except if the rooted tree consists of a single element.

When $b = 2a - 1$, the (a, b)-trees are also known as B-trees, with parameter $m = a - 1$. The most simple case is that of $(2,3)$-trees (see Example 9 and the next exercise).

a) Show that the species A of (a, b)-trees is characterized by the Read–Bajraktarević equation

$$A(X) = X + X^2 + \cdots + X^{a-1} + A(X^a + X^{a+1} + \cdots + X^b). \tag{98}$$

b) Deduce a closed expression for $A(X)$ and calculate the first few terms of $A(X)$.

c) Show that if α is an (a, b)-tree on n leaves and of height h, then
$$\frac{\log n}{\log b} \leq h \leq 1 + \frac{\log(n/2)}{\log a}. \tag{99}$$

d) The (a, b)-trees are used as data structures for the storing (at the leaves) of elements taken from a totally ordered set. Describe algorithms for the insertion, search, or deletion of elements in (a, b)-trees. Analyze the complexity of these algorithms.

10. Let $\mathcal{T}$ be the species of $(2, 3)$-trees (see Example 9).
   a) Show that $\mathcal{T}(x) = \widetilde{\mathcal{T}}(x)$.
   b) Setting $\mathcal{T}(x) = \sum_{n\geq 1} t_n x^n$, establish the recurrence
$$t_1 = 1, \quad t_n = \sum_k \binom{k}{3k-n} t_k \quad (n \geq 2). \tag{100}$$
   c) Generalize to (a, b)-trees.

11. a) Prove that under conditions (44), a) or b), Read–Bajraktarević's equation (45) admits a unique solution $Y = \mathcal{A}$, with $Y(0) = 0$. Deduce the corresponding results in the context of generating and index series.
    b) Prove that any $\mathcal{A}$-structure α has an iterative order k, that is, a minimum integer k (≥ 0) such that (47) is satisfied (U being the underlying set to the structure α). Show moreover that under hypotheses (44), a), $k \leq \log_2 n$, where $n = |U|$.

12. Let G and M be two species of structures such that $G(0) = 0 = M(0)$ and satisfying conditions a) or b) below:
$$\text{a)} \quad G'(0) = 0, \qquad \text{b)} \quad M'(0) = 0. \tag{101}$$

Show that the functional equation
$$Y(X) = X + M(Y(G(X))) \tag{102}$$

has a unique solution $Y = A(X)$, with $Y(0) = 0$, that may be described in the form
$$A(X) = X + M(G + M(G^{(2)} + M(G^{(3)} + \cdots))). \tag{103}$$

13. Show that, under conditions (51), a) or b), Equation (50) has a unique solution $\mathcal{A}$ and that this solution is given by (52) and (53). Deduce the corresponding results for the corresponding generating and index series.

3.3. Quadratic Iterative Methods

14. **STRAHLER ORDER.**
    a) Show that the parameter $s(\beta)$ defined in Example 13 for binary rooted trees β is indeed equal to the minimum number of registers necessary to evaluate the arithmetic expression represented by β.
    HINT: First evaluate the arithmetic expression corresponding to the left or right subtree having maximum Strahler order. In case of equality, it will be necessary to evaluate one of the two branches and save the result in a register.
    b) Use Equation (71) to analyze the average Strahler order of binary rooted trees having n vertices.

15. Consider the species $\mathrm{Jac}(X)$ of Jacobi endofunctions, defined by $\mathrm{Jac}(X) = \mathrm{Jac}(X, X)$ (see Example 3.2.12).
    a) Show that
    $$\mathrm{Jac}(x) = \frac{1}{1 - 2x\mathcal{B}(x)} = \frac{1}{\sqrt{1 - 4x}}. \qquad (104)$$
    b) Deduce the functional equation
    $$\mathrm{Jac}(x) = \frac{1}{1 - 2x} \mathrm{Jac}\left(\left(\frac{x}{1 - 2x}\right)^2\right) \qquad (105)$$
    and calculate the first (rational) approximations given by formula (49) by taking only the n first factors, $n = 1, 2, 3, \ldots$.

16. Extend the results of this section on Read–Bajraktarević functional equations to equations of the form (74) and also, more generally, of the multidimensional form
    $$Y(X) = Q(X) + P(X)Y(G(X)), \qquad (106)$$
    or even
    $$Y(X) = H(X, Y(G_1(X)), \ldots Y(G_n(X))), \qquad (107)$$
    where $Y = (Y_1, \ldots, Y_m)$ is a vector of (unknown) k-sort species $Y_i = Y_i(X_1, \ldots, X_k)$, which is conveniently denoted by $Y : \mathbb{B}^k \longrightarrow \mathbb{B}^m$, and where $G : \mathbb{B}^k \longrightarrow \mathbb{B}^k, G_1 : \mathbb{B}^k \longrightarrow \mathbb{B}^k, \ldots, G_n : \mathbb{B}^k \longrightarrow \mathbb{B}^k, Q : \mathbb{B}^k \longrightarrow \mathbb{B}^m, P : \mathbb{B}^k \longrightarrow \mathbb{B}$, and $H : \mathbb{B}^k \times \mathbb{B}^m \times \cdots \times \mathbb{B}^m = \mathbb{B}^{k+nm} \longrightarrow \mathbb{B}^m$ are given (vectors of) species.

17. Let β be an AVL tree of height h having n leaves (see Example 14). Show that
    $$\log_2 n \leq h \leq \frac{1}{\log_2(\phi)} \log_2(n + 1), \qquad (108)$$
    where ϕ denotes the golden ratio $(1 + \sqrt{5})/2$.
    HINT: Show by induction on h that $n \leq 2^h$ and $n \geq F_{h+2}$, where F_k denotes the kth Fibonacci number ($F_0 = 0$, $F_1 = 1$, $F_{k+2} = F_{k+1} + F_k$).

18. **LEFTIST AND PSEUDO-LEFTIST TREES.** A plane binary rooted tree β is said to be a *leftist* tree (Knuth [172], Kemp [167]) if for any rooted subtree α of β, the shortest branch is the sequence of left children starting from the root; β is said to be a *pseudo-leftist* tree if for all rooted subtrees α, the left height (i.e., the height of the left subtree) is less than or equal to the right height.

a) Show that these two concepts are incomparable (for inclusion).

b) Let Psℓ be the species of pseudo-leftist trees, labeled at the leaves. Show that Ps$\ell(X) = U(X, 0)$, where $U(X, Y)$ is the species characterized by the functional equation

$$U(X, Y) = X + U(X^2 + YX, X + Y). \tag{109}$$

c) Calculate the first few terms of Ps$\ell(X)$.

d) Let Lft$_\lambda$ be the species of leftist trees (labeled at the leaves) whose extreme left branch is of length λ and Lft $= \sum_{\lambda \geq 1}$ Lft$_\lambda$. Show that (see Kemp [167])

i) $\text{Lft}_0(x) = x$,

ii) $\text{Lft}_{\lambda+1}(x) = \text{Lft}_\lambda(x) \cdot \left(\text{Lft}(x) - \sum_{0 \leq \rho < \lambda} \text{Lft}_\rho(x) \right),$ (110)

iii) $\text{Lft}(x) = x + \frac{1}{2}\text{Lft}^2(x) + \frac{1}{2}\sum_{\lambda \geq 0} \text{Lft}_\lambda^2(x).$

e) Deduce the asymptotics for the number $a_n = [x^n]\text{Lft}(x)$ of leftist trees (up to isomorphism) having n leaves.

19. **NEWTON–RAPHSON ITERATION FOR READ–BAJRAKTAREVIĆ EQUATIONS.** Let G be a species of structures with $G(0) = 0$, and $H = H(X, Y)$ a two-sort species satisfying the conditions (44), b), $H(0, 0) = 0$ and $(\partial H/\partial Y)(0, 0) = 0$. Let $Y = \mathcal{A}(X)$ be the solution of the functional equation (45) $Y(X) = H(X, Y(G(X)))$, with $\mathcal{A}(0) = 0$, and let n be a fixed integer ≥ 0. An $\mathcal{A}$-structure on U is called *light* if $|U| \leq n$ and *heavy* if $n + 1 \leq |U| \leq 2n + 1$. Set $\alpha = \mathcal{A}_{\leq n}$ and $\beta = \mathcal{A}_{\leq 2n+1} - \alpha$ for the corresponding species.

a) Show that $\beta =_{2n+1} \gamma$ where $\gamma = \gamma(X)$ is the species solution of the linear Read–Bajraktarević equation

$$\gamma(X) = Q(X) + P(X)\gamma(G(X)), \tag{111}$$

where $Q(X) = H(X, \alpha(G(X))) - \alpha(X)$ and $P(X) = (\partial H/\partial Y)(X, \alpha(G(X)))$, the conditions (51), b) being then satisfied. Deduce an iterative process $\alpha \mapsto \alpha^+ = \alpha + \beta$ having quadratic convergence

b) Adapt Algorithm 2 to calculate the species $\beta_{n+m+1}, 0 \leq m \leq n$, and their generating and index series.

3.4. Elements of Asymptotic Analysis

Let us first recall three classical methods of comparison between numerical sequences, often used in asymptotic analysis.

Definition 1. Let $(a_n)_{n\geq 0}$ and $(b_n)_{n\geq 0}$ be two sequences of complex numbers.
a) One says that a_n is *of order* b_n, for n tending to infinity, if there exists a constant $K > 0$ and an integer N such that

$$|a_n| \leq K|b_n|, \tag{1}$$

for any $n \geq N$. One then writes

$$a_n = O(b_n). \tag{2}$$

b) One says that a_n is *negligible* with respect to b_n, for n tending to infinity, if for any $\epsilon > 0$, there exists an integer N such that

$$|a_n| \leq \epsilon |b_n|, \tag{3}$$

if $n \geq N$. One then writes

$$a_n = o(b_n). \tag{4}$$

c) One says that a_n is *asymptotic* to b_n, for n tending to infinity, if

$$\lim_{n \to \infty} \frac{a_n}{b_n} = 1. \tag{5}$$

One then writes

$$a_n \sim b_n. \tag{6}$$

Observe that if $b_n \neq 0$ for large enough n,

$$\begin{aligned} &\text{a)} \quad a_n = o(b_n) \iff \lim_{n \to \infty} \frac{a_n}{b_n} = 0, \\ &\text{b)} \quad a_n \sim b_n \iff a_n = b_n(1 + o(1)). \end{aligned} \tag{7}$$

Formally, the symbols $O(b_n)$ and $o(b_n)$ denote sets of sequences $(a_n)_{n\geq 0}$ satisfying (1) and (3), respectively. Thus, the equality sign in (2), (4), or in any equation where these symbols appear (as for example (7), b)) should rather be interpreted as set membership.

The following terminology is often used: (2) reads "a_n is big O of b_n" and (4) reads "a_n is little o of b_n". In combinatorial applications, the sequence $(a_n)_{n\geq 0}$ often tends to infinity, since it enumerates structures of a given species. As for b_n, it is often given as an "analytic expression" (or formula), involving the variable

n, whose "computability" is simpler than that of a_n. In particular, when (6) is satisfied, b_n is said to be *asymptotically equivalent* to a_n.

Example 2. a) For a sequence $a_n = \alpha_0 + \alpha_1 n + \cdots + \alpha_\mu n^\mu$, $\alpha_\mu \neq 0$, polynomial in n, and for $b_n = \beta n^\nu$, with $\beta \neq 0$, we have

$$a_n = O(b_n) \quad \text{if } \mu \leq \nu,$$
$$a_n = o(b_n) \quad \text{if } \mu < \nu, \tag{8}$$
$$a_n \sim b_n \quad \text{if } \mu = \nu \text{ and } \alpha_\mu = \beta.$$

b) For any real numbers α and β, with $\beta \neq 0$,

$$n^\alpha = o(\beta^n) \quad \text{if } |\beta| > 1,$$
$$\beta^n = o(n^\alpha) \quad \text{if } |\beta| < 1. \tag{9}$$

c) For $a_n = n!$, we have the following well-known asymptotic equivalence, due to Stirling:

$$n! \sim \left(\frac{n}{e}\right)^n \sqrt{2\pi n}. \tag{10}$$

Example 3. Let $\xi \neq 0$ be a complex number. Then

a) $\left(1 - \frac{x}{\xi}\right)^{-1} = \sum_{n \geq 0} a_n x^n \implies a_n = \frac{1}{\xi^n},$

b) $\left(1 - \frac{x}{\xi}\right)^{-2} = \sum_{n \geq 0} a_n x^n \implies a_n = \frac{n+1}{\xi^n} \sim \frac{n}{\xi^n},$ \quad (11)

c) $\left(1 - \frac{x}{\xi}\right)^{-3} = \sum_{n \geq 0} a_n x^n \implies a_n = \frac{(n+1)(n+2)}{2!\xi^n} \sim \frac{n^2}{2!\xi^n}.$

More generally, if α is an integer ≥ 1, then by Newton's binomial formula,

$$\left(1 - \frac{x}{\xi}\right)^{-\alpha} = \sum_{n \geq 0} a_n x^n \implies a_n = \frac{(n+1)\cdots(n+\alpha-1)}{(\alpha-1)!\xi^n} \sim \frac{n^{\alpha-1}}{(\alpha-1)!\xi^n}. \tag{12}$$

Note that ξ is a pole of order α of the function $(1 - x/\xi)^{-\alpha}$. As the modulus $|\xi|$ of this pole decreases, the coefficient a_n given by (12) gets larger asymptotically.

Polar Singularities

When $f(x)$ is a rational function (i.e., a quotient of polynomials), the asymptotic behavior of $[x^n] f(x)$ is determined by its poles of minimum modulus.

3.4. Elements of Asymptotic Analysis

Proposition 4. *Let $f(x)$ be a rational function analytic at the origin whose poles are*

$$\xi_1, \xi_2, \ldots, \xi_k,$$

the pole ξ_j having multiplicity α_j, $j = 1, \ldots, k$. Then there exists polynomials $\pi_j(n)$, $j = 1, \ldots, k$, and an integer n_0 such that, for $n > n_0$

$$[x^n] f(x) = \sum_{j=1}^{k} \pi_j(n) \frac{1}{\xi_j^n}, \qquad (13)$$

with $\deg \pi_j(n) = \alpha_j - 1$, $j = 1, \ldots, k$. Moreover, if there exists a unique pole ξ_{j_0} with minimum modulus, then

$$[x^n] f(x) \sim \pi_{j_0}(n) \frac{1}{\xi_{j_0}^n}. \qquad (14)$$

Proof. Suppose that $f(x) = p(x)/q(x)$, where p and q are relatively prime polynomials. By the Euclidean algorithm, there exists polynomials $s(x)$ and $r(x)$ such that $p(x) = q(x) s(x) + r(x)$, with $\deg r(x) < \deg q(x)$. Let $n_0 = \deg s(x)$. Then for $n > n_0$, $[x^n] f(x) = [x^n] r(x)/q(x)$. One obtains (13) by decomposing $r(x)/q(x)$ into partial fractions of the form

$$\frac{r(x)}{q(x)} = \sum_{j=1}^{k} \sum_{i=1}^{\alpha_j} \frac{c_{ij}}{(1 - x/\xi_j)^i}, \qquad (15)$$

with complex constants c_{ij}, and by using (12). The asymptotic equivalence (14) follows from the fact that, under the stated condition, $\pi_j(n) \xi_j^{-n} = o(\pi_{j_0}(n) \xi_{j_0}^{-n})$ for any $j \neq j_0$. ∎

A singularity ξ of minimum modulus is often called a *dominant singularity*. When the dominant singularity is not unique, the asymptotic analysis of the coefficient $[x^n] f(x)$ is more delicate (see Exercise 4).

Example 5. The ordinary generating series $f(x)$ for the Fibonacci numbers F_n ($F_n = F_{n-1} + F_{n-2}$, $F_0 = 0$, $F_1 = 1$) is given by $f(x) = x/(1 - x - x^2)$. In this case, $\xi_1 = 2/(1 + \sqrt{5})$, $\xi_2 = 2/(1 - \sqrt{5})$, and $\alpha_1 = \alpha_2 = 1$. Hence, by (13) and (14):

$$F_n = \frac{1}{\sqrt{5}} \left(\frac{1+\sqrt{5}}{2} \right)^n - \frac{1}{\sqrt{5}} \left(\frac{1-\sqrt{5}}{2} \right)^n \sim \frac{1}{\sqrt{5}} \left(\frac{1+\sqrt{5}}{2} \right)^n. \qquad (16)$$

Example 6. Consider an unlimited quantity of red, yellow, and green bricks, of length 2, and blue and orange bricks of length 3. Let a_n be the number of ways to place these bricks edge to edge to have a total length n. Then $a_0 = 1$, $a_1 = 0$, $a_2 = 3$, and $a_n = 3a_{n-2} + 2a_{n-3}$, $n \geq 3$, and we find that $f(x) = \sum_{n \geq 0} a_n x^n = 1/(1 - 3x^2 - 2x^3)$. Thus we obtain $\xi_1 = 1/2$, $\xi_2 = -1$ with $\alpha_1 = 1$ and $\alpha_2 = 2$, and $\pi_1(n) = 4/9$ and $\pi_2(n) = 5/9 + n/3$. It then follows that

$$a_n = \frac{4}{9} 2^n + \left(\frac{5}{9} + \frac{1}{3} n\right)(-1)^n \sim \frac{2^{n+2}}{9}. \tag{17}$$

It is interesting to recall (see Exercise 5) that every sequence $(a_n)_{n \geq 0}$ satisfying a linear recurrence of order k with constant coefficients,

$$a_n = c_1 a_{n-1} + c_2 a_{n-2} + \cdots + c_k a_{n-k}, \quad n \geq k, \tag{18}$$

admits an ordinary generating series which is a rational function, and vice versa.

In the more general case where $f(x)$ is a meromorphic function (i.e., analytic except at isolated pole singularities), for example, $\tan x$, $x/(\exp(x) - 1)$, $\sin(x)/(x-3)^4$, etc., Proposition 4 has the following extension.

Proposition 7. *Let $R > 0$ and $f(x)$ be meromorphic in the disk $|x| < R$ and analytic on the boundary $|x| = R$ as well as at the origin. Suppose that the poles of $f(x)$ in $|x| < R$ are $\xi_1, \xi_2, \ldots, \xi_k$, the pole ξ_j having multiplicity α_j, $j = 1, \ldots, k$. Then there exist polynomials $\pi_1(n), \pi_2(n), \ldots, \pi_k(n)$ such that*

$$[x^n] f(x) = \sum_{j=1}^k \pi_j(n) \frac{1}{\xi_j^n} + o(R^{-n}), \tag{19}$$

with $\deg \pi_j(n) = \alpha_j - 1$. Moreover, if $f(x)$ only has one pole ξ_{j_0} with minimum modulus, then

$$[x^n] f(x) \sim \pi_{j_0}(n) \frac{1}{\xi_{j_0}^n}. \tag{20}$$

Proof. See Exercise 6. ■

Note that when a pole ξ_j is simple ($\alpha_j = 1$), the polynomial $\pi_j(n)$ is constant, and the contribution of ξ_j to (19) is of the form

$$\frac{c_j}{\xi_j^n}, \quad \text{where} \quad c_j = \lim_{x \to \xi_j}(1 - x/\xi_j) f(x). \tag{21}$$

In practice, this situation is frequently encountered.

Example 8. Consider the species Der of derangements, whose generating series is

$$\text{Der}(x) = (1 - x)^{-1} \exp(-x).$$

In this case there is only one simple pole $\xi_1 = 1$. It immediately follows that *for any $R > 1$*,

$$\frac{|\text{Der}[n]|}{n!} = [x^n]\text{Der}(x) = \lim_{x \to 1}(1-x)\text{Der}(x) + o(R^{-n})$$
$$= \frac{1}{e} + o(R^{-n}). \tag{22}$$

In particular, we obtain

$$|\text{Der}[n]| \sim \frac{n!}{e}. \tag{23}$$

More generally (see Exercise 8), the number of permutations all of whose cycles are of length $> k$ is asymptotically given by

$$\frac{n!}{e^{H_k}}, \quad \text{where} \quad H_k = 1 + \frac{1}{2} + \cdots + \frac{1}{k}. \tag{24}$$

Example 9. Consider the species Bal of ballots, whose generating series is

$$\text{Bal}(x) = 1/(2 - e^x).$$

In this case, the poles are the complex numbers ξ such that $\exp(\xi) = 2$, i.e., the numbers

$$\xi_k = \log 2 + 2k\pi i, \quad k \in \mathbb{Z}. \tag{25}$$

These are all simple poles, and the dominant pole is $\xi_0 = \log 2$. After computation, we deduce that

$$\frac{|\text{Bal}[n]|}{n!} = [x^n]\text{Bal}(x) \sim \frac{1}{2}\left(\frac{1}{\log 2}\right)^{n+1}. \tag{26}$$

Moreover, by taking a sequence of circles (of radius R) becoming arbitrarily large and avoiding the poles ξ_k, $k > 0$, one extracts from (19) the following explicit formula

$$|\text{Bal}[n]| = \frac{n!}{2}\sum_{k \in \mathbb{Z}}\left(\frac{1}{\log 2 + 2k\pi i}\right)^{n+1}$$
$$= \frac{n!}{2}\left(\frac{1}{\log 2}\right)^{n+1}\sum_{k \in \mathbb{Z}}\left(1 + \frac{2k\pi i}{\log 2}\right)^{-n-1}. \tag{27}$$

Algebraic Singularities

Euler's gamma function, $\Gamma(\alpha)$, provides the following generalization of (12) (see Exercise 9).

3. Combinatorial Functional Equations

Proposition 10. *Let ξ and α be two complex numbers satisfying $\xi \neq 0$ and $\alpha \notin \{0, -1, -2, \ldots\}$. Then*

$$[x^n]\left(1 - \frac{x}{\xi}\right)^{-\alpha} = \binom{n+\alpha-1}{n}\frac{1}{\xi^n} \sim \frac{n^{\alpha-1}}{\Gamma(\alpha)}\frac{1}{\xi^n}. \tag{28}$$

Example 11. As an application of this proposition, one easily obtains the following asymptotic equivalence for Catalan numbers,

$$c_n \sim \frac{4^n n^{-3/2}}{\sqrt{\pi}}. \tag{29}$$

Indeed, we know that

$$c_n = [x^n]\frac{1 - \sqrt{1-4x}}{2x} = \frac{-1}{2}[x^{n+1}]\sqrt{1-4x}. \tag{30}$$

Hence, from (28) with $\alpha = -1/2$, $\xi = 1/4$, and n replaced by $n+1$, we immediately obtain

$$c_n \sim -\frac{1}{2}\frac{(n+1)^{-1/2-1}}{\Gamma(-1/2)}\frac{1}{(1/4)^{n+1}} \sim \frac{4^n n^{-3/2}}{\sqrt{\pi}}, \tag{31}$$

since $\Gamma(-1/2) = -2\sqrt{\pi}$ and $(n+1)^{-3/2} \sim n^{-3/2}$.

The reader will note that (29) can also be obtained by applying Stirling's formula (10) to the explicit expression $c_n = \binom{2n}{n}/(n+1)$.

Proposition 10 has a very useful extension.

Proposition 12. *Let $\xi \neq 0$ be a complex number, and consider a complex function $f(x)$ which is analytic at the origin.*
a) *If $f(x) = \varphi(x) + (1 - x/\xi)^{-\alpha}\psi(x)$ for two series $\varphi(x)$ and $\psi(x)$ each having a radius of convergence $> |\xi|$, and $\alpha \notin \{0, -1, -2, \ldots\}$, then*

$$[x^n]f(x) \sim \frac{\psi(\xi)}{\xi^n}\frac{n^{\alpha-1}}{\Gamma(\alpha)}. \tag{32}$$

b) *(See Flajolet and Odlyzko [97].) More generally, if $f(x)$ is analytic in an open "Camembert"-shaped region $\mathcal{O}$ (i.e., $\mathcal{O} = \{x \mid x \neq \xi, |x| < R, |\arg((x - \xi)/\xi)| > \omega\}$, where $R > |\xi|$ and $0 < \omega < \pi/2$; see Figure 1) and if, for $x \to \xi$ in $\mathcal{O}$,*

$$f(x) = \sum_{j=0}^{m} c_j(1 - x/\xi)^{-\alpha_j} + O(|1 - x/\xi|^{-A}), \tag{33}$$

3.4. Elements of Asymptotic Analysis

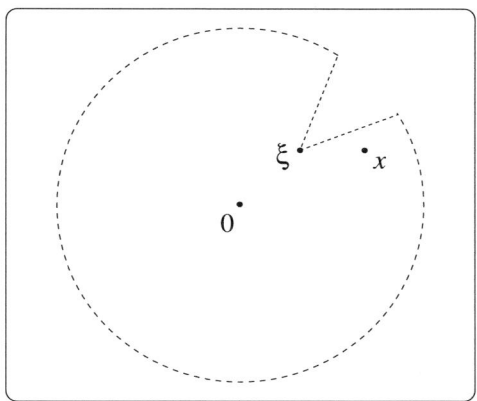

Fig. 1. An open Camembert-shaped region $\mathcal{O}$.

with $\Re\alpha_0 \geq \Re\alpha_1 \geq \cdots \geq \Re\alpha_m > A$, then

$$[x^n]f(x) = \frac{1}{\xi^n}\sum_{j=0}^{m} c_j \binom{n+\alpha_j-1}{n} + O\left(\frac{n^{A-1}}{\xi^n}\right). \tag{34}$$

NOTE: In this proposition, $\Re\alpha$ denotes the real part of α. The notation $O(f(x))$, $x \to \xi$, is the continuous analogue of $O(a_n)$, $n \to \infty$ (see Exercise 3).

The point ξ in (33) or (34) is called an *algebraic singularity* of $f(x)$. The primary interest of Proposition 12, b) comes from the fact that it permits a more detailed analysis of the asymptotic behavior of $[x^n]f(x)$ when used in conjunction with the two following refinements of (28) and (10) (see [97] and [69], and Exercises 17 and 18):

$$[x^n]\left(1-\frac{x}{\xi}\right)^{-\alpha} \sim \frac{n^{\alpha-1}}{\Gamma(\alpha)}\frac{1}{\xi^n}\left(1+\frac{\alpha(\alpha-1)}{2n}+\frac{\alpha(\alpha-1)(\alpha-2)(3\alpha-1)}{24n^2}\right.$$
$$\left.+\frac{\alpha^2(\alpha-1)^2(\alpha-2)(\alpha-3)}{48n^3}+\cdots\right), \tag{35}$$

$$n! \sim \left(\frac{n}{e}\right)^n \sqrt{2\pi n}\left(1+\frac{1}{12n}+\frac{1}{288n^2}-\frac{139}{51840n^3}-\frac{571}{2488320n^4}\right.$$
$$\left.+\frac{163879}{209018880\,n^5}+\cdots\right). \tag{36}$$

Example 13. Consider the function which is analytic at the origin

$$f(x) = \sum_{n\geq 0} f_n \frac{x^n}{n!} = (1-3x)^{-5} e^{x^2} + (1-4x)^{-2/3} e^{x+64x^3}, \tag{37}$$

for which the asymptotic analysis of the coefficient f_n is required. Since the dominant singularity is $\xi = 1/4$, we can neglect the term $(1 - 3x)^{-5} \exp(x^2)$ in the asymptotic analysis of $[x^n] f(x)$. A limited Taylor expansion for $\exp(x + 64x^3)$ around $\xi = 1/4$ gives

$$(1 - 4x)^{-2/3} e^{x+64x^3} = e^{5/4}(1 - 4x)^{-2/3}\left(1 + 13\left(x - \frac{1}{4}\right)\right.$$
$$\left. + \frac{265}{2}\left(x - \frac{1}{4}\right)^2 + \cdots\right)$$
$$= e^{5/4}\left((1 - 4x)^{-2/3} - \frac{13}{4}(1 - 4x)^{1/3}\right.$$
$$\left. + \frac{265}{32}(1 - 4x)^{4/3} + \cdots\right). \tag{38}$$

Using (35) with the following three values of $\alpha: 2/3, -1/3, -4/3$, we find

a) $[x^n](1 - 4x)^{-2/3} \sim \dfrac{4^n}{\Gamma(2/3)}\left(n^{-1/3} - \dfrac{1}{9}n^{-4/3} + \dfrac{1}{81}n^{-7/3} + \cdots\right),$

b) $[x^n](1 - 4x)^{1/3} \sim \dfrac{4^n}{\Gamma(-1/3)}\left(n^{-4/3} + \dfrac{2}{9}n^{-7/3} + \cdots\right),$ \hfill (39)

c) $[x^n](1 - 4x)^{4/3} \sim \dfrac{4^n}{\Gamma(-4/3)}(n^{-7/3} + \cdots).$

An appropriate linear combination of these expressions based on (34) gives, after simplification,

$$\frac{f_n}{n!} = [x^n]f(x) \sim \frac{e^{5/4} 4^n n^{-1/3}}{\Gamma(2/3)}\left(1 + \frac{35}{36n} + \frac{2549}{648n^2} + \cdots\right). \tag{40}$$

Using Stirling's formula (36), we finally find, after some more computation,

$$f_n = n![x^n]f(x) \sim \frac{e^{5/4}\sqrt{2\pi}\, 4^n n^{n+1/6}}{\Gamma(2/3)\, e^n}\left(1 + \frac{19}{18n} + \frac{10415}{2592n^2} + \cdots\right). \tag{41}$$

Example 14. Consider the species $R = L_{\leq 2} = 1 + X + X^2$ of total orderings of cardinality ≤ 2 as well as the species $A = \mathcal{A}_R = X \cdot R(\mathcal{A}_R)$ of the corresponding R-enriched rooted trees, called *unary-binary rooted trees*. Passing to series, we have

$$A(x) = \sum_{n \geq 0} a_n \frac{x^n}{n!} = x(1 + A(x) + A^2(x)). \tag{42}$$

3.4. Elements of Asymptotic Analysis

The asymptotic analysis of the number a_n of *unary-binary* rooted trees having n vertices is carried out as follows. First solve (42) for $A(x)$, to find

$$A(x) = \frac{1 - x - \sqrt{1 - 2x - 3x^2}}{2x}. \tag{43}$$

Thus, for large n

$$\begin{aligned}\frac{a_{n-1}}{(n-1)!} &= [x^{n-1}]\frac{1 - x - \sqrt{(1-3x)(1+x)}}{2x} \\ &= -[x^n]\frac{1}{2}\sqrt{(1-3x)(1+x)}.\end{aligned} \tag{44}$$

The singularities are the roots of the polynomial $(1-3x)(1+x)$; hence $\xi_1 = 1/3$ is the dominant singularity. Similar computations to those of the preceding example give

$$\begin{aligned}\frac{a_{n-1}}{(n-1)!} &= [x^n]\frac{\sqrt{3}}{3}\left(-(1-3x)^{1/2} + \frac{1}{8}(1-3x)^{3/2} + \frac{1}{128}(1-3x)^{5/2} + \cdots\right) \\ &\sim \frac{\sqrt{3}}{6\sqrt{\pi}} 3^n n^{-3/2}\left(1 + \frac{9}{16n} + \frac{265}{512n^2} + \cdots\right).\end{aligned} \tag{45}$$

Substituting $n+1$ for n yields

$$\frac{a_n}{n!} \sim \frac{\sqrt{3}}{2\sqrt{\pi}} 3^n n^{-3/2}\left(1 - \frac{15}{16n} + \frac{505}{512n^2} + \cdots\right), \tag{46}$$

and Stirling's formula finally gives

$$a_n \sim \sqrt{3/2}\,(e/3)^{-n}n^{n-1}\left(1 - \frac{41}{48n} + \frac{4201}{4608n^2} + \cdots\right). \tag{47}$$

Asymptotic Expansions

There exist iterative procedures permitting the computation of as many terms as desired in formulas (35) and (36) (see Exercises 17 and 18). The infinite (formal) sums which result are called *asymptotic expansions* in the sense of the following definition.

Definition 15. Consider a sequence $(\phi_k(n))_{k \geq 0}$ of functions of the integer variable n.
a) The sequence $(\phi_k(n))_{k \geq 0}$ is called an *asymptotic scale*, for $n \to \infty$, if for any $k \geq 0$,

$$\phi_{k+1}(n) = o(\phi_k(n)). \tag{48}$$

b) A formal series $\sum_{k\geq 0} c_k \phi_k(n)$ is called an *asymptotic expansion* of a sequence $(a_n)_{n\geq 0}$, relative to the asymptotic scale $(\phi_k(n))_{k\geq 0}$, if for any $N \geq 0$,

$$a_n = \sum_{k=0}^{N} c_k \phi_k(n) + o(\phi_N(n)). \tag{49}$$

This is summarized by the formula

$$a_n \sim \sum_{k\geq 0} c_k \phi_k(n). \tag{50}$$

Example 16. Considering the asymptotic scale

$$\phi_k(n) = \frac{4^n n^{-3/2}}{\sqrt{\pi}} n^{-k}, \tag{51}$$

we obtain an asymptotic expansion for Catalan numbers, starting with

$$c_n \sim \frac{4^n n^{-3/2}}{\sqrt{\pi}} \left(1 - \frac{9}{8n} + \frac{145}{128n^2} - \frac{1155}{1024n^3} + \frac{36939}{32768n^4} + \cdots \right). \tag{52}$$

Using the explicit formula

$$c_n = \frac{1}{n+1}\binom{2n}{n} \sim \frac{(2n)!}{n!^2}\left(\frac{1}{n} - \frac{1}{n^2} + \frac{1}{n^3} - \frac{1}{n^4} + \cdots \right), \tag{53}$$

Equation (52) is a consequence of Stirling's asymptotic expansion (36) and of the following proposition whose proof is left as an exercise.

Proposition 17. *If $a_n \sim \sum_{k\geq 0} c_k \phi_k(n)$, then*
a) *the coefficients c_k are unique and can be calculated successively by*

$$\begin{aligned} c_0 &= \lim_{n\to\infty} \frac{a_n}{\phi_0(n)}, \\ c_{k+1} &= \lim_{n\to\infty} \frac{a_n - \sum_{i=0}^{k} c_i \phi_i(n)}{\phi_{k+1}(n)}. \end{aligned} \tag{54}$$

b) *If moreover $a'_n \sim \sum_{k\geq 0} c'_k \phi_k(n)$, then for all constants α and α',*

$$\alpha a_n + \alpha' a'_n \sim \sum_{k\geq 0} (\alpha c_k + \alpha' c'_k) \phi_k(n). \tag{55}$$

c) *Asymptotic scales of the form $\phi_k(n) = \omega(n)/n^k$ ($\omega(n)$ not vanishing for large enough n) are closed under product and quotient in the following sense. If*

$a_n \sim \alpha(n)(\alpha_0+\alpha_1 n^{-1}+\alpha_2 n^{-2}+\cdots)$ and $b_n \sim \beta(n)(\beta_0+\beta_1 n^{-1}+\beta_2 n^{-2}+\cdots)$, then

$$a_n b_n \sim \alpha(n)\beta(n)(\gamma_0 + \gamma_1 n^{-1} + \gamma_2 n^{-2} + \cdots),$$

$$\frac{a_n}{b_n} \sim \frac{\alpha(n)}{\beta(n)}(\gamma'_0 + \gamma'_1 n^{-1} + \gamma'_2 n^{-2} + \cdots), \quad \text{if } \beta_0 \neq 0, \tag{56}$$

where

$$\sum \gamma_k n^{-k} = \left(\sum \alpha_k n^{-k}\right)\left(\sum \beta_k n^{-k}\right),$$

$$\sum \gamma'_k n^{-k} = \left(\sum \alpha_k n^{-k}\right)\left(\sum \beta_k n^{-k}\right)^{-1}.$$

An asymptotic expansion may very well be divergent. This is the case in the expansion (36) of Stirling's formula. However, the error caused by truncating an asymptotic expansion is often bounded by the absolute value of the last nonzero term used (as n tends to infinity). It is sometimes possible to obtain a more precise analysis of this error (see Exercise 15). A classical example due to Euler of a divergent asymptotic expansion (see Exercise 19) is

$$\int_0^\infty \frac{e^{-t}}{1+t/n} dt \sim 0! - \frac{1!}{n} + \frac{2!}{n^2} - \frac{3!}{n^3} + \cdots. \tag{57}$$

Singularities at Infinity

W.K. Hayman (see [150]) has introduced a class of power series with real coefficients which he called *admissible*, and whose asymptotic analysis of coefficients is particularly easy. This class has closure properties under certain basic operations. Rather than giving a technical definition, we present here a recursive description of a subclass $\mathcal{H}$ of admissible series which is sufficient for the applications that we have in mind. In particular, all the series appearing in the class $\mathcal{H}$ have an infinite radius of convergence. They then define entire functions, i.e., functions which are analytic on the entire complex plane.

Definition 18. The class $\mathcal{H}$ is the smallest family of entire functions with real coefficients satisfying the following conditions:
i) if $p(x)=c_0+c_1 x+\cdots+c_k x^k$ is a nonconstant polynomial with real coefficients, then
   a) $e^{p(x)} \in \mathcal{H}$ if there exists an n_0 such that $n \geq n_0$ implies $[x^n]e^{p(x)} > 0$,
   b) $p(f(x)) \in \mathcal{H}$ if $c_k > 0$ and $f(x) \in \mathcal{H}$;
ii) if $f(x)$ and $g(x)$ are in $\mathcal{H}$, then $\exp(f(x))$ and $f(x)g(x)$ are in $\mathcal{H}$;
iii) if $q(x)$ is an arbitrary polynomial with real coefficients and if $f(x) \in \mathcal{H}$, then $q(x) + f(x) \in \mathcal{H}$.

Observe that for any function $f(x) = \sum_{n\geq 0}(f_n/n!)x^n \in \mathcal{H}$, $f_n > 0$ for large enough n.

Theorem 19. (HAYMAN [150]). *Let* $f(x) = \sum_{n\geq 0}(f_n/n!)x^n \in \mathcal{H}$ *and put*

$$a(x) = \frac{xf'(x)}{f(x)}, \qquad b(x) = xa'(x). \tag{58}$$

Then

$$\frac{f_n}{n!} \sim \frac{1}{\sqrt{2\pi b(r_n)}} \frac{f(r_n)}{r_n^n}, \tag{59}$$

and

$$f_n \sim \left(\frac{n}{b(r_n)}\right)^{1/2} \left(\frac{n}{er_n}\right)^n f(r_n), \tag{60}$$

where r_n is the smallest positive root of the equation

$$a(x) = n. \tag{61}$$

Example 20. Stirling's formula (10) is a special case of Theorem 19 which, incidentally, has motivated Hayman. We obtain Stirling's formula as follows. Since the exponential power series $\exp(x) = \sum_{n\geq 0}(1/n!)x^n$ is admissible, (59) can be used to estimate $1/n!$. In this case, $a(x) = x$, $b(x) = x$, and we find $r_n = n$. Hence, we immediately obtain

$$\frac{1}{n!} \sim \frac{1}{\sqrt{2\pi n}} \frac{e^n}{n^n}. \tag{62}$$

Example 21. The generating series of the species $\mathrm{Inv} = E(X + E_2)$ of involutions is admissible since $\mathrm{Inv}(x) = \exp(x + x^2/2)$. In this case $a(x) = x + x^2$ and $b(x) = x + 2x^2$. The real positive solution r_n of $x^2 + x = n$ is, for $n \geq 1$,

$$r_n = \frac{1}{2}(1+4n)^{1/2} - \frac{1}{2} = n^{1/2}\left(1 + \frac{1}{4n}\right)^{1/2} - \frac{1}{2}$$

$$= n^{1/2}\left(1 + \frac{1}{8n} - \frac{1}{128n^2} + \cdots\right) - \frac{1}{2} \tag{63}$$

$$= n^{1/2} - \frac{1}{2} + \frac{1}{8}n^{-1/2} - \frac{1}{128}n^{-3/2} + \cdots.$$

Thus

$$b(r_n) = r_n + 2r_n^2 = 2(r_n + r_n^2) - r_n = 2n - r_n$$

$$\sim 2n - n^{1/2} + \frac{1}{2} - \frac{1}{8}n^{-1/2} + \frac{1}{128}n^{-3/2} + \cdots. \tag{64}$$

3.4. Elements of Asymptotic Analysis

Formula (60) gives, after some computation, the following asymptotic equivalence for the number of involutions $|\text{Inv}[n]|$ of a set with n elements (see Exercise 20):

$$|\text{Inv}[n]| \sim \frac{1}{\sqrt{2}} \left(\frac{n}{e}\right)^{n/2} e^{\sqrt{n}-1/4}. \tag{65}$$

Example 22. Consider now the species Par of set partitions. We know that

$$\text{Par}(x) = \exp(\exp(x) - 1).$$

Since $\exp(x) - 1$ is admissible, $\text{Par}(x)$ is also admissible. The following asymptotic equivalence for the number $|\text{Par}[n]|$ of partitions of a set with n elements is deduced from (60):

$$|\text{Par}[n]| \sim (1 + r_n)^{-1/2} \left(\frac{n}{er_n}\right)^n e^{-1+n/r_n}, \tag{66}$$

where r_n is the real positive solution of

$$xe^x = n. \tag{67}$$

In this case, the asymptotic analysis of r_n (and a fortiori of r_n^n) is rather difficult (see de Bruijn [69]). However, for particular numerical values of n, one can efficiently estimate r_n by applying Newton–Raphson's iteration to the equation $xe^x - n = 0$ with the initial value $x_0 = \log n$. For example, for $n = 10$, one finds $r_n \simeq 1.74552800274$ and the right-hand side of (66) approximately equals 118091.956130 (the exact value of $|\text{Par}[10]|$ is 115975), with a relative error less than 0.0183. For $n = 100$, the relative error of the result obtained by the same procedure is less than 0.0033.

The asymptotic analysis of $\log|\text{Par}[n]|$ is easier to carry out. We have the following equivalence (see Exercise 22):

$$\log|\text{Par}[n]| \sim n \log n - n \log \log n - n. \tag{68}$$

Implicitly Defined Series

In Example 14 the asymptotic analysis of the number a_n of R-enriched rooted trees, with $R = 1 + X + X^2$, has been made possible due to the existence of a dominant algebraic singularity of the *explicit* solution

$$y = A(x) = \frac{1 - x - \sqrt{1 - 2x - 3x^2}}{2x} \tag{69}$$

of the implicit equation $y = x(1 + y + y^2)$. Meir and Moon (see [241]) have given very general conditions, on the series $R(x)$, allowing the asymptotic study of R-enriched rooted trees even in the case where the solution $y = A(x)$ of the equation $y = xR(y)$ is not available in *closed* form. The result which follows is an adaptation of their theorem to the context of species of structures.

Theorem 23. (MEIR AND MOON [234] AND [241]). *Let R be a species of structures whose generating series $R(x) = \sum_{n \geq 0} r_k x^k / k!$ has a radius of convergence $\rho > 0$. Let $\mathcal{A}_R$ be the corresponding species of R-enriched rooted trees, and let*

$$\mathcal{A}_R(x) = xR(\mathcal{A}_R(x)) = \sum_{n \geq 1} a_n \frac{x^n}{n!}. \tag{70}$$

Suppose moreover that

    i) $r_0 = 1$, $r_k \geq 0$ *for any* $k \geq 1$,

    ii) $\gcd\{k \mid r_k > 0\} = 1$,   (71)

    iii) $\tau R'(\tau) = R(\tau)$,   *for some* τ, $0 < \tau < \rho$.

Then

$$a_n \sim \alpha \beta^n n^{n-1}, \tag{72}$$

where

$$\alpha = \alpha_R = \sqrt{\frac{R(\tau)}{R''(\tau)}}, \qquad \beta = \beta_R = \frac{R(\tau)}{\tau e}. \tag{73}$$

The asymptotic equivalence (72) is also valid in the purely analytical context when $R(x)$ is an exponential generating series with real (not necessarily integral) coefficients r_k satisfying conditions (71). A sketch of the proof for this theorem is given in Exercise 23.

More generally, Meir and Moon have shown (see also Flajolet, Steyaert [101], and Gardy [123]) under hypotheses (71), for k a fixed integer, that the number $a_n^{\{k\}}$ of forests of k R-enriched rooted trees is given asymptotically by

$$a_n^{\{k\}} \sim \frac{\tau^{k-1}}{(k-1)!} a_n \sim \frac{\tau^{k-1}}{(k-1)!} \alpha \beta^n n^{n-1}, \tag{74}$$

where n is the total number of vertices appearing in the forest. Exercise 25 considers an extension of this result (also due to Meir and Moon) in the case where the number k depends on n.

Example 24. In the case where $R = E$, $\mathcal{A}_R$ is the species of ordinary rooted trees. Then $\tau = 1$ (since $\tau R'(\tau) = R(\tau)$ corresponds to $\tau \exp(\tau) = \exp(\tau)$) and $\alpha = \beta = 1$. Formula (72) then reduces to $a_n \sim n^{n-1}$. In fact, it is an equality, by virtue of Cayley's theorem! Concerning forests with k rooted (ordinary) trees, formula (74) reduces to

$$a_n^{\{k\}} \sim \frac{1}{(k-1)!} n^{n-1}. \tag{75}$$

This last result also follows directly from the explicit formula (3.1.35) for such forests, namely

$$a_n^{\{k\}} = \frac{k}{n}\binom{n}{k} n^{n-k} \sim \frac{k}{n}\frac{n^k}{k!} n^{n-k} = \frac{1}{(k-1)!} n^{n-1}. \tag{76}$$

Example 25. In the cases of ordered rooted trees ($R = L$) and of (plane, nonempty) binary rooted trees ($R = (1 + X)^2$), one finds respectively (see Exercise 24)

$$a_n \sim \frac{1}{2\sqrt{2}}\left(\frac{4}{e}\right)^n n^{n-1}, \qquad a_n \sim \sqrt{2}\left(\frac{4}{e}\right)^n n^{n-1}. \tag{77}$$

Example 26. For *mobiles*, i.e., cyclic rooted trees ($R = 1+C$), the equation $\tau R'(\tau) = R(\tau)$ takes the form of the transcendental equation

$$\frac{\tau}{1-\tau} = 1 - \log(1-\tau). \tag{78}$$

It is possible to approximate τ numerically with the help of classical iterative procedures, and obtain, for example,

$$\tau = .682155567100\cdots. \tag{79}$$

Thus, the number a_n of mobiles with n vertices satisfies $a_n \sim \alpha\beta^n n^{n-1}$, where

$$\alpha = \sqrt{\frac{1-\log(1-\tau)}{(1-\tau)^{-2}}} = \sqrt{\tau(\tau-1)} = .465638646779\cdots,$$

$$\beta = \frac{1-\log(1-\tau)}{e\tau} = \frac{e^{-1}}{1-\tau} = 1.157419803819\cdots. \tag{80}$$

E.A. Bender has given a very general result for the asymptotic analysis of the coefficients of implicitly defined series. Here is a version adapted to the context of exponential generating series.

Theorem 27. (E.A. BENDER [9], [43].) *Let $a(x) = \sum_{n\geq 0} a_n x^n/n!$ be a formal power series with coefficients ≥ 0, which is the implicit solution of the functional equation $f(x, y) = 0$, i.e., $f(x, a(x)) = 0$. Suppose moreover that there exist real numbers $\xi > 0$ and $\tau > a_0$ satisfying the following conditions:*
a) *there exists $\delta > 0$ such that $f(x, y)$ is analytic in the open region (of $\mathbb{C} \times \mathbb{C}$) defined by $|x| < \xi + \delta$ and $|y| < \tau + \delta$,*

b) $f(\xi, \tau) = f_y(\xi, \tau) = 0$,
c) $f_x(\xi, \tau) \neq 0$, $f_{yy}(\xi, \tau) \neq 0$,
d) if $|x| \leq \xi$, $|y| \leq \tau$, and $f(x, y) = f_y(x, y) = 0$, then $x = \xi$ and $y = \tau$.
Then one has

$$a_n \sim \alpha \beta^n n^{n-1}, \tag{81}$$

where

$$\alpha = \alpha_f = \sqrt{\frac{\xi f_x(\xi, \tau)}{f_{yy}(\xi, \tau)}}, \qquad \beta = \beta_f = \frac{1}{\xi e}. \tag{82}$$

The complete proof of Bender's theorem depends on Weierstrass's Preparation theorem and is beyond the scope of this text. The reader is referred to [8] for an exposition of this proof (see also P. Auger [4]).

In fact, Theorem 23 of Meir and Moon follows from Bender's theorem by taking $f(x, y) = y - x R(y)$ (see Exercise 26).

Example 28. Let G be a species of structures satisfying $G(0) = G'(0) = 0$ and $A = {}_G A = X + G({}_G A)$ be the corresponding species of G-rooted trees. Taking $f(x, y) = y - x - G(y)$, equations b) of the theorem can be written in the form
b') $\xi = \tau - G(\tau)$, where $G'(\tau) = 1$.
Thus, for $a_n = n![x^n]A(x)$, we have

$$a_n \sim \alpha \beta^n n^{n-1}, \tag{83}$$

where

$$\alpha = \sqrt{\frac{\tau - G(\tau)}{G''(\tau)}}, \qquad \beta = \frac{1}{e(\tau - G(\tau))}. \tag{84}$$

Example 29. In the case of the implicit species theorem, $A = H(X, A)$, it suffices to take $f(x, y) = y - H(x, y)$. The equations to be satisfied by (ξ, τ) are:
b'') $\tau = H(\xi, \tau)$, $1 = H_y(\xi, \tau)$,
and then

$$a_n \sim \alpha \beta^n n^{n-1}, \tag{85}$$

where

$$\alpha = \sqrt{\frac{\xi H_x(\xi, \tau)}{H_{yy}(\xi, \tau)}}, \qquad \beta = \frac{1}{\xi e}. \tag{86}$$

Theorem 27 can also be used (see Exercise 29) for the asymptotic enumeration of unlabeled R-enriched rooted trees and G-rooted trees, whose ordinary generating series are characterized respectively by

$$\widetilde{A}(x) = x Z_R(\widetilde{A}(x), \widetilde{A}(x^2), \widetilde{A}(x)(x^3), \ldots) \tag{87}$$

and

$$\widetilde{A}(x) = x + Z_G(\widetilde{A}(x), \widetilde{A}(x^2), \widetilde{A}(x^3), \ldots). \tag{88}$$

NOTE: The reader is referred to the texts of Bender [9] or Salvy [291], for more complete expositions on asymptotic analysis methods related to combinatorics, and to Pólya and Read [264], Otter [259], Ford and Uhlenbeck [116], Harary, Robinson and Schenk [116], as well as the work of Meir and Moon listed in the bibliography, for other examples and results related to the asymptotic enumeration of tree-like structures.

Exercises for Section 3.4

1. Show that for all constants α and α', one has
   a) $a_n = O(b_n)$ and $a'_n = O(b_n) \implies \alpha a_n + \alpha' a'_n = O(b_n)$,
   b) $a_n = O(b_n)$ and $a'_n = o(b_n) \implies \alpha a_n + \alpha' a'_n = O(b_n)$,
   c) $a_n = o(b_n)$ and $a'_n = o(b_n) \implies \alpha a_n + \alpha' a'_n = o(b_n)$,
   d) $a_n \sim b_n$ and $a'_n = o(b_n)$, $b_n \neq 0$, $\alpha \neq 0 \implies \alpha a_n + \alpha' a'_n \sim \alpha b_n$,
   e) $a_n = O(b_n)$ and $b_n = O(c_n) \implies a_n = O(c_n)$,
   f) $a_n = O(b_n)$ and $b_n = o(c_n) \implies a_n = o(c_n)$,
   g) $a_n = o(b_n)$ and $b_n = O(c_n) \implies a_n = o(c_n)$,
   h) $a_n = o(b_n)$ and $b_n = o(c_n) \implies a_n = o(c_n)$,
   i) $a_n \sim b_n$ and $a'_n \sim b'_n \implies a_n a'_n \sim b_n b'_n$,
   j) $a_n \sim b_n$ and $a_n > c > 1 \implies \log a_n \sim \log b_n$.
2. Let $\sum_{n \geq 0} a_n x^n$ be a formal power series in x with complex coefficients. The *radius of convergence* R of this series is defined by

$$R = \sup\left\{\rho \geq 0 \,\Big|\, \sum_{n \geq 0} |a_n| \rho^n < \infty\right\}, \tag{89}$$

and one has $0 \leq R \leq \infty$ (i.e., R is an element of $\overline{\mathbb{R}}_+$, the completion of the positive reals). Show that if $0 < R < \infty$, then for any $\epsilon > 0$,

$$(1-\epsilon)^n R^{-n} <_{\text{i.o.}} |a_n| <_{\text{a.e.}} (1+\epsilon)^n R^{-n}, \tag{90}$$

where $u_n <_{\text{i.o.}} v_n$ signifies that $u_n < v_n$ *infinitely often* (i.e., for an infinite number of n) and $u_n <_{\text{a.e.}} v_n$ signifies that $u_n < v_n$ *almost everywhere* (i.e., for all sufficiently large n).

3. a) Consider an accumulation point x_0 of a subset X of the complex plane or of its compactification $\check{\mathbb{C}} = \mathbb{C} \cup \{\infty\}$ (the Riemann sphere). Given two functions $f = f(x)$ and $g = g(x)$,

$$f, g : X \longrightarrow \mathbb{C},$$

define the following notions, analogous to the case of sequences (where $X = \mathbb{N}$ and $x_0 = \infty$):

i) $f = O(g),$ for $x \to x_0,$
ii) $f = o(g),$ for $x \to x_0,$
iii) $f \sim g,$ for $x \to x_0.$

b) The *order* of a formal power series $f(x) = f_0 + f_1 x + f_2 x^2 + \cdots$, denoted ord f, is defined by

$$\operatorname{ord} f = \begin{cases} \min\{k \mid f_k \neq 0\}, & \text{if } f \neq 0, \\ \infty, & \text{if } f = 0. \end{cases} \tag{91}$$

Show that for two formal power series f and g, with radius of convergence > 0, one has, as $x \to 0$,

i) $f = O(g) \Longleftrightarrow \operatorname{ord} f \geq \operatorname{ord} g,$
ii) $f = o(g) \Longleftrightarrow \operatorname{ord} f > \operatorname{ord} g,$ (92)
iii) $f \sim g \Longleftrightarrow \operatorname{ord} f = \operatorname{ord} g = n < \infty$ and $[x^n] f = [x^n] g.$

c) Observe that the right-hand sides of (92) are independent of the radius of convergence of the series and can serve as definitions for O, o, and $\sim$ in the more general case of formal power series f and g with coefficients from any ring. In particular, establish the following connection with the notion of contact of order n (see Section 1.2) for formal power series:

$$f =_n g \Longleftrightarrow f - g = o(x^n) \Longleftrightarrow f - g = O(x^{n+1}). \tag{93}$$

4. Consider the function

$$f(x) = \frac{\rho(\rho - x \cos \theta)}{\rho^2 - 2\rho x \cos \theta + x^2} = \sum_{n \geq 0} f_n x^n, \tag{94}$$

where ρ and θ are two real numbers such that $\rho > 0$ and $|\cos \theta| \neq 1$. Show that $f(x)$ has two poles of minimum modulus and that there does not exist any polynomial $\pi(n)$ and any complex number ξ such that, as $n \to \infty$, $f_n \sim \pi(n)/\xi^n$.

HINT: $\rho^2 - 2\rho x \cos \theta + x^2 = (\rho - xe^{i\theta})(\rho - xe^{-i\theta})$.

3.4. Elements of Asymptotic Analysis

5. a) Let $(a_n)_{n\geq 0}$ be a sequence satisfying a linear recurrence of order k with constant coefficients

$$a_n = c_1 a_{n-1} + c_2 a_{n-2} + \cdots + c_k a_{n-k}, \qquad n \geq k \qquad (95)$$

with initial conditions $a_i = \alpha_i$, $0 \leq i \leq k-1$. Show that there exists a polynomial $p(x)$ such that

$$\sum_{n\geq 0} a_n x^n = \frac{p(x)}{1 - c_1 x - c_2 x^2 - \cdots c_k x^k}. \qquad (96)$$

HINT: $\sum_{n\geq 0} a_n x^n = \sum_{n=0}^{k-1} \alpha_n x^n + \sum_{n\geq k} a_n x^n$.

b) Conversely, show that if $\sum_{n\geq 0} a_n x^n = p(x)/q(x)$ for two polynomials $p(x)$ and $q(x)$, then there exist an integer $k \geq 0$ and constants c_{i+1} and α_i, $0 \leq i \leq k-1$, such that (95) will be valid, with the corresponding initial conditions.

HINT: $q(x) \sum_{n\geq 0} a_n x^n = p(x)$.

6. Prove Proposition 7 by proceding according to the following steps:

a) Use the sum of the principal parts of $f(x)$ at the poles $\xi_1, \xi_2, \ldots, \xi_k$ and deduce the existence of a rational function $p(x)/q(x)$ such that

$$f(x) = \frac{p(x)}{q(x)} + \varphi(x), \qquad (97)$$

where $\varphi(x)$ is analytic at every point in the closed disk $|x| \leq R$.

b) Show, by a compactness argument, that $\varphi(x)$ is in fact analytic on a larger closed disk $|x| \leq R_1$, where $R_1 > R$. Deduce that

$$[x^n]\varphi(x) = o(R^{-n}). \qquad (98)$$

c) Apply Proposition 4 to $p(x)/q(x)$.

7. Show that the polynomials $\pi_j(n)$, $j = 1, \ldots, k$, appearing in Proposition 7 have the explicit form

$$\pi_j(n) = \sum_{i=1}^{\alpha_j} c_{ij} \frac{(n+1)(n+2)\cdots(n+i-1)}{(i-1)!}, \qquad (99)$$

where the constants c_{ij} are given by

$$c_{ij} = \frac{(-\xi_j)^{\alpha_j - i}}{(\alpha_j - i)!} \lim_{x \to \xi_j} \left(\frac{d}{dx}\right)^{\alpha_j - i} \left\{ \left(1 - \frac{x}{\xi_j}\right)^{\alpha_j} f(x) \right\}. \qquad (100)$$

In particular, if ξ_j is a simple pole (i.e., $\alpha_j = 1$), then

$$\pi_j(n) = c_{1j} = \lim_{x \to \xi_j} \left(1 - \frac{x}{\xi_j}\right) f(x). \qquad (101)$$

8. a) Show that
$$\frac{1}{1-x} \exp\left(-x - \frac{1}{2}x^2 - \cdots - \frac{1}{k}x^k\right) \quad (102)$$

is the generating series of the species of permutations whose cycles are of length $> k$. Deduce the asymptotic equivalence (24) for the number of such permutations on a set of n elements.

b) Give an asymptotic equivalence for the number of pointed ballots on a set of n elements.

9. EULER'S GAMMA FUNCTION. Euler's gamma function Γ is defined for $\alpha > 0$ by the integral

$$\Gamma(\alpha) = \int_0^\infty e^{-x} x^{\alpha-1}\, dx. \quad (103)$$

a) Show that $\Gamma(1) = 1$ and that

$$\Gamma(\alpha + 1) = \alpha\, \Gamma(\alpha), \quad (104)$$

by integrating (103) by parts. Deduce that, for any integer $n \geq 0$,

$$\Gamma(n+1) = n!. \quad (105)$$

b) Show that

$$\Gamma(1/2) = \int_{-\infty}^{+\infty} e^{-x^2}\, dx = \sqrt{\pi}. \quad (106)$$

HINT: Compute $(\Gamma(1/2))^2$ by passing to polar coordinates.

c) Starting from the equality

$$\Gamma(\alpha) = \lim_{N \to \infty} \int_0^N \left(1 - \frac{x}{N}\right)^N x^{\alpha-1}\, dx, \quad (107)$$

establish the formula

$$\Gamma(\alpha) = \lim_{N \to \infty} \frac{N^\alpha N!}{\alpha\,(\alpha+1) \cdots (\alpha+N)}. \quad (108)$$

HINT: Let $t = x/N$ in (107), write

$$\Gamma(\alpha) = \lim_{N \to \infty} N^\alpha\, J_N(\alpha), \quad (109)$$

and use integration by parts to show that $J_N(\alpha) = (N/\alpha)\, J_{N-1}(\alpha + 1)$.

d) Formula (108) allows one to define $\Gamma(\alpha)$ for all complex numbers α that are not a negative integer ($\alpha \notin \{0, -1, -2, \ldots, \}$). Show that it also implies

3.4. Elements of Asymptotic Analysis

the formula

$$\frac{1}{\Gamma(\alpha)} = \alpha e^{\alpha \gamma} \prod_{k=1}^{\infty} \left(1 + \frac{\alpha}{k}\right) e^{-\alpha/k}, \tag{110}$$

where γ denotes the *Euler constant*, defined by

$$\gamma = \lim_{N \to \infty} \left(1 + \frac{1}{2} + \frac{1}{3} + \cdots \frac{1}{N} - \log N\right). \tag{111}$$

HINT: $(\alpha + k) = k(1 + \alpha/k) = k(1 + \alpha/k)\exp(-\alpha/k)\exp(\alpha/k)$.

NOTE: The infinite product (110) converges uniformly over every compact set in the complex plane. As a consequence, $1/\Gamma(\alpha)$ is an entire function of α and $\Gamma(\alpha)$ is a meromorphic function on $\mathbb{C}$ whose poles are all simple and of the form $-k$ for $k \in \mathbb{N}$.

e) Show that the asymptotic equivalence (28) is a consequence of (108).

10. MOTZKIN'S PATHS. A *Motzkin path* of length n is a polygonal path in the Cartesian plane $\mathbb{R} \times \mathbb{R}$ whose vertices $p_0, p_1, \ldots, p_n$ belong to $\mathbb{Z} \times \mathbb{Z}$ and which satisfy the following conditions:
   i) $p_0 = (0, 0)$, $p_n = (n, 0)$,
   ii) each step $\overline{p_i\, p_{i+1}}$ is
     – a *northeastern* step, i.e., $p_{i+1} = p_i + (1, 1)$,
     – an *eastern* step, i.e., $p_{i+1} = p_i + (1, 0)$,   or
     – a *southeastern* step, i.e., $p_{i+1} = p_i + (1, -1)$,
   iii) the path does not fall below the x-axis, i.e., for any $p_i = (i, y_i)$ one has $y_i \geq 0$.

Let μ_n be the number of Motzkin paths of length n.

a) Show that $M = M(x) = \sum_{n \geq 0} \mu_n x^n$ satisfies the functional equation

$$M = 1 + xM + x^2 M^2. \tag{112}$$

b) Deduce that

$$M(x) = \frac{1 - x - \sqrt{(1+x)(1-3x)}}{2x^2} \tag{113}$$

and then that

$$\mu_n \sim \left(\frac{3\sqrt{3}}{2\sqrt{\pi}}\right) 3^n\, n^{-3/2}. \tag{114}$$

11. Prove Proposition 17 on asymptotic expansions.
12. a) Let $\omega(n)$ be a function of n which does not vanish for large n and consider a strictly decreasing sequence $\alpha_1 > \alpha_2 > \alpha_3 > \cdots$ of real numbers. Show that

$$\phi_k(n) = \omega(n)\, n^{\alpha_k}, \qquad \psi_k(n) = \omega(n)\, e^{-\alpha_k \cdot n} \tag{115}$$

are two asymptotic scales as $n \to \infty$.

b) Show that two distinct sequences $(a_n)_{n\geq 0}$ and $(b_n)_{n\geq 0}$ can have the same asymptotic expansion according to a given asymptotic scale. For example, verify that

i) $0 \sim 0 + \dfrac{0}{n} + \dfrac{0}{n^2} + \dfrac{0}{n^3} + \cdots$ and $e^{-n} \sim 0 + \dfrac{0}{n} + \dfrac{0}{n^2} + \dfrac{0}{n^3} + \cdots,$

ii) $\dfrac{1}{1+1/n} \sim 1 - \dfrac{1}{n} + \dfrac{1}{n^2} + \cdots$ and $\dfrac{1+e^{-n}}{1+1/n} \sim 1 - \dfrac{1}{n} + \dfrac{1}{n^2} + \cdots,$

(116)

relative to the asymptotic scale $(n^{-k})_{k\geq 0}$.

13. Let $C(t) = \sum_{k\geq 0} c_k t^k$ be an analytic function having a radius of convergence $\rho > 0$. Show that

$$C(1/n) \sim \sum_{k\geq 0} \dfrac{c_k}{n^k}, \qquad (117)$$

according to the asymptotic scale $(n^{-k})_{k\geq 0}$.

14. BERNOULLI NUMBERS. The *Bernoulli numbers* B_n, $n \geq 0$, are rational numbers defined by

$$\dfrac{x}{e^x - 1} = \sum_{n\geq 0} B_n \dfrac{x^n}{n!}. \qquad (118)$$

Here are the first values of B_n, for $n = 0, 1, \ldots, 16$:

$$1, -\dfrac{1}{2}, \dfrac{1}{6}, 0, -\dfrac{1}{30}, 0, \dfrac{1}{42}, 0, -\dfrac{1}{30}, 0, \dfrac{5}{66}, 0, -\dfrac{691}{2730}, 0, \dfrac{7}{6}, 0, -\dfrac{3617}{510}. \qquad (119)$$

a) Show that the B_n satisfy the recurrence

$$B_0 = 1, \qquad B_n = -\sum_{i=1}^{n} \dfrac{1}{i+1} \binom{n}{i} B_{n-i}, \qquad n > 0. \qquad (120)$$

b) Show that for $m > 0$,

$$B_{2m+1} = 0 \qquad (121)$$

and that

$$B_{2m} \sim (-1)^{m-1} \dfrac{(2m)!}{2^{2m-1}\pi^{2m}}. \qquad (122)$$

HINT (for (121)): Show that the function $x/(e^x - 1) + x/2$ is even.

15. EULER–MACLAURIN SUMMATION FORMULA.

a) The *Bernoulli polynomials* $B_n(t)$ are defined by the formula

$$\dfrac{xe^{tx}}{e^x - 1} = \sum_{n\geq 0} B_n(t) \dfrac{x^n}{n!}. \qquad (123)$$

Show for any $n \geq 0$, where the B_n denote the Bernoulli numbers, that

i) $B_n(0) = (-1)^n B_n(1) = B_n,$ ii) $B_n(t) = \sum_{k=0}^{n} \binom{n}{k} B_{n-k} t^k,$

iii) $B_n(t+1) - B_n(t) = nt^{n-1},$ iv) $B'_n(t) = n B_{n-1}(t).$

(124)

b) (de Bruijn [69].) Let k and n be two integers ≥ 1 and $f : [1, n] \longrightarrow \mathbb{C}$ be a function such that $f^{(2k)}$ is continuous. Show that (Euler–Maclaurin summation formula)

$$f(1) + f(2) + \cdots + f(n) = \int_1^n f(t)\,dt + \frac{1}{2}(f(n) + f(1))$$

$$+ \frac{B_2}{2!}(f'(n) - f'(1))$$

$$+ \frac{B_4}{4!}(f'''(n) - f'''(1)) + \cdots$$

$$+ \frac{B_{2k}}{(2k)!}\left(f^{(2k-1)}(n) - f^{(2k-1)}(1)\right) - R_k,$$

(125)

the error term R_k being given by

$$R_k = \int_1^n f^{(2k)}(t) \frac{B_{2k}(\{t\})}{(2k)!}\,dt,$$ (126)

where $\{t\} = t - \lfloor t \rfloor$ denotes the fractional part of t.
HINT: First show by induction on k (by integrating by parts twice) that

$$\frac{g(0) + g(1)}{2} - \int_0^1 g(t)\,dt = \sum_{\nu=1}^{k} \left(g^{(2\nu-1)}(1) - g^{(2\nu-1)}(0)\right) \frac{B_{2\nu}}{(2\nu)!}$$

$$- \int_0^1 g^{(2k)}(t) \frac{B_{2k}(t)}{(2k)!}\,dt,$$ (127)

for any function $g : [0, 1] \longrightarrow \mathbb{C}$ such that $g^{(2k)}$ is continuous. Then take successively $g(x) = f(x+1), g(x) = f(x+2), \ldots, g(x) = f(x+n-1)$.

c) Apply Euler–Maclaurin's formula to obtain the formulas

i) $\sum_{j=1}^{n} j = \frac{n(n+1)}{2},$ ii) $\sum_{j=1}^{n} j^2 = \frac{n(n+1)(2n+1)}{6},$

iii) $\sum_{j=1}^{n} j^3 = \frac{n^2(n+1)^2}{4},$ (128)

and, for any integer $\ell > 1$,

$$\sum_{k=1}^{n} k^\ell = \frac{n^{\ell+1}}{\ell+1} + \frac{n^\ell}{2} + \frac{1}{2}\binom{\ell}{1}B_2 n^{\ell-1} + \frac{1}{4}\binom{\ell}{3}B_4 n^{\ell-3} + \cdots, \quad (129)$$

the last term having the form cn if ℓ is even and cn^2 if ℓ is odd.

NOTE: An analysis of the error R_k in formula (125) shows (see [69]) that if

$$\int_1^\infty |f^{(2k)}(t)|\, dt < \infty,$$

then there exists a constant C, independent of n, such that as $n \to \infty$,

$$\sum_{j=1}^{n} f(j) = \int_1^n f(t)\, dt + C + \frac{1}{2}f(n)$$
$$+ \sum_{\nu=1}^{k} B_{2\nu} f^{(2\nu-1)}(n)/(2\nu)! + O\left(\int_n^\infty |f^{(2k)}(t)|\, dt\right). \quad (130)$$

It is also possible to show (see [40] and [135]) that if, for any $k \geq k_0 \geq 1$, $f^{(2k)}(t)$ and $f^{(2k+2)}(t)$ have the same sign and if the sign remains constant for $t > 1$, and

$$\lim_{t \to \infty} f^{(2k-1)}(t) = 0, \quad (131)$$

then there exists a constant C, independent of n and k, and a θ, with $0 < \theta < 1$, such that

$$\sum_{j=1}^{n} f(j) = C + \int_1^n f(t)\, dt + \frac{1}{2}f(n)$$
$$+ \sum_{\nu=1}^{k-1} B_{2\nu} \frac{f^{(2\nu-1)}(n)}{(2\nu)!} + \theta B_{2k} \frac{f^{(2k-1)}(n)}{(2k)!}. \quad (132)$$

16. Establish the asymptotic expansions

a) $\displaystyle\sum_{j=1}^{n} \frac{1}{j} \sim \log n + \gamma + \frac{1}{2n} - \frac{1}{12n^2} + \frac{1}{120n^4} - \frac{1}{252n^6} + \cdots$

$$-\frac{B_{2k}}{2kn^{2k}} + \cdots,$$

b) $\displaystyle\sum_{j=1}^{n} \frac{1}{j^2} \sim \frac{\pi^2}{6} - \frac{1}{n} + \frac{1}{2n^2} - \frac{1}{6n^3} + \frac{1}{30n^5} - \frac{1}{42n^7} + \cdots \quad (133)$

$$-\frac{B_{2k}}{n^{2k+1}} + \cdots,$$

3.4. Elements of Asymptotic Analysis

c) $\sum_{j=1}^{n} \frac{1}{j^3} \sim \zeta(3) - \frac{1}{2n^2} + \frac{1}{2n^3} - \frac{1}{4n^4} + \frac{1}{12n^6} + \cdots$

$- \frac{2k+1}{2} \frac{B_{2k}}{n^{2k+2}} + \cdots,$

where γ is the Euler constant (see (111)), and, for any real $s > 1$,

d) $\sum_{j=1}^{n} \frac{1}{j^s} \sim \zeta(s) - \frac{1}{(s-1)n^{s-1}} + \frac{1}{2n^s} - \frac{s}{12n^{s+1}} + \cdots$

$- \frac{s(s+1)\cdots(s+2k-2)}{(2k)!} \frac{B_{2k}}{n^{s+2k-2}} + \cdots,$ (134)

where $\zeta(s)$ is the zeta function

$$\zeta(s) = \sum_{k=1}^{\infty} \frac{1}{k^s}. \tag{135}$$

17. Suppose that one is trying to find an asymptotic expansion of the form

$$[x^n]\left(1 - \frac{x}{\xi}\right)^{-\alpha} \sim \frac{n^{\alpha-1}}{\Gamma(\alpha)\xi^n}\left(c_0(\alpha) + \frac{c_1(\alpha)}{n} + \frac{c_2(\alpha)}{n^2} + \cdots\right), \tag{136}$$

where the $c_k(\alpha)$ are polynomials in α, with $c_0(\alpha) = 1$.

a) Show that

$$\sum_{k \geq 0} \frac{c_k(\alpha+1)}{n^k} = \left(1 + \frac{\alpha}{n}\right) \sum_{k \geq 0} \frac{c_k(\alpha)}{n^k} \tag{137}$$

and deduce the recursive scheme

$$\begin{aligned} c_0(\alpha) &= 1, \\ c_k(\alpha+1) &= c_k(\alpha) + \alpha c_{k-1}(\alpha), \qquad k > 0. \end{aligned} \tag{138}$$

b) Show that

$$c_1(\alpha) = \frac{\alpha(\alpha-1)}{2},$$

$$c_2(\alpha) = \frac{\alpha(\alpha-1)(\alpha-2)(3\alpha-1)}{24}, \tag{139}$$

$$c_3(\alpha) = \frac{\alpha^2(\alpha-1)^2(\alpha-2)(\alpha-3)}{48}.$$

18. STIRLING'S FORMULA.
a) By applying Euler–Maclaurin's summation formula (see Exercise 15) to

the function $f(x) = \log x$, show that there exists a constant C such that

$$\log n! \sim C + n\log n - n + \frac{1}{2}\log n + \frac{1}{12n} - \frac{1}{360n^3} + \frac{1}{1260n^5} + \cdots$$

$$+ \frac{B_{2k}}{2k(2k-1)n^{2k-1}} + \cdots . \qquad (140)$$

Show, moreover, that this asymptotic expansion is in fact divergent by using (122).

b) Deduce the existence of Stirling's asymptotic expansion (36):

$$n! \sim C \left(\frac{n}{e}\right)^n \sqrt{n} \left(1 + \frac{1}{12n} + \frac{1}{288n^2} + \cdots\right) \qquad (141)$$

where $C = \sqrt{2\pi}$.

HINT: To determine the constant C in Stirling's formula, set $\alpha = 1/2$ in

$$\Gamma(\alpha) = \lim_{N \to \infty} \frac{N^\alpha N!}{\alpha(\alpha+1)\cdots(\alpha+n)}$$

and use the fact that $\Gamma(1/2) = \sqrt{\pi}$.

19. (Euler.) Show that

$$\int_0^\infty \frac{e^{-t}}{1+t/n}\,dt \sim 0! - \frac{1!}{n} + \frac{2!}{n^2} - \frac{3!}{n^3} + \cdots \qquad (142)$$

and that this asymptotic expansion is divergent.

HINT: Use the identity

$$\frac{1}{1+t/n} = \sum_{k=0}^{N}\left(-\frac{t}{n}\right)^k + \frac{(-t/n)^{N+1}}{1+t/n}. \qquad (143)$$

20. Complete the computations leading to the asymptotic equivalence (65) for the number of involutions of a set with n elements.

HINT: Use the formula $(1+x)^n = \exp(n\log(1+x)) = \exp(nx - nx^2/2 + \cdots)$ to estimate

$$r_n^n = \left(n^{1/2} - \frac{1}{2} + \frac{1}{8}n^{-1/2} - \frac{1}{128}n^{-3/2} + \cdots\right)^n$$

$$= n^{n/2}\left(1 - \frac{1}{2}n^{-1/2} + \frac{1}{8}n^{-1} - \frac{1}{128}n^{-2} + \cdots\right)^n. \qquad (144)$$

21. Consider an integer $k \geq 1$ and a polynomial species $P = P(X)$ of degree k (i.e., for a finite set U, $P[U] \neq \emptyset$ if $|U| = k$, and $P[U] = \emptyset$ if $|U| > k$). Let $F = E \circ P$, the species of assemblies of P-structures and let $F(x) = \sum_{n \geq 0} f_n x^n/n!$. Show that there exist constants $\alpha, \beta_0, \beta_1, \ldots, \beta_{k-1}$ such that

$$f_n \sim \alpha \beta_0^n \beta_1^{n^{1-1/k}} \beta_2^{n^{1-2/k}} \cdots \beta_i^{n^{1-i/k}} \cdots \beta_{k-1}^{n^{1/k}} n^{(1-1/k)n}. \qquad (145)$$

HINT: Use Theorem 19, of Hayman, and show at first that

$$r_n = An^{1/k}\left(1 + \frac{\omega_1}{n^{1/k}} + \frac{\omega_2}{n^{2/k}} + \cdots\right)$$

for certain constants $A, \omega_1, \omega_2, \ldots$.

22. Let r_n be the positive solution of the equation $x \exp(x) = n$, $n \geq 1$.
   a) Show that (see [69])

$$r_n = \log n - \log(\log n) + O\left(\frac{\log(\log(n))}{\log n}\right). \quad (146)$$

HINT: Use the fact that, for $x > 0$, $x \exp(x) = n$ if and only if $x = \log n - \log x$.

   b) Deduce that the logarithm of the number of partitions of a set with n elements satisfies

$$\log |\mathrm{Par}[n]| = n \log n - n \log(\log n) - n + O\left(\frac{n \log(\log(n))}{\log n}\right). \quad (147)$$

HINT: Take the logarithm of both sides of (66), and use the fact that $n/r_n = \exp(r_n)$ as well as (146).

23. a) Show that when the generating series $R(x) = \sum_{n\geq 0} r_n x^n/n!$ of a species R satisfies the conditions of Theorem 23 (Meir and Moon), where ξ and τ satisfy $\tau = \xi R(\tau)$, $1 = \xi R'(\tau)$, then in a neighborhood of the point (ξ, τ) the relation $y = xR(y)$ is equivalent to the relation

$$1 - \frac{x}{\xi} = (y - \tau)^2 \, \frac{\frac{R''(\tau)}{2!} + \frac{R'''(\tau)}{3!}(y-\tau) + \frac{R^{(iv)}(\tau)}{4!}(y-\tau)^2 + \cdots}{R(\tau) + \frac{R'(\tau)}{1!}(y-\tau) + \frac{R''(\tau)}{2!}(y-\tau)^2 + \cdots}. \quad (148)$$

Deduce that, in the same neighborhood,

$$y = \tau + c_1\left(1 - \frac{x}{\xi}\right)^{1/2} + c_2\left(1 - \frac{x}{\xi}\right) + c_3\left(1 - \frac{x}{\xi}\right)^{3/2} + \cdots, \quad (149)$$

where

$$c_1 = \sqrt{\frac{2R(\tau)}{R''(\tau)}}, \qquad c_2 = \frac{3R'(\tau)R''(\tau) - R(\tau)R'''(\tau)}{3(R''(\tau))^2}. \quad (150)$$

HINT: $y = xR(y)$ is equivalent to $\tau + (y-\tau) = (\xi + (x-\xi))\, R(\tau + (y-\tau))$.

   b) Conclude that the number a_n of R-enriched rooted trees on n vertices admits an asymptotic expansion of the form

$$a_n \sim \alpha \beta^n \, n^{n-1}\left(1 + \frac{\omega_1}{n} + \frac{\omega_2}{n^2} + \cdots\right), \quad (151)$$

where α and β are given by (73), and that the ω_k can be calculated recursively.

24. a) Prove the asymptotic equivalences (77) in two ways:
    i) by invoking Theorem 23,
    ii) by explicitly solving $y = xR(y)$, with $R = L$ and $R = (1+X)^2$.
  b) Let $k \geq 1$ be an integer. Show that if $R = L^k$, then

$$a_n \sim \left(\frac{k}{(k+1)^3}\right)^{1/2} \left(\frac{(k+1)^{k+1}}{ek^k}\right)^n n^{n-1}. \qquad (152)$$

  c) Let $k \geq 2$ be an integer. Show that if $R = (1+X)^k$, then

$$a_n \sim \left(\frac{k}{(k-1)^3}\right)^{1/2} \left(\frac{k^k}{e(k-1)^{k-1}}\right)^n n^{n-1}. \qquad (153)$$

25. Meir and Moon (see [241]) have shown, under the hypotheses of Theorem 23, the following result: Let c and d be two constants such that $0 \leq c < 1$, $|d| < \infty$. If $k = cn + dn^{1/2} + o(n^{1/2})$, then the number $a_n^{\{k\}}$ of forests of k R-enriched rooted trees satisfies

$$a_n^{\{k\}} \sim \frac{1}{\sqrt{A}} e^{-d^2/(2A)} k\eta^k \left(\frac{R(\eta)}{e\eta}\right)^n n^{n-1}, \qquad (154)$$

where

$$\eta R'(\eta) = (1-c) R(\eta) \quad \text{and} \quad A = \eta^2 R''(\eta)/R(\eta) + c(1-c). \qquad (155)$$

Illustrate this result in the following cases

  a) $R = E$,   b) $R = L$,
  c) $R = (1+X)^2$, d) $R = 1 + C$.

26. Show that condition d) of Theorem 27 (Bender [9]) is a consequence of the following hypotheses (often much easier to verify):
  d') $a_n > 0$, for n sufficiently large,
  d'') there exists an analytic function $\varphi(x, y)$ in the open set defined by $|x| < \xi + \delta$, $|y| < \tau + \delta$, such that
    i) there exists $c > 0$ such that $\varphi(x, y) = c$ at each point where $F(x, y) = F_y(x, y) = 0$,
    ii) the coefficients of the series expansion of φ around the origin are ≥ 0,
    iii) if φ only depends on x, then the greatest common divisor of the powers of x appearing in its series expansion is 1.
  Deduce Theorem 23 of Meir and Moon.

27. a) Show that if the generating series $G(x)$ of a species G satisfies the conditions of Example 28, then in a neighborhood of the point (ξ, τ), the relation

3.4. Elements of Asymptotic Analysis

$y = x + G(y)$ is equivalent to the relation

$$\xi\left(1 - \frac{x}{\xi}\right) = (y - \tau)^2 \left(\frac{G''(\tau)}{2!} + \frac{G'''(\tau)}{3!}(y - \tau)\right.$$
$$\left. + \frac{G^{(iv)}(\tau)}{4!}(y - \tau)^2 + \cdots\right). \tag{156}$$

Deduce that, in the same neighborhood,

$$y = \tau + d_1\left(1 - \frac{x}{\xi}\right)^{1/2} + d_2\left(1 - \frac{x}{\xi}\right) + d_3\left(1 - \frac{x}{\xi}\right)^{3/2} + \cdots, \tag{157}$$

where

$$d_1 = \sqrt{\frac{2\xi}{G''(\tau)}}, \qquad d_2 = -\frac{G'''(\tau)\xi}{3(G''(\tau))^2}. \tag{158}$$

HINT: $y = x + G(y)$ is equivalent to $\tau + (y - \tau) = (\xi + (x - \xi)) + G(\tau + (y - \tau))$.

b) Conclude that the number a_n of G-rooted trees on n vertices admits an asymptotic expansion of the form

$$a_n \sim \alpha \beta^n n^{n-1}\left(1 + \frac{\gamma_1}{n} + \frac{\gamma_2}{n^2} + \cdots\right), \tag{159}$$

where α and β are given by (84), and that the γ_k can be computed recursively.

28. Give an asymptotic equivalence for the number of G-rooted trees on a set with n elements when

    a) $G = X^3$,   b) $G = X^2 + X^3$,

    c) $G = E_{\geq 2}$,   d) $G = C_{\geq 2}$.

29. Let $\mathcal{A} = \mathcal{A}_R$ be the species of R-enriched rooted trees. The ordinary generating series $\widetilde{\mathcal{A}}(x) = \sum_{n \geq 1} \tilde{a}_n x^n$ enumerating unlabeled $\mathcal{A}$-structures is characterized by the equation

$$\widetilde{\mathcal{A}}(x) = x Z_R(\widetilde{\mathcal{A}}(x), \widetilde{\mathcal{A}}(x^2), \widetilde{\mathcal{A}}(x^3), \ldots), \tag{160}$$

since $\mathcal{A} = X R(\mathcal{A})$. In other words, $y = \widetilde{\mathcal{A}}(x)$ is the solution of the functional equation

$$f(x, y) = y - x Z_R(y, \widetilde{\mathcal{A}}(x^2), \widetilde{\mathcal{A}}(x^3), \ldots) = 0, \tag{161}$$

where $y(0) = 0$. It is often possible to apply Theorem 27 (of Bender) to the equation $f(x, y) = 0$ to obtain an asymptotic equivalence for the coefficients $\tilde{a}_n$.

a) Show that (161) can be written in the form

$$y - xR(y)\omega(x) = 0, \qquad (162)$$

when $R = E$ or $R = S$, or also in the form

$$y - x(R(y) + \omega(x)) = 0, \qquad (163)$$

when $R = 1 + X + E_2$ or $R = 1 + C$, where the functions $\omega(x)$ have radius of convergence greater than ξ. Note that ξ is the radius of convergence of the solution $y = \widetilde{A}(x)$ in Theorem 27. Deduce that for each of these choices for the enriching species R, we have

$$\widetilde{a}_n \sim \alpha \beta^n n^{-3/2}, \qquad n \to \infty, \qquad (164)$$

for well-chosen constants α and β.

HINT: Show that the hypotheses of Theorem 27 are satisfied, assuming the fact that the radius of convergence ξ of $\widetilde{A}(x)$, for this species R, satisfies $0 < \xi < 1$, from which it follows that the radius of convergence of $\widetilde{A}(x^k)$ is $> \xi$, for any $k \geq 2$.

NOTE: In the case $R = E$, Otter [259] has shown that

$$\alpha \approx 0.4399237, \qquad \beta \approx 2.9557649. \qquad (165)$$

b) Analyze the asymptotic enumeration of unlabeled G-rooted trees, by transforming the functional equation

$$y - x - Z_G(y, \widetilde{A}(x^2), \widetilde{A}(x^3), \ldots) = 0 \qquad (166)$$

into an equation of the form

$$y - x - G(y) - \omega(x) = 0 \qquad (167)$$

in the cases $G = E_2$ and $G = C_{\geq 2}$.

4

Complements on Unlabeled Enumeration

4.0. Introduction

In this chapter we study in a deeper manner the enumeration of unlabeled (i.e., isomorphism types of) structures. Traditionally, this question is treated within Pólya theory (see Pólya and Read [264]), in terms of orbits of colored sets. In the context of species, this corresponds to the computation of cycle index series. This task is difficult in general. However, the functional equations which can be established between various species of structures often give an effective reduction tool.

For example, Otter's formula (see [259])

$$t(x) = T(x) + \frac{1}{2}(T(x^2) - T^2(x)), \qquad (1)$$

expressing the generating series $t(x) = \widetilde{\mathfrak{a}}(x)$ of unlabeled trees as a function of that of unlabeled rooted trees, $T(x) = \widetilde{\mathcal{A}}(x)$, can be lifted to a species isomorphism

$$\mathcal{A} + E_2(\mathcal{A}) = \mathfrak{a} + \mathcal{A}^2. \qquad (2)$$

We call this identity the *dissymmetry theorem for trees* (see Section 1). In addition to the automatic passage to the associated enumerative series, this formula suggests extensions to other families of trees and graphs. Hence, in Section 1 we examine in this light R-enriched trees and rooted trees, and, in Section 2, graphs whose *blocks* (i.e., their 2-connected components) are chosen from a fixed set (see Hanlon and Robinson [141]).

One of the goals of this chapter is to give detailed proofs of two fundamental substitution formulas for weighted species, which are used constantly:

$$\widetilde{F_v \circ G_w}(x) = Z_{F_v}(\widetilde{G_w}(x), \widetilde{G_{w^2}}(x^2), \ldots), \qquad (3)$$

$$Z_{F_v \circ G_w} = Z_{F_v}(Z_{G_w}(x_1, x_2, x_3, \ldots), Z_{G_{w^2}}(x_2, x_4, x_6, \ldots), \ldots). \qquad (4)$$

This last formula generalizes both Robinson's composition theorem for graphs (see Harary and Palmer [144], Chapter 8) and Pólya's theorem for the wreath product of two group actions (see Pólya [263] or [264], Section 27). The proofs found in Section 3 are based on the notion of wreath of G-structures, due to Joyal [158], and on the algebraic independence of the *power sum* symmetric functions

$$p_n = t_1^n + t_2^n + t_3^n + \cdots, \tag{5}$$

following Pólya's approach.

In fact, cycle index series of species corresponds to *characters* of permutation representations of the symmetric group. More precisely, a species of structures F is equivalent to a family $(\alpha_n)_{n\in\mathbb{N}}$ of *permutation* representations (group actions $\alpha_n : \mathcal{S}_n \times F_n \longrightarrow F_n$; see Appendix 1) of the symmetric group $\mathcal{S}_n, n = 0, 1, 2, \ldots$. From this point of view, the coefficient $\mathrm{coef}_\sigma Z_F = \mathrm{fix}\, F[\sigma], \sigma \in \mathcal{S}_n$, in the cycle index series Z_F is equal to the character of the linearization of the representation α_n on the conjugacy class of σ. Moreover the symmetric function assigned to this representation under the Frobenius correspondance is given precisely by $Z_F(p_1, p_2, p_3, \ldots)$. The substitution formula (4) gives a direct link (well known in representation theory) between the partitional (functional) composition of species and the operation of plethysm on symmetric functions.

Section 4 is devoted to the study of asymmetric structures, that is, structures whose stabilizer is reduced to the identity permutation. Our main tool for the enumeration of asymmetric F-structures is the *asymmetry index* series, Γ_F, which plays a role analogous to that of cycle index series Z_F for the unlabeled enumeration of F-structures.

4.1. The Dissymmetry Theorem for Trees

Recall that the symbols $\mathfrak{a}$ and $\mathcal{A}$ denote respectively the species of trees and rooted trees. We know that

$$\mathcal{A} = \mathfrak{a}^\bullet = X \cdot \mathfrak{a}' \tag{1}$$

and that the species $\mathcal{A}$ of rooted trees is characterized by the functional equation

$$\mathcal{A} = X \cdot E(\mathcal{A}). \tag{2}$$

This equation facilitates the resolution of many enumeration problems relative to rooted trees since it implies, for example, the relations

a) $\quad A(x) = x \exp(A(x)),$

b) $\quad \widetilde{A}(x) = x \exp \sum_{k\geq 1} \dfrac{\widetilde{A}(x^k)}{k},$ \hfill (3)

c) $\quad Z_\mathcal{A}(x_1, x_2, x_3, \ldots) = x_1 \exp \sum_{k\geq 1} \dfrac{1}{k} Z_\mathcal{A}(x_k, x_{2k}, x_{3k}, \ldots),$

from which explicit or recursive formulas can be deduced for the coefficients of these series (see Example 3.1.3, Exercises 1 and 3 and (3.2.61)).

The molecular decomposition of $\mathcal{A}$ can also be obtained from (2) by a simple iterative procedure or by Newton–Raphson iteration (see Exercise 3.2.3 and Example 3.3.3).

In the case of trees, Cayley's formula is readily deduced from (1) and (2): $\alpha_n = |\mathfrak{a}[n]| = \frac{1}{n}|\mathcal{A}[n]| = n^{n-2}$, but the enumeration of isomorphism types (unlabeled trees) is more delicate. One possible approach (see Robinson [282]) consists of computing the series $Z_{\mathfrak{a}}$ by integration, from the relation

$$Z_{\mathcal{A}}(x_1, x_2, x_3, \ldots) = x_1 \frac{\partial}{\partial x_1} Z_{\mathfrak{a}}(x_1, x_2, x_3, \ldots). \quad (4)$$

On the other hand, Otter [259] has given in 1948 an explicit formula for $t(x) = \widetilde{\mathfrak{a}}(x)$ as a function of $T(x) = \widetilde{\mathcal{A}}(x)$:

$$t(x) = T(x) + \frac{1}{2}(T(x^2) - T^2(x)). \quad (5)$$

The first terms of these series are

a) $T(x) = x + x^2 + 2x^3 + 4x^4 + 9x^5 + 20x^6 + 48x^7 + 115\,x^8$
$+ 286x^9 + 719x^{10} + 1842x^{11} + 4766x^{12} + 12486x^{13}$
$+ 32973x^{14} + \cdots,$

b) $t(x) = x + x^2 + x^3 + 2x^4 + 3x^5 + 6x^6 + 11x^7 + 23x^8 + 47x^9$
$+ 106x^{10} + 235x^{11} + 551x^{12} + 1301x^{13}$
$+ 3159x^{14} + \cdots.$

Otter's formula admits the following analogues in the context of generating and cycle index series:

a) $\mathfrak{a}(x) = \mathcal{A}(x) - \frac{1}{2}\mathcal{A}^2(x),$

b) $Z_{\mathfrak{a}} = Z_{\mathcal{A}} + \frac{1}{2}(x_2 \circ Z_{\mathcal{A}} - Z_{\mathcal{A}}^2),$ $\quad (6)$

where, by virtue of the plethystic substitution, $x_2 \circ Z_{\mathcal{A}}(x_1, x_2, x_3, \ldots) = Z_{\mathcal{A}}(x_2, x_4, x_6, \ldots)$.

In fact, formulas (5) and (6), a) and b), all follow the same natural species isomorphism (see (7) below). We call this relation the *dissymmetry theorem for trees*, in analogy with the *Dissimilarity Characteristic Theorem*, used to prove Otter's formula (see [144]), but also because this theorem expresses the dissymmetry which occurs whenever a tree is pointed outside of its center. The idea of this theorem goes back to Robinson [282] and Norman [257], and relies upon the concept of *center* of a tree, i.e., the subgraph generated by the points of minimum eccentricity. Here, the *eccentricity* of a vertex is defined as being the maximum distance from this vertex to any other vertex of the tree. It is easy to see that the center always consists of a single point or of a pair of points connected by an edge. One way to obtain the center is to recursively prune the leaves from the tree. A different concept of center, *the center of mass*, where the *eccentricity* of a vertex is defined as the maximum cardinality (i.e., the number of vertices) of any branch attached to this vertex, also leads to a valid proof (see Nickel [254]). The main point is that the concept of center should be *canonical*, that is, preserved under isomorphism.

Theorem 1. DISSYMMETRY THEOREM FOR TREES. *The species of structures $\mathfrak{a}$, of trees, and $\mathcal{A}$, of rooted trees, are related by the natural isomorphism*

$$\mathcal{A} + E_2(\mathcal{A}) = \mathfrak{a} + \mathcal{A}^2, \tag{7}$$

where E_2 denotes the species of sets of cardinality 2.

Proof. The left-hand side of (7) represents those trees which are pointed either at a vertex (the term $\mathcal{A} = \mathfrak{a}^\bullet$) or at an edge, the structures obtained in this case being identified with unordered pairs of rooted trees whose roots are the extremities of the distinguished edge, whence the term $E_2(\mathcal{A})$. On the right-hand side, the term $\mathfrak{a}$ may be identified with trees which have been pointed in a canonical manner, at their center, be it a vertex or an edge. It then remains to find a natural isomorphism between the species of trees pointed at a vertex or an edge, distinct from the center, and the ordered (left, right) pairs of rooted trees, represented by $\mathcal{A}^2$.

1st *case:* The tree is pointed at a vertex s, different from the center. Let p be the vertex adjacent to s, in the direction of the center (if the center is an edge to which s belongs, p is the other vertex of this edge). By cutting the edge joining s and p, one obtains an ordered pair of rooted trees, the left one with root s, and the right one with root p, and thus an $\mathcal{A}^2$-structure. See Figure 1.

2nd *case:* The tree is pointed at an edge, different from the center. In this case, let p be the vertex of the distinguished edge nearest to the center. Then cut the edge and place on the left the rooted tree with root p, and on the right the other part. See Figure 2.

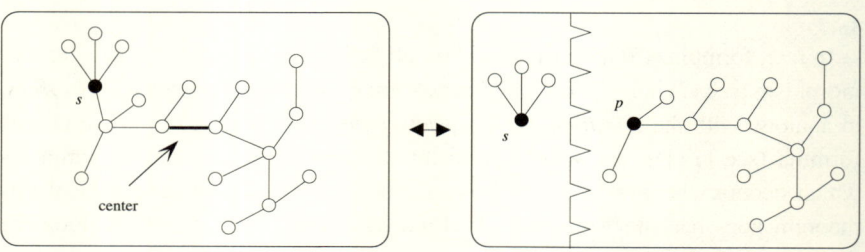

Fig. 1.

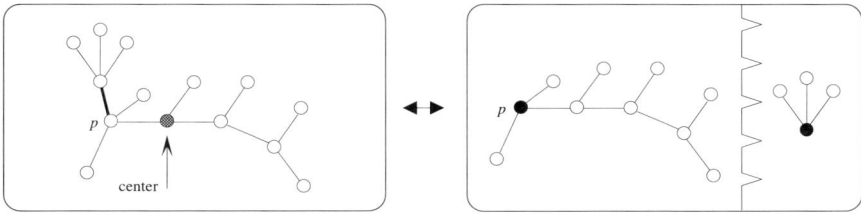

Fig. 2.

For the inverse mapping, given an $\mathcal{A}^2$-structure, join with an edge the roots of the two rooted trees and find the center of the tree thus obtained. If the center comes from the right rooted tree (or if the center is precisely the added edge), then we are in the first case and the root of the left tree is distinguished. If the center comes from the left rooted tree, we are in the second case and the added edge is distinguished. Finally, it is clear that this bijection is a natural isomorphism of species, that is, it commutes with any relabeling of the structures. ∎

In the context of virtual species (see Section 2.5), the dissymmetry theorem can be rewritten in the form

$$\mathfrak{a} = \mathcal{A} + E_2(\mathcal{A}) - \mathcal{A}^2$$
$$= (X + E_2 - X^2) \circ \mathcal{A}. \tag{8}$$

It is readily seen that identities (5) and (6), a) are consequences of (8). By passing to cycle index series, using $Z_{E_2}(x_1, x_2, x_3, \ldots) = (1/2)(x_1^2 + x_2)$, we obtain the formula

$$Z_\mathfrak{a} = \left(x_1 + \frac{1}{2}(x_2 - x_1^2)\right) \circ Z_\mathcal{A}, \tag{9}$$

clearly equivalent to (6), b). From (4) and (9) we immediately deduce (see Robinson [282])

$$Z_\mathfrak{a}(x_1, x_2, x_3, \ldots) = \int_0^{x_1} \frac{1}{\xi_1} Z_\mathcal{A}(\xi_1, x_2, x_3, \ldots) d\xi_1 + Z_\mathfrak{a}(x_1, x_2, x_3, \ldots)\bigg|_{x_1=0}$$

$$= \int_0^{x_1} \frac{1}{\xi_1} Z_\mathcal{A}(\xi_1, x_2, x_3, \ldots) d\xi_1 + \frac{1}{2} Z_\mathcal{A}(x_2, x_4, x_6, \ldots). \tag{10}$$

This identity permits the expression of the coefficients $\mathfrak{a}_\sigma = |\text{Fix } \mathfrak{a}[\sigma]|$ of the index series $Z_\mathfrak{a}$ in terms of those of $Z_\mathcal{A}$ (see Exercise 3).

The dissymmetry theorem can easily be extended to R-enriched trees and rooted trees. Recall that in Section 3.1 we have described the species $\mathcal{A}_R$ of R-enriched rooted trees, characterized by the equation

$$\mathcal{A}_R = X \cdot R(\mathcal{A}_R), \tag{11}$$

and also the species $\mathfrak{a}_R$ of *R-enriched trees*, i.e., of trees together with an *R*-structure on the set of *neighbors* of any vertex (see Figure 3.1.13). Some examples are given by plane trees (see Example 3.1.14 and Example 4 below) and by the homeomorphically irreducible trees (see Example 3). We remind the reader that Equation (11) permits the explicit computation of the coefficients of the series $F(\mathcal{A}_R)(x)$ and $Z_{F(\mathcal{A}_R)}(x_1, x_2, x_3, \ldots)$ for any species structures F (see Section 3.1 and Proposition 3.2.14).

One verifies directly (see Proposition 3.1.15) that

$$\mathfrak{a}_R^{\bullet} = XR(\mathcal{A}_{R'}). \tag{12}$$

This means that the study of *R*-enriched trees is related to R'-enriched rooted trees rather than to *R*-enriched rooted trees. The same is true for the species $\mathfrak{a}_R^{\bullet-\bullet}$, of *R-enriched trees with a distinguished edge*, and for $\mathfrak{a}_R^{\bullet\to\bullet}$, where the distinguished edge is oriented, since (see Figure 3)

$$\mathfrak{a}_R^{\bullet-\bullet} = E_2(\mathcal{A}_{R'}) \quad \text{and} \quad \mathfrak{a}_R^{\bullet\to\bullet} = \mathcal{A}_{R'}^2. \tag{13}$$

Taking these observations into account, the dissymmetry theorem for *R*-enriched trees has the following form and its proof remains essentially the same.

Theorem 2. DISSYMMETRY THEOREM FOR *R*-ENRICHED TREES [187]. *Let R be a species such that* $R'(0) \neq 0$. *One then has*

$$\mathfrak{a}_R^{\bullet} + \mathfrak{a}_R^{\bullet-\bullet} = \mathfrak{a}_R + \mathfrak{a}_R^{\bullet\to\bullet} \tag{14}$$

or equivalently

$$\mathfrak{a}_R = X \cdot R(\mathcal{A}_{R'}) + E_2(\mathcal{A}_{R'}) - \mathcal{A}_{R'}^2. \tag{15}$$

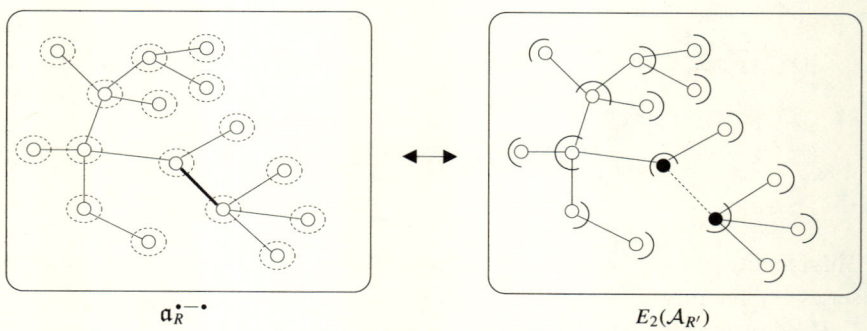

Fig. 3.

Together with equation $\mathcal{A}_{R'} = X \cdot R'(\mathcal{A}_{R'})$, this theorem permits the computation of all the series associated to $\mathfrak{a}_R$. For example, for the unlabeled structures, one finds the equations

a) $\widetilde{\mathcal{A}_{R'}}(x) = x Z_{R'}\big(\widetilde{\mathcal{A}_{R'}}(x), \widetilde{\mathcal{A}_{R'}}(x^2), \ldots\big),$

b) $\widetilde{\mathfrak{a}_R}(x) = x Z_R\big(\widetilde{\mathcal{A}_{R'}}(x), \widetilde{\mathcal{A}_{R'}}(x^2), \ldots\big) + \dfrac{1}{2}\big(\widetilde{\mathcal{A}_{R'}}(x^2) - \widetilde{\mathcal{A}_{R'}}^2(x)\big).$ (16)

See also Exercise 5 and Appendix 2, which contains tables giving the first few terms of various series associated to the species $\mathcal{A}_R$ and $\mathfrak{a}_R$ as a function of the generating series associated to R.

Example 3. HOMEOMORPHICALLY IRREDUCIBLE TREES (Harary and Prins [146], [144], p. 62). Denote by $\mathfrak{h}$ the species of *homeomorphically irreducible trees*, i.e., trees without vertices of degree 2. These are R-enriched trees, with $R = E_{\neq 2} = E - E_2$. Let $R' = E_{\neq 1} = E - X$, and consider the species $\mathcal{H} = \mathcal{A}_{R'}$ of R'-enriched rooted trees, i.e., without fibers of cardinality 1, also called *homeomorphically irreducible rooted trees*. We underline that $\mathcal{H} \neq \mathfrak{h}^\bullet$. Equations (11), (12), and (15) give

$$\mathcal{H} = XE(\mathcal{H}) - X\mathcal{H} \quad \text{and} \quad \mathfrak{h} = XE_{\neq 2}(\mathcal{H}) + E_2(\mathcal{H}) - \mathcal{H}^2. \quad (17)$$

In any rooted tree, the chains of consecutive vertices having fibers of cardinality 1 can be grouped together (see Figure 4), whence the identity

$$\mathcal{A} = \mathcal{H}\left(\dfrac{X}{1-X}\right). \quad (18)$$

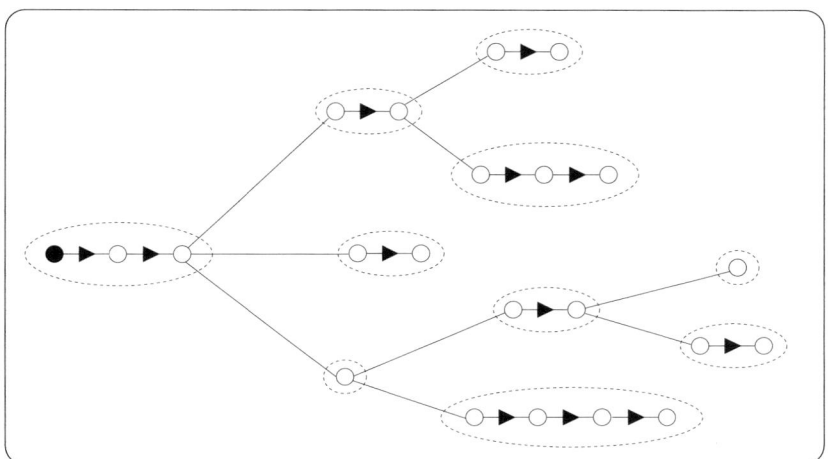

Fig. 4.

Inverting $X/(1-X)$ (for substitution), the equation

$$\mathcal{H} = A\left(\frac{X}{1+X}\right) \tag{19}$$

follows. Thus, using (8) and the fact that $XE_{\neq 2}(\mathcal{H}) = \mathcal{H} + X\mathcal{H} - XE_2(\mathcal{H})$, the following relation is obtained:

$$\mathfrak{h} = \mathfrak{a}\left(\frac{X}{1+X}\right) + XA\left(\frac{X}{1+X}\right) - XE_2\left(A\left(\frac{X}{1+X}\right)\right). \tag{20}$$

After computation, one finds for unlabeled homeomorphically irreducible trees (see Exercise 6)

$$\widetilde{\mathfrak{h}}(x) = x + x^2 + x^4 + x^5 + 2x^6 + 2x^7 + 4x^8 + 5x^9 + 10x^{10}$$
$$+ 14x^{11} + 26x^{12} + 42x^{13} + 78x^{14} + 132x^{15} + \cdots.$$

Example 4. PLANE TREES (Harary, Prins, and Tutte [147], [144], p. 67). Denote by $\mathfrak{p}$ the species of *plane trees*, i.e., trees topologically embedded in the plane or, more precisely, homeomorphism classes of such embeddings. Fixing an orientation of the plane, the set of neighbors of every vertex is seen to have a structure of circular permutation and this characterizes the embedding of the tree in the plane. In other words, $\mathfrak{p} = \mathfrak{a}_R$ with $R = 1 + C$, where C denotes the species of cyclic permutations. Note that $R' = L$, the species of linear orders, and recall the index series of these species:

$$Z_C(x_1, x_2, x_3, \ldots) = \sum_{n \geq 1} \frac{\phi(n)}{n} \log \frac{1}{1 - x_n}, \quad Z_L(x_1, x_2, x_3, \ldots) = \frac{1}{1 - x_1}. \tag{21}$$

Denote by $\mathcal{P} = \mathfrak{p}^{\bullet}$, the species of rooted (i.e., pointed) plane trees, and by $\mathcal{A}_L$, the species of ordered rooted trees (see Example 3.1.4), defined by the functional equation

$$\mathcal{A}_L = XL(\mathcal{A}_L). \tag{22}$$

The species $\mathcal{A}_L$ is asymmetric (i.e., the automorphism group of any ordered rooted tree is reduced to the identity) and it follows from (22) that the series $y = \mathcal{A}_L(x) = \widetilde{\mathcal{A}}_L(x)$ satisfies the quadratic equation $y^2 - y + x = 0$. Solving this equation gives

$$\mathcal{A}_L(x) = \widetilde{\mathcal{A}}_L(x) = \frac{1 - \sqrt{1 - 4x}}{2} = \sum \frac{1}{n}\binom{2n-2}{n-1} x^n. \tag{23}$$

Now Equation (12) gives $\mathcal{P} = X + X C(\mathcal{A}_L)$ and the dissymmetry theorem implies the identity

$$\mathfrak{p} = \mathcal{P} + E_2(\mathcal{A}_L) - \mathcal{A}_L^2, \tag{24}$$

which yields

a) $p(x) = x + x \log \dfrac{1}{1 - \mathcal{A}_L(x)} - \dfrac{1}{2} \mathcal{A}_L^2(x),$

b) $\widetilde{p}(x) = x + x \displaystyle\sum_{n \geq 1} \dfrac{\phi(n)}{n} \log \dfrac{1}{1 - \widetilde{\mathcal{A}}_L(x^n)} + \dfrac{1}{2}(\widetilde{\mathcal{A}}_L(x^2) - \widetilde{\mathcal{A}}_L^2(x)).$ (25)

After computation, one finds

$$\widetilde{p}(x) = x + x^2 + x^3 + 2x^4 + 3x^5 + 6x^6 + 14x^7 + 34x^8 + 95x^9 + 280x^{10}$$
$$+ 854x^{11} + 2694x^{12} + 8714x^{13} + 28640x^{14} + 95640x^{15} + \cdots. \quad (26)$$

An explicit formula can be given for the coefficients $[x^n]\widetilde{p}(x)$, enumerating unlabeled plane trees of order n, as well as a more refined formula, taking into account the degree distribution (see Exercise 4).

Certain classes of trees and rooted trees do not enter as such in the setting of R-enrichments, but lend themselves to an analogous treatment. This is the case for oriented trees which are studied below, as well as for mixed, signed, or other trees, proposed in the exercises.

Example 5. ORIENTED TREES (Harary and Prins [146] and [144]). An *oriented* tree is a tree in which each edge pq is given one of two orientations, $\overrightarrow{pq}$ or $\overrightarrow{qp}$. Denote by $\mathfrak{o}$, the species of *oriented trees* and by $\mathcal{O} = \mathfrak{o}^{\bullet}$, the species of rooted oriented trees. Evidently the functional equation

$$\mathcal{O} = XE^2(\mathcal{O}) \quad (27)$$

holds (see Figure 5), from which are deduced

a) $\mathcal{O}(x) = x \exp(2\mathcal{O}(x)),$ \qquad b) $\widetilde{\mathcal{O}}(x) = x \exp\left(2 \displaystyle\sum_{k \geq 1} \dfrac{\widetilde{\mathcal{O}}(x^k)}{k}\right).$ (28)

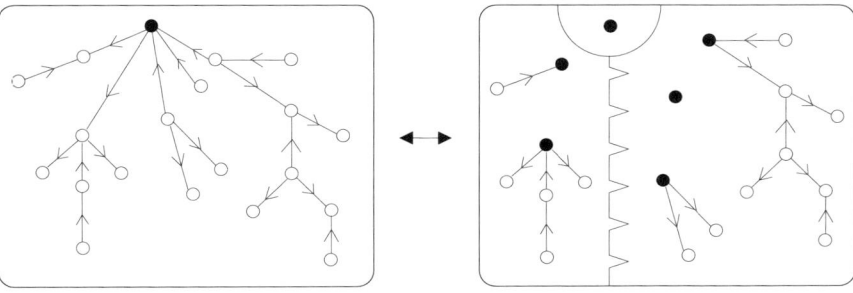

Fig. 5.

The labeled enumeration of these structures is quite simple. Indeed since each of the $n-1$ edges can be oriented in two different fashions, we have

$$|\mathcal{O}[n]| = (2n)^{n-1} \quad \text{and} \quad |\mathfrak{o}[n]| = 2(2n)^{n-2}. \tag{29}$$

On the other hand, from (28), b) one finds, by recursive computation,

$$\widetilde{\mathcal{O}}(x) = x + 2x^2 + 7x^3 + 26x^4 + 107x^5 + 458x^6 + 2058x^7 + 9498x^8$$
$$+ 44947x^9 + 216598x^{10} + 1059952x^{11} + 5251806x^{12}$$
$$+ 26297238x^{13} + 132856766x^{14} + 676398395x^{15} + \cdots. \tag{30}$$

For oriented trees, the dissymmetry theorem takes a particularly simple form:

Proposition 6. *The species $\mathfrak{o}$ of oriented trees and $\mathcal{O}$ of rooted oriented trees are related by*

$$\mathcal{O} = \mathfrak{o} + \mathcal{O}^2. \tag{31}$$

Proof. Observe first that the species $\mathcal{O}^2$ can be identified to that of oriented trees where one (oriented) edge has been distinguished. Redefine the center of an oriented tree as being the vertex s which is either the center of the underlying tree if it is a vertex, or, if it is an edge, the origin of this oriented edge. To prove (31), consider a rooted oriented tree, with root s. If this distinguished point is the center of the oriented tree, an $\mathfrak{o}$-structure is obtained directly. Otherwise distinguish the oriented edge emanating from s and leading to the center of the oriented tree. This gives, as mentioned above, an $\mathcal{O}^2$-structure (see Figure 6). It is easy to see that this correspondence is invertible and natural. ∎

From (31) follows immediately the identity

$$\widetilde{\mathfrak{o}}(x) = \widetilde{\mathcal{O}}(x) - \widetilde{\mathcal{O}^2}(x), \tag{32}$$

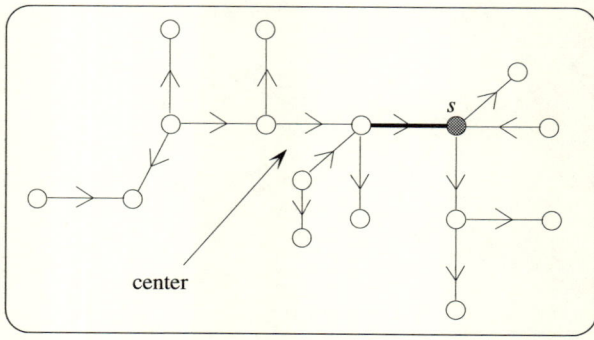

Fig. 6.

and, after computation,

$$\tilde{o}(x) = x + x^2 + 3x^3 + 8x^4 + 27x^5 + 91x^6 + 350x^7 + 1376x^8$$
$$+ 5743x^9 + 24635x^{10} + 108968x^{11} + 492180x^{12}$$
$$+ 2266502x^{13} + 10598452x^{14} + 50235931x^{15} + \cdots. \quad (33)$$

It is often useful to consider the leaves of trees and of rooted trees as elements of a different sort. For example, this is the case in chemistry where the molecular structures of alkanes (paraffins), of alcohols, and their isomers are determined by trees and rooted trees whose internal vertices correspond to carbon atoms of valence (degree) 4, and the leaves to hydrogen atoms of valence 1. See Figure 7 which represents an alkane isomer of molecular formula C_7H_{16}. Other examples come from tree-like data structures in computer science, where the data are often stored in the leaves.

Consider the species of two sorts $B_R(X, Y)$ of R-enriched rooted trees with internal vertices of sort X and external vertices (leaves or isolated vertices) of sort Y; suppose here that $R(0) = 1$ and set $R_+ = R - 1$. This species is defined by the functional equation

$$B_R(X, Y) = Y + X R_+(B_R(X, Y)). \quad (34)$$

Likewise, denote by $b_R(X, Y)$ the species of R-enriched trees with internal vertices of sort X and external vertices of sort Y. Here the hypothesis on R is that $R(0) = 0$, which implies that the tree reduced to one vertex is excluded, and that $|R[1]| = 1$ (or $R'(0) = 1$); in other words, the enrichment is only meaningful for internal vertices. Denoting by $b_R^{\bullet X}(X, Y)$ the species of such R-enriched trees pointed at an internal

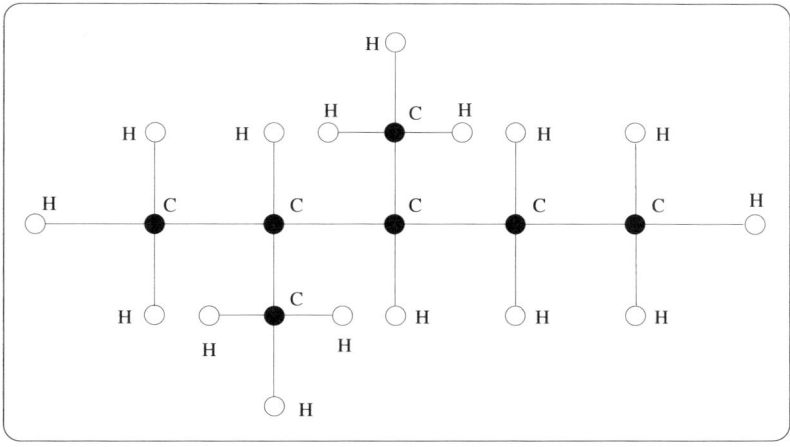

Fig. 7.

point, we have

$$\mathfrak{b}_R^{\cdot X}(X, Y) = X R_\diamond(B_{R'}(X, Y)), \tag{35}$$

where $R_\diamond = R - X$. The dissymmetry theorem for R-enriched trees is then generalized as follows.

Theorem 7. DISSYMMETRY THEOREM FOR R-ENRICHED TREES WITH LEAVES. *Let R be a species such that $R(0) = 0$ and $R'(0) = 1$ and let $R_\diamond = R - X$. Then the species $\mathfrak{b}_R(X, Y)$ of R-enriched trees with leaves is expressed as a function of R'-enriched rooted trees with leaves $B_{R'}(X, Y)$ in the following manner:*

$$\mathfrak{b}_R(X, Y) = X R_\diamond(B_{R'}(X, Y)) + E_2(B_{R'}(X, Y)) - (B_{R'}(X, Y) - Y) B_{R'}(X, Y). \tag{36}$$

Proof. The equation may be written in the form

$$X R_\diamond(B_{R'}(X, Y)) + E_2(B_{R'}(X, Y)) = \mathfrak{b}_R(X, Y) + (B_{R'}(X, Y) - Y) B_{R'}(X, Y)$$

and is shown essentially in the same manner as the dissymmetry theorems for trees. One should consider R-enriched trees with leaves which are pointed either at an internal vertex or at an edge, and then localize this pointing with respect to the center. Details are left to the reader. ∎

Example 8. ALKANES (PARAFFINS) AND ALCOHOL MOLECULES (Pólya [264]). Denote by E_4, C_4, and $E_4^\pm$, the species of sets, cyclic permutations, and oriented sets (see Example 2.6.9, b)), respectively, all of which are restricted to cardinality 4. Recall that the corresponding cycle index series of these species and their derivatives are

$$\begin{aligned}
\text{a)} \quad & Z_{E_4} = \frac{1}{24}(x_1^4 + 6x_1^2 x_2 + 8x_1 x_3 + 3x_2^2 + 6x_4), \\
\text{b)} \quad & Z_{C_4} = \frac{1}{4}(x_1^4 + x_2^2 + 2x_4), \\
\text{c)} \quad & Z_{E_4^\pm} = \frac{1}{12}(x_1^4 + 8x_1 x_3 + 3x_2^2), \\
\text{d)} \quad & Z_{E_3} = \frac{1}{6}(x_1^3 + 3x_1 x_2 + 2x_3), \\
\text{e)} \quad & Z_{X^3} = x_1^3, \\
\text{f)} \quad & Z_{E_3^\pm} = \frac{1}{3}(x_1^3 + 2x_3).
\end{aligned} \tag{37}$$

For $R = X + E_4$, $X + C_4$, and $X + E_4^\pm$, respectively, the R-enriched trees with leaves correspond to isomers of alkane molecules, to plane isomers, and to stereoisomers, that is, to isomers embedded in three-dimensional space, respectively, with molecular formula $C_n H_{2n+2}$, $n \geq 0$ (see Figure 7). Denote these species by $\varphi(X, Y)$, $\gamma(X, Y)$, and $\eta(X, Y)$, respectively, and by $F(X, Y)$, $G(X, Y)$, and

$H(X, Y)$, the corresponding species of R'-enriched rooted trees, where $R' = 1 + E_3$, $1 + X^3$, and $1 + E_3^{\pm}$, respectively. These rooted trees correspond to isomers (respectively, plane isomers and stereoisomers) of alcohol molecules of formula $C_n H_{2n+1} OH$, $n \geq 0$. These last species then satisfy the equations

a) $F(X, Y) = Y + X E_3(F(X, Y))$,

b) $G(X, Y) = Y + X G^3(X, Y)$, (38)

c) $H(X, Y) = Y + X E_3^{\pm}(H(X, Y))$,

and the following equations for the corresponding unlabeled rooted trees enumeration, due to Pólya, are immediately deduced:

a) $\widetilde{F}(x, y) = y + \dfrac{x}{6}(\widetilde{F}^3(x, y) + 3\widetilde{F}(x, y)\widetilde{F}(x^2, y^2) + 2\widetilde{F}(x^3, y^3))$,

b) $\widetilde{G}(x, y) = y + x\widetilde{G}^3(x, y)$, (39)

c) $\widetilde{H}(x, y) = y + \dfrac{x}{3}(\widetilde{H}^3(x, y) + 2\widetilde{H}(x^3, y^3))$.

For reasons of homogeneity, the variable y can be removed from these three functions $\Phi(x, y)$: Letting $\Phi(x) = \Phi(x, 1)$, one recovers $\Phi(x, y) = y\Phi(xy^2)$. This reflects the fact that any such rooted tree with n internal vertices always has $2n + 1$ leaves, whence the molecular formula $C_n H_{2n+1} OH$, the supplementary link OH being attached automatically to the root. Setting $f(x) = \widetilde{F}(x, 1)$, $g(x) = \widetilde{G}(x, 1)$, and $h(x) = \widetilde{H}(x, 1)$, the following equations are equivalent to (39):

a) $f(x) = 1 + \dfrac{x}{6}(f^3(x) + 3f(x)f(x^2) + 2f(x^3))$,

b) $g(x) = 1 + xg^3(x)$, (40)

c) $h(x) = 1 + \dfrac{x}{3}(h^3(x) + 2h(x^3))$.

The expansion at the origin of these series can be computed, recursively for (40), a) and c), and explicitly for (40), b) (see Exercise 10).

It is now possible to enumerate the corresponding families of paraffin isomers, $\varphi(X, Y)$, $\gamma(X, Y)$, and $\eta(X, Y)$, with the aid of the dissymmetry theorem for R-enriched trees with leaves, with $R_\diamond = E_4$, C_4, and $E_4^{\pm}$, respectively. Setting $\varphi(x) = \widetilde{\varphi}(x, 1)$, $\gamma(x) = \widetilde{\gamma}(x, 1)$, and $\eta(x) = \widetilde{\eta}(x, 1)$, for unlabeled trees, Equation (36) gives the following Otter type formulas:

a) $\varphi(x) = \dfrac{x}{24}[f^4(x) + 6f^2(x)f(x^2) + 8f(x)f(x^3) + 3f^2(x^2) + 6f(x^4)]$

$+ \dfrac{1}{2}[f(x^2) - f^2(x)] + f(x)$,

(41)

b) $\gamma(x) = \dfrac{x}{4}[g^4(x) + g^2(x^2) + 2g(x^4)] + \dfrac{1}{2}[g(x^2) - g^2(x)] + g(x)$,

c) $\eta(x) = \dfrac{x}{12}[h^4(x) + 3h^2(x^2) + 8h(x)h(x^3)] + \dfrac{1}{2}[h(x^2) - h^2(x)] + h(x)$.

After computations, one obtains

$$\varphi(x) = 1 + x + x^2 + x^3 + 2x^4 + 3x^5 + 5x^6 + 9x^7 + 18x^8 + 35x^9 + 75x^{10}$$
$$+ 159x^{11} + 355x^{12} + 802x^{13} + 1858x^{14} + 4347x^{15} + \cdots,$$

$$\gamma(x) = 1 + x + x^2 + 2x^3 + 7x^4 + 25x^5 + 108x^6 + 492x^7 + 2431x^8$$
$$+ 12371x^9 + 65169x^{10} + 350792x^{11} + 1926372x^{12} \qquad (42)$$
$$+ 10744924x^{13} + 60762760x^{14} + 347653944x^{15} + \cdots,$$

$$\eta(x) = 1 + x + x^2 + x^3 + 2x^4 + 3x^5 + 5x^6 + 11x^7 + 24x^8 + 55x^9$$
$$+ 136x^{10} + 345x^{11} + 900x^{12} + 2412x^{13} + 6563x^{14} + \cdots.$$

Exercises for Section 4.1

1. Denote respectively by $T_n = [x^n]\widetilde{\mathcal{A}}(x)$ and $t_n = [x^n]\widetilde{a}_n(x)$ the number of unlabeled rooted trees and trees with n vertices.
   a) Show that the functional equation (3), b) is equivalent to Cayley's identity

$$\sum_{n \geq 1} T_n x^n = x \prod_{n \geq 1} (1 - x^n)^{-T_n}. \qquad (43)$$

   b) Show that the numbers T_n satisfy the recurrence

$$T_{n+1} = \frac{1}{n} \sum_{k=1}^{n} \left(\sum_{d \mid k} d T_d \right) T_{n-k+1}. \qquad (44)$$

   c) Show that

$$t_n = T_n + \frac{1}{2} \chi_{\text{even}}(n) T_{\frac{n}{2}} - \frac{1}{2} \sum_{k=1}^{n-1} T_k T_{n-k}, \qquad (45)$$

   where χ_{even} is the characteristic function of even integers.

2. Establish the identity

$$2(n-1)n^{n-2} = \sum_{k=1}^{n-1} \binom{n}{k} k^{k-1} (n-k)^{n-k-1}. \qquad (46)$$

3. Let $\sigma = (\sigma_1, \sigma_2, \ldots)$. Show that the number of trees left fixed by a permutation σ having σ_k cycles of length k (i.e., of type $1^{\sigma_1} 2^{\sigma_2} \cdots$) is given by

$$\text{fix } \mathfrak{a}[\sigma] = \begin{cases} \frac{1}{\sigma_1} \text{fix } \mathcal{A}[\sigma], & \text{if } \sigma_1 > 0, \\ \frac{1}{2} 2^{\sigma_2 + \sigma_4 + \cdots} \text{fix } \mathcal{A}[\sigma_2, \sigma_4, \ldots], & \text{if } \sigma_1 = \sigma_3 = \cdots = 0, \\ 0, & \text{otherwise.} \end{cases} \qquad (47)$$

NOTE: The number fix $\mathcal{A}[\sigma]$ is given by (3.2.61).

4.1. The Dissymmetry Theorem for Trees

4. a) Show that the number $\tilde{p}_n$ of unlabeled plane trees on $n \geq 2$ vertices is given by

$$\tilde{p}_n = \frac{1}{2(n-1)} \sum_{d|(n-1)} \phi\left(\frac{n-1}{d}\right)\binom{2d}{d} - \frac{1}{2}c_{n-1} + \frac{1}{2}\chi_{\text{even}}(n)c_{\frac{n}{2}-1}, \quad (48)$$

where ϕ denotes Euler's function and $c_n = \frac{1}{n+1}\binom{2n}{n}$ denotes the nth Catalan number.

b) Show that the dissymmetry theorem for R-enriched trees is still valid in the weighted case, i.e., when $R = R_w(X)$ is a weighted species (see Exercise 3.1.23).

c) Prove the dissymmetry theorem for plane trees, weighted by degrees (see Exercise 3.1.24 for the notation):

$$\mathfrak{p}_r = \mathcal{P}_r + E_2(\mathcal{A}_{L_{r'}}) - \mathcal{A}^2_{L_{r'}}. \quad (49)$$

d) Show that the number $\tilde{p}(\mathbf{i})$ of unlabeled plane trees with degree distribution

$$\mathbf{i} = (i_0, i_1, i_2, \ldots),$$

under condition (3.1.109), satisfies

$$\tilde{p}(\mathbf{i}) = \tilde{\mathcal{P}}(\mathbf{i}) - \frac{1}{n}\binom{n}{\mathbf{i}} + \frac{1}{n}\chi_{\text{even}}(\mathbf{i})\binom{n/2}{\mathbf{i}/2}, \quad (50)$$

where $\tilde{\mathcal{P}}(\mathbf{i})$ is given by (3.1.113) and $\chi_{\text{even}}(\mathbf{i}) = \prod_{k\geq 0}\chi_{\text{even}}(i_k)$.

5. Let R be a species of structures such that $R'(0) \neq 0$. Show that

a) $Z_{\mathfrak{a}_R}(x_1, x_2, \ldots) = \int_0^{x_1} Z_{R(\mathcal{A}_{R'})}(\xi_1, x_2, \ldots) d\xi_1 + \frac{1}{2}Z_{\mathcal{A}_{R'}}(x_2, x_4, \ldots),$

b) $\text{fix } \mathfrak{a}_R[\sigma_1, \sigma_2, \ldots] = \begin{cases} \text{fix }_{R(\mathcal{A}_{R'})}[\sigma_1-1, \sigma_2, \ldots], & \text{if } \sigma_1 > 0, \\ \frac{1}{2}2^{\sigma_2+\sigma_4+\cdots}\text{ fix }_{\mathcal{A}_{R'}}[\sigma_2, \sigma_4, \ldots], & \text{if } \sigma_1 = \sigma_3 = \cdots = 0, \\ 0, & \text{otherwise.} \end{cases}$

(51)

6. Let $\mathfrak{h}$ be the species of homeomorphically irreducible trees. Show that

$$\tilde{\mathfrak{h}}(x) = (1-x)u(x) + xv(x)(2 - v(x)), \quad (52)$$

where

$$u(x) = Z_\mathfrak{a}\left(\frac{x}{1+x}, \frac{x^2}{1+x^2}, \frac{x^3}{1+x^3}, \ldots\right),$$

$$v(x) = Z_\mathcal{A}\left(\frac{x}{1+x}, \frac{x^2}{1+x^2}, \frac{x^3}{1+x^3}, \ldots\right),$$

(53)

$\mathfrak{a}$ and $\mathcal{A}$ denoting respectively the species of trees and that of rooted trees.

7. Denote by $\widetilde{O}_n = [x^n]\widetilde{O}(x)$, and by $\widetilde{o}_n = [x^n]\widetilde{o}(x)$, the number of unlabeled oriented rooted trees and trees, respectively, with n vertices. Show that
a) functional equation (28), b) is equivalent to

$$\sum_{n\geq 1} \widetilde{O}_n x^n = x \prod_{n\geq 1}(1-x^n)^{-2\widetilde{O}_n}, \qquad (54)$$

b) the numbers $\widetilde{O}_n$ satisfy the recurrence

$$\widetilde{O}_n = \frac{2}{n}\sum_{k=1}^{n}\left(\sum_{d|k} d\widetilde{O}_d\right)\widetilde{O}_{n-k+1}, \qquad (55)$$

c) one has

$$\widetilde{o}_n = \widetilde{O}_n - \sum_{\substack{i+j=n \\ i,j\geq 1}} \widetilde{O}_i \widetilde{O}_j, \qquad (56)$$

d) and

$$Z_O(x_1, x_2, x_3, \ldots) = \frac{1}{2}Z_A(2x_1, 2x_2, 2x_3, \ldots). \qquad (57)$$

8. DIRECTED TREES (Harary and Palmer [144], Leroux and Miloudi [215]). Let us call *directed* (or *mixed*) trees those trees where each edge is assigned one direction or both directions. Denote this species by $\mathfrak{d}$ and by $\mathcal{D} = \mathfrak{d}^{\bullet}$, the species of rooted directed trees.
a) Prove the relations

$$\begin{aligned}\text{i)} \quad & \mathcal{D} = XE^3(\mathcal{D}), \\ \text{ii)} \quad & \mathcal{D} + E_2(\mathcal{D}) = \mathfrak{d} + 2\mathcal{D}^2.\end{aligned} \qquad (58)$$

b) Deduce the corresponding relations for cycle index and type generating series.
c) Invoking the fact that $E^3 = E(3X)$, show that one has

$$3\mathcal{D} = \mathcal{A}(3X). \qquad (59)$$

Deduce the value of the 15 first terms of the series $\widetilde{\mathcal{D}}(x)$ and $\widetilde{\mathfrak{d}}(x)$, as well as explicit formulas for the coefficients of the series $Z_\mathcal{D}$ and $Z_\mathfrak{d}$.

9. SIGNED AND MARKED TREES ([144], [215]). Denote by $\mathfrak{s}$, the species of *signed* trees, that is, those trees where each edge is assigned a sign $+$ or $-$, and by $S = \mathfrak{s}^{\bullet}$, the species of rooted signed trees. Likewise, denote by $\mathfrak{m}$ and $\mathcal{M} = \mathfrak{m}^{\bullet}$, respectively, the species of trees and of rooted trees which are *marked*, that is, in which each edge and each vertex are assigned a sign $+$ or $-$. Establish the following relations and deduce the corresponding equalities for the generating

4.1. The Dissymmetry Theorem for Trees

and index series:

a) $S = XE^2(S)$ and $S = \mathcal{O}$,

b) $S + 2E_2(S) = \mathfrak{s} + 2S^2$, (60)

c) $M = 2XE^2(M)$,

d) $M + 2E_2(M) = \mathfrak{m} + 2M^2$.

e) Establish the combinatorial equalities

$$2S = A(2X), \qquad 2M = A(4X), \tag{61}$$

where $\mathcal{A}$ is the species of ordinary rooted trees. Deduce explicit formulas for the coefficients of the generating series Z_S and Z_M.

10. a) Calculate, starting with functional equations (40), a) and c), the first 15 terms of the series $f(x)$ and $h(x)$.

b) Show that relation (40), b) implies (see Exercise 3.2.4)

$$g(x) = 1 + \sum_{n \geq 1} \frac{1}{n} \binom{3n}{n-1} x^n. \tag{62}$$

c) For $n \geq 0$, set $f_n = [x^n] f(x)$ and $h_n = [x^n] h(x)$. Show that these coefficients satisfy the recurrences

$$f_0 = 1, \quad f_{n+1} = \frac{1}{6}\left(\sum_{i+j+k=n} f_i f_j f_k + 3 \sum_{i+2j=n} f_i f_j + 2 \sum_{3i=n} f_i \right), \tag{63}$$

$$h_0 = 1, \quad h_{n+1} = \frac{1}{3}\left(\sum_{i+j+k=n} h_i h_j h_k + 2 \sum_{3i=n} h_i \right). \tag{64}$$

11. Let $\mathfrak{a}$ be the species of trees and $\mathcal{C}$ that of cyclic permutations.

a) Show that there exist two series $w_1(x)$ and $w_2(x)$ such that

$$Z_{\mathfrak{a} \times \mathcal{C}} = w_1(x_1) + w_2(x_2). \tag{65}$$

HINT: Use (2.1.39) and the fact that the center of a tree consists of either 1 or 2 vertices and is invariant under every automorphism.

b) Show moreover that the series $w_1(x)$ and $w_2(x)$ are given explicitly by the formulas

i) $w_1(x_1) = (\mathfrak{a} \times \mathcal{C})(x_1) = \sum_{k \geq 1} k^{k-3} x_1^k$,

ii) $w_2(x_2) = \sum_{k \geq 1} 2^{k-1} k^{k-2} x_2^k$. (66)

12. a) Let t_n be the number of unlabeled trees with n vertices and f_n the number of unlabeled forests of trees of order n, that is, containing n vertices in total.

Show that
$$f_n = \sum_{j_1+2j_2+\cdots=n} \prod_{k\geq 1} \binom{t_k + j_k - 1}{j_k}. \qquad (67)$$

b) Let G be an arbitrary species such that $G(0) = 0$. Generalize the preceding result to the enumeration of unlabeled assemblies of G-structures.

13. a) Prove formula (3.2.54), i.e., the dissymmetry theorem for R, S-enriched bicolored trees.

    HINT: Fix an order between the two colors thus allowing an edge orientation to be defined for any bicolored tree, and adapt the proof of Proposition 6.

    b) Show that the basic concepts of R, S-enriched bicolored trees and relations (3.2.52)–(3.2.54) are also valid in a weighted context, i.e., when the species R and S are weighted.

14. c-CHROMATIC TREES AND ROOTED TREES (Riordan [281]). Let c be an integer ≥ 2 and consider the set $C = \{0, 1, \ldots, c-1\}$ of c colors. Let $\mathcal{a}_c(X)$ be the species of c-chromatic trees, i.e., where each vertex is given a color from C so that adjacent vertices are colored differently (proper coloring), and let $\mathcal{A}_c$ be the species of rooted (pointed) c-chromatic trees where the root is of color 0.

    a) Show, using addition modulo c, that

    i) $\mathcal{a}_c^\bullet = c\,\mathcal{A}_c$,

    ii) $\mathcal{A}_c = X E^{c-1}(\mathcal{A}_c)$, \hfill (68)

    iii) $\mathcal{a}_c^\bullet = \mathcal{a}_c + \binom{c}{2}\mathcal{A}_c^2$. \quad (dissymmetry formula)

    HINT: Each proper coloring of a tree defines a natural orientation of its edges.

    b) Let $T_c(x) = \widetilde{\mathcal{a}}_c^\bullet(x)$ and $t_c(x) = \widetilde{\mathcal{a}}_c(x)$. Show that

    i) $T_c(x) = cx\,\exp\left(\dfrac{c-1}{c}\sum_{k\geq 1}\dfrac{T_c(x^k)}{k}\right)$,

    \hfill (69)

    ii) $t_c(x) = T_c(x) - \dfrac{c-1}{2c}T_c^2(x)$.

    c) Show that there exist two sequences of polynomials $P_i(c)$ and $p_i(c)$, $i = 1, 2, 3, \ldots$, such that

    $$T_c(x) = \sum_{i\geq 1} P_i(c)x^i, \qquad t_c(x) = \sum_{i\geq 1} p_i(c)x^i \qquad (70)$$

    and compute the first terms of these sequences. For example, $P_1(c) = c$, $P_2(c) = 2\binom{c}{2}$, $P_3(c) = 4\binom{c}{2} + 9\binom{c}{3}$, $P_4(c) = 8\binom{c}{2} + 54\binom{c}{3} + 64\binom{c}{4}$.

15. EXPECTED NUMBER OF LEAVES IN A ROOTED TREE LEFT FIXED BY A GIVEN PERMUTATION (Bergeron, G. Labelle, and Leroux [24]).

a) Let $\mathcal{B}(X, Y)$ denote the species of rooted trees, with internal vertices of sort X and leaves (including the case where the tree is reduced to a vertex) of sort Y, characterized by Equation (34) with $R = E$. Show that $\mathcal{B}(X, Y)$ is related to the species $\mathcal{A}$ of rooted trees by

$$\mathcal{B}(X, X + T) = T + \mathcal{A}(X \cdot E(T)), \tag{71}$$

where T denotes another sort of elements.

HINT: Either use a purely computational reasoning based on the functional equations satisfied by these species or give a constructive combinatorial proof.

b) Deduce that the weighted species $\mathcal{A}_w = \mathcal{B}(X, X_t)$ of rooted trees, where the variable t is a leaf counter (see Exercise 2.4.5), has the virtual expression

$$\mathcal{A}_w(X) = X_t - X + \mathcal{A}(X \cdot E(X_t - X)), \tag{72}$$

and that its cycle generating series is given by

$$Z_{\mathcal{A}_w} = (t-1)x_1 + Z_\mathcal{A}\left(x_1 \exp \sum_{k \geq 1} \frac{(t^k - 1)x_k}{k}, x_2 \exp \sum_{k \geq 1} \frac{(t^{2k} - 1)x_{2k}}{k}, \ldots\right). \tag{73}$$

c) Let U be a set of cardinality $n \geq 1$ and σ a permutation of U of cycle type $(\sigma_1, \sigma_2, \ldots)$. Let

$$a_\sigma = \text{fix } \mathcal{A}[\sigma] = \text{coeff}_\sigma Z_\mathcal{A} \text{ (see (3.2.61))} \quad \text{and} \tag{74}$$
$$\text{Supp}(\sigma) = \{k \in [n] \mid \sigma_k \neq 0\}.$$

Show that the expected number of leaves $\text{Exp}(\mathcal{A}, \sigma)$ of a random rooted tree on U left fixed by σ is given by 1 if $n = 1$ and, if $n > 1$, by

$$\text{Exp}(\mathcal{A}, \sigma) = \frac{1}{a_\sigma} \sum_{k \in \text{Supp}(\sigma)} k\sigma_k(\phi_k - k)a_{\sigma - \delta_k}, \tag{75}$$

where $\phi_k = \phi_k(\sigma) = \sum_{d \mid k} d\sigma_d$ and $\delta_k = (\delta_{k,1}, \delta_{k,2}, \ldots, \delta_{k,n})$.

HINT: This number is the quotient b_σ/a_σ, where $b_\sigma = \text{coeff}_\sigma (\partial/\partial t)(Z_{\mathcal{A}_w})|_{t=1}$.

d) Consider the particular cases where $\sigma = (n, 0, 0, \ldots)$ (see (3.1.69)) and $\sigma = (n-2, 1, 0, \ldots)$. Show moreover that the expected number of leaves of unlabeled random rooted trees of order n is given by λ_n/T_n, where $T_n = [x^n] \widetilde{\mathcal{A}}(x)$ and

$$\lambda_n = [x^n]\left(x + \sum_{k \geq 1} \frac{kx^{2k}}{1 - x^k} \frac{\partial Z_\mathcal{A}}{\partial x_k}(x, x^2, \ldots)\right). \tag{76}$$

e) Denote by $\mathfrak{b}(X, Y)$, the species of trees with internal vertices of sort X and leaves of sort Y, and by $\mathfrak{f}(X, Y)$, the species of endofunctions with internal vertices (i.e., with nonempty fiber) of sort X and leaves (i.e., elements with an empty fiber) of sort Y. Show that these species are related to those of trees and endofunctions by the formulas

$$\mathfrak{b}(X, X+T) = T + E_2(T) + \mathfrak{a}(X \cdot E(T)),$$
$$\mathfrak{f}(X, X+T) = \text{End}(X \cdot E(T)),$$
(77)

and extend the results of parts b), c), and d) to trees and endofunctions.

16. (Bouchard, Chiricota, and G. Labelle [35].) Consider a total order

$$1 < X < E_2 < X^2 < E_3 < C_3 < XE_2 < X^3 < \cdots$$
(78)

on the set $\mathfrak{M}$ of molecular species. For a virtual species

$$G = \sum_{M \in \mathfrak{M}} g_M M,$$
(79)

with arbitrary constant term $g_1 \in \mathbb{Z}$, the virtual species $E_2(G)$ is expressed by the formula

$$E_2(G) = \sum_{M \in \mathfrak{M}} g_M E_2(M) + \sum_{M \in \mathfrak{M}} \binom{g_M}{2} M^2 + \sum_{\substack{P,Q \in \mathfrak{M} \\ P < Q}} g_P g_Q P \cdot Q.$$
(80)

We emphasize that $E_2(1) = 1$ (see (2.4.31)). If $F = E_2(G)$, F is said to be the *symmetric square* of G or that G is the *symmetric square root* of F.

a) Show that the species X of singletons possesses a unique symmetric square root, of which the first terms of the molecular decomposition are given by

$$-1 - X - E_2(X) + X^2 + XE_2(X) - X^3 - E_2(E_2(X))$$
$$+ E_2(X)^2 - 2X^2 E_2(X) + E_2(X^2) + \cdots.$$
(81)

b) Show that for any virtual species F and G,

$$E_2(F + G) = E_2(F) + E_2(G) + F \cdot G.$$
(82)

c) Show that the dissymmetry formula for trees is equivalent to

$$E_2(1 - \mathcal{A}) = 1 - \mathfrak{a}.$$
(83)

In other words, the virtual species $1 - \mathfrak{a}$ is the symmetric square of $1 - \mathcal{A}$.

17. a) Show that the number $\tilde{\mathfrak{a}}_n$ of unlabeled trees having n vertices is asymptotically given by

$$\tilde{\mathfrak{a}}_n \sim ab^n n^{-5/2},$$
(84)

for certain constants a and b.

HINT: Use the dissymmetry theorem for trees to express the series $\tilde{\mathfrak{a}}(x)$ as a function of the series $\tilde{\mathcal{A}}(x)$. Then appeal to the fact, implied by Bender's theorem (Theorem 3.4.27), that in a neighborhood of the dominating

singularity $\xi < 1$, there is an expansion of the form
$$\widetilde{A}(x) = 1 + c_1(1 - x/\xi)^{1/2} + c_2(1 - x/\xi) + \cdots. \tag{85}$$
NOTE: Otter [259] showed that
$$a \approx 0.5349485 \quad \text{and} \quad b \approx 2.9557649. \tag{86}$$
b) Show how Theorem 2 (of dissymmetry for R-enriched trees) can be used, in conjunction with Exercise 3.4.29, to enumerate asymptotically unlabeled R-enriched trees for other species R.

4.2. Connected Graphs and Blocks

In this section we consider particular classes of graphs, for example connected, nonseparable (2-connected) graphs, graphs without end-vertices, etc. Many functional relations between these species are given, permitting their enumeration. They generalize earlier results of Robinson, Hanlon, and others (see, for example, [140], [141], [144], and [282]) on cycle index series of these species.

We first consider the species $\mathcal{G} = \mathcal{G}_w$ of simple graphs, weighted by an edge counter y, that is where, for $g \in \mathcal{G}[U]$, we set $e(g)$ equal to the number of edges of g and $w(g) = y^{e(g)}$; the number of vertices $n = |U|$ is called the *order* of the graph g. Many species of graphs correspond to subclasses of graphs closed under isomorphisms and weighted in the same manner. Here are some examples:
- 0, the empty class,
- 1, the class of empty graphs,
- X, the class of graphs consisting of one point,
- E, the class of discrete graphs,
- $K_2 \simeq E_{2,y}$, the class of complete graphs of order 2,
- $K_n \simeq E_{n,w}$, the class of complete graphs of order n, with weight $w = y^{\binom{n}{2}}$,
- $C = \mathcal{G}_w^c$, the class of connected graphs,
- $P_n = P_{n,y^n}$, the class of polygons of order n, $n \geq 3$,
- $\text{Cha}_n = \text{Cha}_{n,y^{n-1}}$, the class of (nonoriented) chains of order n, with $n \geq 1$.

The enumeration of unlabeled simple graphs uses the cycle index series $Z_{\mathcal{G}_w}$ of the weighted species $\mathcal{G}_w$ of simple graphs, or equivalently the cycle index polynomial $P_{\mathcal{S}_n^{(2)}}$ of the symmetric group $\mathcal{S}_n$ acting on the set $\wp^{[2]}[n]$ of pairs of elements of $[n]$, $n \geq 0$ (see Section 2.1 and Appendix 1). In fact, we have $\mathcal{G}_w = \wp_w \Box \wp^{[2]}$ (see Exercise 2.3.14) and

$$Z_{\mathcal{G}_w}(x_1, x_2, x_3, \ldots) = \sum_{n \geq 0} \frac{1}{n!} \sum_{\sigma \in \mathcal{S}[n]} \left| \text{Fix } \wp_w[\wp^{[2]}[\sigma]] \right| x_1^{\sigma_1} x_2^{\sigma_2} x_3^{\sigma_3} \cdots$$

$$= \sum_{n \geq 0} \frac{1}{n!} \sum_{\sigma \in \mathcal{S}[n]} (1+y)^{c_1} (1+y^2)^{c_2} (1+y^3)^{c_3} \cdots$$

$$x_1^{\sigma_1} x_2^{\sigma_2} x_3^{\sigma_3} \cdots, \tag{1}$$

where c_k denotes the number of cycles $(\wp^{[2]}[\sigma])_k$ of length k of $\wp^{[2]}[\sigma]$ and is given by (see Proposition 2.2.7)

$$\left(\wp^{[2]}[\sigma]\right)_k = \frac{1}{2}\sum_{[i,j]=k}(i,j)\sigma_i\sigma_j + \sigma_{2k} - \sigma_k + \frac{1}{2}(k \bmod 2)\sigma_k. \tag{2}$$

It follows that

$$\begin{aligned}\widetilde{\mathcal{G}}_w(x) &= Z_{\mathcal{G}_w}(x_1, x_2, x_3, \ldots) \\ &= \sum_{n\geq 0}\frac{1}{n!}\sum_{\sigma\in S[n]}(1+y)^{c_1}(1+y^2)^{c_2}(1+y^3)^{c_3}\cdots x^n \\ &= \sum_{n\geq 0} P_{S_n^{(2)}}(1+y, 1+y^2, 1+y^3, \ldots)x^n,\end{aligned} \tag{3}$$

in accordance with Pólya–Redfield's Theorem (see Example A.1.20).

We can compute the molecular decomposition of the species $\mathcal{G}_{n,w}$ of simple graphs of order n, for small values of n (see Exercise 1), through an analysis of isomorphism types of graphs. A more refined problem, interesting for its applications in chemistry, is the enumeration of graphs according to degree distribution. See Exercise 2.6.24.

Connected Graphs

The following fundamental relation is used for the enumeration of connected graphs according to the number of edges:

$$\mathcal{G}_w = E(C), \tag{4}$$

where $C = \mathcal{G}_w^c$ denotes the weighted subspecies of $\mathcal{G}_w$ formed by connected graphs. Set

$$C(x) = \sum_{n\geq 1}c_n(y)\frac{x^n}{n!} \quad \text{and} \quad \widetilde{C}(x) = \sum_{n\geq 1}\widetilde{c}_n(y)x^n.$$

From (4) we immediately deduce

$$\mathcal{G}'_w = \mathcal{G}_w \cdot C', \tag{5}$$

which implies the recurrence

$$c_{n+1}(y) = (1+y)^{\binom{n+1}{2}} - \sum_{k=1}^{n}\binom{n}{k-1}(1+y)^{\binom{n-k+1}{2}}c_k(y). \tag{6}$$

Note also the following recurrence, related to the Mallows and Riordan tree inversion polynomial (see [232] and Exercise 5.2.10),

$$c_{n+1}(y) = \sum_{k=1}^{n}\binom{n-1}{k-1}((1+y)^k - 1)c_k(y)c_{n-k+1}(y), \tag{7}$$

which follows from the relation

$$C'' = \left(C \times \wp_w^+\right)' \cdot C', \tag{8}$$

where $\wp_w^+ = E \cdot (E_+)_{(y)}$ denotes the weighted species of nonempty subsets (see Exercise 2).

Möbius inversion can then be used to determine the cycle index series and type generating series of C in terms of those of $\mathcal{G}_w$ (see (2.3.71)). We find

$$Z_C(x_1, x_2, x_3, \ldots) = \sum_{k \geq 1} \frac{\mu(k)}{k} \log Z_{\mathcal{G}_{wk}}(x_k, x_{2k}, x_{3k}, \ldots) \tag{9}$$

and

$$\widetilde{c}_n(y) = \sum_{d \mid n} \frac{\mu(d)}{d} a_{n/d}(y^d), \tag{10}$$

where we have set $\log \widetilde{\mathcal{G}}_w(x) = \sum_{n \geq 1} a_n(y) x^n$.

In [282] Robinson proposes an iterative procedure simplifying the computation of the cycle index series for C, starting from that of $\mathcal{G}_w$. This method can be applied to the computation of the molecular decomposition of C and generalized to arbitrary species of structures F and F^c for which $F = E(F^c)$. Here is a description of this procedure in the case of the species $\mathcal{G} = \mathcal{G}_w$ and $C = \mathcal{G}_w^c$. For $n \geq 1$, let C_n be the restriction of C to sets of cardinality n, that is, the species of connected graphs of order n, and denote by $\mathcal{G}_{\geq n}$ the species of graphs whose connected components are all of order $\geq n$, that is

$$\mathcal{G}_{\geq n} = E\left(\sum_{j \geq n} C_j\right). \tag{11}$$

We then have $\mathcal{G}_{\geq 1} = \mathcal{G}_w$ and, for $n \geq 1$, we find

$$C_n = (\mathcal{G}_{\geq n})_n, \tag{12}$$

$$\mathcal{G}_{\geq n} = \mathcal{G}_{\geq n+1} \cdot E(C_n) \quad \Longrightarrow \quad \mathcal{G}_{\geq n+1} = \mathcal{G}_{\geq n} \cdot E^{-1}(C_n). \tag{13}$$

See Example 2.5.10 for a description of the virtual species E^{-1}. For example,
- $C_1 = (\mathcal{G}_{\geq 1})_1 = (\mathcal{G}_w)_1 = X$,
- $\mathcal{G}_{\geq 2} = \mathcal{G}_{\geq 1} \cdot E^{-1}(X)$, the species of graphs without isolated points,
- $C_2 = (\mathcal{G}_{\geq 2})_2 = K_{2,w} = E_{2,y}$ (an edge),
- $\mathcal{G}_{\geq 3} = \mathcal{G}_{\geq 2} \cdot E^{-1}(E_{2,y})$, the species of graphs without isolated points or edges,
- $C_3 = (\mathcal{G}_{\geq 3})_3 = K_3 + Cha_3 = E_{3,y^3} + (XE_2)_{y^2}$,
- $C_4 = E_2(X^2)_{y^3} + XE_3(X_y) + P_{4,y^4} + \left(X^2 E_2\right)_{y^4} + \left(E_2^2\right)_{y^5} + (E_4)_{y^6}$,
- etc.

Nonseparable Graphs

A *cut-point* of a connected graph is a vertex of the graph whose removal destroys connectivity. A *nonseparable* (or *2-connected*) graph is a connected graph with no

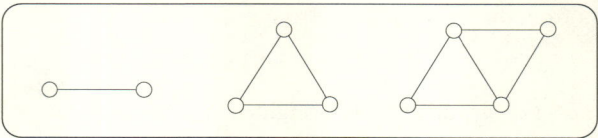

Fig. 1. Three nonseparable graphs.

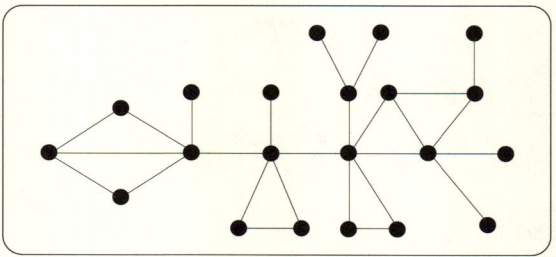

Fig. 2. A connected graph g.

cut-point (this excludes the graph consisting of a single point). Figure 1 illustrates three types of nonseparable graphs.

A *block* of a graph is a maximal nonseparable subgraph. The blocks of a graph g do not form a partition of the vertices; many blocks can be joined together by the same cut-point. The bc-*tree*, bc(g), of a connected graph g is a bicolored (white-black) tree describing precisely the links between the blocks of g (the white vertices of bc(g)) and the cut-points of g (the black vertices of bc(g)). Figure 3 represents the bc-tree bc(g) of the connected graph illustrated in Figure 2. The letters "bc" stand for *block-cut-point tree* (see Harary [142], Chapter 4, and Tutte [314], Chapter 9). Since all the leaves of a bicolored tree bc(g) are of the same color (white), its center is always reduced to a single point. The concept of bc-center can then be defined for any connected graph: It consists of either the block or the cut-point of g corresponding to the center of its bc-tree bc(g). This permits the extension of the dissymmetry theorem to all connected graphs (see Theorem 3 below).

If B denotes a family of unlabeled nonseparable graphs, for example those in Figure 1, we also denote by B the weighted species of graphs isomorphic to an element of B. We denote more simply by C_B the weighted species of connected graphs all of whose blocks are in B. For example, all the blocks of the graph g in Figure 2 are in the class B described by Figure 1, that is, $g \in C_B$. This class B is also used for the other illustrations in this section.

Example 1.
a) If $B = B_a$, the family of all blocks, then $C_B = C$ is the species of all connected graphs.
b) If $B = K_2$, then $C_B = \mathfrak{a}$ is the species of trees.

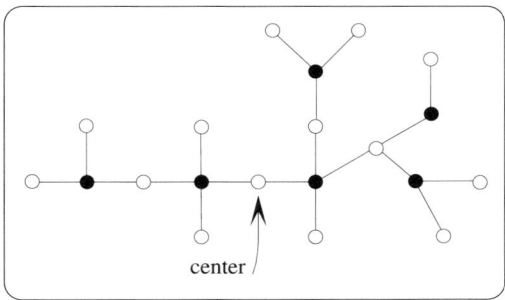

Fig. 3. The bc-tree bc(g).

c) If $B = K_3$, the complete graph on three vertices, then C_B is the species of triangular cacti (see Example 7).
d) If $B = \{P_n \mid n \geq 3\}$, the family of polygons, then C_B is the species of cacti or of Husimi graphs (see [144], p. 71; [143], [149]).
e) If $B = B_a - K_2$, then C_B is the species of 2-edge-connected graphs, that is, connected graphs not having any edge whose removal separates the graphs into two connected components.

The following relation is well known (see Harary and Palmer [144], 1.33 and 8.71; Robinson [282], Theorem 4; J. Labelle [199], 2.10).

Proposition 2. *Let B be a class of nonseparable graphs and C_B be the weighted species of connected graphs all of whose blocks are in B. Then we have*

$$C'_B = E(B'(C^{\bullet}_B)). \tag{14}$$

Proof. Given a graph $g \in C_B[U + \{*\}]$, consider the blocks b to which the vertex $*$ belongs. The other vertices of these blocks are the roots of $C^{\bullet}_B$-structures (see Figure 4). This canonical decomposition yields the desired weight-preserving isomorphism with assemblies of $B'(C^{\bullet}_B)$-structures. ∎

Theorem 3. DISSYMMETRY THEOREM FOR GRAPHS. *Let B be a species of nonseparable graphs, and let C_B be the species of connected graphs all of whose blocks are in B. Then we have*

$$C_B = C^{\bullet}_B + B(C^{\bullet}_B) - C^{\bullet}_B \cdot B'(C^{\bullet}_B). \tag{15}$$

Proof. When written in the form

$$C^{\bullet}_B + B(C^{\bullet}_B) = C_B + C^{\bullet}_B \cdot B'(C^{\bullet}_B), \tag{16}$$

this result generalizes the dissymmetry formula for trees (4.1.7) and the proof is similar. The left-hand side of (16) represents those graphs in C_B which are pointed either at a vertex ($C^{\bullet}_B$) or at a block ($B(C^{\bullet}_B)$). In the right-hand side, the term C_B

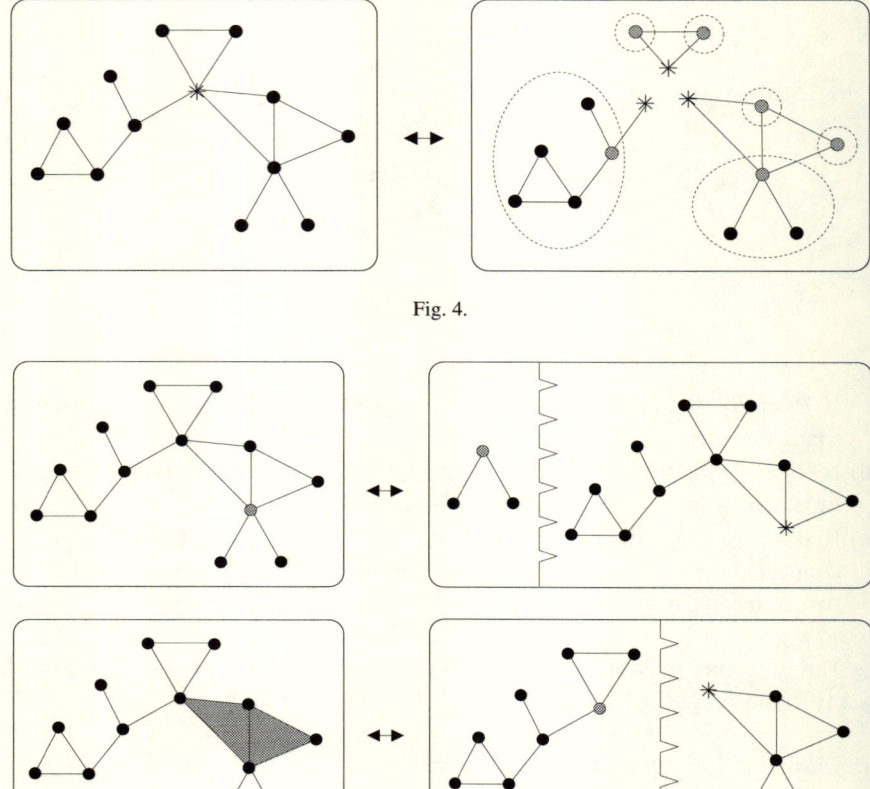

Fig. 4.

Fig. 5.

corresponds to the case where the pointing has been made in a canonical fashion, at the bc-center of the graph. In the other cases, for which the pointing is done out of the center (either on a point or a block), a natural decomposition illustrated in Figure 5 gives a $C_B^\bullet B'(C_B^\bullet)$-structure in a bijective manner. ∎

There is another remarkable identity in this context, associated to the species N_B of connected graphs not in B containing at least two vertices and having no end-block (i.e., of degree 1 in the bc-tree) belonging to B (see Robinson [282], Theorem 8, for the case $B = K_2$, and Hanlon [140], Theorem 4).

Proposition 4. Let B be a species of nonseparable graphs. Then we have

$$C = C_B + N_B(C_B^\bullet). \tag{17}$$

Proof. To a graph g in the class $C - C_B$ we canonically associate a subgraph h of g, belonging to the class N_B, arising from the bc-tree $\mathrm{bc}(g)$. The subgraph h is

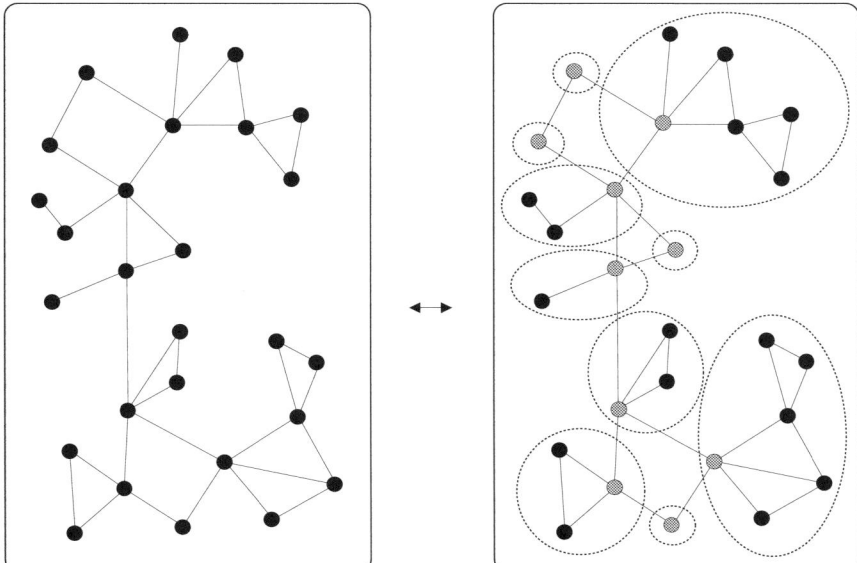

Fig. 6.

characterized by the fact that bc(h) is the maximal bc-subtree of bc(g) containing no leaves of block type $\in B$. The graph g is recovered by replacing the vertices of h by the $C_B^{\bullet}$-structures which are attached to it in g. See Figure 6. ∎

Equations (14), (15), and (17) are used to determine C_B and N_B, when B is known, or to determine B when C_B and N_B are known. It is then necessary to invert some of these relations; this may be done in an elegant manner in the framework of virtual species.

Example 5. For $B = K_2 = E_{2,y}$, we have $C_B = \mathfrak{a}_w$, the weighted species of trees, and $C_B^{\bullet} = \mathcal{A}_w$, that of rooted trees. Note that $w(\alpha) = y^{n-1}$, for any tree or rooted tree α of order n. Thus we have

$$\mathcal{A}_w = XE((\mathcal{A}_w)_y), \tag{18}$$

which yields, by substitution,

$$\mathcal{A}_w^{\langle -1 \rangle} = XE^{-1}(X_y), \tag{19}$$

and

$$\widetilde{\mathcal{A}_w^{\langle -1 \rangle}}(x) = x \exp\left(-\sum_{k \geq 1} \frac{y^k x^k}{k}\right)$$
$$= x(1 - yx). \tag{20}$$

Moreover, N_B is the species of connected graphs not consisting of a single point and without end-points (i.e., vertices of degree 1). Denote by $W_w = E(X + N_B)$, the species of all graphs without end-points. Equations (16) and (17) give, since $B' = X_y$,

$$\text{a)} \quad \mathcal{A} + K_2(\mathcal{A}) = \mathfrak{a} + \mathcal{A} \cdot \mathcal{A}_y,$$
$$\text{b)} \quad C = \mathfrak{a} + N_B(\mathcal{A}), \tag{21}$$

the weighting w being implicit. We deduce

$$N_B(\mathcal{A}) = C - \mathfrak{a} = C - (X + K_2 - X_y^2)(\mathcal{A}),$$
$$N_B = C(\mathcal{A}^{\langle -1 \rangle}) - (X + K_2 - X_y^2), \tag{22}$$

and

$$W_w = E(X + N_B) = E(C(\mathcal{A}^{\langle -1 \rangle}) - K_2 + X_y^2)$$
$$= \mathcal{G}_w(\mathcal{A}^{\langle -1 \rangle}) \cdot E(X_y^2 - K_2). \tag{23}$$

Considering the type generating series for these weighted species, observing that $(X_y^2 - K_2)\widetilde{}(x) = 0$ and taking (20) into account, we obtain the following result, due to Robinson ([282], Corollary 4.1):

$$\widetilde{W}_w(x) = Z_{\mathcal{G}_w}(x - yx^2, x^2 - y^2x^4, x^3 - y^3x^6, \ldots). \tag{24}$$

Example 6. NONSEPARABLE GRAPHS. (See Robinson [282] and Hanlon [140].) Take for B the class B_a of all nonseparable graphs. We then have $C_{B_a} = C$, $N_{B_a} = 0$, as well as the following fundamental relations, deduced from (14) and (15):

$$\text{a)} \quad C^\bullet = XE(B'_a(C^\bullet)),$$
$$\text{b)} \quad C = (X + B_a - XB'_a)(C^\bullet). \tag{25}$$

Using these equations we can determine, by inversion, the weighted species B_a as well as its associated series. For example, we deduce from (25), a) and from (2.3.70),

$$\text{a)} \quad E(B'_a) = \frac{X}{C^{\bullet \langle -1 \rangle}},$$
$$\text{b)} \quad Z_{B'_a} = \sum_{k \geq 1} \frac{\mu(k)}{k} \log \frac{x_k}{Z_{C^{\bullet \langle -1 \rangle}}(x_k, x_{2k}, x_{3k}, \ldots)}, \tag{26}$$

and from (25), b),

$$B_a = C(C^{\bullet \langle -1 \rangle}) + XB'_a - X. \tag{27}$$

Let
$$M(x, y) = \widetilde{C^{\bullet(-1)}}(x). \tag{28}$$

From $C^{\bullet} = XC'$, we get $X = C^{\bullet(-1)} \cdot C'(C^{\bullet(-1)})$, from which we extract the equation
$$x = M(x, y) Z_{C'_w}(M(x, y), M(x^2, y^2), M(x^3, y^3), \ldots), \tag{29}$$

allowing the recursive computation of $M(x, y)$. Taking into account (26), b) and (27), we finaly have
$$\widetilde{B_a}(x) = Z_{C_w}(M(x, y), M(x^2, y^2), M(x^3, y^3), \ldots)$$
$$+ x \sum_{k \geq 1} \frac{\mu(k)}{k} \log \frac{x^k}{M(x^k, y^k)} - x. \tag{30}$$

Observe that for any class B of nonseparable graphs, the species $A = C_B^{\bullet}$ satisfies the functional equation
$$A = XE(B')(A), \tag{31}$$

so that Lagrange inversion methods can be used to determine A and its associated series. The inversion method of Hanlon (see Exercise 3) can also be applied.

Example 7. TRIANGULAR CACTI (Harary and Palmer [144]). Consider the class $B = K_3$ of graphs isomorphic to a triangle. As species, we have $B = E_3$ and $B' = E_2$. Put $\delta = C_B$, the species of *triangular cacti*, that is, connected graphs all of whose blocks are triangles, and $\Delta = \delta^{\bullet}$, the species of *pointed triangular cacti*. An application of (14) and (15) immediately gives
$$\Delta = XE(E_2(\Delta)) \tag{32}$$

and the dissymmetry formula
$$\delta = \Delta + E_3(\Delta) - \Delta E_2(\Delta). \tag{33}$$

These equations are solved as before (see Exercise 4) and we find
$$\widetilde{\Delta}(x) = x \exp\left(\sum_{n \geq 1} \frac{\widetilde{\Delta}^2(x^n) + \widetilde{\Delta}(x^{2n})}{2n} \right)$$
$$= x + x^3 + 2x^5 + 5x^7 + 13x^9 + 37x^{11} + 111x^{13} + 345x^{15} + 1105x^{17}$$
$$+ 3624x^{19} + 12099x^{21} + 41000x^{23} + 140647x^{25} + 487440x^{27}$$
$$+ 1704115x^{29} + \cdots, \tag{34}$$

$$\widetilde{\delta}(x) = \widetilde{\Delta}(x) + \frac{1}{3}(\widetilde{\Delta}(x^3) - \widetilde{\Delta}^3(x))$$
$$= x + x^3 + x^5 + 2x^7 + 4x^9 + 8x^{11} + 19x^{13} + 48x^{15} + 126x^{17} + 355x^{19}$$
$$+ 1037x^{21} + 3124x^{23} + 9676x^{25} + 30604x^{27} + 98473x^{29} + \cdots, \quad (35)$$

as well as the relations

$$\widetilde{\Delta^{\langle -1\rangle}}(x) = x(1-x^2) \quad \text{and} \quad Z_\delta(x(1-x^2), x^2(1-x^4), x^3(1-x^6), \ldots) = x. \tag{36}$$

Note that a triangular cactus consisting of n triangles has $2n+1$ vertices (and $3n$ edges). There will be an economy of scale if we consider the labels to be on the triangles rather than on the vertices. This gives the species $\mathfrak{k}$ of *triangular cacti* constructed on the set of triangles and $\mathcal{K}$, the species of these triangular cacti pointed at an arbitrary vertex. Note that $\mathcal{K} \neq \mathfrak{k}^\bullet$. The proof of the following identities is left as an exercise:

a) $\mathcal{K} = E(XE_2(\mathcal{K}))$,   b) $\mathfrak{k} = \mathcal{K} + X(E_3(\mathcal{K}) - \mathcal{K}E_2(\mathcal{K}))$,
c) $\widetilde{\Delta}(x) = x\widetilde{\mathcal{K}}(x^2)$,   d) $\widetilde{\delta}(x) = x\widetilde{\mathfrak{k}}(x^2)$. $\quad(37)$

Exercises for Section 4.2

1. Refine Table 6 of Appendix 2 giving the molecular decomposition of the species of graphs up to order 6 by introducing an edge counter y. Deduce the first few terms of the series associated to the species $\mathcal{G}_w$. Also consider the weighting of graphs according to degree distribution (see Exercise 2.6.24) up to order 6.
   HINT: Use a table of unlabeled graphs, for example, Appendix 1 of Harary [142].

2. a) Show that the recursive formula (7) follows from the relation

$$C''_w = (C_w \times \wp^+)' \cdot C'_w. \tag{38}$$

   b) Prove this relation.
      HINT: Show at first the simpler relation (Gessel [128], Theorem 1)

$$C'_w = E(C_w \times \wp^+). \tag{39}$$

3. (Hanlon [140].) Let F be a species of structures such that $F(0) = 0$, and let A be the species of structures defined by the equation

$$A = XE(F(A)). \tag{40}$$

   Show that

   a) $A^{\langle -1\rangle} = XE^{-1}(F)$. $\tag{41}$

b) If the sequence $(f_n)_{n\geq 1}$ is defined by $\widetilde{F}(x) = \sum_{n\geq 1} f_n x^n$, then

$$\widetilde{A^{(-1)}}(x) = x \prod_{k\geq 1} (1 - x^k)^{f_k}. \tag{42}$$

c) For any species of structures G and H such that $H = G(A)$,

$$\widetilde{G}(x) = Z_H(M(x), M(x^2), M(x^3), \ldots), \tag{43}$$

where $M(x) = \widetilde{A^{(-1)}}(x)$.

d) Generalize these results to the case where $F = F_w$ is a weighted species.

4. TRIANGULAR CACTI. Consider the species δ of triangular cacti, as well as the species $\Delta = \delta^\bullet$, introduced in Example 7.

a) Check that the first few terms of the series $\Delta(x)$ are given by

$$\Delta = x + \frac{x^3}{2} + \frac{5}{8}x^5 + \frac{49}{48}x^7 + \frac{243}{128}x^9 + \frac{14641}{3840}x^{11} + \frac{371293}{46080}x^{13}$$
$$+ \frac{253125}{14336}x^{15} + \frac{410338673}{10321920}x^{17} + \frac{16983563041}{185794560}x^{19}$$
$$+ \frac{1400846643}{6553600}x^{21} + \frac{41426511213649}{81749606400}x^{23} + \cdots. \tag{44}$$

b) Prove identity (34), take its logarithmic derivative, and deduce a recursion expressing $\widetilde{\Delta}_n = [x^n] \widetilde{\Delta}(x)$ as a function of $\widetilde{\Delta}_1, \widetilde{\Delta}_2, \ldots, \widetilde{\Delta}_{n-1}$, $n > 1$.

c) Prove identity (35) and deduce the following formula for the coefficients $\widetilde{\delta}_n = [x^n] \widetilde{\delta}(x)$ as a function of the coefficients $\widetilde{\Delta}_k$, $k \leq n$,

$$\widetilde{\delta}_n = \widetilde{\Delta}_n + \frac{1}{3}\left(\sum_{3i=n} \widetilde{\Delta}_i - \sum_{i+j+k=n} \widetilde{\Delta}_i \widetilde{\Delta}_j \widetilde{\Delta}_k \right). \tag{45}$$

d) Prove identities (37).

5. 2-EDGE-CONNECTED GRAPHS. (See Hanlon and Robinson [141], Joyal [158], Labelle [199], and Leroux [213].) A *bridge* of a graph g is a block of g of type K_2, that is, an edge of g whose removal increases the number of connected components. A *2-edge-connected graph* is a connected graph without any bridges. A *lump* is a maximal 2-edge-connected subgraph of a graph.

a) Show that the lumps of a connected graph g determine a partition of the vertices of g inducing a quotient lump-edges graph which is a tree.

b) Let M be a weighted species of 2-edge-connected graphs. Denote by $\mathcal{G}_M$ the weighted species of graphs whose lumps are in M, and by $\mathcal{G}_M^c$, the subclass of connected graphs. Establish the following relations:

i) $\mathcal{G}_M^{c\bullet} = M^\bullet\left(XE\left((\mathcal{G}_M^{c\bullet})_y\right)\right),$

ii) $M\left(XE\left((\mathcal{G}_M^{c\bullet})_y\right)\right) + K_2(\mathcal{G}_M^{c\bullet}) = \mathcal{G}_M^c + \mathcal{G}_M^{c\bullet} \cdot (\mathcal{G}_M^{c\bullet})_y.$ (dissymmetry) (46)

c) Consider the particular case where $M = X$ and where $M = M_a$, the class of all 2-edge-connected graphs (see [141]).

6. Given a subspecies G of the species $\mathcal{G}_w$ of simple graphs, with edge counter y, denote by $G^{(y)}$, the species composed of the same class of graphs G but with weight function $w'(s) = \frac{d}{dy} w(s)$.

a) Show that $G^{(y)}$ can be interpreted as the weighted species of graphs in G where an edge has been removed and that, for $G = \mathcal{G}$, the class of all graphs, we have (see Exercises 2.4.16 and 2.4.17)

$$\mathcal{G}^{(y)} = E_2 \left\langle\!\left\langle X \frac{d}{dX} \right\rangle\!\right\rangle (\mathcal{G}) - y \mathcal{G}^{(y)}, \tag{47}$$

where y denotes the species 1_y, that is, the species 1 weighted by y.

b) Prove the identity (see Hanlon and Robinson [141], 6.12)

$$\mathcal{C}^{(y)} = E_2 \left\langle\!\left\langle X \frac{d}{dX} \right\rangle\!\right\rangle (\mathcal{C}) - y \mathcal{C}^{(y)} + E_2(\mathcal{C}^\bullet). \tag{48}$$

HINT: If an edge of a connected graph is removed, the resulting graph is either connected or an assembly of two pointed connected graphs.

c) Prove Temperley's identity (see [330])

$$B_a^{(y)} = E_2 \left\langle\!\left\langle X \frac{d}{dX} \right\rangle\!\right\rangle B_a - y B_a^{(y)} + E_2(X) + \left(X L_{\geq 2} \left(X \frac{d^2}{dX^2} B_a \right) \right) / \mathbb{Z}_2, \tag{49}$$

where B_a denotes the weighted species of nonseparable graphs.

HINT: If, after having removed an edge, the resulting graph is no longer nonseparable, it consists either of two isolated points or of a (nonordered) chain of at least two nonseparable graphs joined by cut-points.

7. Let $\mathcal{M}_w = T_{Y;y} \Gamma_{\text{mult}}(X, Y)$ be the species of multigraphs without loops, with edge counter y (see Exercise 2.4.4).

a) Give a formula for the cycle index series $Z_{\mathcal{M}_w}$ and show that

$$\widetilde{\mathcal{M}_w}(x) = \sum_{n \geq 0} P_{S_n^{(2)}} \left(\frac{1}{1-y}, \frac{1}{1-y^2}, \frac{1}{1-y^3}, \ldots \right) x^n. \tag{50}$$

b) Calculate the two-sort cycle index series

i) $Z_\Gamma(x_1, x_2, \ldots; y_1, y_2, \ldots)$, ii) $Z_{\Gamma_{\text{mult}}}(x_1, x_2, \ldots; y_1, y_2, \ldots)$.

(51)

8. ITERATIVE PROCEDURE OF ROBINSON. Let $F = F_w$ be a virtual weighted species such that $F(0) = 1$. We know that there exists a unique weighted virtual species $F^c = F_w^c$ satisfying $F = E(F^c)$, namely $F^c = E_+^{\langle -1 \rangle}(F_+)$ (see (2.5.56)).

a) Adapt Robinson's iterative procedure, described in (11)–(13), to calculate the successive restrictions $(F^c)_n$, $n = 1, 2, 3, \ldots$. Describe a corresponding algorithm for the computation of associated series of F^c.

b) Verify that the expressions given in the text for $(F^c)_n$, $n \leq 4$, are recovered when $F = \mathcal{G}_w$, the species of simple graphs with edge counter y. Extend these computations to the case $n = 5$ and compute the corresponding series.

c) Considering the case $F = 1 + X$, apply the procedure described in a) to compute the substitutional inverse $E_+^{\langle -1 \rangle}$ of the species E_+ of nonempty sets, and obtain its first few terms:

$$E_+^{\langle -1 \rangle} = X - E_2 + XE_2 - E_3 - X^2 E_2 + XE_3 - E_4 + E_2 \circ E_2$$
$$- XE_2^2 + E_2 E_3 + X^3 E_2 - X^2 E_3 + XE_4 - E_5 + \cdots. \quad (52)$$

d) Deduce the first few terms of the molecular decomposition of the weighted virtual species

$$\Lambda^{(\alpha)} = E \circ X_\alpha \circ E_+^{\langle -1 \rangle}$$

described in Exercise 2.3.12.

e) Show that $Y = F^c$ is the solution of the functional equation

$$Y = F_+ - G(Y), \quad Y(0) = 0, \quad (53)$$

where $G = E_{\geq 2}$. Develop an iterative procedure with quadratic convergence (in the sense of contact order), adapted from Proposition 3.3.7, to compute F^c and the associated series.

4.3. Proof of the Substitution Formulas

The goal of this section is to prove the following fundamental formula, already used in the preceding chapters:

$$(\widetilde{F_w \circ G_v})(x) = Z_{F_w}(\widetilde{G_v}(x), \widetilde{G_{v^2}}(x^2), \widetilde{G_{v^3}}(x^3), \ldots). \quad (1)$$

As we will see, this formula is equivalent to the Pólya–Redfield Theorem. However we will first give a direct proof, following an approach of Joyal [158]. The general substitution formula for cycle index series

$$Z_{F_w \circ G_v} = Z_{F_w} \circ Z_{G_v} \quad (2)$$

will follow from (1), using the fact that the "power sum" symmetric functions are algebraically independent.

Recall that to each species of structures F there corresponds a family of group actions $S_n \times F[n] \longrightarrow F[n]$, $n \geq 0$, defined by transport of structures. For any bijection $\sigma : U \longrightarrow V$ and any F-structure $s \in F[U]$, we use the notation $\sigma \cdot s$ for $F[\sigma](s)$, or sometimes $\sigma \cdot_F s$ to avoid ambiguity.

To each weighted species of structures $F = F_w$, we associate an auxiliary species $\widetilde{F}$ in such a way that the generating series $\widetilde{F}(x)$ coincides with the ordinary type generating series of the species F, in agreement with the notation already in use.

Intuitively, an $\widetilde{F}$-structure is an F-structure together with an automorphism. Here is the definition.

Definition 1. Let $F = F_w$ be a weighted species of structures. The weighted species $\widetilde{F} = \widetilde{F}_w$ is defined by the formulas

$$\widetilde{F}[U] = \{(s, \alpha) \in F[U] \times \mathcal{S}[U] \mid \alpha \cdot s = s\}, \tag{3}$$

with $w(s, \alpha) = w(s)$, and, for $\sigma : U \longrightarrow V$,

$$\widetilde{F}[\sigma](s, \alpha) = (\sigma \cdot s, \sigma \circ \alpha \circ \sigma^{-1}). \tag{4}$$

It is easily verified that this indeed defines a weighted species of structures $\widetilde{F}_w$ (see Exercise 1). In fact, $\widetilde{F}$ is a subspecies of the Cartesian produce $F \times \mathcal{S}$. Moreover, by virtue of the weighted Cauchy–Frobenius Theorem (alias Burnside's Lemma; see Proposition A.1.7) applied to the action of the symmetric group $\mathcal{S}_n$ on $F[n]$, one finds, as generating series for $\widetilde{F}_w$,

$$\widetilde{F}_w(x) = \sum_{n \geq 0} |\widetilde{F}[n]|_w \frac{x^n}{n!}$$

$$= \sum_{n \geq 0} \frac{1}{n!} |\{(s, \alpha) \in F[n] \times \mathcal{S}[n] \mid \alpha \cdot s = s\}|_w x^n$$

$$= \sum_{n \geq 0} \frac{1}{n!} \sum_{\alpha \in \mathcal{S}[n]} |\{s \in F[n] \mid \alpha \cdot s = s\}|_w x^n$$

$$= \sum_{n \geq 0} |F[n]/\mathcal{S}_n|_w x^n$$

$$= \sum_{n \geq 0} |\mathrm{T}(F_n)|_w x^n.$$

That is, $\widetilde{F}(x)$ is the ordinary generating series for types of (or unlabeled) F-structures, agreeing with the previous notation. With this approach, the cycle index series of F can be considered as a refinement of the generating series of $\widetilde{F}$. This gives a new and very useful expression for Z_F as a sum of cycle index polynomials of stabilizers of (types of) F-structures:

Proposition 2. *For any weighted species $F = F_w$,*

$$Z_F(x_1, x_2, x_3, \ldots) = \sum_{t \in T(F)} w(t) P_{\mathcal{G}_t}(x_1, x_2, x_3, \ldots), \tag{5}$$

where t ranges over a system of representatives of types of F-structures and where $P_{\mathcal{G}_t}$ denotes the cycle index polynomial of the stabilizer $\mathcal{G}_t$ of the F_w-structure t acting on the underlying set of t (see Appendix 1).

Proof. By definition,

$$\begin{aligned}
Z_F(x_1, x_2, x_3, \ldots) &= \sum_{n \geq 0} \frac{1}{n!} \sum_{\sigma \in S_n} |\text{Fix } F[\sigma]|_w x_1^{\sigma_1} x_2^{\sigma_2} \cdots \\
&= \sum_{n \geq 0} \frac{1}{n!} \sum_{(s,\sigma) \in \widetilde{F}[n]} w(s) x_1^{\sigma_1} x_2^{\sigma_2} \cdots \\
&= \sum_{n \geq 0} \frac{1}{n!} \sum_{s \in F[n]} w(s) \sum_{\sigma \in \mathcal{G}_s} x_1^{\sigma_1} x_2^{\sigma_2} \cdots \\
&= \sum_{n \geq 0} \sum_{t \in T(F_n)} \frac{|\mathcal{O}(t)|}{n!} w(t) \sum_{\sigma \in \mathcal{G}_t} x_1^{\sigma_1} x_2^{\sigma_2} \cdots \\
&= \sum_{n \geq 0} \sum_{t \in T(F_n)} \frac{w(t)}{|\mathcal{G}_t|} \sum_{\sigma \in \mathcal{G}_t} x_1^{\sigma_1} x_2^{\sigma_2} \cdots \\
&= \sum_{t \in T(F)} w(t) P_{\mathcal{G}_t}(x_1, x_2, x_3, \ldots).
\end{aligned}$$

Here we have used Proposition A.1.6 which gives the cardinality of an orbit: $|\mathcal{O}(t)| = n!/|\mathcal{G}_t|$. ∎

In particular, if F is a molecular species and if $s \in F[U]$, then

$$Z_F(x_1, x_2, x_3, \ldots) = P_{\mathcal{G}_s:U}(x_1, x_2, x_3, \ldots). \tag{6}$$

Observe that, for $\sigma \in S_n$, the substitution $x_k \mapsto x^k$ in $x_1^{\sigma_1} x_2^{\sigma_2} \cdots$ gives x^n. The following formula is then recovered immediately:

Corollary 3. *For any weighted species $F = F_w$,*

$$\widetilde{F}_w(x) = Z_{F_w}(x, x^2, x^3, \ldots). \tag{7}$$

Proof. We have

$$\begin{aligned}
\widetilde{F}_w(x) &= \sum_{n \geq 0} |T(F_n)|_w x^n \\
&= \sum_{n \geq 0} \sum_{t \in T(F_n)} w(t) x^n \\
&= \sum_{n \geq 0} \sum_{t \in T(F_n)} \frac{w(t)}{|\mathcal{G}_t|} \sum_{\sigma \in \mathcal{G}_t} x^n \\
&= Z_F(x, x^2, x^3, \ldots),
\end{aligned}$$

which completes the proof. ∎

To establish formula (1), we now analyze a generic $(F_w \circ G_v)\widetilde{}$-structure on a set U. Such a structure is a pair (s, σ), where $\sigma \in \mathcal{S}[U]$ and s is an $(F \circ G)$-structure on U left fixed by σ. We then have $s = (\pi, f, g)$, where
- π is a partition of U,
- $f \in F[\pi]$, and
- $g = (g_p)_{p \in \pi}$, with $g_p \in G[p]$,

with $\sigma \cdot \pi = \pi$, which induces a bijection $\sigma_\pi : \pi \longrightarrow \pi$, sending a block p of π on $\sigma(p)$ so that $\sigma_\pi \cdot {}_F f = f$; moreover, for each $p \in \pi$, one has $\sigma|_p \cdot {}_G g_p = g_{\sigma(p)}$ (see Figure 1). Observe that the overall weight of such an $(F \circ G)\widetilde{}$-structure is

$$w(s, \sigma) = w(f) \prod_{p \in \pi} v(g_p). \tag{8}$$

Denote by $(F \widetilde{\circ} G)_N$ the subspecies of $(F \widetilde{\circ} G)$-structures for which the induced permutation σ_π is of cycle type $N = (n_1, n_2, n_3, \ldots)$, where n_k is the number of cycles of length k of σ_π. We have

$$(F \widetilde{\circ} G) = \sum_N (F \widetilde{\circ} G)_N, \tag{9}$$

the sum being taken over the set of sequences $N = (n_1, n_2, n_3, \ldots)$ with finite support, that is, with $|N| := n_1 + 2n_2 + \cdots < \infty$.

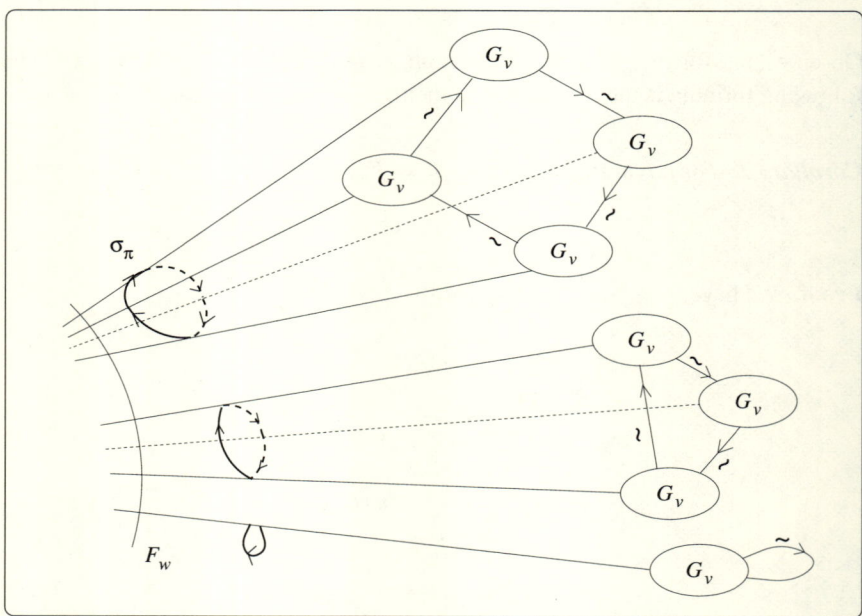

Fig. 1.

Following [158, p. 26], we call *wreath of G-structures* an $(E \widetilde{\circ} G)$-structure whose induced permutation σ_π consists of one single cycle and denote by $K_m(G)$ the species of such wreaths of G-structures with cycle length m. The analysis we have just made shows that if the permutation σ_π is of type $N = (n_1, n_2, n_3, \ldots)$, the $(F \circ G)\widetilde{\,}$-structure (s, σ) is entirely determined by the assembly of wreaths of G-structures of the form $E_{n_1}(K_1(G))E_{n_2}(K_2(G))\cdots$ and by an F-structure left fixed by a permutation of type N. Hence we have

$$(F \widetilde{\circ} G)_N(x) = |\text{Fix } F[N]|_w \frac{(K_1(G)(x))^{n_1}}{n_1!} \frac{(K_2(G)(x))^{n_2}}{n_2!} \cdots . \tag{10}$$

Lemma 4. *The generating series of the species of wreaths of G_v-structures, of length m, is given by*

$$K_m(G_v)(x) = \frac{\widetilde{G_{v^m}}(x^m)}{m}, \tag{11}$$

Proof. We will rather show that

$$m K_m(G_v)(x) = \widetilde{G_{v^m}}(x^m). \tag{12}$$

A $K_m(G_v)$-structure on a set U is a triple $(\pi, (g_p)_{p\in\pi}, \sigma)$, where π is a partition of U, $g_p \in G[p]$ for each $p \in \pi$, and σ is a permutation of U inducing a cyclic permutation of order m on π and isomorphisms of G_v-structures: $\sigma|_p \cdot_G g_p = g_{\sigma(p)}$. Clearly $m K_m(G_v)(x)$ enumerates wreaths of G-structures where one of the m classes has been distinguished. Call this class p_1. As σ_π is a cyclic permutation of π, the restrictions $\sigma|_p$, $p \in \pi$ determine a cycle of G_v-structure isomorphisms $g_p \longrightarrow g_{\sigma(p)}$. These m structures are all isomorphic and essentially determined by the G-structure g_{p_1}, and the overall weight is then

$$\prod_{p\in\pi} v(g_p) = v^m(g_{p_1}). \tag{13}$$

Moreover, the permutation $\sigma^m|_{p_1}$ is an automorphism of the G-structure g_{p_1}. Finally, the set p_1 can be identified with the set of chains

$$u \mapsto \sigma(u) \mapsto \sigma^2(u) \mapsto \cdots \mapsto \sigma^{m-1}(u)$$

for $u \in p_1$, which partitions the underlying set. We have thus constructed a $(\widetilde{G_{v^m}} \circ X^m)$-structure and the reader can easily verify that this correspondence is bijective. ∎

In conclusion, we have the following fundamental result.

Theorem 5. *The generating series of unlabeled $(F_w \circ G_v)$-structures is given by*

$$(F_w \widetilde{\circ} G_v)(x) = Z_{F_w}\big(\widetilde{G_v}(x), \widetilde{G_{v^2}}(x^2), \ldots\big). \tag{14}$$

Proof. By virtue of Equations (9) and (10), and recalling Definition (2.3.18) of the cycle index series, we immediately deduce from the proceeding lemma that

$$(F_w \widetilde{\circ} G_v)(x) = \sum_{|N|<\infty} (F_w \widetilde{\circ} G_v)_N(x)$$

$$= \sum_{n_1+2n_2+\cdots<\infty} |\text{Fix } F[n_1, n_2, \ldots]|_w \frac{(\widetilde{G_v}(x))^{n_1}}{1^{n_1} n_1!} \frac{(\widetilde{G_{v^2}}(x^2))^{n_2}}{2^{n_2} n_2!} \cdots$$

$$= Z_{F_w}(\widetilde{G_v}(x), \widetilde{G_{v^2}}(x^2), \ldots),$$

which completes the proof. ∎

The analysis that we have made of an $(F \widetilde{\circ} G)$-structure allows us to establish a direct link between the composition $F \circ G$ of two molecular species and the wreath product $(K \wr H : C \times D)$ of two finite group actions $(K : C)$ and $(H : D)$. Recall that this wreath product consists of elements $(\gamma, (\lambda_x)_{x \in C}) \in K \times H^C$ acting on the pairs $(c, d) \in C \times D$ by the formula

$$(\gamma, (\lambda_x)_{x \in C}) \cdot (c, d) = (\gamma \cdot c, \lambda_c \cdot d). \quad (15)$$

Hence, as a set, $K \wr H = K \times H^C$ and the multiplication in $K \wr H$ is given by (see Exercise 5)

$$(\gamma', (\lambda'_x))(\gamma, (\lambda_x)) = (\gamma'\gamma, (\lambda'_{\gamma(x)}\lambda_x)). \quad (16)$$

Also recall that a species is called molecular if it contains only one isomorphism type of structures (see Section 2.6).

Proposition 6. *Let F and G be molecular species. Then the composite species $F \circ G$ is also molecular. Moreover, if $s = (\pi, f, (g_p)_{p \in \pi})$ is an arbitrary $(F \circ G)$-structure on a set U and if $\mathcal{G}_s$, $\mathcal{G}_f$, and $\mathcal{G}_g$ denote respectively the automorphism groups of the structures s, $f \in F[\pi]$, and $g \in G[D]$, then we have a group action isomorphism*

$$(\mathcal{G}_s : U) \simeq (\mathcal{G}_f \wr \mathcal{G}_g : \pi \times D).$$

Proof. To show that $F \circ G$ is molecular, suppose that $s = (\pi, f, (g_p)_{p \in \pi}) \in (F \circ G)[U]$ and $s' = (\pi', f', (g'_{p'})_{p' \in \pi'}) \in (F \circ G)[U']$ are two $(F \circ G)$-structures. Let $\gamma : \pi \longrightarrow \pi'$ be a bijection such that $\gamma \cdot_F f = f'$ and, for each $p \in \pi$. Let $\alpha_p : p \longrightarrow \gamma(p)$ be a bijection such that $\alpha_p \cdot_G g_p = g'_{\gamma(p)}$. The existence of these isomorphisms is ensured by the hypothesis that F and G are molecular. Then it is easy to see that the union $\alpha = \bigcup_{p \in \pi} \alpha_p : U \longrightarrow U'$ defines an isomorphism between s and s'.

Now, if $\sigma \in S[U]$ is an automorphism of $s \in (F \circ G)[U]$, then $(s, \sigma) \in (F \widetilde{\circ} G)[U]$ and, in particular, $\sigma_\pi : \pi \longrightarrow \pi$ is an automorphism of the F-structure f, that is, $\sigma_\pi \in \mathcal{G}_f$. For each $p \in \pi$, fix an isomorphism $\beta_p : p \longrightarrow D$ between

$g_p \in G[p]$ and a fixed G-structure $g \in G[D]$. We know that for each class $p \in \pi$, $\sigma|_p : p \longrightarrow \sigma(p) = \sigma_\pi(p)$ determines an isomorphism $g_p \simeq g_{\sigma(p)}$ of G-structures. Set

$$\lambda_p = \beta_{\sigma(p)} \circ \sigma|_p \circ \beta_p^{-1}, \qquad p \in \pi. \tag{17}$$

Then λ_p, as a composition of isomorphisms, is an automorphism of the G-structure g, that is, $\lambda_p \in \mathcal{G}_g$, for any $p \in \pi$. Thus, we have constructed a function

$$\mathcal{G}_s \longrightarrow \mathcal{G}_f \wr \mathcal{G}_g, \quad \text{sending} \quad \sigma \mapsto (\sigma_\pi, (\lambda_p)_{p \in \pi}). \tag{18}$$

Moreover, as π is a partition of U, each element $u \in U$ appears in a unique class $p = p(u) \in \pi$. We then define a function

$$U \longrightarrow \pi \times D, \quad \text{sending} \quad u \mapsto (p(u), \beta_{p(u)}(u)), \tag{19}$$

and we leave to the reader the tasks of verifying that (18) and (19) are bijections, that (18) is a group isomorphism, and that, globally, this pair of functions establishes the desired group action isomorphism. ∎

Characterization of Cycle Index Series

Let $\mathbb{K}$ be a field, with $\mathbb{Q} \subseteq \mathbb{K} \subseteq \mathbb{C}$. We will consider species of structures with weight in rings $\mathbb{A}$ of formal power series over $\mathbb{K}$ in an arbitrary number of indeterminates and show that the substitution formula (1) completely characterizes the index series $Z_{F_w}(x_1, x_2, x_3, \ldots)$.

Theorem 7. Let $F = F_w$ be a weighted species of structures. Then the cycle index series of F_w is the unique formal power series $f(x_1, x_2, x_3, \ldots) \in \mathbb{A}[\![x_1, x_2, x_3, \ldots]\!]$ such that for any weighted species $G = G_v$,

$$\widetilde{F_w \circ G_v}(x) = f\big(\widetilde{G_v}(x), \widetilde{G_{v^2}}(x^2), \widetilde{G_{v^3}}(x^3), \ldots\big). \tag{20}$$

Proof. It is clear that the series $f = Z_{F_w}$ satisfies condition (20). To show uniqueness, choose variables $t_1, t_2, t_3, \ldots$ not appearing in $\mathbb{A}$ and consider the particular species $G_v = X_{\mathbf{t}} = X_{t_1} + X_{t_2} + X_{t_3} + \cdots$, introduced in Example 2.3.15, for which

$$\widetilde{G_{v^k}}(x^k) = \big(t_1^k + t_2^k + t_3^k + \cdots\big) x^k = p_k x^k, \tag{21}$$

where p_k is the power sum symmetric function (in the t_i). It suffices to show that for $f \in \mathbb{A}[\![\mathbf{x}]\!]$, we have

$$f(p_1 x, p_2 x^2, p_3 x^3, \ldots) = 0 \implies f(x_1, x_2, x_3, \ldots) = 0. \tag{22}$$

If $f(x_1, x_2, x_3, \ldots) = \sum_{n_1, n_2, \ldots} a_{n_1, n_2, \ldots} x^{n_1} x^{n_2} \cdots$, set, for $n \geq 0$,

$$f_n(x_1, x_2, x_3, \ldots) = \sum_{n_1 + 2n_2 + \cdots = n} a_{n_1, n_2, \ldots} x_1^{n_1} x_2^{n_2} \cdots.$$

Then f_n is a polynomial in the x_i and

$$f(p_1x, p_2x^2, p_3x^3, \ldots) = 0 \implies \sum_{n \geq 0} f_n(p_1, p_2, p_3, \ldots) x^n = 0$$

$$\implies f_n(p_1, p_2, p_3, \ldots) = 0, \quad n \geq 0$$

$$\implies f_n(x_1, x_2, x_3, \ldots) = 0, \quad n \geq 0$$

$$\implies f(x_1, x_2, x_3, \ldots) = 0,$$

since the symmetric functions $p_k(t_1, t_2, \ldots)$ ($k \geq 0$) are algebraically independent over $\mathbb{A}$ (see Macdonald [228] and Exercise 4.4.4). ∎

Remark 8. This characterization can be used to define the cycle index series Z_F of an $\mathbb{A}$-weighted species F_w. In fact, the power sum symmetric functions $p_k(t_1, t_2, t_3, \ldots)$ form an algebraic basis in the ring Λ of symmetric functions with coefficients in $\mathbb{A}$. It is then possible to express the symmetric function $F_w(X_\mathbf{t})(x)|_{x=1}$, giving the inventory of unlabeled colored F_w-structures, in terms of power sums. This is exactly what the cycle index series Z_{F_w} does:

$$\widetilde{F_w(X_\mathbf{t})}(x)|_{x=1} = Z_{F_w}(p_1, p_2, p_3, \ldots). \tag{23}$$

Following an approach similar to that of Pólya (see [263] and [264], Section 27), we now use Theorem 7 to prove the substitution formula for cycle index series.

Theorem 9. *For any weighted species F_w and G_v, with $G(0) = 0$, one has*

$$Z_{F_w \circ G_v} = Z_{F_w} \circ Z_{G_v}$$
$$= Z_{F_w}\big((Z_{G_v})_1, (Z_{G_v})_2, (Z_{G_v})_3, \ldots\big), \tag{24}$$

where, for $k \geq 1$,

$$(Z_{G_v})_k(x_1, x_2, x_3, \ldots) = Z_{G_{v^k}}(x_k, x_{2k}, x_{3k}, \ldots). \tag{25}$$

Proof. By virtue of the preceding theorem, it suffices to show that for any weighted species H_u, with $H_u(0) = 0$,

$$\widetilde{((F_w \circ G_v) \circ H_u)}(x) = \big(Z_{F_w} \circ Z_{G_v}\big)\big(\widetilde{H_u}(x), \widetilde{H_{u^2}}(x^2), \widetilde{H_{u^3}}(x^3), \ldots\big). \tag{26}$$

But we have

$$\widetilde{((F_w \circ G_v) \circ H_u)}(x) = \widetilde{(F_w \circ (G_v \circ H_u))}(x)$$

$$= Z_{F_w}\big(\widetilde{(G_v \circ H_u)}(x), \widetilde{(G_{v^2} \circ H_{u^2})}(x^2), \ldots\big)$$

$$= Z_{F_w}\Big(Z_{G_v}\big(\widetilde{H_u}(x), \widetilde{H_{u^2}}(x^2), \ldots\big), Z_{G_{v^2}}\big(\widetilde{H_{u^2}}(x^2),$$
$$\widetilde{H_{u^4}}(x^4), \ldots\big), \ldots\Big)$$

$$= Z_{F_w}\big(Z_{G_v}(x_1, x_2, \ldots), Z_{G_{v^2}}(x_2, x_4, \ldots), \ldots\big)\big|_{x_i = \widetilde{H_{u^i}}(x^i)}$$

$$= \big(Z_{F_w} \circ Z_{G_v}\big)\big(\widetilde{H_u}(x), \widetilde{H_{u^2}}(x^2), \ldots\big). \qquad\blacksquare$$

Connections with Pólya Theory

We now associate to any group action $\alpha : G \times D \longrightarrow D$, where G and D are finite, a species of structures, denoted $M_{G:D}$, defined by

$$M_{G:D}[U] = \mathrm{Bij}[D, U]/G, \tag{27}$$

where $\mathrm{Bij}[D, U]/G$ denotes the set of G-reduced labelings (see Appendix 1), that is, the set of orbits of bijections $\beta : D \longrightarrow U$ under the induced action of G, defined, for $g \in G$, by

$$g \cdot \beta = \beta \circ \varphi(g^{-1}), \tag{28}$$

where $\varphi : G \longrightarrow \mathcal{S}[D]$ is the group homomorphism associated to the action α. Hence, the orbit $\mathcal{O}(\beta)$ of a bijection $\beta : D \longrightarrow U$ has an explicit description, analogous to the coset of a subgroup:

$$\mathcal{O}(\beta) = \beta \circ \varphi(G) = \{\beta \circ \varphi(g) \mid g \in G\}. \tag{29}$$

Obviously $M_{G:D}[U] \neq \emptyset$ if and only if $|U| = |D|$. Now if $\sigma : U \xrightarrow{\sim} V$ is a bijection and $\mathcal{O}(\beta)$ is the orbit of a labeling $\beta : D \longrightarrow U$ then we set

$$M_{G:D}[\sigma](\mathcal{O}(\beta)) = \mathcal{O}(\sigma \circ \beta). \tag{30}$$

It is easily verified (see Exercise 3) that this definition is coherent and indeed defines a species of structures. Moreover, $M_{G:D}$ is a molecular species, that is, there exists only one isomorphism type, since if $\beta_1 \in \mathrm{Bij}[D, U_1]$ and $\beta_2 \in \mathrm{Bij}[D, U_2]$ then $\sigma = \beta_2 \circ \beta_1^{-1}$ is a bijection of U_1 to U_2 such that $M_{G:D}[\sigma](\mathcal{O}(\beta_1)) = \mathcal{O}(\sigma \circ \beta_1) = \mathcal{O}(\beta_2)$. Finally, the stabilizer $\mathcal{G}_{\mathcal{O}(\beta)}$ of an $M_{G:D}$-structure $\mathcal{O}(\beta)$ on U is isomorphic to the subgroup $\varphi(G)$ of $\mathcal{S}[D]$:

$$\mathcal{G}_{\mathcal{O}(\beta)} = \beta\varphi(G)\beta^{-1}. \tag{31}$$

Indeed, for $\sigma \in \mathcal{S}[U]$ and $\beta \in \mathrm{Bij}[D, U]$, we have

$$\sigma \in \mathcal{G}_{\mathcal{O}(\beta)} \iff \sigma \cdot_{M_{G:U}} \mathcal{O}(\beta) = \mathcal{O}(\beta)$$
$$\iff \mathcal{O}(\sigma \circ \beta) = \mathcal{O}(\beta)$$
$$\iff \exists g \in G : g \cdot (\sigma \circ \beta) = \beta$$
$$\iff \exists g \in G : \sigma \circ \beta \circ \varphi(g^{-1}) = \beta$$
$$\iff g \in G : \sigma \circ \beta = \beta \circ \varphi(g)$$
$$\iff \beta^{-1} \circ \sigma \circ \beta \in \varphi(G)$$
$$\iff \sigma \in \beta\varphi(G)\beta^{-1}.$$

In fact, this yields an isomorphism of group actions

$$(\mathcal{G}_{\mathcal{O}(\beta)} \times U \longrightarrow U) \simeq (\varphi(G) \times D \longrightarrow D),$$

and it follows that (see Exercise A.1.7)

$$
\begin{aligned}
P_{\mathcal{G}_{\mathcal{O}(\beta)}:U}(y_1, y_2, y_3, \ldots) &= P_{\varphi(G):D}(y_1, y_2, y_3, \ldots) \\
&= P_{G:D}(y_1, y_2, y_3, \ldots).
\end{aligned} \tag{32}
$$

Observe that if the action $H \times [n] \longrightarrow [n]$ is given by a subgroup H of $\mathcal{S}_n$, then the species $M_{H:[n]}$ coincides with the molecular species $M_H = X^n/H$ introduced in Definition 2.6.8.

Proposition 10. *Let $\alpha : G \times D \longrightarrow D$ be an action of a finite group G on a finite set D. Then the associated species of structures $M_{G:D}$ is molecular and, moreover,*

$$Z_{M_{G:D}}(x_1, x_2, x_3, \ldots) = P_{G:D}(x_1, x_2, x_3, \ldots). \tag{33}$$

Proof. It has already been established that $M_{G:D}$ is molecular. Formula (33) is then a consequence of (6) and (32). ∎

Now let $F = F_v$ be a weighted species of structures. We associate to it the weighed set $(W, w) = (T(F), w)$, where $T(F)$ denotes the set of isomorphism types of F-structures (i.e., unlabeled F-structures), the weights $w(t)$ of a type of F_v-structure t being defined by

$$w(t) = v(t) x^{n(t)}, \tag{34}$$

where $n(t)$ denotes the order of t, that is, the cardinality of the underlying set of a structure of type t. The inventory of this weighted set is evidently

$$|T(F)|_w = \widetilde{F}_v(x). \tag{35}$$

Given a group action $\alpha : G \times D \longrightarrow D$, we can use the types of F_v-structures as colors and construct the set $\Phi[D, T(F)]/G$ of G-reduced colorings (see Appendix 1), that is, the orbits of functions $f : D \longrightarrow T(F)$ under the action of the group G defined, for $g \in G$, $f \in \Phi[D, T(F)]$, and $d \in D$, by

$$(g \cdot f)(d) = f(g^{-1} \cdot d). \tag{36}$$

Recall that the weight $w(f)$ of a coloring f is defined by $w(f) = \prod_{d \in D} w(f(d))$. Then there is an isomorphism of weighted sets

$$\Phi[D, T(F)]/G \xrightarrow{\sim} TM_{G:D} \circ F_v, \tag{37}$$

between G-reduced colorings and types of $(M_{G:D} \circ F_v)$-structures, defined in the following manner. For a coloring $f : D \longrightarrow T(F)$, choose pairwise disjoint sets U_d, $d \in D$, and structures $s_d \in F_v[U_d]$, of type $f(d)$, for each $d \in D$. Set $U = \bigcup_{d \in D} U_d$ and let π be the partition of U whose blocks are the sets U_d and $\beta : D \longrightarrow \pi$ the induced bijection $d \mapsto U_d$.

4.3. Proof of the Substitution Formulas

Let $\sigma = \varphi(g^{-1}) \in \mathcal{S}[D]$, the permutation of D induced by the action of an element $g \in G$. It is clear that to a coloring $f \circ \sigma : D \longrightarrow T(F)$ there corresponds a family $s_{\sigma(d)} \in F_v[U_{\sigma(d)}]$, $d \in D$, that is, the same assembly of F_v-structures on the set U, together with the bijection $\beta \circ \sigma : D \longrightarrow \pi$. Thus to an orbit $\mathcal{O}(f) \in \Phi[D, T(F)]/G$ corresponds a $(M_{G:D} \circ F)$-structure $s = (\mathcal{O}(\beta), \pi, \{s_d\}_{d \in D})$ on U and

$$w(s) = \left(\prod_{d \in D} v(s_d)\right) x^{|U|} = \prod_{d \in D} v(s_d) \, x^{|U_d|} = \prod_{d \in D} w(f(d)) = w(f).$$

The reader can verify that this correspondence determines an isomorphism of the form (37). The following result is immediately deduced.

Proposition 11. *Let $\alpha : H \times D \longrightarrow D$ be an action of a finite group H on a finite set D, and let $F = F_v$ be a weighted species of structures. Then*

$$(M_{H:D} \circ F_v)\widetilde{\ }(x) = |\Phi[D, T(F)]/H|_w, \tag{38}$$

where $T(F)$ denotes the set of types of F-structures, weighted by the formula (34).

As an application, we show the equivalence between the substitution formula (1) and Pólya–Redfield's Theorem (Theorem A.1.17). To prove Pólya–Redfield's Theorem starting from (1), it suffices to consider the case where $W = [\infty] = \{1, 2, 3, \ldots\}$ and $w(i) = t_i x$. Indeed, every weighted set of colors is obtained from the latter by a specialization of the t_i and of x (for example $x = 1$). Take the weighted species $F_v = X_{\mathbf{t}} = X_{t_1} + X_{t_2} + \cdots$, whose weighted set of types coincides with W. Recall that $X_{\mathbf{t}}\widetilde{\ }(x^k) = p_k(t_1, t_2, \ldots) x^k = |W|_{w^k}$. We then have, for any finite action $H \times D \longrightarrow D$,

$$\begin{aligned}
|\Phi[D, W]/H|_w &= |\Phi[D, T(X_{\mathbf{t}})]/H|_w \\
&= (M_{H:D} \circ X_{\mathbf{t}})\widetilde{\ }(x) & \text{by (38)} \\
&= Z_{M_{H:D}}(p_1 x, p_2 x^2, p_3 x^3, \ldots) & \text{by (1)} \\
&= P_{H:D}(|W|_w, |W|_{w^2}, |W|_{w^3}, \ldots), & \text{by (33)}
\end{aligned}$$

whence formula (A.1.36).

Conversely, to prove the substitution formula

$$(F_w \widetilde{\circ} G_v)(x) = Z_{F_w}\left(\widetilde{G_v}(x), \widetilde{G_{v^2}}(x^2), \ldots\right) \tag{39}$$

beginning from the Pólya–Redfield Theorem, one can suppose, by linearity, that the species F_w is molecular, $F_w = M_{H:D}$, with constant weight $w = 1$, and then

$$(F_w \widetilde{\circ\, G_v})(x) = (M_{H:D} \circ G_v \widetilde{)}(x)$$
$$= |\Phi[D, T(G)]/H|_w \qquad \text{by (38)}$$
$$= P_{H:D}(|T(G)|_w, |T(G)|_{w^2}, \ldots) \qquad \text{by (A.1.36)}$$
$$= P_{H:D}(\widetilde{G_v}(x), \widetilde{G_{v^2}}(x^2), \ldots)$$
$$= Z_{M_{H:D}}(\widetilde{G_v}(x), \widetilde{G_{v^2}}(x^2), \ldots). \qquad \text{by (33)}$$

Exercises for Section 4.3

1. Show that $\widetilde{F_w}$ is indeed a weighted species of structures, by verifying that the "transport" (4) is well defined and satisfies the functoriality conditions (1.1.11) and (1.1.12).
2. Complete the proof of Proposition 6 by showing that
   a) functions (18) and (19) are bijective (in each case describe the inverse function),
   b) function (18) is a group isomorphism,
   c) functions (18) and (19) establish a group action isomorphism

   $$(\mathcal{G}_s : U) \xrightarrow{\sim} (\mathcal{G}_f \wr \mathcal{G}_g : \pi \times D)$$

   in the sense of Definition A.1.3.
3. Let $\alpha : G \times D \longrightarrow D$ be a finite set action of a finite group G.
   a) Show that formulas (27) and (30) indeed define a species of structures $M_{G:D}$, by verifying, in particular, that the transports are well defined (i.e., independent of the orbit representatives β) and satisfy the functoriality conditions (1.1.11) and (1.1.12).
   b) Let $\varphi : G \longrightarrow S(D)$ be the group homomorphism associated to the action α and let $\beta : D \longrightarrow U$ be a bijection, of orbit $\mathcal{O}(\beta)$. Show that there is a group action isomorphism

   $$(\mathcal{G}_{\mathcal{O}(\beta)} \times U \longrightarrow U) \simeq (\varphi(G) \times D \longrightarrow D)$$

   and that

   $$P_{\mathcal{G}_{\mathcal{O}(\beta)}:U}(y_1, y_2, y_3, \ldots) = P_{\varphi(G):D}(y_1, y_2, y_3, \ldots)$$
   $$= P_{G:D}(y_1, y_2, y_3, \ldots).$$

4. a) Verify that if an action $H \times [n] \longrightarrow [n]$ is given by a subgroup H of S_n, then the species $M_{H:[n]}$ coincides with the species $M_H = X^n/H$ introduced in Section 2.6.

b) Establish the following species isomorphisms, in relation with Example 2.6.9:

i) $M_{1:[n]} = L_n$, ii) $M_{S_n:[n]} = E_n$, iii) $M_{A_n:[n]} = E_n^\pm$, (40)

iv) $M_{C_n:[n]} = C_n$, v) $M_{D_n:[n]} = P_n$,

where A_n, C_n and D_n denote respectively the alternating group, the cyclic group and the dihedral group.

5. a) Show that the multiplication rule in $K \wr H \simeq K \times H^C$ given by (16) defines a group structure on $K \wr H$ whose identity element is given by $(1_K, (1_H))$ and that the inverse of an arbitrary element $(\gamma, (\lambda_x))$ is given by

$$(\gamma, (\lambda_x))^{-1} = (\gamma^{-1}, (\lambda^{-1}_{\gamma^{-1} \cdot x})). \tag{41}$$

Show also that (15) defines a group action.

b) Establish the following relation between the cycle index polynomials:

$$P_{K \wr H : C \times D} = P_{K:C} \circ P_{H:D} \tag{42}$$

6. a) Prove the substitution formulas (2.4.21), b) and c), for the 2-sort species.
b) Generalize to the case of the substitution of k m-sort species into a k-sort species.

4.4. Asymmetric Structures

The goal of this section is to describe the necessary tools for the classification and enumeration of discrete asymmetric structures, that is, combinatorial structures whose automorphism group is reduced to the identity permutation of the underlying set. Of the two trees represented in Figure 1, the first admits a nontrivial *symmetry* (an automorphism different from the identity), whereas the second is totally *asymmetric*. A nontrivial symmetry of the first tree is given by the permutation $\sigma = (c, d, x)(b, y)$ (cyclic notation). On the contrary, only the identity permutation leaves the second tree invariant.

Asymmetric trees are often called *identity trees* in the literature. The problem of enumerating them has already been considered and solved by Harary and Prins

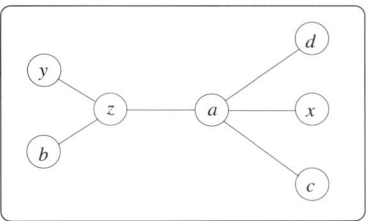 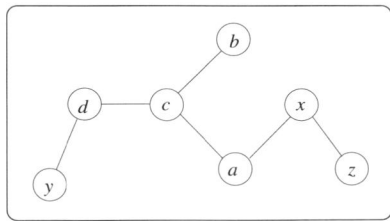

Fig. 1.

(see [144], [146], and [215]). In 1968, Rota [285] produced an analog of the cycle indicator for the enumeration of asymmetric structures which is based on a Galois connection between subgroups of a group acting on a set S and partitions of S. Rota's acyclic indicator is defined in terms of Möbius inversion in the lattice of *periods* (closed partitions of the Galois connection). See also Rota–Sagan [288] and Rota–Smith [289]. Other authors (see Stockmeyer [307], White [325], Kerber and Thurlings [171]) have attacked the enumeration of structures according to stabilizers using Möbius inversion in the lattice of subgroups of a permutation group. This section presents a general method for the solution of this problem in the framework of species of structures (see G. Labelle [186–187], Yang [331]).

Definition 1. Let F be a species of structures, and s, an F-structure on a finite set U (i.e., $s \in F[U]$). The structure s is called *asymmetric* if its stabilizer $\mathcal{G}_s$ is reduced to the identity, that is, if

$$\text{for any } \sigma \in \mathcal{S}[U], \ F[\sigma](s) = s \implies \sigma = \text{Id}_U. \tag{1}$$

It is equivalent to say that the size of the orbit of an asymmetric F-structure s on U under the action of $\mathcal{S}[U]$ is $|U|!$ (see Proposition A.1.6).

Definition 2. Let F be a species of structures. The *flat part* of F is the subspecies $\overline{F}$ of F defined, for any finite set U, by

$$\overline{F}[U] = \{s \in F[U] \mid s \text{ is asymmetric}\}. \tag{2}$$

One says that the species F is *flat* (or that F is *asymmetric*) if $\overline{F} = F$.

For the species X, E, L, $\mathcal{C}$, and $\mathcal{S}$ of singletons, sets, linear orders, oriented cycles, and permutations, respectively, it can be verified easily that

$$\overline{X} = X, \quad \overline{E} = 1 + X, \quad \overline{L} = L, \quad \overline{\mathcal{C}} = X, \quad \overline{\mathcal{S}} = 1 + X. \tag{3}$$

In particular, the species X and L are flat. The same holds for the species X^n, $n \geq 0$, since every subspecies of a flat species is flat. Also note that the only molecular species M which are flat are those whose atomic structure is X^n. Hence, a flat species of structures $G = G(X)$ can always be written in the form

$$G(X) = g_0 + g_1 X + g_2 X^2 + g_3 X^3 + \cdots \in \mathbb{N}[\![X]\!]. \tag{4}$$

Example 3. ASYMMETRIC ROOTED TREES. Consider the species $\mathcal{A} = XE(\mathcal{A})$ of rooted trees. Starting from the list of unlabeled rooted trees of order ≤ 5, we obtain the first few terms of the atomic decomposition of the species $\mathcal{A}$ (see

Example 2.6.5):

$$A = X + X^2 + X^3 + XE_2 + 2X^4 + X^2E_2 + XE_3$$
$$+ 3X^5 + 3X^3E_2 + X^2E_3 + X(E_2 \circ X^2) + XE_4 + \cdots. \quad (5)$$

This atomic decomposition can also be calculated by an iterative combinatorial procedure (see Sections 3.2 and 3.3). It follows that the decomposition of the flat part $\overline{A}$ of A is given by the series

$$\overline{A} = X + X^2 + X^3 + 2X^4 + 3X^5 + 6X^6 \cdots \in \mathbb{N}[\![X]\!]. \quad (6)$$

In fact, Harary and Prins (see [146]) have shown that the coefficients $a_n = [x^n]\overline{A}(x)$ satisfy the equation

$$\sum_{n \geq 1} a_n x^n = x \prod_{k \geq 1} (1 + x^k)^{a_k}, \quad (7)$$

which permits their recursive computation (see Example 13 and Exercise 14).

More generally, we propose to compute $\overline{F}$ when the species F is given explicitly in terms of simpler species or implicitly by a combinatorial functional equation. The main difficulty lies in the fact that the operation $F \mapsto \overline{F}$ does not commute with the usual combinatorial operations of the theory of species (except for sum and product). Recall (see Proposition 2.6.14) that the semiring Spe of species of structures is isomorphic to the semiring $\mathbb{N}[\![\mathfrak{A}]\!] = \mathbb{N}[\![X, E_2, E_3, C_3, \ldots]\!]$ of formal power series with coefficients in $\mathbb{N}$, on the set $\mathfrak{A}$ of atomic species. We then see that for any species of structures F, the flat part $\overline{F}$ of F is given by the sum of terms of the form $f_n X^n$ in the atomic decomposition of F.

The proof of the following proposition is left as an exercise.

Proposition 4. *The operator $F \mapsto \overline{F}$ defines an idempotent and surjective semiring homomorphism*

$$\mathbb{N}[\![\mathfrak{A}]\!] \twoheadrightarrow \mathbb{N}[\![X]\!],$$

that is, the following equalities are satisfied:

a) $\overline{F + G} = \overline{F} + \overline{G},$ b) $\overline{F \cdot G} = \overline{F} \cdot \overline{G},$ c) $\overline{\overline{F}} = \overline{F}.$ (8)

Moreover, the following relations are valid:

a) $\overline{F'} \supseteq \overline{F}',$ b) $\overline{F \times G} \supseteq \overline{F} \times \overline{G},$ c) $\overline{F \circ G} = \overline{F \circ \overline{G}} \supseteq \overline{F} \circ \overline{G},$
(9)

the inclusions being strict in general. ∎

Examples showing that inequalities (9) are strict in general are: $F = E_2$ for a), $F = E_2$ and $G = X^2$ for b), and $F = E_2$ and $G = X + X^2$ for c).

In order to make up for the noncommutativity of $F \mapsto \overline{F}$ for the combinatorial operations $\circ$ and $'$, we introduce a new tool: the *asymmetry index series* $\Gamma_F = \Gamma_F(x_1, x_2, \ldots)$ of a species F. This series has an infinite number of variables and plays, for asymmetric structures, a role analogous to that of the cycle index series $Z_F = Z_F(x_1, x_2, \ldots)$ for the enumeration of unlabeled structures.

Observe first that the flat part $\overline{F_w}$ of a weighted species can be defined as the subspecies of F_w formed of asymmetric F-structures. Since for any flat species of structures H, one has $H(x) = \widetilde{H}(x)$ (see Exercise 1), the *asymmetry generating series* $\overline{F_w}(x)$ can be interpreted as the type generating series of asymmetric F-structures. In other words, if $T(F)$ denotes the set of isomorphism types of F-structures, we have $T(F) \supseteq T(\overline{F})$ and

$$\widetilde{\overline{F_w}}(x) = \sum_{n \geq 0} |T(F_n)|_w x^n, \qquad \overline{F_w}(x) = \sum_{n \geq 0} |T(\overline{F_n})|_w x^n, \tag{10}$$

where F_n denotes the restriction of the species F to the cardinality n.

Let $X_\mathbf{t} = X_{t_1} + X_{t_2} + X_{t_3} + \cdots$ be the species of singletons colored by a color i and given weight t_i, for $i \in \{1, 2, 3, \ldots\}$. Recall (see Remark 4.3.8) that the cycle index series $Z_{F_w}(x_1, x_2, x_3, \ldots)$ is characterized by the equality

$$\widetilde{F_w(X_\mathbf{t})}(x)|_{x=1} = Z_{F_w}(p_1(\mathbf{t}), p_2(\mathbf{t}), p_3(\mathbf{t}), \ldots), \tag{11}$$

the "power sum" functions $p_k = p_k(\mathbf{t}) = t_1^k + t_2^k + t_3^k + \cdots$ forming an algebraic basis of the ring of symmetric functions. By analogy, since the series $\overline{F_w(X_\mathbf{t})}(x)$ is also a symmetric function in the variables $t_1, t_2, t_3, \ldots$, we give the following definition.

Definition 5. The *asymmetry index series* Γ_{F_w} of a weighted species $F = F_w$ is the unique series $\Gamma_{F_w}(x_1, x_2, x_3, \ldots)$ such that

$$\overline{F_w(X_\mathbf{t})}(x)|_{x=1} = \Gamma_{F_w}(p_1(\mathbf{t}), p_2(\mathbf{t}), p_3(\mathbf{t}), \ldots). \tag{12}$$

To better understand this definition, it is necessary to analyze the asymmetric $F(X_\mathbf{t})$-structures. An $F(X_\mathbf{t})$-structure on a set U is an F-structure $s \in F[U]$ supplied with a coloring $f : U \longrightarrow [\infty] = \{1, 2, 3, \ldots\}$, the weight of the structure (s, f) being given by

$$\gamma(s, f) = w(s) \prod_{u \in U} t_{f(u)}. \tag{13}$$

An automorphism of this structure (s, f) is a permutation $\sigma \in \mathcal{S}[U]$ which is an automorphism of s ($F[\sigma](s) = s$) and which leaves the coloring f fixed ($f = f \circ \sigma$), that is to say, that the coloring f must be constant on all the cycles of σ. The asymmetry of the structure (s, f) is then expressed by saying that if σ is an automorphism of s different from the identity, then there exists an element $u \in U$

such that $f(u) \neq f(\sigma^j(u))$, for some integer $j \geq 1$. In this case, we say that f is an F-*injective coloring* of the structure s. Denote by $F^{\text{inj}}\langle X_\mathbf{t}\rangle$ the subspecies of $F(X_\mathbf{t})$ formed by the $F(X_\mathbf{t})$-structures (s, f), where f is an F-injective coloring. We then have $\overline{F^{\text{inj}}\langle X_\mathbf{t}\rangle} = \overline{F(X_\mathbf{t})}$. In particular, the (isomorphism) type inventories are the same and we get

$$|T(F^{\text{inj}}\langle X_\mathbf{t}\rangle)|_\gamma = F^{\text{inj}}\langle X_\mathbf{t}\rangle\widetilde{\ }(x)|_{x=1} = \overline{F(X_\mathbf{t})}(x)|_{x=1} = \Gamma_F(p_1, p_2, p_3, \ldots). \quad (14)$$

The concept of F-injective coloring f of an F-structure s makes sense for any summable weighted set of colors (W, v). We use the same notation $F^{\text{inj}}\langle W\rangle$ to denote the species of pairs (s, f), where s is an F-structure on the set U and $f : U \longrightarrow W$ is an F-injective coloring, weighted by

$$\gamma(s, f) = w(s) \prod_{u \in U} v(f(u)). \quad (15)$$

Let $c = |W| \leq \infty$ be the cardinality of W. Take a bijection $\beta : [c] \longrightarrow W$ and set, for $i \leq c$, $v_i = v(\beta(i))$, the weight of the ith color of this bijection and, for $i > c$, if W is finite, $v_i = 0$. Denote by $t_i \mapsto v_i$, the simultaneous substitution of the v_i in place of the t_i, $i \in \{1, 2, 3, \ldots\}$. For example, we have

$$p_k(\mathbf{t})|_{t_i \mapsto v_i} = \sum_{i \leq c} v_i^k = |W|_{v^k}. \quad (16)$$

Composition with β induces an isomorphism $(s, f) \mapsto (s, \beta \circ f)$ of weighted species and we obtain by virtue of (14), for isomorphism types,

$$|T(F^{\text{inj}}\langle W\rangle)|_\gamma = |T(F^{\text{inj}}\langle X_\mathbf{t}\rangle)|_{\gamma, t_i \mapsto v_i}$$
$$= \Gamma_{F_w}(p_1(\mathbf{t}), p_2(\mathbf{t}), p_3(\mathbf{t}), \ldots)|_{t_i \mapsto v_i}$$
$$= \Gamma_{F_w}(|W|_v, |W|_{v^2}, |W|_{v^3}, \ldots). \quad (17)$$

Now let $G = G_v$ be another weighted species. Our objective is to compute $\overline{F_w \circ G_v}(x)$. In order to do this, consider $W = T_v(G)$ as the set of colors weighted by:

$$v(\mathcal{O}(s)) = v(s) x^{|U|}. \quad (18)$$

We have, for $k = 1, 2, 3, \ldots$,

$$|W|_{v^k} = \widetilde{G_{v^k}}(x^k). \quad (19)$$

Denote by $F_w^{\text{inj}}\langle G_v\rangle$, the weighted subspecies of $F_w \circ G_v$ formed of F-injective F-assemblies of G-structures, that is, for which the coloring induced by the types of G-structures that appear is F-injective. In other words, the only automorphism of an $F^{\text{inj}}\langle G\rangle$-structure are those which originate from isolated automorphisms of

G-structures in the F-assembly. Note that this operation is not to be confused with a substitution of species. Indeed, $F_w^{\text{inj}}\langle\ \rangle$ is not defined as a species of structures, but rather as an operator on species. This is what we wish to emphasize with the special notation $F_w^{\text{inj}}\langle G_v\rangle$.

Lemma 6. *Let $F = F_w$ and $G = G_v$ be two weighted species. Then the type generating series of F-injective assemblies of G-structures is given by*

$$\left(F_w^{\text{inj}}\langle G_v\rangle\right)\widetilde{}(x) = \Gamma_{F_w}\left(\widetilde{G_v}(x), \widetilde{G_{v^2}}(x^2), \widetilde{G_{v^3}}(x^3), \dots\right). \tag{20}$$

Proof. For isomorphism types, it is clear that

$$\mathrm{T}\left(F_w^{\text{inj}}\langle G_v\rangle\right) = \mathrm{T}\left(F_w^{\text{inj}}\langle \mathrm{T}_v(G)\rangle\right)$$

and that, taking into account the weightings (15) and (18) and using (17) and (19),

$$\left(F_w^{\text{inj}}\langle G_v\rangle\right)\widetilde{}(x) = \left|\mathrm{T}\left(F_w^{\text{inj}}\langle W\rangle\right)\right|_\gamma$$
$$= \Gamma_{F_w}\left(\widetilde{G_v}(x), \widetilde{G_{v^2}}(x^2), \widetilde{G_{v^3}}(x^3), \dots\right). \qquad\blacksquare$$

Observe now that an $F_w \circ G_v$-structure is asymmetric if and only if it is an F-injective F-assembly of G-structures which are themselves all asymmetric. In other words,

$$\overline{F_w \circ G_v} = F_w^{\text{inj}}\langle\overline{G_v}\rangle \tag{21}$$

and it follows, since $\overline{G}(x) = \overline{G}\widetilde{}(x)$, that

$$\overline{F_w \circ G_v}(x) = \left(F_w^{\text{inj}}\langle\overline{G_v}\rangle\right)\widetilde{}(x)$$
$$= \Gamma_{F_w}\left(\overline{G_v}(x), \overline{G_{v^2}}(x^2), \overline{G_{v^3}}(x^3), \dots\right).$$

Hence we obtain a characterization of asymmetry index series analogous to that of the cycle index series (see Theorem 4.3.7).

Theorem 7. *Let F_w be a species of $\mathbb{A}$-weighted structures. Then there exists a unique formal series $\Gamma_{F_w}(x_1, x_2, x_3, \dots) \in \mathbb{A}[\![x_1, x_2, x_3, \dots]\!]$ such that, for any weighted species G_v,*

$$\overline{F_w \circ G_v}(x) = \Gamma_{F_w}\left(\overline{G_v}(x), \overline{G_{v^2}}(x^2), \overline{G_{v^3}}(x^3), \dots\right). \tag{22}$$

In particular, for $G_v = X$, we obtain

$$\overline{F_w}(x) = \Gamma_{F_w}(x, x^2, x^3, \dots). \tag{23}$$

4.4. Asymmetric Structures

Moreover we have the following formula (see Exercise 7):

$$F_w(x) = \Gamma_{F_w}(x, 0, 0, \ldots), \tag{24}$$

in analogy with Theorem 1.2.8 for the cycle index series. The asymmetry index series also has the following properties.

Corollary 8. *For any weighted species* $F = F_w$ *and* $G = G_v$,

a) $\Gamma_{F+G} = \Gamma_F + \Gamma_G$,   b) $\Gamma_{F \cdot G} = \Gamma_F \cdot \Gamma_G$,

c) $\Gamma_{F'} = \dfrac{\partial}{\partial x_1} \Gamma_F$,   d) $\Gamma_{F \circ G} = \Gamma_F \circ \Gamma_G$ ($G(0) = 0$). $\tag{25}$

Proof. These properties follow from the characterization (23) of asymmetry index series. For a), it suffices to use (8), a) and b). The proof of c) is left as an exercise as well as that of d), which is similar to the proof of Theorem 4.3.9. We point out that the same notation is used for plethystic substitution, that is,

$$\Gamma_{F_w} \circ \Gamma_{G_v} = \Gamma_{F_w}\left((\Gamma_{G_v})_1, (\Gamma_{G_v})_2, (\Gamma_{G_v})_3, \ldots\right), \tag{26}$$

where we set

$$(\Gamma_{G_v})_k = \Gamma_{G_{v^k}}(x_k, x_{2k}, x_{3k}, \ldots). \blacksquare$$

It is interesting to observe that the series Γ_F and Z_F are independent in the sense that Γ_F is not a function of Z_F and vice versa. Indeed (see Exercise 9), it is possible to have $Z_F = Z_G$ without having $\Gamma_F = \Gamma_G$, and to have $\Gamma_F = \Gamma_G$ without having $Z_F = Z_G$. This can be partially explained by the recently observed fact that two nonisomorphic group actions can have the same lattice of periods (see Doran [79]).

Example 9. For the species X, E, L, C, and S, we have the following explicit formulas whose proofs are left as exercises:

a) $\Gamma_X(x_1, x_2, x_3, \ldots) = x_1$,

b) $\Gamma_L(x_1, x_2, x_3, \ldots) = \dfrac{1}{1 - x_1}$,

c) $\Gamma_E(x_1, x_2, x_3, \ldots) = \exp\left(\sum_{k \geq 1} (-1)^{k-1} \dfrac{x_k}{k}\right)$, $\tag{27}$

d) $\Gamma_S(x_1, x_2, x_3, \ldots) = \dfrac{1 - x_2}{1 - x_1}$,

e) $\Gamma_C(x_1, x_2, x_3, \ldots) = \sum_{n \geq 1} \dfrac{\mu(n)}{n} \log \dfrac{1}{1 - x_n}$.

Invoking Theorem 7 and Möbius inversion, we deduce the following results expressing the asymmetry-generating series of many species of enriched assemblies in terms of their members, and conversely:

— Asymmetric assemblies. If $F_w = E \circ G_v$, then

a) $\overline{F_w}(x) = \exp\left(\overline{G_v}(x) - \frac{1}{2}\overline{G_{v^2}}(x^2) + \frac{1}{3}\overline{G_{v^3}}(x^3) - \cdots\right),$ (28)

b) $\overline{G_v}(x) = \Lambda_w(x) + \Lambda_{w^2}(x^2) + \Lambda_{w^4}(x^4) + \Lambda_{w^8}(x^8) + \cdots,$

where $\Lambda_w(x) = \sum_{k \geq 1} \frac{\mu(k)}{k} \log \overline{F_{w^k}}(x^k)$.

— Asymmetric circular assemblies. If $F_w = \mathcal{C} \circ G_w$, then

a) $\overline{F_w}(x) = \sum_{k \geq 1} \frac{\mu(k)}{k} \log \frac{1}{1 - \overline{G_{v^k}}(x^k)},$

b) $\overline{G_v}(x) = 1 - \exp\left(-\sum_{k \geq 1} \frac{1}{k}\overline{F_{w^k}}(x^k)\right).$ (29)

— Asymmetric permuted assemblies. If $F_w = \mathcal{S} \circ G_v$, then

a) $\overline{F_w}(x) = \frac{1 - \overline{G_{v^2}}(x^2)}{1 - \overline{G_v}(x)},$

b) $\overline{G_v}(x) = 1 - \frac{1}{\overline{F_w}(x)\overline{F_{w^2}}(x^2)\overline{F_{w^4}}(x^4)\overline{F_{w^8}}(x^8)\cdots}.$ (30)

Example 10. Consider the species End of endofunctions and $\mathcal{A}$ of rooted trees. Since End $= \mathcal{S} \circ \mathcal{A}$, the following relations are immediately obtained:

a) $\overline{\text{End}}(x) = \frac{1 - \overline{\mathcal{A}}(x^2)}{1 - \overline{\mathcal{A}}(x)},$ b) $\overline{\mathcal{A}}(x) = 1 - \frac{1}{\overline{\text{End}}(x)\overline{\text{End}}(x^2)\overline{\text{End}}(x^4)\cdots}.$ (31)

The computation of the asymmetry index series Γ_M of a molecular species $M = X^n/H$, where H is a subgroup of $\mathcal{S}_n$, can be done using the following result (see G. Labelle [186]).

Proposition 11. Let $M = X^n/H$ be a molecular species, with $H \leq \mathcal{S}_n$. Then

$$\overline{M(X_t)}(x) = \sum_{n_1 + n_2 + \cdots = n} c_{n_1, n_2, \ldots}(H) \, t_1^{n_1} t_2^{n_2} \cdots x^n,$$ (32)

where

$$c_{n_1, n_2, \ldots}(H) = |\{\tau \epsilon \mathcal{S}_{n_1, n_2, \ldots} \backslash \mathcal{S}_n / H \mid \tau H \tau^{-1} \cap \mathcal{S}_{n_1, n_2, \ldots} = \{Id\}\}|,$$ (33)

the notation ϵ indicating that τ runs over a system of double coset representatives (see Exercise 2.6.16).

4.4. Asymmetric Structures

Table 4. *Asymmetry index polynomials for atomic species supported by $n \leq 4$ points.*

	F	$\Gamma_F(x_1, x_2, x_3, \ldots)$
$n = 1$	X	x_1
$n = 2$	E_2	$\frac{1}{2}x_1^2 - \frac{1}{2}x_2$
$n = 3$	E_3	$\frac{1}{6}x_1^3 - \frac{1}{2}x_1 x_2 + \frac{1}{3}x_3$
	C_3	$\frac{1}{3}x_1^3 - \frac{1}{3}x_3$
$n = 4$	E_4	$\frac{1}{24}x_1^4 - \frac{1}{4}x_1^2 x_2 + \frac{1}{3}x_1 x_3 + \frac{1}{8}x_2^2 - \frac{1}{4}x_4$
	$E_4^\pm$	$\frac{1}{12}x_1^4 - \frac{1}{3}x_1 x_3 - \frac{1}{4}x_2^2 + \frac{1}{2}x_4$
	$E_2 \circ E_2$	$\frac{1}{8}x_1^4 - \frac{1}{4}x_1^2 x_2 - \frac{1}{8}x_2^2 + \frac{1}{4}x_4$
	P_4^{bic}	$\frac{1}{4}x_1^4 - \frac{3}{4}x_2^2 + \frac{1}{2}x_4$
	C_4	$\frac{1}{4}x_1^4 - \frac{1}{4}x_2^2$
	$E_2 \circ X^2$	$\frac{1}{2}x_1^4 - \frac{1}{2}x_2^2$

Table 4 gives the polynomial Γ_F for any atomic species F of cardinality $n \leq 4$ (see also Appendix 2).

The asymmetry index series $\Gamma_{\mathcal{A}_R}$ of the species $Y = \mathcal{A}_R$ of R-enriched rooted trees can be computed explicitly, starting from the functional equation $Y = XR(Y)$, in a manner analogous to the cycle index series (see Proposition 3.2.14).

Proposition 12. *Let R be a species of structures and $\mathcal{A}_R = XR(\mathcal{A}_R)$ the corresponding species of R-enriched rooted trees. Then for any species of structures F and for any finite sequence $\mathbf{n} = (n_1, n_2, n_3, \ldots)$ of nonnegative integers,*

$$[\mathbf{x}^\mathbf{n}]\Gamma_{F(\mathcal{A}_R)}(\mathbf{x}) = [\mathbf{t}^\mathbf{n}]\Gamma_F(\mathbf{t}) \prod_{k \geq 1}\left(1 - \frac{t_1 \frac{\partial}{\partial t_1}\Gamma_R(\mathbf{t})}{\Gamma_R(\mathbf{t})}\right)_k (\Gamma_R(\mathbf{t}))_k^{\sigma_k}, \qquad (34)$$

where $\mathbf{x} = (x_1, x_2, x_3, \ldots)$, $\mathbf{t} = (t_1, t_2, t_3, \ldots)$, and $\mathbf{x}^\mathbf{n} = x_1^{n_1} x_2^{n_2} x_3^{n_3} \cdots$.

Example 13. ASYMMETRIC TREES AND ROOTED TREES. (See [144], [146], [187], and [215].) The species $\mathfrak{u} = \overline{\mathfrak{a}}$ and $U = \overline{\mathcal{A}}$ of trees and asymmetric rooted trees satisfy the functional equations

a) $U = XE^{\text{inj}}\langle U \rangle$,

b) $\mathfrak{u} = U + E_2^{\text{inj}}\langle U \rangle - U^2$, (dissymmetry formula)

(35)

obtained from the equation $\mathcal{A} = XE(\mathcal{A})$ and the dissymmetry formula for trees (see (4.0.2)). We deduce, for the enumeration of asymmetric trees and rooted trees, the relations

$$U(x) = x \exp\left(\sum_{k \geq 1} (-1)^{k-1} \frac{U(x^k)}{k}\right), \quad (36)$$

from which (7) follows, and

$$\mathfrak{u}(x) = U(x) - \frac{1}{2}(U(x^2) + U^2(x)). \quad (37)$$

After computation, we find

a) $U(x) = x + x^2 + x^3 + 2x^4 + 3x^5 + 6x^6 + 12x^7 + 25x^8 + 52x^9$
$+ 113x^{10} + 247x^{11} + 548x^{12} + 1226x^{13} + 2770x^{14}$
$+ 6299x^{15} + \cdots$, (38)

b) $\mathfrak{u}(x) = x + x^7 + x^8 + 3x^9 + 6x^{10} + 15x^{11} + 29x^{12} + 67x^{13}$
$+ 139x^{14} + 310x^{15} + \cdots$.

The asymmetry index series satisfy

a) $\Gamma_\mathcal{A} = x_1 \exp\left(\sum_{k \geq 1} (-1)^{k-1} \frac{(\Gamma_\mathcal{A})_k}{k}\right)$,

b) $\Gamma_\mathfrak{a} = \Gamma_\mathcal{A} - \frac{1}{2}(\Gamma_\mathcal{A})^2 - \frac{1}{2}(\Gamma_\mathcal{A})_2$,

(39)

and an application of the preceding proposition yields

$$\Gamma_\mathcal{A}(x_1, x_2, x_3, \ldots) = \sum_{\sigma_1 + 2\sigma_2 + \cdots < \infty} \gamma_\sigma \frac{x_1^{\sigma_1} x_2^{\sigma_2} x_3^{\sigma_3} \cdots}{1^{\sigma_1} \sigma_1! \, 2^{\sigma_2} \sigma_2! \, 3^{\sigma_3} \sigma_3! \cdots},$$

with $\gamma_\sigma = 0$, if $\sigma_1 = 0$, and, for $\sigma_1 \geq 0$,

$$\gamma_\sigma = \sigma_1^{\sigma_1 - 1} \prod_{k \geq 2} (\theta_k^{\sigma_k} - k\sigma_k \theta_k^{\sigma_k - 1}), \quad (40)$$

where

$$\theta_k = \sum_{d \mid k} (-1)^{(k/d)-1} d\sigma_d, \quad k \geq 1. \quad (41)$$

4.4. Asymmetric Structures

Moreover, if the asymmetry index series of a species R is given by

$$\Gamma_R(x_1, x_2, x_3, \ldots) = r_0 + r_1 x_1 + r_2 \frac{x_1^2}{2} + r_{01} \frac{x_2}{2} + \cdots$$

$$+ r_{n_1, n_2, \ldots} \frac{x_1^{n_1} x_2^{n_2} \cdots}{1^{n_1} n_1! 2^{n_2} n_2! \cdots} + \cdots,$$

then by solving equation $\Gamma_{A_R} = x_1 \Gamma_R(\Gamma_{A_R})$, we find

$$\Gamma_{A_R} = r_0 x_1 + r_0 r_1 x_1^2$$

$$+ \left(3r_0^2 r_2 + 6r_0 r_1^2\right) \frac{x_1^3}{6} + r_0 r_{01} \frac{x_1 x_2}{2}$$

$$+ \left(4r_0^3 r_3 + 36r_0^2 r_1 r_2 + 24r_0 r_1^3\right) \frac{x_1^4}{24}$$

$$+ \left(2r_0^2 r_{11} + 2r_0 r_1 r_{01}\right) \frac{x_1^2 x_2}{4} + r_0 r_{001} \frac{x_1 x_3}{3}$$

$$+ \left(5r_0^4 r_4 + 80r_0^3 r_1 r_3 + 60r_0^3 r_2^2 + 360r_0^2 r_1^2 r_2 + 120r_0 r_1^4\right) \frac{x_1^5}{120}$$

$$+ \left(3r_0^3 r_{21} + 12r_0^2 r_1 r_{11} + 6r_0^2 r_2 r_{01} + 6r_0 r_1 r_{01}\right) \frac{x_1^3 x_2}{12}$$

$$+ \left(2r_0^2 r_{101} + 2r_0 r_1 r_{001}\right) \frac{x_1^2 x_3}{6}$$

$$+ \left(r_0^2 r_{02} + 4r_0 r_1 r_{01}\right) \frac{x_1 x_2^2}{8} + r_0 r_{0001} \frac{x_1 x_4}{4}$$

$$+ \left(6r_0^5 r_5 + 150r_0^4 r_1 r_4 + 300r_0^4 r_2 r_3 + 1200r_0^3 r_1^2 r_3 \right.$$

$$\left. + 1800r_0^3 r_1 r_2^2 + 3600r_0^2 r_1^3 r_2 + 720r_0 r_1^5\right) \frac{x_1^6}{720}$$

$$+ \left(4r_0^4 r_{31} + 36r_0^3 r_1 r_{21} + 36r_0^3 r_2 r_{11} + 12r_0^3 r_{01} r_3 + 72r_0^2 r_1^2 r_{11}\right.$$

$$\left. + 72r_0^2 r_1 r_2 r_{01} + 24r_0 r_1^3 r_{01}\right) \frac{x_1^4 x_2}{48}$$

$$+ \left(3r_0^3 r_{201} + 12r_0^2 r_1 r_{101} + 6r_0^2 r_2 r_{001} + 6r_0 r_1^2 r_{001}\right) \frac{x_1^3 x_3}{18}$$

$$+ \left(2r_0^3 r_{12} + 8r_0^2 r_1 r_{11} + 2r_0^2 r_1 r_{02} + 4r_0^2 r_{01} r_{11} + 8r_0 r_1^2 r_{01}\right) \frac{x_1^2 x_2^2}{16}$$

$$+ \left(2r_0^2 r_{1001} + 2r_0 r_1 r_{0001}\right) \frac{x_1^2 x_4}{8} + r_0^2 r_{011} \frac{x_1 x_2 x_3}{6} + r_0 r_{00001} \frac{x_1 x_5}{5}$$

$$+ \cdots. \tag{42}$$

Substituting in this expression the explicit coefficients of the asymmetry index series of a given species, we obtain the asymmetry index series for any R-enriched rooted trees. The expression above is thus the generic asymmetric index series for R-enriched rooted trees. Appendix 2 also contains the generic asymmetric index series $\overline{\Gamma_{a_R}}$ for R-enriched trees as well as the generic series $\overline{\mathcal{A}_R}(x)$ and $\overline{a_R}(x)$ up to order 6.

Exercises for Section 4.4

1. a) Show that for any species of structures H, the following properties are equivalent:
      i) The species H is flat, i.e., all the H-structures are asymmetric.
      ii) $H(x) = \widetilde{H}(x)$.
      iii) $Z_H(x_1, x_2, x_3, \ldots) = H(x_1)$.
   b) Show moreover that if the species H is flat and if $H(x) = \sum_{n \geq 0} a_n x^n$, then the molecular decomposition of H is $H(X) = \sum_{n \geq 0} a_n X^n$.

2. Prove Proposition 4 and verify the examples which are given, showing that inequalities (9) are strict in general.

3. INJECTIVE ASSEMBLIES. Let $G = G_v$ be a weighted species.
   a) Show that $E^{\mathrm{inj}}\langle G \rangle$ is the species of *injective* assemblies of G-structures, that is, assemblies of pairwise nonisomorphic G-structures.
   b) (See [263], and [144], Section 2.6.) Show that, for $n \geq 2$,

   $$E_n^{\mathrm{inj}}\langle G \rangle \widetilde{}(x) = \left(E_n^{\pm} \circ G\right)\widetilde{}(x) - \left(E_n \circ G\right)\widetilde{}(x). \tag{43}$$

   HINT: Show that $(E_n^{\pm} \circ G)\widetilde{}(x)$ counts unlabeled injective assemblies of G-structures twice, but the noninjective ones only once, since in this case there would be an isomorphism between the two orientations.

   c) Deduce from b) and formula (2.6.44), iii), expression (27), c) for the asymmetric index series Γ_E.
   d) Show that

   $$E^{\mathrm{inj}}\langle G_v \rangle\widetilde{}(x) = \prod_{s \in T(G)} (1 + v(s) x^{n(s)}), \tag{44}$$

   where s runs over a system of representatives of types of G-structures and $n(s)$ denotes the order of s. In particular, if G is not weighted and $\widetilde{G}(x) = \sum_{n \geq 1} \widetilde{g}_n x^n$, then

   $$E^{\mathrm{inj}}\langle G_v \rangle\widetilde{}(x) = \prod_{k \geq 1} (1 + x^k)^{\widetilde{g}_k}. \tag{45}$$

   e) Setting $G_v = X_t = X_{t_1} + X_{t_2} + X_{t_3} + \cdots$, deduce from (44) the following identity which implies (27), c):

   $$\sum_{n \geq 0} e_n x^n = \exp\left(\sum_{k \geq 1} (-1)^{k-1} \frac{p_k x^k}{k}\right), \tag{46}$$

where $e_n = e_n(\mathbf{t})$ denotes the elementary symmetric function of degree n:

$$e_n = \sum_{j_1 < j_2 < \cdots < j_n} t_{j_1} t_{j_2} \cdots t_{j_n}. \tag{47}$$

4. Show that the *power sum* symmetric functions

$$p_k = t_1^k + t_2^k + t_3^k + \cdots, \qquad k \geq 1, \tag{48}$$

form an algebraic basis over $\mathbb{Q}$ in the ring Λ of symmetric functions.
HINT: A partition λ of an integer n is a decreasing sequence $\lambda_1 \geq \lambda_2 \geq \cdots \geq \lambda_k$ (for arbitrary k), of positive integers, such that $\lambda_1 + \lambda_2 + \cdots + \lambda_k = n$. One denotes by $\lambda \vdash n$ the fact that λ is a partition of n. For some families (multiplicative basis) of symmetric functions $(f_k)_{k=1,2,\ldots}$, it is convenient to denote by f_λ the symmetric function $f_{\lambda_1} f_{\lambda_2} \cdots f_{\lambda_k}$, for each partition λ of an integer n. Use (46) of the preceding exercise to show that the families $(p_\lambda)_{\lambda \vdash n}$ linearly generate the space Λ_n of homogeneous symmetric functions of degree n, given that the family $(e_\lambda)_{\lambda \vdash n}$ generates it (see Macdonald [228], Chapter 1). Then use the fact that the dimension of Λ_n is equal to the number of partitions of n. Conclude that the p_k are algebraically independent.

5. Prove that the transformation $F \mapsto \Gamma_F$ preserves the differentiation and substitution operations (see (25), c) and d)).
HINT: For (25), c), add an auxiliary variable t_0 to the variables $t_1, t_2, \ldots$. Then introduce the limited Taylor expansion

$$F(X_{t_0} + X_{t_1} + X_{t_2} + \cdots) = F(X_{t_1} + X_{t_2} + \cdots)$$
$$+ X_{t_0} F(X_{t_1} + X_{t_2} + \cdots) + O(t_0^2), \tag{49}$$

where $O(t_0^2)$ denotes a species whose structures have a weight divisible by t_0^2. Finally, take the type generating series of the two sides of (49) and identify the terms which are linear in t_0. For d), invoke Theorem 7 and proceed as in the proof of $Z_{F \circ G} = Z_F \circ Z_G$ (see Theorem 4.3.9).

6. a) Prove formulas (27), a), b), d), and e).
   HINT: For d) and e), invoke respectively c) and g) of Exercise 2.3.7. Formula e) can also be established by using c) and d) in the formula $\Gamma_S = \Gamma_E \circ \Gamma_C$ and appealing to property (2.3.84) of the Möbius function.
   b) Prove formulas (28), (29), and (30).
   HINT: For (28), b), use formula (2.3.84) to first show that

$$\Lambda_w(x) = \overline{G_v}(x) - \overline{G_{v^2}}(x^2). \tag{50}$$

7. Let $M = X^n/H$, be a molecular species, where H is a subgroup of S_n.
   a) Prove Proposition 11.
      HINT: Invoke Exercise 2.6.16 with $X_i := X_{t_i}$.
   b) (See Kerber [168], G. Labelle, J. Labelle, and Pineau [190], Rota [285], and Stockmeyer [307].) Let $\mu = \mu(U, V)$ be the Möbius function of the lattice

of subgroups of the group H. Show that

$$\Gamma_M(x_1, x_2, x_3, \ldots) = \frac{1}{|H|} \sum_{V \leq H} \mu(\{1\}, V) x_1^{V_1} x_2^{V_2} \cdots \qquad (51)$$

where $V \leq H$ denotes that V is a subgroup of H, 1 is the identity element of H, and V_i, for $i \geq 1$, denotes the number of orbits with i elements with respect to the natural action of V on $[n]$.

   c) Prove formula (24).
   HINT: Assume first that F_w is molecular and use (51).

8. Starting from the combinatorial description of atomic species of order $n \leq 4$ and from (12),
   a) Verify Table 4 giving the asymmetry index series of atomic species of order $n \leq 4$.
   b) Compute the asymmetry index series of the molecular species of order $n \leq 4$.
   HINT: For $M = P \cdot Q$, invoke the formula $\Gamma_M = \Gamma_P \cdot \Gamma_Q$.

9. INDEPENDENCE BETWEEN Z_F AND Γ_F.
   a) Show that the species

$$F = 2E_3 + X^3, \qquad G = 2XE_3 + C_3 \qquad (52)$$

satisfy

$$Z_F = Z_G \quad \text{and} \quad \Gamma_F \neq \Gamma_G. \qquad (53)$$

   b) Show that the species

$$F = XE_2, \qquad G = E_3 + C_3 \qquad (54)$$

satisfy

$$Z_F \neq Z_G \quad \text{and} \quad \Gamma_F = \Gamma_G. \qquad (55)$$

10. Show that two *nonisomorphic* species F and G can have the same cycle index series ($Z_F = Z_G$) and the same asymmetry index series ($\Gamma_F = \Gamma_G$).
    HINT: Take, for example, $F = E_4 + E_2 \circ E_2 + X^2 E_2$ and $G = XE_3 + E_2^2 + C_4$.

11. (G. Labelle and Pineau [197].) Show that the asymmetry index series of the species $E^\pm$ of oriented sets is given by

   i) $\Gamma_{E^\pm}(x_1, x_2, \ldots) = (2 + x_2 - x_3 + x_4 - x_5 + \cdots) \Gamma_E(x_1, x_2, \ldots) - 1 - x_1,$

   ii) $\Gamma_{E^\pm}(x_1, x_2, \ldots) = 1 + x_1 + \sum_{n \geq 2} \frac{1}{n!} \sum_{\sigma \in S_n} (2 + \sigma_1 - n) \cdot \text{sgn}(\sigma) x_1^{\sigma_1} x_2^{\sigma_2} \cdots.$

$$\qquad (56)$$

HINT: Invoke Proposition 11 and note that, for $n \geq 2$, $n_1 + n_2 + \cdots = n$, one has $A_n \cap S_{n_1, n_2, \ldots} = \{1\}$ if and only if, for any i, $n_i \leq 1$ or there exists a unique j such that $n_j = 2$ and, for $i \neq j$, $n_i \leq 1$ (A_n denotes the alternating group).

12. a) Show that for species F and G, it is possible to have

$$\Gamma_{F \times G} \neq \Gamma_F \times \Gamma_G, \tag{57}$$

by taking, for example, $F = G = C_3$.

b) Show that there does not exist a function $\omega(u, v)$ such that, for any species F and G one has $\Gamma_{F \times G} = \omega(\Gamma_F, \Gamma_G)$.
HINT: Show that for $F = G = XE_2$ and $\Phi = \Psi = E_3 + C_3$, one has $\Gamma_F = \Gamma_\Phi$ and $\Gamma_G = \Gamma_\Psi$, but $\Gamma_{F \times G} \neq \Gamma_{\Phi \times \Psi}$.

c) Show that for any molecular species $M = X^n/H$ and $N = X^n/K$, where H and K are subgroups of S_n, one has (see (2.6.50))

$$\overline{M \times N}(x) = \gamma(H, K) x^n, \tag{58}$$

where

$$\gamma(H, K) = |\{\tau \underline{\in} H \backslash S_n / K \mid |H \cap \tau K \tau^{-1}| = 1\}|. \tag{59}$$

13. a) Prove Proposition 12.
HINT: Invoke Good–Lagrange multidimensional inversion, as in the proof of Proposition 3.2.14.

b) Show that

$$\Gamma_{\mathfrak{a}_R} = \int_0^{x_1} \Gamma_{R(A_{R'})}(\xi_1, x_2, x_3, \ldots) d\xi_1 - \frac{1}{2} \Gamma_{A_{R'}}(x_2, x_4, x_6, \ldots). \tag{60}$$

c) Let $\sigma = (\sigma_1, \sigma_2, \ldots, \sigma_n)$ be a vector of integers. Deduce from (60) the following formula for the coefficients $\alpha_{\sigma_1, \sigma_2, \ldots} = \text{coeff}_\sigma \Gamma_{\mathfrak{a}_R}$ of the asymmetry index series of the species $\mathfrak{a}_R$ of R-enriched trees:

$$\alpha_{\sigma_1, \sigma_2, \ldots} = \begin{cases} \omega_{\sigma_1 - 1, \sigma_2, \ldots}, & \text{if } \sigma_1 \neq 0, \\ -2\left(\sum_k \sigma_{2k}\right)^{-1} b_{\sigma_2, \sigma_4, \ldots}, & \text{if } 0 = \sigma_1 = \sigma_3 = \cdots, \\ 0, & \text{otherwise}, \end{cases} \tag{61}$$

where $\omega_\sigma = \text{coeff}_\sigma \Gamma_R \Delta$ and $b_\sigma = \text{aut}(\sigma)[\mathbf{x}^\sigma] x_1 \Delta$, with

$$\Delta = \prod_{k \geq 1} \left(1 - \frac{x_1 \partial^2 \Gamma_R / \partial x_1^2}{\partial \Gamma_R / \partial x_1}\right)_k (\partial \Gamma_R / \partial x_1)^{\sigma_k}. \tag{62}$$

14. a) Let $U(x) = \overline{A}(x) = \sum_n a_n x^n$. Show that relations (7) and (36) are equivalent. Deduce a recurrence relation for the coefficients of $U(x)$.

b) Use (37) to express the number of types of asymmetric trees of order n as a function of a_n (see Exercise 4.1.1).

c) Prove formula (40) for γ_σ and express the coefficient $\alpha_\sigma = \text{coeff}_\sigma \Gamma_\mathfrak{a}$ as a function of γ_σ.

d) Show that $\varepsilon_\sigma = \text{coeff}_\sigma \Gamma_{\text{End}}$ is given by the formula

$$\varepsilon_\sigma = \sigma_1^{\sigma_1}\left(\theta_2^{\sigma_2} - 4\sigma_2\theta_2^{\sigma_2-1} + 4\sigma_2(\sigma_2-1)\theta_2^{\sigma_2-2}\right)\prod_{k\geq 3}\left(\theta_k^{\sigma_k} - k\sigma_k\theta_k^{\sigma_k-1}\right), \quad (63)$$

where θ_k is defined by (41).

e) Extend the results of Exercise 4.1.15 to trees, rooted trees, and asymmetric endofunctions.

15. ASYMMETRIC PLANE TREES AND ROOTED TREES.

a) Starting from identities (3.1.18) and (4.1.24), deduce the fundamental relations for the flat parts $\overline{\mathcal{P}}$ and $\overline{p}$ and the asymmetry index series $\Gamma_\mathcal{P}$ and Γ_p of the species $\mathcal{P}$ of rooted plane trees, and p, plane trees.

b) Conclude, in analogy with (3.1.87) and (4.1.48), that the numbers $\overline{p}_n$ and $\overline{\mathfrak{p}}_n$, of unlabeled asymmetric plane rooted trees and asymmetric plane trees, respectively, are given, for $n \geq 2$, by

i) $\displaystyle \overline{p}_n = \frac{1}{2(n-1)}\sum_{d|(n-1)}\mu((n-1)/d)\binom{2d}{d},$

ii) $\displaystyle \overline{\mathfrak{p}}_n = \overline{p}_n - \frac{1}{2}c_{n-1} - \frac{1}{2}\chi_{\text{even}}(n)c_{\frac{n}{2}-1}.$

(64)

c) (G. Labelle and Leroux [194].) Give formulas analogous to (3.1.113) and (4.1.50) for unlabeled asymmetric plane trees and rooted trees according to degree distribution.

d) [194] Give a formula analogous to (3.2.105) for unlabeled asymmetric bicolored plane trees according to the double distribution of degrees.

16. Let F_v be a weighted species of structures. Denote by $E_{=2}\langle F_v\rangle$, the subspecies of $E_2\langle F_v\rangle$ formed of pairs of isomorphic F_v-structures, and for $k \geq 1$, by $\mathcal{C}_{\geq k}\langle F_v\rangle$, the species of oriented cycles of F_v-structures having a circular symmetry of order k, and by $\mathcal{C}_{=k}\langle F_v\rangle$ the subspecies of $\mathcal{C}_{\geq k}\langle F_v\rangle$ containing oriented cycles of F_v-structures whose cyclic automorphism group is exactly of order k. These three species are weighted by defining the weight w as the product of the v-weights of the F-structures that appear.

a) Show that (Stockmeyer [307])

i) $E_{=2}\langle F_v\rangle^\sim(x) = \widetilde{F_{v^2}}(x^2),$

ii) $\mathcal{C}_{\geq k}\langle F_v\rangle^\sim(x) = Z_C\left(\widetilde{F_{v^k}}(x^k), \widetilde{F_{v^{2k}}}(x^{2k}), \ldots\right),$

(65)

iii) $\mathcal{C}_{=k}\langle F_v\rangle^\sim(x) = \Gamma_C\left(\widetilde{F_{v^k}}(x^k), \widetilde{F_{v^{2k}}}(x^{2k}), \ldots\right).$

HINT: Analyze unlabeled structures of the species $F_{v^2}(X^2)$, $C(F_{v^k}(X^k))$, and $C^{\text{inj}}\langle F_{v^k}(X^k)\rangle$.

b) [194] Apply these results to the enumeration of unlabeled plane uni- or bi-colored trees and rooted trees (according to degree distributions) when the automorphism group, necessarily cyclic, is of order h equal to (or a multiple of) an integer $k \geq 1$.

17. Let $\mathfrak{h}$ be the species of homeomorphically irreducible trees (topological trees, see Example 4.1.3). Show that

$$\overline{\mathfrak{h}}(x) = \Gamma_\mathfrak{a}\left(\frac{x}{1+x}, \frac{x^2}{1+x^2}, \ldots\right) + x\Gamma_\mathcal{A}\left(\frac{x}{1+x}, \frac{x^2}{1+x^2}, \ldots\right)$$
$$- \frac{x}{2}\Gamma_\mathcal{A}^2\left(\frac{x}{1+x}, \frac{x^2}{1+x^2}, \ldots\right) + \frac{x}{2}\Gamma_\mathcal{A}\left(\frac{x^2}{1+x^2}, \frac{x^4}{1+x^4}, \ldots\right),$$
(66)

where $\mathcal{A}$ and $\mathfrak{a}$ denote respectively the species of rooted trees and that of trees.
HINT: Use formula (4.1.20).

18. Let $\mathcal{A}_S$ be the species of S-enriched rooted trees, where S is the species of permutations, and set $\overline{\mathcal{A}_S}(x) = \sum_n \overline{a}_n x^n$. Show that $\overline{a}_0 = 0$, and that, for $n \geq 0$,

$$\overline{a}_{n+1} = (\overline{a}_1\overline{a}_n + \overline{a}_2\overline{a}_{n-1} + \cdots + \overline{a}_n\overline{a}_1) - X(n \text{ even})\overline{a}_{n/2}. \tag{67}$$

19. a) Let Inv be the species of involutions and set $J = \text{Inv} \times \text{Inv}$. Show that

$$\overline{J}(x) = \prod_{k \geq 0}(1 + x^{2k+1}) = \sum_{n \geq 0} p^*(n)x^n, \tag{68}$$

where $p^*(n)$ is the number of partitions of n into distinct odd parts.

20. a) (G. Labelle and Pineau [197].) Show that the species A, of even permutations, admits the expression

$$\text{A} = E(\mathcal{C}_1 + \mathcal{C}_3 + \mathcal{C}_5 + \cdots)E_{\text{even}}(\mathcal{C}_2 + \mathcal{C}_4 + \mathcal{C}_6 + \cdots). \tag{69}$$

b) Deduce that

$$\Gamma_\text{A} = \frac{2 - x_1^2 - x_2}{2(1-x_1)}. \tag{70}$$

NOTE: Formula (70) shows that Γ_A only depends on x_1 and x_2, just as $\Gamma_S = (1-x_2)/(1-x_1)$.

21. Let H be a virtual species whose molecular decomposition contains a single copy of the species X. If $G = H \circ F$, describe the computation procedures for Γ_F in terms of Γ_G and Γ_H,

a) by using the formula $F = H^{(-1)} \circ G$, with

$$H^{(-1)} = X - \Delta_H X + \Delta_H^2 X - \Delta_H^3 X + \cdots, \tag{71}$$

the operator Δ_H being defined, for a species K, by $\Delta_H K = K \circ H - K$ (see Proposition 2.5.19).

b) by adapting Newton–Raphson iteration to the context of index series applied to the equation $(h \circ f)(x_1, x_2, x_3, \ldots) - g(x_1, x_2, x_3, \ldots) = 0$, with $f = \Gamma_F$ as unknown and $g = \Gamma_G$ and $h = \Gamma_H$ two given series.

22. HEREDITARY FINITE SETS AND ASYMMETRIC ROOTED TREES. (Meir, Moon and Mycielsky [243].) The family V_ω of hereditary finite sets is defined by $V_\omega = \bigcup_{n \geq 0} V_n$, where the V_n are defined recursively by $V_0 = \emptyset$, and, for $n \geq 0$,

$$V_{n+1} = \{U \mid U \subseteq V_n\}. \tag{72}$$

a) Show that the function $\theta : V_\omega \longrightarrow \mathbb{N}$ defined by

$$\theta(U) = \sum_{W \in U} 2^{\theta(W)} \tag{73}$$

is, in fact, a bijection (in particular, $\theta(\emptyset) = 0$).

b) Construct a bijection β between the set V_ω and the set of unlabeled asymmetric rooted trees.

HINT: For $n \geq 0$, set $H_n = \theta^{-1}(n)$. Show that $V_\omega = \{H_0, H_1, H_2, \ldots\}$ with $H_0 = \emptyset$, and, for $H_m = \{H_{j_1}, \ldots, H_{j_k}\}$, with $0 \leq j_1 < j_2 < \cdots < j_k$, set $\beta(H_m)$ equal to the unlabeled rooted tree obtained by joining the sequence of $\beta(H_{j_i})$, $i = 1, \ldots, k$, to a common root.

23. Show that the number $\bar{a}_n$ of unlabeled asymmetric rooted trees on n vertices has an asymptotic equivalent of the form

$$\bar{a}_n \sim \alpha \beta^n n^{-3/2}, \tag{74}$$

for certain constants α and β.
HINT: See Exercise 3.4.29, a).

NOTE: Harary, Robinson, and Schwenk [148] have shown that

$$\alpha \approx 0.3625, \qquad \beta \approx 2.5176. \tag{75}$$

Meir, Moon, and Mycielski [243] have also established that the limiting expected value, as $n \longrightarrow \infty$, of the degree of the root of a random asymmetric rooted tree with n vertices is given by

$$2 - \sum_{k=2}^{\infty} (-1)^k \overline{A}(\xi^k) \approx 1.8602 \tag{76}$$

4.4. Asymmetric Structures

and that the limiting probability that the root is of degree k is

$$\xi = \Gamma_{E_{k-1}}(\overline{A}(\xi), \overline{A}(\xi^2)\overline{A}(\xi^3), \ldots), \qquad k \geq 1. \tag{77}$$

24. **q-SERIES ASSOCIATED TO A SPECIES.** (Décoste [70–71].) Let q and x be two formal variables. A q-series in x is a series of the form (see (5.1.16) for the definition of $n!_q$)

$$f(x;q) = \sum_{n \geq 0} f_n(q) \frac{x^n}{n!_q}, \tag{78}$$

where the coefficients $f_n(q)$ are functions of q. If $\lim_{q \to 1} f(x;q) = f(x) = \sum_n f_n x^n/n!$, the series $f(x;q)$ is said to be a q-analogue of $f(x)$, and that the sequence $(f_n(q))_{n \geq 0}$ is a q-analogue of the sequence $(f_n)_{n \geq 0}$. Two classical q-analogues of the exponential series are:

$$E(x;q) = \sum_{n \geq 0} \frac{x^n}{n!_q}, \qquad E\langle x;q\rangle = \sum_{n \geq 0} q^{\binom{n}{2}} \frac{x^n}{n!_q}. \tag{79}$$

Every species of structures F determines two q-series as follows:

$$F(x;q) = Z_F\left(\frac{(1-q)x}{(1-q)}, \frac{(1-q)^2 x^2}{(1-q^2)}, \frac{(1-q)^3 x^3}{(1-q^3)}, \ldots\right), \tag{80}$$

$$F\langle x;q\rangle = \Gamma_F\left(\frac{(1-q)x}{(1-q)}, \frac{(1-q)^2 x^2}{(1-q^2)}, \frac{(1-q)^3 x^3}{(1-q^3)}, \ldots\right). \tag{81}$$

a) Show that

$$\begin{array}{ll} \text{i)} \quad \lim_{q \to 1} F(x;q) = F(x), & \text{ii)} \quad \lim_{q \to 0} F(x;q) = \widetilde{F}(x), \\ \text{iii)} \quad \lim_{q \to 1} F\langle x;q\rangle = F(x), & \text{iv)} \quad \lim_{q \to 0} F\langle x;q\rangle = \overline{F}(x). \end{array} \tag{82}$$

In particular, if we set

$$F(x;q) = \sum_{n \geq 0} f_n(q) x^n/n!_q, \qquad F\langle x;q\rangle = \sum_{n \geq 0} f_n\langle q\rangle x^n/n!_q,$$

then the sequences $(f_n(q))_{n \geq 0}$ and $(f_n\langle q\rangle)_{n \geq 0}$ give two q-analogues of the sequence $(f_n)_{n \geq 0}$ enumerating F-structures.

b) Show that the two q-series associated to the species E of sets coincide with the two classical q-analogues (79) of the exponential.

c) Show that taking the species $F = E_k E$ of subsets of cardinality k gives

$$f_n(q) = \begin{bmatrix} n \\ k \end{bmatrix}_q, \qquad f_n\langle q\rangle = \begin{bmatrix} n \\ k \end{bmatrix}_q q^{\binom{k}{2}+\binom{n-k}{2}}, \tag{83}$$

where

$$\begin{bmatrix} n \\ k \end{bmatrix}_q = \frac{n!_q}{k!_q (n-k)!_q}. \qquad (84)$$

d) Show that for $F = L$, $f_n(q) = f_n\langle q \rangle = n!_q$.

e) Show that the operations $F \mapsto F(x; q)$ and $F \mapsto F\langle x; q \rangle$ preserve the sum and the product, and that

$$(F \circ G)(x; q) = Z_F(G(x_1; q), G(x_2; q^2), \ldots)\big|_{x_i := (1-q)^i x^i / (1-q^i)},$$
$$(F \circ G)\langle x; q \rangle = \Gamma_F(G\langle x_1; q \rangle, G\langle x_2; q^2 \rangle, \ldots)\big|_{x_i := (1-q)^i x^i / (1-q^i)}. \qquad (85)$$

f) Show that, for any species F, the functions $f_n(q)$ and $f_n\langle q \rangle$ are polynomials in q with integer coefficients (positive for $f_n(q)$), and of degree $\leq \binom{n}{2}$.
NOTE: Moreover, for each partition $\lambda \vdash n$, let $t_\lambda = t_\lambda(F)$ be equal to the number of orbits of the canonical action $S_\lambda \times F[n] \longrightarrow F[n]$ of the Young subgroup $S_\lambda = S_{\lambda_1} \times S_{\lambda_2} \times \cdots$ of S_n on the set $F[n]$ of F-structures on $[n]$. Then each coefficient of q^k in $f_n(q)$ is a linear combination of the t_λ, $\lambda \vdash n$ with *universal* coefficients independent of F (see Décoste and G. Labelle [72]).

25. (G. Labelle and Leroux [195].) Recall the definition of the connected components weighting species as $\Lambda^{(\alpha)} = E \circ X_\alpha \circ E_+^{\langle -1 \rangle}$ (see Exercices 2.3.10 and 2.3.12).

a) Show that

i) $\overline{\Lambda^{(\alpha)}}(x) = \prod_{k \geq 1} (1 + x^k)^{\gamma_k(\alpha)}$,

ii) $\Gamma_{\Lambda^{(\alpha)}}(x_1, x_2, x_3, \ldots) = \prod_{k \geq 1} (1 + x_k)^{\gamma_k(\alpha)}$, $\qquad (86)$

where

$$\gamma_k(\alpha) = -\lambda_k(-\alpha) - \lambda_{k/2}(-\alpha) - \lambda_{k/4}(-\alpha) - \cdots. \qquad (87)$$

b) Deduce explicit formulas for the series $\overline{F_{w^{(\alpha)}}}(x)$ and $\Gamma_{F_{w^{(\alpha)}}}(x_1, x_2, x_3, \ldots)$.

5

Species on Totally Ordered Sets

5.0. Introduction

In many applications of enumerative and algebraic combinatorics, the structures depend on a given total (linear) order on the underlying set. Good examples are provided by alternating permutations or standard Young tableaux. Similarly, in computer science, data structures often make use of some explicit order on the set of data, as in the case of searching or sorting algorithms. For instance, in a *binary search tree*, it is required that each vertex (identified here with its label) be larger (respectively smaller) than all the vertices of its left subtree (respectively right subtree). See Figure 1.

In order to take into account such constructions, we now develop a variant of the theory of species of structures, the $\mathbb{L}$-*species*, where the structures are supported by sets with a total order. This theory takes its source in the work of Foata [106], [112], Schützenberger [293], and Viennot [317, 318]. In contrast with ordinary species (henceforth called $\mathbb{B}$-species), any $\mathbb{L}$-species admits a canonical primitive (anti-derivative), and differential equations admit unique solutions, following more closely the properties of formal power series (see Leroux and Viennot [217, 218]). For example, the differential equation

$$Y' = R(Y), \quad Y(0) = 0, \qquad (1)$$

for any $\mathbb{L}$-species R, admits the canonical solution $Y = \mathcal{A}_R^\uparrow$ of *increasing R-enriched rooted trees*, that is, whose vertex (labels) are increasing on any path from the root to the leaves. See Figure 5.1.1. More generally, any differential equation, or system, of the form

$$Y' = H(X, Y), \quad Y(0) = Z, \qquad (2)$$

has a unique solution which can be described in terms of increasing enriched rooted trees.

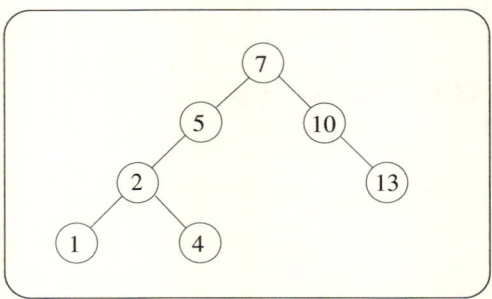

Fig. 1.

Another major difference with $\mathbb{B}$-species is the transport of structures of $\mathbb{L}$-species which is carried out only along increasing bijections. As a consequence, the structures have no symmetries and the notions of automorphism or of isomorphism type (nonlabeled structures) become meaningless. However, the theory is well adapted to the analysis of combinatorial structures on totally ordered sets.

Essentially, an $\mathbb{L}$-species is a functor from the category $\mathbb{L}$ of totally ordered sets and increasing bijections to the category of weighted sets. In Section 1 are found the basic concepts of the theory, including the combinatorial operations of addition, multiplication, (partitional) composition, and derivation of $\mathbb{L}$-species, as well as two new operations: integration and convolution. This development follows closely the classical algebraic, differential and integral calculus on formal series, while presenting it in a new light.

Section 2 develops a general combinatorial theory of (systems of) differential equations in the context of $\mathbb{L}$-species. In particular, the classical results concerning autonomous and nonautonomous differential equations are recovered, including the Lie–Gröbner–Taylor expansions, the linear differential equations, and the method of separation of variables, from a combinatorial point of view.

The exercises contain some simple integrals easy to deduce combinatorially, a combinatorial Taylor expansion with integral remainder, an extension of the method of separation of variables with an application to equations appearing in control theory, and some results on asymptotic analysis of increasing enriched rooted trees. Some of these applications are particularly interesting for the development of computer algebra methods in nonlinear dynamical systems.

5.1. $\mathbb{L}$-Species

Totally Ordered Sets

A *totally ordered set* is a pair $\ell = (U, \preceq)$, where U is a finite set and $\preceq$ is a total order relation on U. In other words, $\preceq$ is an L-structure on U. For u and v in U, we write $u \preceq v$ or $u \preceq_\ell v$ when u is smaller than v with respect to the order of ℓ,

5.1. 𝕃-Species

and $u \prec v$ if $u \preceq v$ and $u \neq v$. An element $u \in \ell$ is, more precisely, an element of the underlying set U of ℓ. The smallest element of a nonempty totally ordered set ℓ is denoted by $\min \ell$:

$$\min \ell = \operatorname{minimum}\{x \mid x \in \ell\}. \tag{1}$$

Let $\ell = (U, \preceq)$ be a totally ordered set. Each subset V of U is naturally ordered according to the order *induced* on V by $\preceq$. The corresponding totally ordered set is denoted by $\ell|_V$, and $\ell|_V$ is said to be the *restriction* of ℓ to V. For subsets $U_1, U_2, \ldots, U_k$ of U such that $U_1 + U_2 + \cdots + U_k = U$ (i.e., $U_i \cap U_j = \emptyset$, if $i \neq j$, and $\bigcup_{i=1}^k U_i = U$), we write

$$\ell_1 + \ell_2 + \cdots + \ell_k = \ell, \tag{2}$$

if ℓ_i is the restriction $\ell|_{U_i}$ of ℓ to U_i, for $i = 1, 2, \ldots, k$.

Let π be a partition of a totally ordered set $\ell = (U, \preceq)$. Each block p of π inherits a total order $\ell|_p$. It is also the case for the set π, of blocks, ordered according to the minimum element of each block, that is, for $p_1, p_2 \in \pi$, by

$$p_1 \preceq p_2 \iff \min p_1 \preceq \min p_2. \tag{3}$$

This induced total order is denoted by ℓ_π.

The *ordinal sum* of two ordered sets $\ell_1 = (U_1, \preceq_1)$ and $\ell_2 = (U_2, \preceq_2)$ is the ordered set $\ell = (U, \preceq)$, denoted by

$$\ell = \ell_1 +_O \ell_2, \tag{4}$$

where $U = U_1 + U_2$, and, for $u, v \in U$, one has

$$u \preceq_\ell v \iff \begin{cases} u \preceq_1 v, & \text{when } u, v \in U_1, \\ u \in U_1 \text{ and } v \in U_2, \\ u \preceq_2 v, & \text{when } u, v \in U_2. \end{cases} \tag{5}$$

In other words, $\ell_1 = \ell|_{U_1}$, $\ell_2 = \ell|_{U_2}$ and all the elements of ℓ_1 are smaller than those of ℓ_2. If ℓ_1 and ℓ_2 are totally ordered, then $\ell_1 +_O \ell_2$ also is. In particular, denote by $1 +_O \ell$, the totally ordered set obtained by adding a new minimum element to ℓ.

Between two ordered sets ℓ_1 and ℓ_2, an *increasing (nondecreasing)* function $f : \ell_1 \longrightarrow \ell_2$ is such that, for u and v in ℓ_1,

$$u \prec_{\ell_1} v \implies f(u) \preceq_{\ell_2} f(v). \tag{6}$$

The *empty* ordered set, denoted by $\emptyset$, corresponds to the unique total order relation on the empty set. Likewise, a set U of cardinality 1 admits only one total order, and such an ordered set is also denoted by 1. Finally the cardinality $|\ell|$ of an ordered set is the cardinality of its underlying set.

For $n \geq 1$, we denote by $[n]$, the ordered set $\{1, 2, \ldots, n\}$ with the usual order, with $[0] = \emptyset$. Note that for any ordered set ℓ of cardinality n, there exists a unique increasing bijection $\gamma : [n] \longrightarrow \ell$.

$\mathbb{L}$-Species

An $\mathbb{L}$-species involves constructions that can be performed on totally ordered sets. As motivation, let us first mention a classical example due to D. André [2], going back to 1879. An *alternating permutation* is a permutation (list) σ of $[n]$ such that

$$\sigma(1) > \sigma(2) < \sigma(3) > \sigma(4) < \cdots . \tag{7}$$

Another example is that of increasing binary rooted trees (see Figure 1) or, more generally, increasing rooted trees enriched in various ways.

Let $\mathbb{K}$ be a ring with $\mathbb{Q} \subseteq \mathbb{K} \subseteq \mathbb{C}$ and $\mathbb{A} = \mathbb{K}[\![\alpha, t, q, \ldots]\!]$ be a ring of formal power series on $\mathbb{K}$. We consider the possibility that the structures s are given weights $w(s) \in \mathbb{A}$. The reader is referred to Section 2.3 for notions relating to weighted sets.

Definition 1. A (weighted) $\mathbb{L}$-*species* is a rule F which
i) to each totally ordered set ℓ, *associates* a (weighted) set $F[\ell]$,
ii) to each increasing bijection $\gamma : \ell_1 \longrightarrow \ell_2$, *associates* a function (morphism of weighted sets)

$$F[\gamma] : F[\ell_1] \longrightarrow F[\ell_2];$$

these functions $F[\gamma]$ must satisfy the following functoriality properties:

$$F[\mathrm{Id}_\ell] = \mathrm{Id}_{F[\ell]}, \qquad F[\beta \circ \gamma] = F[\beta] \circ F[\gamma]. \tag{8}$$

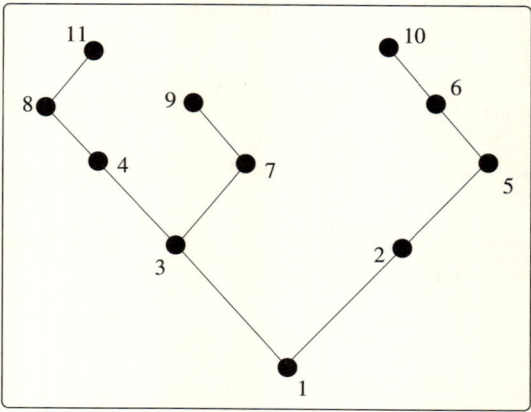

Fig. 1.

5.1. $\mathbb{L}$-Species

In other words, an $\mathbb{L}$-species is a functor

$$F : \mathbb{L} \longrightarrow \mathbb{E}_\mathbb{A}, \tag{9}$$

where $\mathbb{L}$ denotes the category of totally ordered sets and increasing bijections and $\mathbb{E}_\mathbb{A}$, that of $\mathbb{A}$-weighted (finite or) summable sets. Observe that the category $\mathbb{L}$ has a simple structure, since increasing bijections between totally ordered sets are unique.

An element $s \in F[\ell]$ is said to be an *F-structure* on ℓ, and the function $F[\gamma]$ is the *transport of F-structures* along γ.

Remark 2. The functoriality properties (8) imply that every transport function $F[\gamma]$ is a bijection (isomorphism of weighted sets).

Example 3. a) Alternating permutations, as defined by condition (1), make sense for any finite totally ordered set ℓ, and the transport (relabeling) along any increasing bijection is clear. We denote this $\mathbb{L}$-species by Alt.
b) The notion of *increasing rooted tree*, that is, those rooted trees whose vertices are increasing along any path from the root to the leaves, gives rise to an $\mathbb{L}$-species, denoted $\mathcal{A}^\uparrow$. Likewise, we denote by $\mathcal{B}^\uparrow$ (respectively $\mathcal{B}_c^\uparrow$), the $\mathbb{L}$-species of binary increasing (respectively complete) rooted trees, and more generally, by $\mathcal{A}_R^\uparrow$, the $\mathbb{L}$-species of increasing R-enriched rooted trees.

Definition 4. An *isomorphism of* (weighted) $\mathbb{L}$-*species* $\varphi : F \longrightarrow G$ is a family of (weight-preserving) bijections

$$\varphi_\ell : F[\ell] \longrightarrow G[\ell],$$

one for each totally ordered set ℓ, which commutes with the transports of structures; that is, for any increasing bijection $\gamma : \ell \longrightarrow m$, one should have

$$G[\gamma] \circ \varphi_\ell = \varphi_m \circ F[\gamma]. \tag{10}$$

We write $F = G$ to indicate that F and G are isomorphic. In categorical terms, an isomorphism of $\mathbb{L}$-species $\varphi : F \longrightarrow G$ is an invertible *natural transformation*.

Remark 5. Any weighted $\mathbb{B}$-species F (in the sense of previous chapters) gives rise to a weighted $\mathbb{L}$-species, also denoted by F, defined by setting, for any totally ordered set $\ell = (U, \preceq)$,

$$F[U, \preceq] = F[U], \tag{11}$$

the transport of structures being obtained by restriction to increasing bijections.

Note however that two nonisomorphic $\mathbb{B}$-species can become isomorphic when considered as $\mathbb{L}$-species. A typical example is given by the species S of permutations (bijections), and L of lists (or total orders), the transition being naturally made from $\sigma \in S[n]$ to the "word" $\sigma(1)\,\sigma(2)\cdots\sigma(n) \in L[n]$.

We thus obtain an arsenal of $\mathbb{L}$-species which can be readily used:
- 0, the empty species,
- 1, the species of (totally ordered) empty sets,
- X, the species of singletons,
- X^n, the species of lists of length n,
- E, the species of (totally ordered) sets,
- S, the species of permutations,
- L, the species of lists,
- $\mathcal{C}$, the species of cyclic permutations, etc.

Example 6. DESCENT, INVERSION, AND MAJOR INDEX STATISTICS OF PERMUTATIONS. Here are some examples of weighted $\mathbb{L}$-species obtained using parameters which depend on the underlying order. Let $\ell = (U, \preceq)$ be a totally ordered set of cardinality $n \geq 0$, let $\sigma = \sigma(1)\sigma(2)\cdots\sigma(n)$ be a permutation (list) on ℓ, and let $i, j \in [n]$. One says that i is a *descent* of σ if $1 \leq i \leq n-1$ and $\sigma(i) > \sigma(i+1)$. Denote by $\mathrm{Des}(\sigma)$, the descent set of σ and $\mathrm{des}(\sigma) = |\mathrm{Des}(\sigma)|$. Also set $\mathrm{maj}(\sigma) = \sum_{i \in \mathrm{Des}(\sigma)} i$. This last parameter is called the *major index* of σ. Finally, a pair (i, j) is an *inversion* of σ if $i < j$ and $\sigma(i) > \sigma(j)$; the inversion set of σ is denoted by $\mathrm{Inv}(\sigma)$, and we set $\mathrm{inv}(\sigma) = |\mathrm{Inv}(\sigma)|$. Denote by $L_{(q)}^{\mathrm{des}}$, $L_{(q)}^{\mathrm{maj}}$, and $L_{(q)}^{\mathrm{inv}}$ the weighted species of lists with the respective weightings

$$w(\sigma) = q^{\mathrm{des}(\sigma)}, \qquad w(\sigma) = q^{\mathrm{maj}(\sigma)}, \qquad w(\sigma) = q^{\mathrm{inv}(\sigma)}. \tag{12}$$

These concepts can be extended to R-enriched rooted trees, where R is an arbitrary $\mathbb{L}$-species (see Exercise 5.2.10). Lists correspond to the enrichment $R = 1 + X$.

Let F be an $\mathbb{L}$-species. Since the only increasing bijection $\ell \longrightarrow \ell$ from a totally ordered set to itself is the identity function Id_ℓ, any F-structure on ℓ has only one (trivial) automorphism. Moreover, two distinct F-structures s and t on ℓ are never isomorphic. The notion of isomorphism type is thus meaningless in this context.

Definition 7. Let F be a (weighted) $\mathbb{L}$-species. The *generating series* of F, denoted by $F(x)$ or sometimes by $\mathrm{card}(F)$, is defined by

$$F(x) = \sum_{n \geq 0} f_n \frac{x^n}{n!}, \tag{13}$$

where $f_n = |F[n]|$ (or $|F[n]|_w$ in the weighted case).

Note that two $\mathbb{L}$-species F and G, with trivial weight (i.e., $w = 1$), are isomorphic if and only if $F(x) = G(x)$. This statement is false in the case of general weights (see Exercises 3, b)).

Example 8. a) (André [2].) As we will see the $\mathbb{L}$-species Alt is characterized by a differential equation (see Example 15); solving this differential equation for the associated generating series we get

$$\text{Alt}(x) = \tan x + \sec x. \tag{14}$$

b) Likewise, one finds, solving associated differential equations,

a) $\mathcal{A}^{\uparrow} = \log \dfrac{1}{1-x}$, b) $\mathcal{B}^{\uparrow}(x) = \dfrac{1}{1-x}$, c) $\mathcal{B}_c^{\uparrow}(x) = \tan x.$ (15)

c) We introduce the classical q-analogues: $0_q = 1, 0!_q = 1$ and, for $n \geq 1$,

$$n_q = 1 + q + q^2 + \cdots + q^{n-1} \quad \text{and} \quad n!_q = 1_q \cdot 2_q \cdot 3_q \cdots n_q. \tag{16}$$

For the species L of permutations (lists), weighted by the parameters des, maj, or inv, one finds (see Exercises 1 and 2)

a) $\left|L_{(q)}^{\text{maj}}[n]\right| = \left|L_{(q)}^{\text{inv}}[n]\right| = n!_q,$

b) $L_{(q)}^{\text{des}}(x) = \dfrac{1-q}{\exp(x(q-1)) - q}.$ (17)

The operations defined on $\mathbb{L}$-species reflect the usual operations on exponential generating series and extend those of $\mathbb{B}$-species interpreted as $\mathbb{L}$-species. Thus the theory follows closely what has already been done for weighted $\mathbb{B}$-species, with some necessary adaptations. In particular, functorial composition no longer plays a role, but new operations become possible, such as integration, ordinal product (corresponding to the product of ordinary generating series; see Joyal [158]), and convolution.

Definition 9. Let F and G be two weighted $\mathbb{L}$-species. For $\ell = (U, \preceq)$, we define a number of operations by the following rules; in each case the transport of structures is obvious:
– The *sum*, $F + G$:

$$(F + G)[\ell] = F[\ell] + G[\ell]. \tag{18}$$

– The *product*, $F \cdot G$:

$$(F \cdot G)[\ell] = \sum_{\ell_1 + \ell_2 = \ell} F[\ell_1] \times G[\ell_2]. \tag{19}$$

– The *substitution* (partitional composition of Foata [45]), $F(G) = F \circ G$ ($G(0) = 0$):

$$(F \circ G)[\ell] = \sum_{\pi \in \text{Par}[\ell]} F[\ell_\pi] \times \prod_{p \in \pi} G[\ell|_p]. \tag{20}$$

– The *Cartesian product*, $F \times G$:

$$(F \times G)[\ell] = F[\ell] \times G[\ell]. \tag{21}$$

– The *derivative*, $F' = \frac{d}{dX} F(X)$:

$$F'[\ell] = F[1 +_\mathcal{O} \ell]. \tag{22}$$

– The *integral*, $\int F = \int_0^X F(T) \, dT$:

$$\left(\int F\right)[\ell] = \begin{cases} \emptyset, & \text{if } \ell = \emptyset, \\ F[\ell \setminus \{\min \ell\}], & \text{if } \ell \neq \emptyset. \end{cases} \tag{23}$$

– The *ordinal product*, $F \cdot_\mathcal{O} G$:

$$(F \cdot_\mathcal{O} G)[\ell] = \sum_{\ell_1 +_\mathcal{O} \ell_2 = \ell} F[\ell_1] \times G[\ell_2]. \tag{24}$$

– The *convolution*, $F * G$:

$$F * G = F \cdot_\mathcal{O} X \cdot_\mathcal{O} G. \tag{25}$$

Example 10. Let E_{even} and E_{odd} be the species of sets of even and odd cardinality, respectively, with generating series

$$E_{\text{even}}(x) = \cosh x, \qquad E_{\text{odd}}(x) = \sinh x. \tag{26}$$

The analytic identity $\cosh^2 x - \sinh^2 x = 1$ is reflected by the $\mathbb{L}$-species isomorphism

$$E_{\text{even}}^2 = 1 + E_{\text{odd}}^2. \tag{27}$$

Indeed, a species of structures E_{even}^2 on $\ell = (U, \preceq)$ is a pair of complementary subsets of U, each with even cardinality. If U is nonempty, it suffices to remove the element $\min \ell$ from the subset where it occurs and to add it to the complementary subset. This gives an involutive bijection with the pairs of complementary subsets of U, each of odd cardinality. Note that identity (27) is false in the context of $\mathbb{B}$-species; to see this, it suffices to compare their cycle index series.

Example 11. If L^c denotes the species of *connected* lists, that is, nonempty lists which begin with the minimum element of the underlying set, then we have

$$L = E(L^c), \tag{28}$$

since cutting up a list according to its *left to right minimum* elements (see Figure 2) gives a factorization into connected lists. This is analogous to the cyclic decomposition of a permutation (see Example 1.2.13).

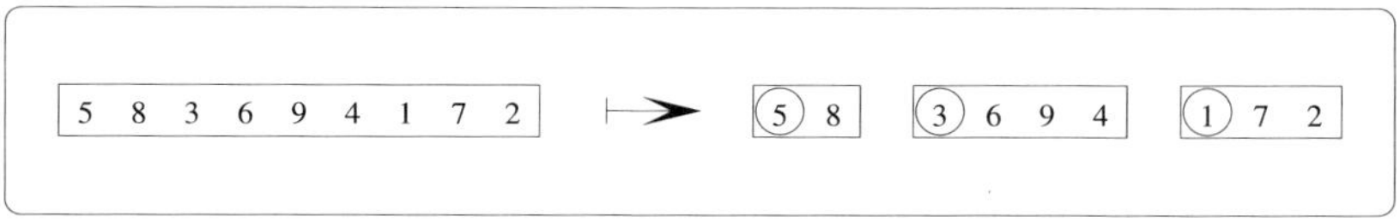

5.1. L-Species

| 5 | 8 | 3 | 6 | 9 | 4 | 1 | 7 | 2 | ⟼ | ⑤ 8 | ③ 6 9 4 | ① 7 2 |

Fig. 2.

More generally, for any $\mathbb{L}$-species G, an $E(G)$-structure, that is, an assembly of G-structures, can always be presented in a canonical fashion by ordering its parts according to the order (increasing or decreasing) of their minimum elements. As in the case of $\mathbb{B}$-species, we denote by F_n the restriction of an $\mathbb{L}$-species F to (totally ordered) sets of cardinality n. Note that $F_0 = F(0)$ and that $F[\emptyset] = \emptyset \iff F(0) = 0$.

It is easy to verify that the operations defined on $\mathbb{L}$-species satisfy the usual properties of associativity, commutativity, and linearity (up to isomorphism), in analogy with formal power series. The following proposition, whose proof is left as an exercise, asserts that the passage to generating series preserves these operations.

Recall the definition of the Hadamard product $f \times g$ of two exponential formal power series $f(x) = \sum a_n x^n / n!$ and $g(x) = \sum b_n x^n / n!$,

$$(f \times g)(x) = \sum_{n \geq 0} a_n b_n \frac{x^n}{n!}, \tag{29}$$

and the convolution $f * g$ of two continuous functions $f = f(x)$ and $g = g(x)$,

$$(f * g)(x) = \int_0^x f(x-t)g(t)\,dt. \tag{30}$$

Proposition 12. *Let F and G be two weighted $\mathbb{L}$-species. Then*

a) $(F + G)(x) = F(x) + G(x)$,

b) $(F \cdot G)(x) = F(x)G(x)$,

c) $(F \circ G)(x) = F(G(x))$, $(G(0) = 0)$

d) $(F \times G)(x) = F(x) \times G(x)$, \hfill (31)

e) $F'(x) = \dfrac{d}{dx} F(x)$,

f) $\left(\int_0^X F(T)\,dT \right)(x) = \int_0^x F(t)\,dt$,

g) $(F * G)(x) = F(x) * G(x)$.

The ordinal product $F \cdot_\mathrm{O} G$ of two species F and G, is defined by (24). A structure of species $F \cdot_\mathrm{O} G$ on a total order ℓ is obtained by cutting ℓ into an *initial* segment ℓ_1 and a *terminal segment* ℓ_2, and then giving ℓ_1 an F-structure and ℓ_2 a G-structure. The ordinal product corresponds to the product of *ordinary generating series*, $\sum_{n\geq 0} |F[n]| x^n$, of $\mathbb{L}$-species. The latter are obtained from exponential generating series by the *Laplace–Carson transform*, $\mathcal{L}_c$ (see Comtet [58], Section 1.13), defined simply, for $F(x) = \sum_{n\geq 0} a_n x^n/n!$ and $f(x) = \sum_{n\geq 0} a_n x^n$, by

$$\mathcal{L}_c(F(x)) = f(x). \tag{32}$$

It is a slight modification of the classical Laplace transform since, from an analytical point of view, one has, for $f = \mathcal{L}_c(F)$,

$$f(1/s) = s \int_0^\infty e^{-st} F(t)\, dt. \tag{33}$$

The convolution $F * G$ of F and G is defined by $(F * G) = F \cdot_O X \cdot_O G$, and formula (31), g) corresponds to the classical theorem in analysis saying that the Laplace transform of the convolution of two functions is the product of two transforms. See Exercise 8 for a combinatorial proof of this property.

Combinatorial Integrals

Recall the definition of the integral $\int F = \int_0^X F(T)\, dT$ of an $\mathbb{L}$-species F: For any totally ordered set ℓ, one has

$$\int F[\ell] = \begin{cases} \emptyset, & \text{if } \ell = \emptyset, \\ F[\ell \setminus \{\min \ell\}], & \text{if } \ell \neq \emptyset. \end{cases} \tag{34}$$

Figure 2 gives a generic representation of an $\int F$-structure. We observe that the minimum element of the underlying total order ℓ of an $\int F$-structure does not appear in the corresponding F-structure. This fact is emphasized by the position of min $= \min \ell$ in Figure 2. It can be verified that

$$\int F' = F_+ = F - F(0) \quad \text{and} \quad \left(\int F\right)' = F \tag{35}$$

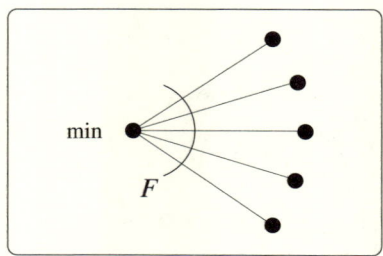

Fig. 2.

5.1. 𝕃-Species

and that the usual properties of the integral are valid, among which are additivity and change of variables (see Exercise 6):

a) $\quad \int_0^X (F(T)+G(T))\, dT = \int_0^X F(T)\, dT + \int_0^X G(T)\, dT,$

b) $\quad \int_0^X F(G(T))G'(T)\, dT = \int_0^{G(X)} F(Y)\, dY, \qquad (G(0)=0).$

(36)

Example 13. Let $\mathcal{G}$ be the 𝕃-species of simple graphs and $\mathcal{G}_p$, the subspecies of even graphs, that is, graphs whose vertices all have even degree. Since in any simple graph the number of vertices of odd degree is even, we can, in any simple graph on $\ell \setminus \{\min \ell\}$, attach vertices of odd degree to $\min \ell$ to obtain a nonempty even graph. An 𝕃-species isomorphism is thus obtained, that is, the equality

$$\int_0^X \mathcal{G}(T)\, dT = \mathcal{G}_p - 1. \tag{37}$$

There does not exist an analytic counterpart of formula (13) since the series considered have a zero radius of convergence. Here are now some examples corresponding to analytic formulas.

Example 14. The following combinatorial integrals are immediate.

a) $\quad \int_0^X T^n\, dT = L^c_{n+1}(X), \quad (n \geq 0),\qquad$ b) $\quad \int_0^X E(T)\, dT = E(X),$

c) $\quad \int_0^X E_{\text{odd}}(T)\, dT = E_{\text{even}}(X) - 1,\qquad$ d) $\quad \int_0^X E_{\text{even}}(T)\, dT = E_{\text{odd}}(X),$

e) $\quad \int_0^X L(T)\, dT = L^c(X).$

(38)

These identities correspond to the elementary integrals

a) $\quad \int_0^x t^n\, dt = \dfrac{x^{n+1}}{n+1}, \quad (n \geq 0),\qquad$ b) $\quad \int_0^x e^t\, dt = e^x - 1,$

c) $\quad \int_0^x \sinh t\, dt = \cosh x - 1.\qquad$ d) $\quad \int_0^x \cosh t\, dt = \sinh x,$

e) $\quad \int_0^x \dfrac{dt}{1-t} = \log \dfrac{1}{1-x}.$

(39)

Example 15. We denote respectively by $Z = \text{Alt}_{\text{even}}$ and $Y = \text{Alt}_{\text{odd}}$, the subspecies of even (i.e., on an even set of elements) and odd alternating permutations (lists). Note that by convention, for $|\ell| \geq 2$, an alternating permutation always starts with a descent ($\sigma(1) > \sigma(2)$). If we remove the minimum element from an alternating permutation σ, the latter is cut into two alternating permutations (σ_1, σ_2), where σ_1 is odd and σ_2 is of the same parity as σ. We thus have the relations

$$Y' = 1 + Y^2 \quad (Y(0) = 0), \qquad Z' = YZ \quad (Z(0) = 1), \tag{40}$$

or equivalently

$$Y(X) = \int_0^X (1 + Y^2(T)) \, dT, \qquad Z(X) = 1 + \int_0^X Y(T)Z(T) \, dT. \tag{41}$$

The analytic solution of this system of differential equations is given by

$$Y(x) = \tan x \quad \text{and} \quad Z(x) = \sec x. \tag{42}$$

Hence alternating permutations constitute a combinatorial model (due to André [2]) for *tangent* and *secant* numbers (also called the *Euler numbers*), which is the coefficients of the Taylor–Maclaurin (exponential) series of these functions. Taking into consideration the combinatorial relation (see Exercise 5.2.8)

$$Z^2 = 1 + Y^2, \tag{43}$$

which corresponds to the trigonometric identity $\sec^2 x = 1 + \tan^2 x$, we have found combinatorial interpretations of the integrals

$$\int_0^x \sec^2 t \, dt = \tan(x), \qquad \int_0^x \tan t \, \sec t \, dt = \sec x - 1. \tag{44}$$

Example 16. Other trigonometric, algebraic, or transcendental integrals can be deduced from combinatorial models, for example (see Exercise 7 and Section 5.2),

a) $\int_0^x \tan t \, dt = \log \sec x$, \qquad b) $\int_0^x \sec t \, dt = \log(\sec x + \tan x)$,

c) $\int_0^x \dfrac{dt}{\sqrt{1-2t}} = 1 - \sqrt{1-2x}$, \qquad d) $\int_0^x \dfrac{dt}{1-t^2} = \log\sqrt{\dfrac{1+x}{1-x}}$, \qquad (45)

e) $\int_0^x \dfrac{dt}{\sqrt{1-t^2}} = \arcsin x$.

The formula for integration by parts,

$$\int g \, df = fg - \int f \, dg, \tag{46}$$

reflects the Leibnitz rule for the derivative of a product. Setting $df = F(T) \, dT$ and $dg = G(T) \, dT$, it can be expressed as

$$\int_0^X F(T) \, dT \int_0^X G(Y) \, dY = \int_0^X F(T) \int_0^T G(Y) \, dY \, dT$$

$$+ \int_0^X \int_0^Y F(T) \, dT \, G(Y) \, dY. \tag{47}$$

5.1. $\mathbb{L}$-Species

Written in this manner, it admits another analytic interpretation, that is, the evaluation of the double integral

$$\int\int_R F(T)G(Y)\,dT\,dY, \tag{48}$$

where R is the square region (of points (T, Y)) determined by the points $(0, 0)$, $(0, X)$, $(X, 0)$, and (X, X). The right-hand side of (47) corresponds to partitioning this region into two triangular subregions, one where $Y < T$ and the other where $Y > T$, respectively. A similar dichotomy occurs in the combinatorial interpretation of the left-hand side of (47), according to $m_1 < m_2$ or $m_1 > m_2$, where m_1 and m_2 are minimum elements of the $\int F-$ and $\int G-$structures in this product. In the first case, m_1 is the absolute minimum and we have, in fact, a structure of species $\int (F \cdot \int G)$. In the second case we obtain an $\int (\int F \cdot G)$-structure.

Multisort $\mathbb{L}$-Species

It is also appropriate to introduce the weighted k-sort $\mathbb{L}$-species as functors in k variables

$$F : \mathbb{L} \times \mathbb{L} \times \cdots \times \mathbb{L} \longrightarrow \mathbb{E}_\mathbb{A}. \tag{49}$$

In particular, a two-sort $\mathbb{L}$-species is a rule which
i) to each pair (ℓ_1, ℓ_2) of totally ordered (disjoint) sets associates a finite (or summable) $\mathbb{A}$-weighted set $F[\ell_1, \ell_2]$,
ii) to each pair $(\gamma_1, \gamma_2) : (\ell_1, \ell_2) \longrightarrow (m_1, m_2)$ of increasing bijections associates a function (morphism of weighted sets)

$$F[\gamma_1, \gamma_2] : F[\ell_1, \ell_2] \longrightarrow F[m_1, m_2]$$

such that the following functoriality properties are satisfied:

$$F[\mathrm{Id}_{\ell_1}, \mathrm{Id}_{\ell_2}] = \mathrm{Id}_{F[\ell_1, \ell_2]}, \qquad F[\tau_1 \circ \gamma_1, \tau_2 \circ \gamma_2] = F[\tau_1, \tau_2] \circ F[\gamma_1, \gamma_2]. \tag{50}$$

We consider the elements of ℓ_1 as belonging to a first sort associated with the variable X, and those of ℓ_2 as being of a second sort associated with Y. We then write $F = F(X, Y)$. We agree that "every element of the first sort is smaller than those of the second sort," or equivalently that any pair of totally ordered sets (ℓ, m) gives rise to a totally ordered set consisting of the (disjoint) ordinal sum $\ell +_\mathcal{O} m$.

With these definitions and conventions, it is easy to extend the usual operations on $\mathbb{L}$-species to two (or many) sorts, similarly to operations on k-sort $\mathbb{B}$-species. For example, for a two-sort $\mathbb{L}$-species F and two unisort $\mathbb{L}$-species G and H, with $G(0) = 0 = H(0)$, the (partitional) composition $F(G, H)$ is the one-sort $\mathbb{L}$-species defined by

$$F(G, H)[\ell] = \sum_{\substack{\pi \in \text{Par}[\ell] \\ \chi: \pi \to \{1,2\}}} F[\chi^{-1}(1), \chi^{-1}(2)] \times \prod_{p \in \chi^{-1}(1)} G[p] \times \prod_{q \in \chi^{-1}(2)} H[q], \quad (51)$$

where, for $\pi \in \text{Par}[\ell]$, $\chi : \pi \longrightarrow \{1, 2\}$, and $i \in \{1, 2\}$, the set of blocks $\chi^{-1}(i)$ is (totally) ordered according to increasing order of the minimum element of each block.

We also note that the partial derivatives $\frac{\partial}{\partial X} F$ and $\frac{\partial}{\partial Y} F$ of a two-sort $\mathbb{L}$-species $F = F(X, Y)$ are defined by

$$\frac{\partial F}{\partial X}[\ell, m] = F[1 +_O \ell, m], \qquad \frac{\partial F}{\partial Y}[\ell, m] = F[\ell, 1 +_O m]. \quad (52)$$

These operations are preserved when passing to the exponential generating series, defined by

$$F(x, y) = \sum_{k,n} F[k, n] \frac{x^k}{k!} \frac{y^n}{n!}, \quad (53)$$

where $[k, n]$ denotes the pair of totally ordered sets $([k], [n])$.

Exercises for Section 5.1

1. EULERIAN POLYNOMIALS AND NUMBERS.
   a) Consider the weighted species $\text{Bal}_{(q)}$ of ballots (see (1.1.24)) with weighting $w(\pi) = q^{n-k}$, where $\pi \in \text{Bal}[n]$ and k denotes the number of blocks of π. Show that

$$\text{Bal}_{(q)}(x) = \left(1 - \frac{\exp(qx) - 1}{q}\right)^{-1}. \quad (54)$$

   b) Establish the equipotence

$$L^{\text{des}}_{(q+1)}(x) = \text{Bal}_{(q)}(x) \quad (55)$$

and deduce formula (17) b), as well as Frobenius' formula for the *Eulerian polynomials*:

$$A_n(q) = \sum_{k=0}^{n-1} (n - k)! S(n, n - k)(q - 1)^k, \quad (56)$$

where $S(n, k)$ denotes the Stirling numbers of the second kind and where we have set

$$L^{\text{des}}_{(q)}(x) = \sum_{n \geq 0} A_n(q) \frac{x^n}{n!}. \quad (57)$$

HINT: A list σ weighted by $(q + 1)^{\text{des}\,\sigma}$ is equivalent to a weighted list where certain descents have been distinguished and marked by the counter q. One

5.1. $\mathbb{L}$-Species

constructs a ballot by introducing divisions between the elements of the list σ except for marked descents.

c) The *Eulerian numbers* $A_{n,k}$ are the coefficients of the Eulerian polynomials:

$$A_{n,k} = [q^k]A_n(q) = n![q^k x^n]L_{(q)}^{\text{des}}(x). \tag{58}$$

In other words, $A_{n,k}$ is the number of lists $\sigma \in L[n]$ having k descents. Establish the following:

i) $A_{n+1,k} = (n-k+1)A_{n,k-1} + (k+1)A_{n,k},$ (59)

ii) $\displaystyle\sum_{m \geq 1} m^n t^m = \frac{t A_n(t)}{(1-t)^{n+1}}.$

d) An *ascent* of a list $\sigma = \sigma(1)\cdots\sigma(n) \in L[n]$ is an $i \in [n-1]$ such that $\sigma(i) < \sigma(i+1)$, and an *excedance* of a permutation $\sigma \in S[n]$ is an $i \in [n]$ such that $i < \sigma(i)$. Denote by $\text{asc}(\sigma)$, the number of ascents of a list σ and by $\text{exc}(\sigma)$, the number of excedances of a permutation σ. Prove the equipotences

$$L_{(q)}^{\text{des}} = L_{(q)}^{\text{asc}} = S_{(q)}^{\text{exc}}. \tag{60}$$

HINT: For the first equipotence, use vertical symmetry or else the transformation $L[\iota]$, where ι is the unique order-reversing bijection $\ell \longrightarrow \ell$. For the second, use the fundamental transformation (see Example 1.2.13).

e) Show that (Foata and Schutzenberger [112])

i) $A_{2n}(-1) = 0, \quad (n > 0),$

ii) $(-1)^n A_{2n+1}(-1) = (2n+1)![x^{2n+1}]\tan(x),$ (61)

iii) $B_{2n+1}(-1) = 0,$

iv) $(-1)^n B_{2n}(-1) = (2n)![x^{2n}]\sec(x),$

where $B_n(q) = |B_{(q)}^{\text{exc}}([n])|$ and B denotes the class of permutations without fixed points.

HINT: For i) and iii) define suitable involutions on the structures; for ii) and iv), use (17) b).

2. Prove the following formulas involving the weighted L-species L^{maj} and L^{inv} (see Example 6):

a) $\left|L_q^{\text{inv}}[n]\right| = n!_q, \quad n \geq 0.$ (62)

HINT: Given a list $\sigma \in L[n]$ and k with $1 \leq k \leq n$, denote by $\sigma\mid_k$ the restriction of σ to the set $[k]$, that is, the list obtained from σ by erasing all elements larger than k. Construct the *inversion table* of σ: $\lambda = (\lambda(1), \lambda(2), \ldots, \lambda(n))$, where $\lambda(k) \leq k-1$ is the number of inversions created by inserting k in $\sigma\mid_{k-1}$ (that is,

$\lambda(k) = \mathrm{inv}(\sigma|_k) - \mathrm{inv}(\sigma|_{k-1}))$. Show that the list σ can be reconstructed from λ. Such a sequence $\lambda = (\lambda(1), \lambda(2), \ldots, \lambda(n))$, with $0 \leq \lambda(k) \leq k-1$, is called a *Lehmer code*.

$$\text{b)} \quad |L_q^{\mathrm{maj}}[n]| = n!_q, \quad n \geq 0. \tag{63}$$

HINT: Show that the parameter maj also gives rise to a Lehmer code.

3. a) Show that two (nonweighted) $\mathbb{L}$-species F and G are isomorphic if and only if $F(x) = G(x)$.
   b) Give an example of a nontrivial weighted $\mathbb{L}$-species F such that $F(x) = 0$.

4. Let F be a $\mathbb{B}$-species. To be more precise, we denote by $\pi(F)$ the associated $\mathbb{L}$-species (see Remark 5). Conversely, if F is an $\mathbb{L}$-species, a $\mathbb{B}$-species $\rho(F)$ is defined by setting

$$\rho(F)[U] = \sum_{\preceq \in L[U]} F[U, \preceq]. \tag{64}$$

a) Show that π preserves addition, product, and substitution.
b) Complete the definition of the $\mathbb{B}$-species $\rho(F)$ for transport of structures and show that for two $\mathbb{L}$-species F and G and a $\mathbb{B}$-species H,

$$\text{i)} \quad \rho(F+G) = \rho(F) + \rho(G),$$

$$\text{ii)} \quad \rho\left(\int F\right) = X\rho(F), \tag{65}$$

$$\text{iii)} \quad \rho\pi(H) = L \times H.$$

5. Let $\varphi: G \longrightarrow H$ be an $\mathbb{L}$-species isomorphism. Show how to infer the following $\mathbb{L}$-species isomorphisms, where F is an arbitrary $\mathbb{L}$-species:

a) $F + \varphi : F + G \longrightarrow F + H$,

b) $F \cdot \varphi : F \cdot G \longrightarrow F \cdot H$,

c) $F \circ \varphi : F \circ G \longrightarrow F \circ H$ $\quad (G(0) = 0 = H(0))$,

d) $\varphi \circ F : G \circ F \longrightarrow H \circ F$ $\quad (F(0) = 0)$, $\tag{66}$

e) $\varphi' : G' \longrightarrow H'$,

f) $\int \varphi : \int G \longrightarrow \int H$,

g) $F \times \varphi : F \times G \longrightarrow F \times H$,

h) $F \cdot_O \varphi : F \cdot_O G \longrightarrow F \cdot_O H$.

HINT: For b), if $s = (\pi, f, (g_p)_{p \in \pi}) \in (F \circ G)[\ell]$, where $\pi \in \mathrm{Par}[\ell]$, $f \in F[\ell_\pi]$, and $g_p \in G[\ell|_p]$, set $(F \circ \varphi)(s) = (\pi, f, (h_p)_{p \in \pi}) \in (F \circ H)[\ell]$, where $h_p = \varphi_{\ell|_p}(g_p) \in H[\ell|_p]$.

6. Establish the usual properties of the combinatorial integral, in particular additivity and change of variables (see (36)).
   NOTE: Interpret the right-hand side of (36), b) as the composition $(\int_0^X F(T)\, dT) \circ G$.
7. a) Starting from the relation
$$L = L_{odd} + L_{even}, \tag{67}$$
   show that the integral $\mathbb{L}$-species $\int L_{odd}$ and $\int L_{even}$ respectively form the even and odd parts of the $\mathbb{L}$-species $\int L$. Deduce the integrals
$$\text{i)} \int_0^x \frac{t}{1-t^2}\, dt = \ln \frac{1}{\sqrt{1-t^2}}, \quad \text{ii)} \int_0^x \frac{1}{1-t^2}\, dt = \ln\sqrt{\frac{1+t}{1-t}}. \tag{68}$$
   b) Show that the generating series of the species F and G of permutations all of whose cycles have even and odd length, respectively, are given by (see J. Labelle [199])
$$\text{i)} \quad F(x) = \frac{1}{\sqrt{1-x^2}}, \quad \text{ii)} \quad G(x) = \sqrt{\frac{1+x}{1-x}}. \tag{69}$$
8. a) Establish the associativity, commutativity (up to isomorphism), and distributivity with respect to the addition of the ordinal product and of the convolution of $\mathbb{L}$-species (see (24) and (25)).
   b) Show that the ordinal product corresponds to the product of ordinary generating series of $\mathbb{L}$-species.
   c) CONVOLUTION OF $\mathbb{L}$-SPECIES. Given two $\mathbb{L}$-species F and G, introduce a new $\mathbb{L}$-species H defined by
$$H(X) = \int_0^X F(Y - T)\, G(T)\, dT \bigg|_{Y=X}, \tag{70}$$
   where Y denotes a second kind of points and where the species $-T$ (or $-X$) is interpreted as the species of singletons weighted by -1. Show that the species H and $F * G$ are equipotent (that is, have the same generating series).
   HINT: An H-structure on ℓ ($\neq \emptyset$) is a sextuplet (x, A, B, C, f, g) such that $\{x\}$, A, B, and C are pairwise disjoint sets whose union gives ℓ, $x = \min(\{x\} \cup B \cup C)$, $f \in F[A \cup B]$ and $g \in G[C]$, the weight of such a structure being $(-1)^{|B|}$. Define the element y to be the minimum of the elements of $A \cup B$ which are larger than x, if it exists, and construct an involution on the H-structures by exchanging the position of this element y between A and B. Show that the sole survivors of this involution are the H-structures where y does not exist, that is, the $F * G$-structures.
   d) Deduce property (31), g) for the convolution of $\mathbb{L}$-species.

9. TAYLOR'S FORMULA WITH INTEGRAL REMAINDER. Show that for any $\mathbb{L}$-species F and any $n \geq 0$, one has

$$F(X+Y) = F(Y) + E_1(X)F'(Y) + E_2(X)F^{(2)}(Y) + \cdots + E_n(X)F^{(n)}(Y) + R_n, \tag{71}$$

where

$$R_n = E_n(X) * F^{(n+1)}(X+Y) = \int_0^X E_n(T-X) \cdot F^{(n+1)}(T+Y) \, dT. \tag{72}$$

5.2. Combinatorial Differential Equations

There is a great deal of interaction between enumerative combinatorics and differential equations. A first direction is the use of differential equations to obtain information about the generating series of certain classes of combinatorial structures. Examples abound in the literature (see, for example, Goulden and Jackson [133]). A second consists in finding natural combinatorial models for solutions of differential equations (or systems of differential equations) and using combinatorial methods (constructions, bijections) to establish the properties of these solutions.

We first present a general method, developed by Leroux and Viennot [217, 219] (see also Bergeron and Reutenauer [26, 27], Fliess [103], and G. Labelle [183]), to solve combinatorially systems of differential equations. In this approach, the solutions appear naturally as increasing enriched rooted trees.

Example 1. As a motivating example, consider the system of differential equations

$$\begin{aligned} Y' &= 1 + Y^2, & Y(0) &= 0, \\ Z' &= YZ, & Z(0) &= 1, \end{aligned} \tag{1}$$

where Y and Z are unknown $\mathbb{L}$-species. Integrating, these equations become equivalent to

$$Y = X + \int_0^X Y^2(T) \, dT, \qquad Z = 1 + \int_0^X Y(T)Z(T) \, dT \tag{2}$$

and may be interpreted combinatorially, using the definition of integral of $\mathbb{L}$-species, in the manner indicated in Figure 1. Iterating this decomposition for Y and for Z, we see that the $\mathbb{L}$-species Y can be identified canonically with the class of complete binary rooted trees which are increasing (on any path from the root to the leaves; see Figure 2(a)). This $\mathbb{L}$-species is denoted by $Y = \mathcal{B}_c^\uparrow$. Likewise the species Z can be identified with the $\mathbb{L}$-species of increasing binary rooted trees which are complete with the exception of the extreme right branch, where there always remains an empty leaf (see Figure 2(b)). In this case the number of vertices is always even and this species is denoted by $Z = \mathcal{B}_{ce}^\uparrow$.

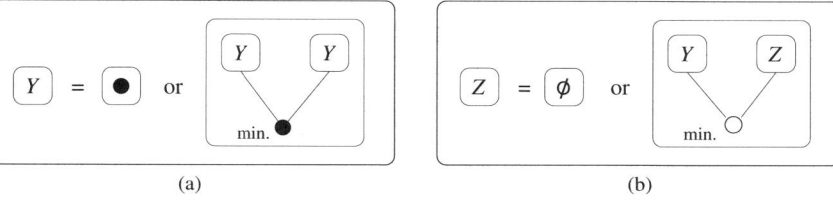

Fig. 1.

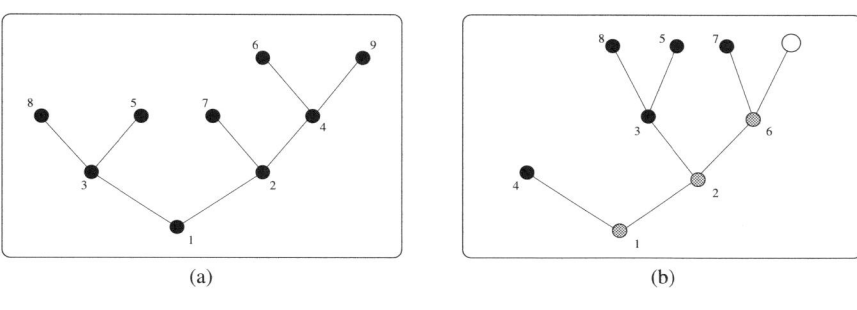

Fig. 2.

Using the uniqueness principle (established in Theorem 3), these canonical solutions of (1), as odd and even complete increasing binary rooted trees, must be respectively isomorphic to the $\mathbb{L}$-species of odd and even alternating permutations (lists), since these are also solutions of system (1) (see Example 5.1.15). In fact the canonical isomorphism $\mathcal{B}_c^\uparrow \simeq \mathrm{Alt}_\mathrm{odd}$, $\mathcal{B}_{ce}^\uparrow \simeq \mathrm{Alt}_\mathrm{even}$, is a well-known bijection, due to Foata and Schützenberger (see [113], Françon and Viennot [118, 121], and Goulden and Jackson [133], 5.2.14), which associates to each increasing binary rooted tree (not necessarily complete) the permutation (list) obtained by "projecting on the x-axis." For example, the permutations associated to the increasing binary rooted trees of Figure 2(a) and (b) are, respectively, 835172649 and 41835276.

Since the analytical solutions of system (1) are $Y(x) = \tan x$, $Z(x) = \sec x$, we obtain two combinatorial models for the *tangent* and *secant numbers*

$$\mathcal{B}_c^\uparrow(x) = \mathrm{Alt}_\mathrm{odd}(x) = \tan x, \qquad \mathcal{B}_{ce}^\uparrow(x) = \mathrm{Alt}_\mathrm{even}(x) = \sec x. \qquad (3)$$

We first consider the case of a single autonomous differential equation.

Example 2. a) Recall that the species $Y = L$ of lists satisfies the differential equation

$$Y' = Y^2, \qquad Y(0) = 1, \qquad (4)$$

since the removal of the minimum element of a permutation (list) breaks it into two lists. The canonical combinatorial solution of (4) is the $\mathbb{L}$-species $Y = \mathcal{B}^\uparrow$ of increasing binary rooted trees and one has

$$\mathcal{B}^{\uparrow}(x) = L(x) = \frac{1}{1-x}. \tag{5}$$

The $\mathbb{L}$-species isomorphism $B \xrightarrow{\sim} L$ is precisely the Foata–Schützenberger bijection described in Example 1.

b) We also see, by removing the minimum element, that the $\mathbb{L}$-species $Y = \mathcal{A}^{\uparrow}$ of increasing rooted trees satisfies the differential equation

$$Y' = E(Y), \qquad Y(0) = 0. \tag{6}$$

The analytic solution of this equation is $Y(x) = -\log(1-x)$, which comes to saying that $\mathcal{A}^{\uparrow}(x) = -\log(1-x)$ or also, in an equivalent manner,

$$E(\mathcal{A}^{\uparrow}(x)) = \frac{1}{1-x} = L(x). \tag{7}$$

Once more an $\mathbb{L}$-species isomorphism $L \simeq E(\mathcal{A}^{\uparrow})$ explains (7) combinatorially: It is the bijection (Burge [41]) obtained by iterating the decomposition of lists according to their left to right records (minimum elements; see Example 5.1.11). For example the assembly of increasing rooted trees corresponding to the list (8, 11, 4, 3, 9, 7, 1, 2, 10, 6, 5), as well as to the increasing binary rooted tree of Figure 5.1.1, is given in Figure 3, where the assembly is presented in decreasing order of their minimum elements.

More generally, let R be an $\mathbb{L}$-species and consider the autonumuous differential equation

$$Y' = R(Y), \qquad Y(0) = Z, \tag{8}$$

where Z is a variable which represents the initial condition of the equation. This variable can take on particular values, for instance $Z = 0$ or more generally (when

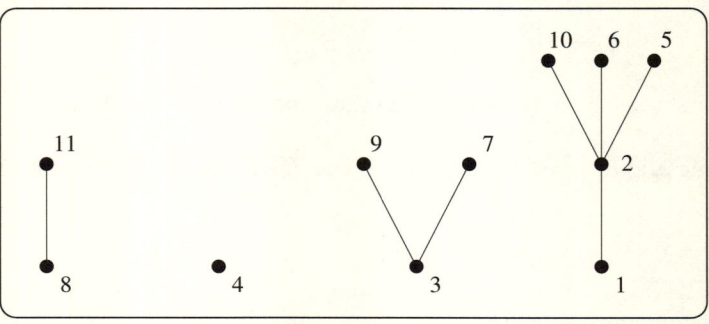

Fig. 3.

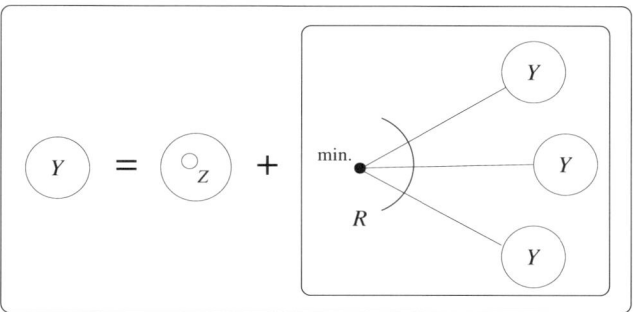

Fig. 4.

possible) $Z = \gamma$ a constant $\in \mathbb{A}$. Separating variables, we see that the analytical solution of (8), with $Z = \gamma$, is given by $Y = \Phi^{\langle -1 \rangle}(x)$, where Φ is determined by the relation

$$\Phi(Y) = \int_\gamma^Y \frac{du}{R(u)} = x. \tag{9}$$

Hence the canonical solution which we will construct constitutes a combinatorial model for this function $\Phi^{\langle -1 \rangle}(x)$.

To preserve generality, we consider the variable Z as representing a second sort of elements. It follows that the solution is an $\mathbb{L}$-species of two sorts $Y = Y(X, Z)$. Written in integral form, Equation (8) becomes

$$Y(X, Z) = Z + \int_0^X R(Y(T, Z)) \, dT \tag{10}$$

and is interpreted combinatorially by Figure 4. Iterating this decomposition we obtain in a canonical manner, for any pair of totally ordered sets (ℓ, m) corresponding to the sorts X and Z respectively, increasing R-enriched rooted trees like that of Figure 5, where $\ell = [8]$ and $m = \{a, b, \ldots, h\}$. Observe that to each vertex of sort X (the black points) is attached an R-assembly (possibly empty) of $Y(X, Z)$-structures. These are the *internal* (or *fertile*) vertices of the rooted tree. Moreover the vertices of sort Z (the white points) are found at the leaves of the rooted tree and do not carry R-structures. These vertices are called *buds*. With the convention that every element of sort X is smaller than the elements of sort Z, the vertices of the rooted tree are overall increasing from the root to the leaves.

These properties determine the two-sort $\mathbb{L}$-species of R-enriched increasing rooted trees, denoted by $\mathcal{A}_R^\uparrow(X, Z)$. It is clear that $\mathcal{A}_R^\uparrow(0, Z) = Z$ and that the decomposition of any structure $s \in \mathcal{A}_R^\uparrow[\ell, m]$, where $\ell \neq \emptyset$, into a minimum element

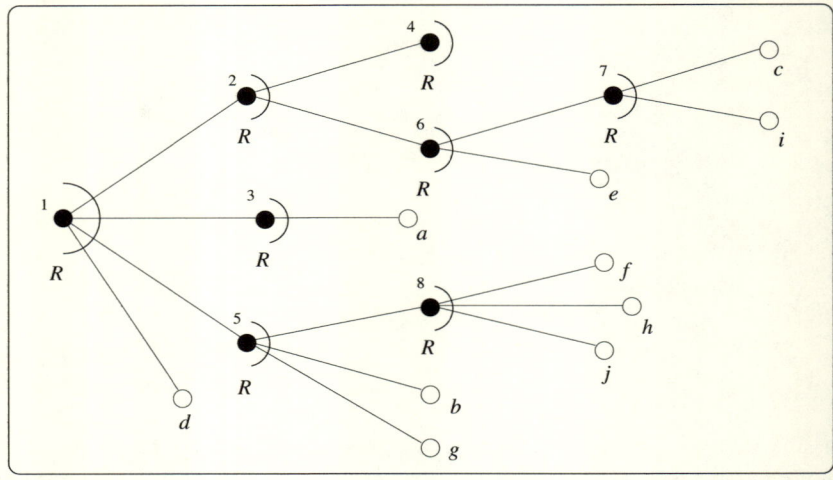

Fig. 5.

followed by an R-assembly of $\mathcal{A}_R^\uparrow$-structures gives an isomorphism

$$\alpha : \frac{\partial}{\partial X} \mathcal{A}_R^\uparrow(X, Z) \xrightarrow{\sim} R(\mathcal{A}_R^\uparrow(X, Z)). \tag{11}$$

In this sense, the species $Y = \mathcal{A}_R^\uparrow$, or more precisely the pair $(\mathcal{A}_R^\uparrow, \alpha)$, is the solution of differential equation (8). Now if a pair $(\mathcal{B}, \beta)$ is another solution of (8), that is, $\mathcal{B} = \mathcal{B}(X, Z)$ is a two-sort $\mathbb{L}$-species with $\mathcal{B}(0, Z) = Z$, and β is an isomorphism

$$\beta : \frac{\partial}{\partial X} \mathcal{B}(X, Z) \xrightarrow{\sim} R(\mathcal{B}(X, Z)), \tag{12}$$

then there exists a unique isomorphism of solutions $\Phi : (\mathcal{A}_R^\uparrow, \alpha) \longrightarrow (\mathcal{B}, \beta)$, that is, an $\mathbb{L}$-species isomorphism $\Phi : \mathcal{A}_R^\uparrow \longrightarrow \mathcal{B}$ such that the following diagram commutes:

$$\begin{array}{ccc} \frac{\partial}{\partial X}\mathcal{A}_R^\uparrow & \xrightarrow{\alpha} & R(\mathcal{A}_R^\uparrow) \\ \frac{\partial}{\partial X}\Phi \downarrow & & \downarrow R(\Phi) \\ \frac{\partial}{\partial X}\mathcal{B} & \xrightarrow{\beta} & R(\mathcal{B}) \end{array} \tag{13}$$

Here the induced natural transformations $\frac{\partial}{\partial X}\Phi$ and $R(\Phi)$ are defined in an obvious manner (see Exercise 5.1.5). This is proved by induction on the cardinality of ℓ, that is on the number of fertile points of the $\mathcal{A}_R^\uparrow$-structure. First, one has $\mathcal{A}_R^\uparrow(0, Z) = Z = \mathcal{B}(0, Z)$ and the unique choice of Φ_0 is the identity $Z \longrightarrow Z$. Now, for $n \geq 0$, suppose that the natural bijection $\Phi_{(h,s)}$ has already been defined in a unique manner for any pair of totally ordered sets (h, s), with $|h| \leq n$, and let (ℓ, m) be a pair of totally ordered sets, with $|\ell| = n + 1$. We can assume, without loss of generality, that $\ell = 1 +_O \ell^-$, where $\ell^- = \ell \setminus \min \ell$. We then have, since $A_R^\uparrow$ and $\mathcal{B}$ are solutions of (8), bijections

$$A_R^\uparrow[\ell, m] = A_R^\uparrow[1 +_O \ell^-, m] = \frac{\partial}{\partial X} A_R^\uparrow[\ell^-, m] \xrightarrow{\alpha_{(\ell^-, m)}} R(A_R^\uparrow)[\ell^-, m], \qquad (14)$$

$$\mathcal{B}[\ell, m] = \mathcal{B}[1 +_O \ell^-, m] = \frac{\partial}{\partial X} \mathcal{B}[\ell^-, m] \xrightarrow{\beta_{(\ell^-, m)}} R(\mathcal{B})[\ell^-, m].$$

Now an $R(A_R^\uparrow)$-structure on (ℓ^-, m) is an R-assembly of $A_R^\uparrow$-structures on pairs (h, s), with $|h| \leq n$, and we can use the induction hypothesis, which ensures the existence of bijections $\Phi_{(h,s)} : A_R^\uparrow[h, s] \longrightarrow \mathcal{B}[h, s]$ on each member of the R-assembly, to induce a bijection $R(\Phi)_{(\ell^-, m)} : R(A_R^\uparrow)[\ell^-, m] \longrightarrow R(\mathcal{B})[\ell^-, m]$. We need only set

$$\Phi_{(\ell, m)} = \frac{\partial}{\partial X} \Phi_{(\ell^-, m)} = \beta_{(\ell^-, m)}^{\langle -1 \rangle} \circ R(\Phi)_{(\ell^-, m)} \circ \alpha_{(\ell^-, m)} \qquad (15)$$

to obtain a bijection $A_R^\uparrow[\ell, m] \longrightarrow \mathcal{B}[\ell, m]$, and this is the only possible choice which makes diagram (13) commutative.

In summary, we have established the following result.

Theorem 3. *Let R be an $\mathbb{L}$-species. Then the two-sort $\mathbb{L}$-species $Y = A_R^\uparrow(X, Z)$, with the natural isomorphism $\alpha : \partial A_R^\uparrow / \partial X \longrightarrow R(A_R^\uparrow)$, is a solution of the differential equation (8). Moreover, this solution is canonical in the sense that for any other solution $(\mathcal{B}, \beta)$ of (8), there exists a unique isomorphism $\Phi : A_R^\uparrow \longrightarrow \mathcal{B}$ making Diagram (13) commutative.*

The simplest case for the initial condition is when $Z = 0$. This means that buds (the leaves of sort Z) are forbidden and that increasing R-enriched rooted trees live only on vertices of sort X. We set $A_R^\uparrow(X) = A_R^\uparrow(X, 0)$. It is then natural to suppose that $R(0) \neq 0$, for otherwise (i.e., if $R(0) = 0$), we have $A_R^\uparrow = 0$.

Example 4. INCREASING ORDERED ROOTED TREES. The species $Y = A_L^\uparrow$ of increasing ordered rooted trees is the solution of the differential equation

$$Y' = L(Y), \qquad Y(0) = 0, \qquad (16)$$

L denoting as usual the species of total orders (lists). This means that the children of each internal vertex v (the fiber of v) are linearly ordered and, the rooted tree being increasing, larger than v (see Figure 6(a)). These structures are often called increasing planted *planar* (or *plane*) trees. To be more precise, the $\mathbb{L}$-species $\mathcal{P}^\uparrow$

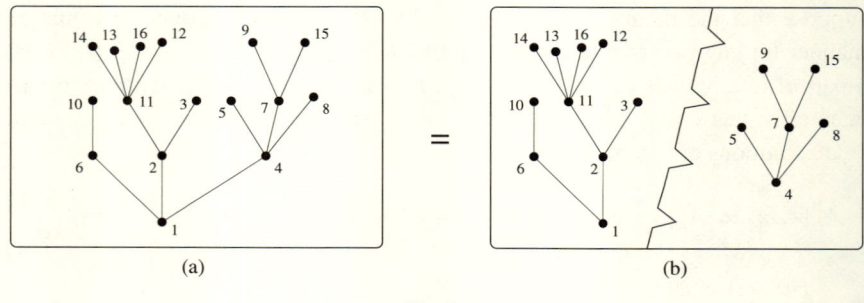

Fig. 6.

of increasing plane (i.e., embedded in the plane) rooted trees is defined by

$$\mathcal{P}^\uparrow = X + \int \mathcal{C}(\mathcal{A}_L^\uparrow), \tag{17}$$

where $\mathcal{C}$ denotes the species of cyclic permutations.

The $\mathbb{L}$-species $Y = \mathcal{A}_L^\uparrow$ also satisfies the functional equation

$$Y = X + E_2(Y) \tag{18}$$

which says that an $\mathcal{A}_R^\uparrow$-structure is either a singleton or a pair of $\mathcal{A}_L^\uparrow$-structures. One way to realize this assertion is to cut the rightmost sub-rooted tree attached to the root of any increasing ordered rooted tree which does not consist of a single point (see Figure 6(b)).

It follows that the generating series $y = \mathcal{A}_L^\uparrow(x)$ satisfies the equation $y^2 - 2y + 2x = 0$, whose resolution gives

$$\mathcal{A}_L^\uparrow(x) = 1 - \sqrt{1 - 2x}. \tag{19}$$

The substitution $Z := \gamma$, where γ is a nonzero element of $\mathbb{A}$ (for example $\gamma = 1$) in the species $\mathcal{A}_R^\uparrow(X, Z)$ is more delicate. It is equivalent to making the buds (the elements of sort Z) indistinguishable, nonlabeled, as if one passed to the isomorphism types with respect to the sort Z. This last point of view can be formalized by considering, when R is in fact a $\mathbb{B}$-species, that the solution $Y = \mathcal{A}_R^\uparrow(X, Z)$ is a mixed species, $\mathbb{L}$-species for the sort X and $\mathbb{B}$-species for the sort Z (see Exercise 5). The question of finiteness, for $Z := 1$, or of summability, for $Z := \gamma$, a formal variable acting as a bud counter, must then be examined. Special care must also be taken when passing to series.

A case where everything works well is when the species R is asymmetric and polynomial, as in the following example.

Example 5. Let $a_0, a_1, \ldots, a_m$ be scalar parameters or formal variables. Consider the $\mathbb{L}$-species

$$R = a_0 + a_1 X + a_2 X^2 + \cdots + a_m X^m \tag{20}$$

of lists of length $\leq m$, weighted by a_k where k is the length of the list. The solution $Y = Y(X, Z)$ of the differential equation

$$Y' = a_0 + a_1 Y + a_2 Y^2 + \cdots + a_m Y^m, \qquad Y(0) = Z \qquad (21)$$

is given by increasing ordered rooted trees such that if an internal vertex has k children, then $k \leq m$ and this vertex has weight a_k. The weight of a rooted tree is then equal to the product of the weights of its internal vertices. If moreover we set $Z := \gamma$ then the weight of a rooted tree a is multiplied by $\gamma^{b(a)}$ where $b(a)$ denotes the number of buds of a. For example, the weight of the rooted tree represented by Figure 5 is $a_0 a_1 a_2^3 a_3^2 a_4 \gamma^{10}$.

Many other choices of enriching species R are worthy of interest. Note the following classes of enriched increasing rooted trees:
– cyclic (the *mobiles*): $R = 1 + C$,
– alternating: $R = \text{Alt}$,
– complete m-ary: $R = 1 + E_m$ ($m \geq 1$),
– complete ordered m-ary: $R = 1 + X^m$ ($m \geq 1$),
– ordered m-ary: $R = (1 + X)^m$.

Note in the last case the equivalence of the two differential equations

$$Y' = (1 + Y)^m, \quad Y(0) = 0 \qquad \text{and} \qquad V' = V^m, \quad V(0) = 1. \qquad (22)$$

The Lie–Gröbner–Taylor Formula

It is possible to adapt to $\mathbb{L}$-species the method of combinatorial eclosions introduced in G. Labelle [179, 180] for the study of implicit functions and of differential equations in the context of $\mathbb{B}$-species (see Exercises 3.2.9 and 3.2.10). We first consider the case of the autonomous differential equation

$$Y' = R(Y), \quad Y(0) = Z. \qquad (23)$$

Theorem 6. (See Gröbner and Knapp [138].) *Let R be an $\mathbb{L}$-species. Then the canonical solution of (23), the $\mathbb{L}$-species $Y = A_R^\uparrow(X, Z)$, of increasing R-enriched rooted trees can be expressed in the form*

$$A_R^\uparrow(X, Z) = \exp(X R(Z) \partial / \partial Z) Z. \qquad (24)$$

More generally, for any $\mathbb{L}$-species F, one has

$$F\bigl(A_R^\uparrow(X, Z)\bigr) = \exp(X R(Z) \partial / \partial Z) F(Z). \qquad (25)$$

Proof. Call *points* and *buds* the elements of sort X and Z respectively, and let $\Psi = \Psi(X, Z)$ be a two-sort $\mathbb{L}$-species. Following the definition of the derivative and product of $\mathbb{L}$-species, we can interpret the application of the differential operator

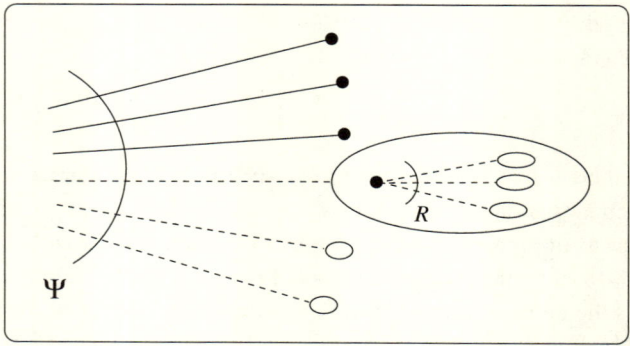

Fig. 7.

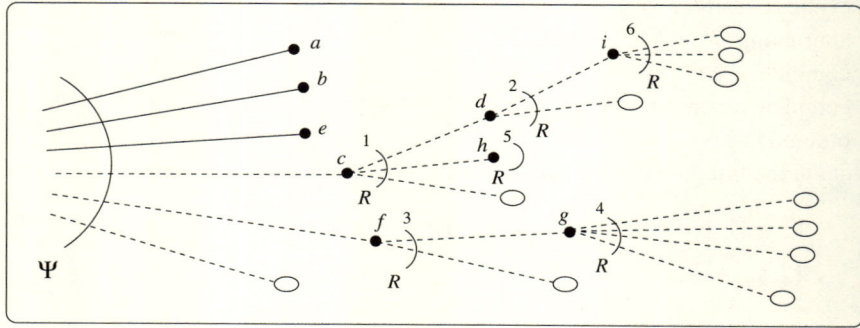

Fig. 8.

$X\mathcal{D}$ on Ψ, where $\mathcal{D} = R(Z)\partial/\partial Z$, in the following fashion: An $X\mathcal{D}\Psi$-structure can be viewed as a Ψ-structure having undergone an "eclosion," that is, where the minimum (phantom) bud has been replaced by an $XR(Z)$-structure, that is, by an arbitrary point (called the *eclosion point*) followed by an R-assembly of buds. Figure 7 represents an $X\mathcal{D}\Psi$-structure. For $n \geq 0$, $(X\mathcal{D})^n\Psi$ is obtained by successive applications (n times) of the eclosion operator $X\mathcal{D}$. The order of eclosions is important and must be noted at the eclosion points. However, this numbering is distinct from the labeling of the black points. In fact, after n eclosions, there are $n!$ ways to label the n eclosion points in an $(X\mathcal{D})^n\Psi$-structure, since their appearance comes from n products with X. Among these $n!$ structures, there is one canonical representative, where the order of the labels coincides with the order of the eclosions; we call this an *orderly labeling* of the $(X\mathcal{D})^n\Psi$-structure. See Figure 8 representing an $(X\mathcal{D})^n\Psi$-structure, where $n = 6$, and where the labeling of the eclosion points is orderly, that is, respects the order of the eclosions.

We can then interpret the quotient $(X\mathcal{D})^n/n!$ as follows: An $((X\mathcal{D})^n/n!)\Psi$-structure is identified with an $(X\mathcal{D})^n\Psi$-structure whose labeling is orderly. Finally, since

$$\exp(X\mathcal{D}) = \sum_{n\geq 0} \frac{(X\mathcal{D})^n}{n!}, \qquad (26)$$

an $\exp(X\mathcal{D})\Psi$-structure is an $((X\mathcal{D})^n/n!)\Psi$-structure for some $n \geq 0$. If we take $\Psi(X, Z) = Z$, the beginning structure is a simple bud and after one eclosion we find an $XR(Z)$-structure. After additional eclosions, we obtain an R-enriched rooted tree with (internal) points and buds. Then all the points are eclosion points, and in the case of an orderly labeling, that is, for an $\exp(X\mathcal{D})\Psi$-structure, the point labels are in increasing order, from the root to the leaves, since the eclosions occur in this manner. We then have (24).

If, at the start, we have $\Psi(X, Z) = F(Z)$, that is, an F-assembly of buds, it is clear that an $\exp(X\mathcal{D})\Psi$-structure will be simply an F-assembly of $A_R^{\uparrow}(X, Z)$-structures, whence (25). ∎

We remark that formulas (24) and (25) can be written in the form

$$A_R^{\uparrow}(X, Z) = \sum_{n\geq 0} \mathcal{D}^n(Z) \cdot E_n(X), \qquad (27)$$

$$F\big(A_R^{\uparrow}(X, Z)\big) = \sum_{n\geq 0} \mathcal{D}^n(F(Z)) \cdot E_n(X), \qquad (28)$$

since the differential operator $\mathcal{D} = R(Z)\partial/\partial Z$ commutes with multiplication by X and the factor $X^n/n!$ corresponds precisely to $E_n(X)$, that is, an assembly of n internal points arranged in an increasing manner. In other words, the right-hand side of (27) tells us to first perform the eclosions with buds only (that is, apply the operator $\mathcal{D}^n = (R(Z)\partial/\partial Z)^n$ to Z) keeping track of the order of the eclosions (see Figure 9, (a), where $n = 8$) and then to put the internal points in their natural position (compare with Figure 5). Moreover, each $(R(Z)\partial/\partial Z)^n(Z)$-structure can be transformed in a bijective manner into a structure represented generically by Figure 9, (b). It is, in fact, a (free, that is, nonstructured) rooted tree on $[n]$, where, in addition, an $R^{(\delta_i)}(Z)$-structure is attached to each vertex $i \in [n]$ of indegree $d^-(i) = \delta_i$ (only counting the vertices in $[n]$). Here, for $j \geq 0$, $R^{(j)}$ denotes the jth derivative of the species R. From this observation we deduce the following proposition (see [26]).

Proposition 7. *For $n \geq 1$, the $\mathbb{L}$-species H defined by $H(Z) = (G(Z)\frac{\partial}{\partial Z})^n Z$ can be written in the form*

$$H = \sum_{\mathbf{m}} c(\mathbf{m}) R^{m_0} R^{(1)m_1} R^{(2)m_2} \cdots, \qquad (29)$$

where the sum ranges over sequences $\mathbf{m} = (m_0, m_1, m_2, \ldots)$ of integers ≥ 0 (the multiplicities) such that $n = \sum_{i\geq 0} m_i$ and $c(\mathbf{m})$ denotes the number of increasing rooted trees on $[n]$ having, for $j = 0, 1, 2, \ldots, m_j$ vertices of indegree j.

Note that the condition

$$m_0 = 1 + \sum_{j \geq 1}(j-1) m_j \tag{30}$$

is necessary and sufficient so that the coefficient $c(\mathbf{m})$ is nonzero (see Exercise 12, e) and f), where this coefficient is computed explicitly). More generally, a similar interpretation can be given to powers of a differential operator, $(R(Z)\,d/dZ)^n$, applied to any species $F(Z)$. Indeed, writing

$$\mathcal{D}^n = \left(R(Z)\frac{d}{dZ}\right)^n = \sum_{k=1}^{n} H_k(Z)\left(\frac{d}{dZ}\right)^k, \tag{31}$$

we see, since $(d/dZ)^k E_j(Z) = E_{j-k}(Z)$, with $E_h = 0$ for $h < 0$, that

$$\mathcal{D}^n Z = H_1(Z) \quad (= H(Z)),$$
$$\mathcal{D}^n E_2(Z) = H_1(Z)Z + H_2(Z), \tag{32}$$
$$\mathcal{D}^n(E_j(Z)) = H_1(Z)E_{j-1}(Z) + H_2(Z)E_{j-2}(Z) + \cdots + H_j(Z), \quad 1 \leq j \leq n.$$

Using (28) and the preceding observation (the transition from Figure 9(a) to Figure 9(b)), a combinatorial interpretation can be given to $H_k(Z)$, similar to that of $H(Z)$, but using assemblies of k increasing rooted trees on $[n]$.

Another observation which follows from (27), or from Figure 9(a), is the fact that, for $n \geq 0$,

$$\left(G(Z)\frac{d}{dZ}\right)^n (Z) = \left.\left(\frac{\partial}{\partial X}\right)^n \mathcal{A}_R^\dagger(X, Z)\right|_{X=0}. \tag{33}$$

This shows that Gröbner's formula is in fact a form of Taylor expansion of $\mathcal{A}_R^\dagger(X, Z)$ around $X = 0$.

Example 8. (INCREASING TERNARY ROOTED TREES). Here is a simple example illustrating the use of combinatorial eclosions. Let $Y = T^\uparrow(X)$ be the $\mathbb{L}$-species of increasing ternary (plane) rooted trees, solution of the differential equation

$$Y' = Y^3, \qquad Y(0) = 1, \tag{34}$$

where the leaves are indistinguishable buds (see Figure 11(a)). Theorem 6 then permits us to write

$$T^\uparrow(X) = \exp\left(XZ^3\frac{\partial}{\partial Z}\right)(Z) \tag{35}$$

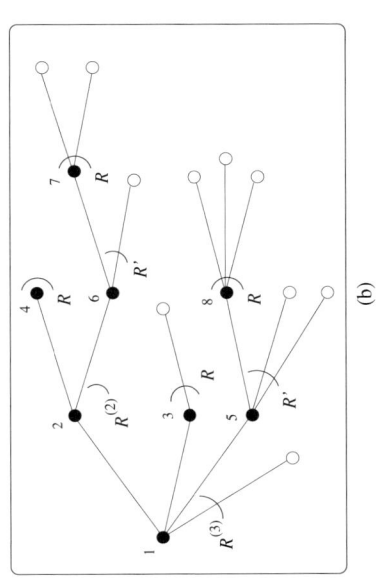
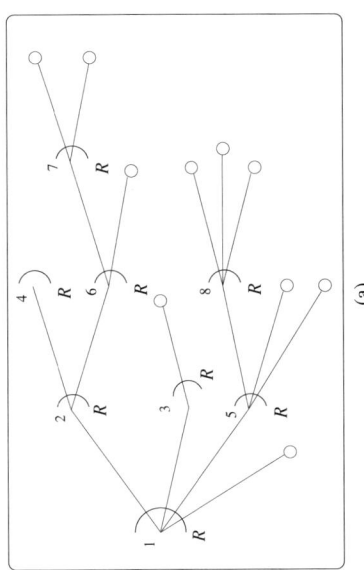

Fig. 9.

370   5. Species on Totally Ordered Sets

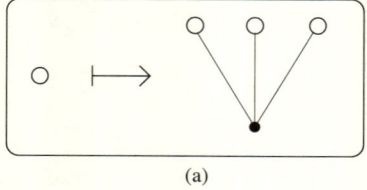
(a)

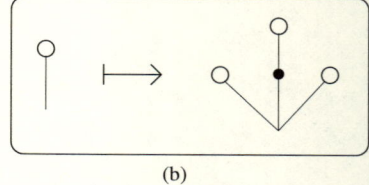
(b)

Fig. 10.

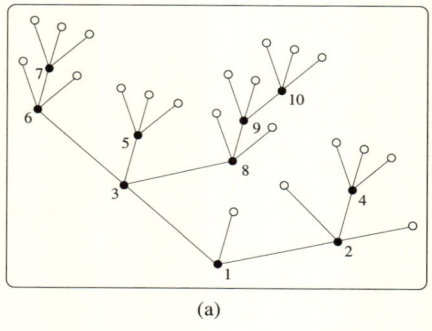
(a)

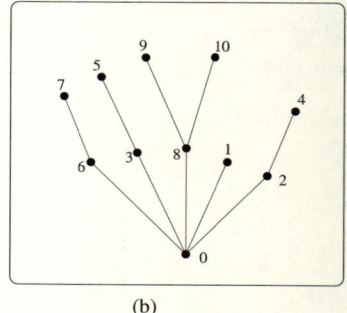
(b)

Fig. 11.

and two options are offered to us for the combinatorial interpretation of the eclosion operator $XZ^3 \partial/\partial Z$. The first (see Figure 10(a)) is the natural interpretation given by the general theory. The second supposes a base (like a flowerpot) to which the bud is attached before giving rise to an XZ^3 structure of the kind indicated in Figure 10(b). Thus the same sequence of eclosions, applied to the same buds and in the same order, will give rise, according to option (a), to a traditional ternary rooted tree such as that of Figure 11(a), or, according to option (b), to an apparently strange structure but which yields, after pruning the unnecessary buds $(Z = 1)$, quite simply an increasing ordered rooted tree, with a new minimum element (0) serving as a base for the initial bud. For example, the ordered rooted tree of Figure 11(b) corresponds to the ternary rooted tree of Figure 11(a) under this correspondence. We conclude that

$$T^\uparrow = \left(\mathcal{A}_L^\uparrow\right)' \tag{36}$$

(also see Exercise 3) and, consequently, for the generating series, using (19),

$$\begin{aligned} T(x) &= \frac{d}{dx}\mathcal{A}_L^\uparrow(x) \\ &= \frac{1}{\sqrt{1-2x}}. \end{aligned} \tag{37}$$

Systems of Differential Equations

Let us now study systems of differential equations, of the form

$$\frac{\partial Y_i}{\partial X} = Y'_i = R_i(Y_1, \ldots, Y_k), \quad Y_i(0) = Z_i, \quad i = 1, \ldots, k. \tag{38}$$

It is often advantageous to use vectorial notation: $\mathbf{Y} = (Y_1, \ldots, Y_k)$, $\mathbf{Y}' = (Y'_1, \ldots, Y'_k)$, $\mathbf{R} = (R_1, \ldots, R_k)$, $\mathbf{Z} = (Z_1, \ldots, Z_k)$, etc. The system of differential equations (38) is then more simply written as

$$\mathbf{Y}' = \mathbf{R}(\mathbf{Y}), \qquad \mathbf{Y}(0) = \mathbf{Z}. \tag{39}$$

This case is not only important by itself but also because other types of differential equations reduce to such a system by adding auxillary unknown functions (or species). Classical examples include the higher order equation

$$Y^{(k)} = F(Y, Y', Y^{(2)}, \ldots, Y^{(k-1)}) \tag{40}$$

and the nonautonomous first-order equation

$$Y' = H(X, Y), \qquad Y(0) = Z. \tag{41}$$

In (38), the R_i are given k-sort $\mathbb{L}$-species. The initial conditions $Z_1, \ldots, Z_k$ are considered as k supplementary sorts (of buds), on which the solutions also depend. However, to simplify the notation, this dependency is not always made explicit.

A solution of (38) is a family (A_i, ψ_i), for $i = 1, \ldots, k$, where A_i is an $\mathbb{L}$-species such that $A_i(0) = Z_i$ and ψ_i is an isomorphism

$$A'_i \longrightarrow R_i(A_1, \ldots, A_k). \tag{42}$$

As in the case of one equation, system (38) is equivalent to the system of integral equations

$$A_i(X) = Z_i + \int_0^X R_i(A_1(T), \ldots, A_k(T)) \, dT, \quad i = 1, \ldots, k. \tag{43}$$

For each i, the corresponding equation can be combinatorially represented as in Figure 12. It asserts that an A_i-structure is either a bud of type i, or consists of a minimum element of sort X, which is given the "color" i, and to which is attached an R_i-structure on a multiset of A_j-structures of the same kind, $j = 1, \ldots, k$. Repeating this process, we obtain, for $i = 1, \ldots, k$, the canonical solution $A_i = A^\dagger_{\mathbf{R},i}$ of increasing $\mathbf{R}$-enriched rooted trees (see Section 3.2) α, where the root is of "color" i, and where all the internal vertices u (of sort X) are first given a "color" j ($1 \leq j \leq k$) and then an R_j-structure supported by the fiber $\alpha^{-1}(u)$ (see Figure 13). The proof of the following theorem is similar to the case of one equation and is left as an exercise.

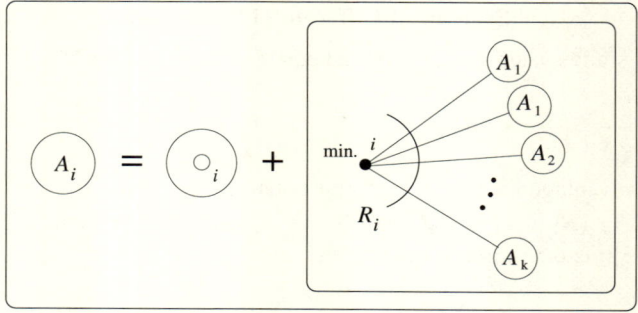

Fig. 12.

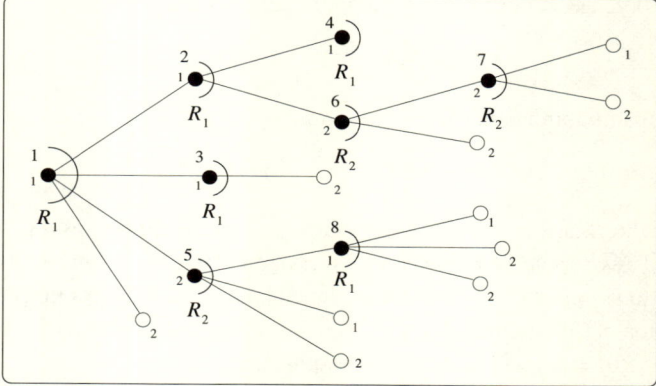

Fig. 13.

Theorem 9. *Let* $\mathbf{R} = (R_1, \ldots, R_k)$ *be a vector of k-sort* $\mathbb{L}$*-species and* $\mathbf{Z} = (Z_1, \ldots, Z_k)$, *a vector of supplementary sorts (of buds). Then the vector of* $\mathbb{L}$*-species* $\mathbf{Y} = \mathcal{A}_{\mathbf{R}}^{\uparrow} = (\mathcal{A}_{\mathbf{R},1}^{\uparrow}, \ldots, \mathcal{A}_{\mathbf{R},k}^{\uparrow})$, *of increasing* $\mathbf{R}$*-enriched rooted trees is, up to isomorphism, the unique solution of the system of differential equations (39). Moreover, for any k-sort* $\mathbb{L}$*-species* $F(X_1, \ldots, X_k)$, *one has*

$$F(\mathcal{A}_{\mathbf{R}}^{\uparrow}(X)) = \exp(X\mathcal{D})F(\mathbf{Z})$$
$$= \sum_{n \geq 0} \mathcal{D}^n F(\mathbf{Z}) E_n(X), \qquad (44)$$

where $\mathcal{D}$ *is the differential operator defined by*

$$\mathcal{D} = \sum_{j=1}^{k} R_j(Z_1, \ldots, Z_k) \frac{\partial}{\partial Z_j}. \qquad (45)$$

Note that the eclosions can be of k different sorts. For $j = 1, \ldots, k$, the operator

$$XR_j(Z_1, \ldots, Z_k)\partial/\partial Z_j$$

is interpreted by the following instructions: "Replace the minimum (phantom) bud of sort j by a point (marked with the color j) together with an $R_j(Z_1, \ldots, Z_k)$ structure." We also note that

$$A_{\mathbf{R},i}^\uparrow(X) = \sum_{n\geq 0} \mathcal{D}^n(Z_i) E_n(X). \tag{46}$$

Example 10. a) The system of differential equations (1) considered at the beginning of this section can be written

$$\begin{aligned} Y_1' &= 1 + Y_1^2, & Y_1(0) &= 0, \\ Y_2' &= Y_1 Y_2, & Y_2(0) &= 1. \end{aligned} \tag{47}$$

Its canonical solution, such as described there, is given by the increasing complete binary rooted trees, more precisely by $Y_1 = \mathcal{B}_c^\uparrow$ and $Y_2 = \mathcal{B}_{ce}^\uparrow$. See Figure 2(a) and (b), where such structures are illustrated, the "colors" 1 and 2 being represented by the colors black and white, respectively.

b) The tangent and secant functions are also solutions of the following slightly different system:

$$\begin{aligned} Y_1' &= Y_2^2, & Y_1(0) &= 0, \\ Y_2' &= Y_1 Y_2, & Y_2(0) &= 1. \end{aligned} \tag{48}$$

In this case the two sorts of eclosions are as follows (see Figure 14):

$$\text{a)} \quad XZ_2^2 \frac{\partial}{\partial Z_1}, \qquad \text{b)} \quad XZ_1 Z_2 \frac{\partial}{\partial Z_2}. \tag{49}$$

The canonical solution $Y_1 = A_1(X)$ and $Y_2 = A_2(X)$ is then obtained by applying these eclosion operators a certain number of times, starting with the buds Z_1 and Z_2, respectively, then setting $Z_1 = 0$, $Z_2 = 1$. We then obtain two subspecies A_1 and A_2 of the $\mathbb{L}$-species $\mathcal{B}^\uparrow$ of increasing binary rooted trees. We call a *left branch* of a binary rooted tree b, any maximal sub-rooted trees of b containing only left edges. The species A_1 and A_2 are the classes of increasing binary rooted trees,

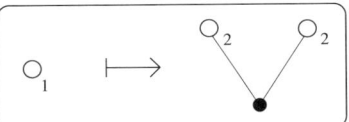 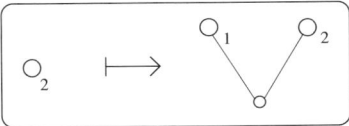

Fig. 14.

characterized by the following properties:

- A_2: all the left branches are even (contain an even number of points, not counting the buds),
- A_1: all the left branches are even, except the leftmost branch (containing the root), which is odd.

(50)

This combinatorial model of tangent and secant functions is well adapted to the establishment of integral (5.1.45), b), which can be written in an equivalent manner as

$$\tan x + \sec x = \exp\left(\int_0^x \sec t\, dt\right). \tag{51}$$

Indeed, by cutting the edges from the extreme left branch of an A_1- or of an A_2-structure, we obtain an assembly of $\int A_2$-structures.

We remark that this model can take other interesting forms. For example, Foata–Schützenberger's bijection (see Example 1) transforms these increasing binary rooted trees into particular lists called *Jacobi permutations* by Viennot (see [317]), who uses them to obtain a combinatorial model of Jacobi's elliptic functions. Another possibility is the bijection of Burge (see Example 2, b) to obtain odd or even assemblies of increasing rooted trees with even fibers.

The Nonautonomous Equation

It is worthwhile to consider nonautonomous first-order differential equations, of the form

$$\frac{\partial Y}{\partial X} = Y' = H(X, Y), \qquad Y(0) = Z, \tag{52}$$

where $H = H(X, Y)$ is a given two-sort $\mathbb{L}$-species. This equation can be replaced by a system of autonomous equations

$$\begin{aligned} Y' &= H(T, Y), & Y(0) &= Z, \\ T' &= 1, & T(0) &= 0, \end{aligned} \tag{53}$$

or directly by the integral equation

$$Y(X, Z) = Z + \int_0^X H(T, Y(T, Z))\, dT \tag{54}$$

represented by Figure 15. In any case, interpreting Z as a sort variable (the buds) and T as an auxiliary sort variable (the minibuds which will be transformed into points of sort X), we obtain the canonical solution $Y = \mathcal{A}_H^\uparrow(X, Z)$, of increasing H-enriched rooted trees (see Section 3.2), with buds. See Figure 16, which represents a generic $\mathcal{A}_H^\uparrow$-structure. We then have the following theorem, whose proof is left as an exercise.

5.2. Combinatorial Differential Equations

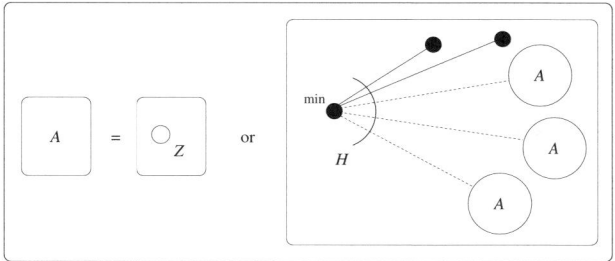

Fig. 15.

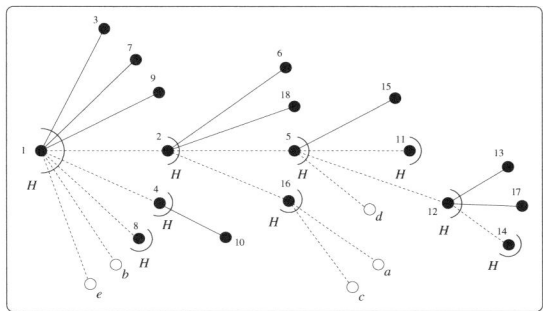

Fig. 16. An $\mathcal{A}_H^\uparrow(X, Z)$-structure.

Theorem 11. *Let H be a two-sort $\mathbb{L}$-species. Then the $\mathbb{L}$-species $Y = \mathcal{A}_H^\uparrow(X, Z)$, of increasing H-enriched rooted trees is, up to canonical isomorphism, the unique solution of the differential equation (52). Moreover, for any $\mathbb{L}$-species F,*

$$\begin{aligned} F\left(\mathcal{A}_H^\uparrow(X, Z)\right) &= \exp(X\mathcal{D})(F(Z))|_{T=0} \\ &= \sum_{n \geq 0} \mathcal{D}^n(F(Z))\bigg|_{T=0} E_n(X), \end{aligned} \qquad (55)$$

where $\mathcal{D}$ is the differential operator defined by

$$\mathcal{D} = \frac{\partial}{\partial T} + H(T, Z)\frac{\partial}{\partial Z}. \qquad (56)$$

Example 12. Consider the differential equation

$$Y' = XY^3, \quad Y(0) = 1, \qquad (57)$$

where $H(X, Y) = XY^3$ and $Z := 1$, whose analytic solution is $Y(x) = 1/\sqrt{1 - x^2}$. This equation takes the integral form

$$Y(X) = 1 + \int_0^X TY^3(T)\, dT \qquad (58)$$

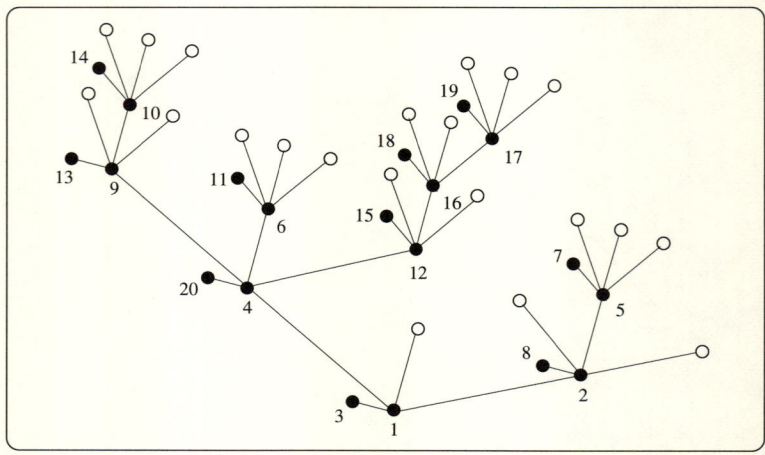

Fig. 17.

and we see the combinatorial solution of the increasing XY^3-enriched rooted trees appear, as illustrated in Figure 17. A close look at this figure and at Figure 11(a) allows us to state that

$$Y(X) = T^\uparrow(E_2(X)), \tag{59}$$

where $T^\uparrow$ denotes the species of increasing ternary rooted trees (see Example 8). We conclude that, as anticipated,

$$Y(x) = T^\uparrow\left(\frac{x^2}{2}\right) = \frac{1}{\sqrt{1-x^2}}. \tag{60}$$

Separation of Variables

The preceding example may be generalized to the case of the differential equation with separable variables

$$Y' = F(X)R(Y), \qquad Y(0) = Z. \tag{61}$$

The method of separation of variables enables us to write, for the analytic solution $Y = y(x)$ of (61):

$$\Phi(y) = \int_Z^y \frac{du}{R(u)} = \int_0^x F(t)\,dt$$
$$\implies y(x) = \Phi^{\langle-1\rangle}\left(\int_0^x F(t)\,dt\right), \tag{62}$$

where, as we have already observed at the start of this section, $y(x) = \Phi^{\langle -1 \rangle}(x)$ is a solution of the autonomous equation

$$Y' = R(Y), \qquad Y(0) = Z. \tag{63}$$

A separation of variables also occurs at the combinatorial level.

Proposition 13. *For any* $\mathbb{L}$*-species F and R, the canonical combinatorial solution* $Y = Y(X, Z)$ *of the differential equation* (61) *is the* $\mathbb{L}$*-species given by*

$$Y(X, Z) = \mathcal{A}_R^\uparrow \left(\int_0^X F(T) \, dt, Z \right), \tag{64}$$

where $\mathcal{A}_R^\uparrow(X, Z)$ *denotes the* $\mathbb{L}$*-species of increasing R-enriched rooted trees (with buds of sort Z), solution of the autonomous differential equation* (63). *Moreover, for any* $\mathbb{L}$*-species G, one has*

$$G(Y(X, Z)) = \sum_{n \geq 0} \mathcal{D}^n(G(Z)) E_n \left(\int_0^X F(T) \, dT \right), \tag{65}$$

where $\mathcal{D} = R(Z)\frac{\partial}{\partial Z}$.

Proof. Taking $H(X, Y) = F(X)R(Y)$, we immediately see from Figure 16 that an increasing $H(X, Y)$-enriched rooted tree is always obtained by substituting $\int F$-structures for the vertices of sort X in an increasing R-enriched rooted tree (see Figure 5). This is precisely what (64) asserts. Formula (65) is then an immediate consequence of (28). ∎

This result can be generalized to differential equations (and to systems of differential equations) of the form

$$Y' = \sum_j F_j(T) R_j(Y), \qquad Y(0) = Z, \tag{66}$$

where j runs over a finite set J of indices and $\{F_j\}_{j \in J}$, $\{R_j\}_{j \in J}$ are given families of $\mathbb{L}$-species (see Exercise 14 and [219, 220]).

Example 14. LINEAR EQUATIONS. Consider the first-order *linear* differential equation

$$Y' = F(X)Y + G(X), \qquad Y(0) = Z, \tag{67}$$

where F and G are given $\mathbb{L}$-species.

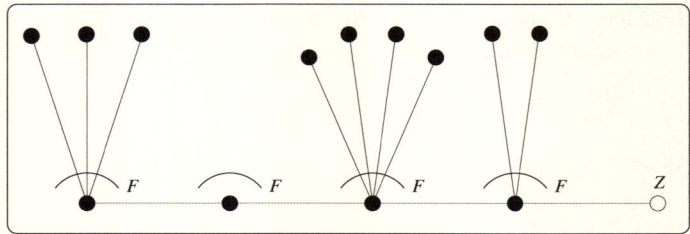

Fig. 18.

a) The homogeneous case. If $G(X) = 0$, we have the *homogeneous* linear differential equation

$$Y' = F(X)Y, \qquad Y(0) = Z. \tag{68}$$

Taking $H(X, Y) = F(X)Y$, the combinatorial solution of (68) is given by increasing $F(X)Y$-enriched rooted trees, that is, increasing rooted trees of threadlike shape illustrated in Figure 18. We denote this species by $A(X, Z)$. It is essentially an assembly (given in the increasing order of the minimum elements) of $\int F$-structures, followed by a bud. We thus have

$$A = E\left(\int_0^X F(T)\,dT\right) \cdot Z. \tag{69}$$

b) For the general case, we denote by $Y = B(X)$, the solution of the differential equation

$$Y' = F(X)Y + G(X), \qquad Y(0) = 0, \tag{70}$$

that is, (67), with $Z = 0$. Then it is easy to verify that the general solution $Y = Y(X, Z)$ of (67) is given by

$$Y(X, Z) = A(X, Z) + B(X). \tag{71}$$

The $\mathbb{L}$-species $B(X)$ also has a simple combinatorial description. Essentially (see Figure 19) a B-structure is an assembly (increasing list) of structures which are all of species $\int F$ except the last, which is an $\int G$-structure. Analytically, the method of variation of the constant further suggests the expression

$$Y = C(X) = E\left(\int_0^X F(T)\,dT\right) \cdot \int_0^X G(T) \cdot E\left(-\int_0^T F(U)\,dU\right) dT \tag{72}$$

as a solution of (70). It is left as an exercise to the reader (see [217]) to give a combinatorial interpretation of the expression (72) in terms of weighted $\mathbb{L}$-species, and to find a sign-reversing involution of the C-structures whose fixed points are B-structures, proving that $B(X)$ and $C(X)$ are equipotent $\mathbb{L}$-species.

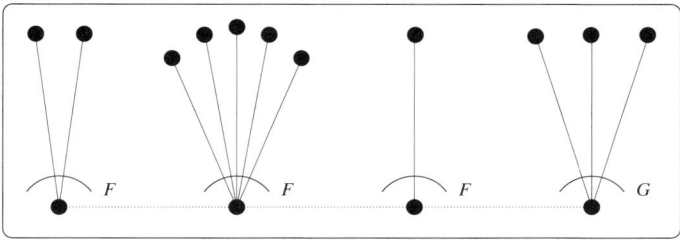

Fig. 19.

Exercises for Section 5.2

1. For the species $Y = \mathcal{A}_L^\uparrow$ of increasing ordered rooted trees set $\mathcal{A}_L^\uparrow(x) = \sum_{n \geq 1} a_n x^n / n!$.

   a) Show that $a_1 = 1$ and, for $n \geq 2$,
   $$a_n = \prod_{k=1}^{n-1}(2k-1). \tag{73}$$

   b) Let α be an ordered rooted tree on $[n]$ ($n \geq 1$), labeled in a canonical manner, for example according to the prefix order (depth first search). For $i \in [n]$, denote by n_i the cardinality (number of vertices) of the maximal sub-rooted tree α_i of α whose root is the vertex i. Show that the number of ways to label the rooted tree α in an increasing manner is equal to
   $$n! \Big/ \prod_{i \in [n]} n_i. \tag{74}$$

   HINT: First show that for any labeling of α, there is for each vertex i a unique cyclic permutation of the labels of the sub-rooted tree α_i (according to the prefix order of its vertices) which places the minimum element at the root.

   c) Show that formula (74) is valid more generally for any list of ordered rooted trees.

   HINT: Add a new root and use b) or else proceed by induction.

2. a) Show that the species $Y = \mathcal{A}_L$ of (not necessarily increasing) ordered rooted trees satisfies the functional equation $Y = X + Y^2$. Deduce an explicit isomorphism (in the form of two algorithms) $\mathcal{A}_L \simeq \mathcal{P}$, where $\mathcal{P}$ denotes the species of parenthesizations (see Example 3.2.5).

   HINT: If an ordered rooted tree is not reduced to a point, detach from the root the sub-rooted tree located at the extreme right.

   b) Likewise show that the $\mathbb{L}$-species $\mathcal{A}_L^\uparrow$, of increasing ordered rooted trees, and $\mathcal{P}$, of commutative parenthesizations (see Exercise 3.3.5), are isomorphic

and that the restriction of the isomorphism obtained in a) gives a simple encoding of the commutative parenthesizations on a totally ordered set.
3. Let $\mathcal{T}^\uparrow$ be the $\mathbb{L}$-species of increasing ternary rooted trees (see Example 8).
   a) Show that the species $Y = \int \mathcal{T}^\uparrow$ satisfies the differential equation
   $$Y' = \mathcal{B}^\uparrow(Y), \qquad Y(0) = 0, \qquad (75)$$
   where $\mathcal{B}^\uparrow$ denotes the species of increasing binary rooted trees.
   b) Using the $\mathbb{L}$-species isomorphism $\mathcal{B}^\uparrow \simeq L$, deduce an isomorphism
   $$\int \mathcal{T}^\uparrow \simeq \mathcal{A}_L^\uparrow \qquad (76)$$
   and give a fast algorithm to realize this isomorphism. Compare with the isomorphism (36) described in Example 8.
   c) Conclude that the number $\mathcal{T}_n^\uparrow$ of increasing ternary rooted trees on $[n]$ is given by
   $$\mathcal{T}_n^\uparrow = \prod_{k=1}^n (2k-1), \quad n \geq 1. \qquad (77)$$
4. By considering the extreme right branch of the even complete increasing binary rooted trees (the $\mathcal{B}_e^\uparrow$-structures; see Example 1), show that
   $$\mathcal{B}_e^\uparrow \simeq E \circ \left(\int \mathcal{B}^\uparrow \right) \qquad (78)$$
   and deduce integral (5.1.45), a): $\int_0^x \tan t \, dt = \log \sec x$.
5. MIXED SPECIES. Introduce the concept of weighted mixed species $F = F(X, Z)$ as functors
   $$F : \mathbb{L} \times \mathbb{B} \longrightarrow \mathbb{E}_\mathbb{A} \qquad (79)$$
   and develop the theory of these species (operations, associated series,...). Show how, by analogy with Definitions 2.4.12 and 2.4.13, one can pass to types according to the second sort. Consider the example of the species $Y = F(X, Z)$, solution of the differential equation
   $$\frac{\partial Y}{\partial X} = R(Y), \qquad Y(0, Z) = Z, \qquad (80)$$
   where R is a weighted $\mathbb{B}$-species and justify the specialization of the initial condition to $Z = y$, a formal variable, as well as $Z = 1$, under the hypothesis of polynomiality. Generalize to the case of systems of differential equations.
6. Prove Theorem 9 and show that Theorem 11 is a corollary.
7. Let $U = \mathcal{A}_{E_{\text{even}}}^\uparrow$ be the species of increasing rooted trees with even fibers, determined by the differential equation
   $$U' = E_{\text{even}}(U), \qquad U(0) = 0. \qquad (81)$$

a) Show that the species $Y = E_{\text{odd}}(U)$ and $Z = E_{\text{even}}(U) = U'$ satisfy the system of differential equations (48), that is,

$$Y' = Z^2, \qquad Y(0) = 0, \qquad Z' = YZ, \qquad Z(0) = 1. \qquad (82)$$

b) Illustrate with examples the isomorphism between this combinatorial model of the tangent and secant numbers and the increasing binary rooted trees of type A_1 and A_2 described in Example 10.

8. Use the various proposed combinatorial models for tangent and secant trigonometric functions to prove the following identities:

a) $\sec^2(x) = 1 + \tan^2(x)$,

b) $\tan(x+y) = \tan(x) + \tan(y) + \tan(x)\tan(y)\tan(x+y)$, \qquad (83)

c) $\sec(x+y) = \sec(x)\sec(y) + \tan(x)\tan(y)\sec(x+y)$.

9. Let $U = \mathcal{A}^{\uparrow}_{L_{\text{even}}}$ be the species of increasing ordered rooted trees with even fibers, determined by the differential equation

$$U' = L_{\text{even}}(U), \qquad U(0) = 0, \qquad (84)$$

and V, the species of increasing ordered rooted trees with even fibers, except that of the root which is odd, determined by

$$V' = L_{\text{odd}}(U), \qquad V(0) = 0. \qquad (85)$$

a) Prove the functional relations

$$\text{i)} \quad U = X + L_3^c(U), \qquad \text{ii)} \quad V = E_2(U). \qquad (86)$$

HINT: For ii), imitate the bijection illustrated in Figure 6 to establish the functional equation (18). For i), if the rooted tree is not reduced to a point, remove the last two sub-rooted trees (at the right) which are attached to the root to obtain a triplet of U-structures whose minimum is in the first position.

b) Deduce the functional equations

$$\text{i)} \quad U(x) = \frac{x}{1 - U^2(x)/3}, \qquad \text{ii)} \quad V(x) = \frac{U^2(x)}{2}, \qquad (87)$$

and use Lagrange inversion to evaluate

$$\text{i)} \quad |U[2m+1]| = \frac{(3m)!}{3^m m!}, \qquad \text{ii)} \quad |V[2m]| = \frac{(3m-2)!}{3^{m-1}(m-1)!}. \qquad (88)$$

10. INVERSIONS IN TREES AND ROOTED TREES (Mallows and Riordan [232]). An *inversion* in a rooted tree α on a totally ordered set ℓ is defined as a pair of

vertices (x, y) of α, with $x > y$ in ℓ, such that the unique path from the root to y passes through x (including the root). Denote by $\mathrm{inv}(\alpha)$ the number of inversions of α and by $\mathcal{A}_{(q)}^{\mathrm{inv}}$ (or more simply $\mathcal{A}_{(q)}$) the $\mathbb{L}$-species of rooted trees, weighted by the function $w(\alpha) = q^{\mathrm{inv}(\alpha)}$. An *inversion in a tree* α is an inversion of α considered as a rooted tree whose root is the minimum element. This defines a sub-weighted species $\mathcal{J}_{(q)} = \mathcal{J}_{(q)}^{\mathrm{inv}}$ of $\mathcal{A}_{(q)}^{\mathrm{inv}}$.

For $n \geq 1$, the polynomials $A_n(q)$ and $J_n(q)$ are defined as the generating polynomial of rooted trees and of trees on $[n]$ according to the number of inversions. In other words, we have

$$A_{(q)}(x) = \sum_{n \geq 1} A_n(q) \frac{x^n}{n!} \quad \text{and} \quad J_{(q)}(x) = \sum_{n \geq 1} J_n(q) \frac{x^n}{n!}. \tag{89}$$

a) Show that, for $n \geq 1$,

$$A_n(q) = n_q J_n(q). \tag{90}$$

b) Establish the identity

$$J'_{(q)} = E(A_{(q)}) \tag{91}$$

as well as the mutually equivalent relations

$$\begin{aligned} &\text{i)} \quad J''_{(q)} = J'_{(q)} A'_{(q)}, \\ &\text{ii)} \quad J_{n+1}(q) = \sum_{k=1}^{n} \binom{n-1}{k-1} k_q J_k(q) J_{n-k+1}(q). \end{aligned} \tag{92}$$

Conclude (Gessel and Wang [131], Kreweras [175]) that $y^{n-1} J_n(1+y)$ is the generating polynomial of connected graphs on $[n]$ according to the number of edges and, in particular, that $J_n(2)$ is equal to the number of connected graphs on $[n]$.

HINT: See recurrence formula (4.2.7).

c) Show that

$$J_n(0) = A_n(0) = (n-1)!. \tag{93}$$

HINT: For $q = 0$, $\mathcal{A}_{(0)} = \mathcal{J}_{(0)} = \mathcal{A}^{\uparrow}$ is the $\mathbb{L}$-species of increasing rooted trees.

d) For $q = -1$, decompose the species $\mathcal{J}_{(-1)}$ into its even and odd subspecies and set

$$\mathcal{E} = \mathcal{J}_{(-1),\mathrm{even}}, \quad \mathcal{O} = \mathcal{J}_{(-1),\mathrm{odd}}. \tag{94}$$

Deduce from (90) and (91) the differential equations

$$\mathcal{O}' \equiv E_{\mathrm{even}}(\mathcal{O}), \quad \mathcal{E}' \equiv E_{\mathrm{odd}}(\mathcal{O}), \tag{95}$$

where we write $F \equiv G$ to denote that two weighted $\mathbb{L}$-species F and G are equipotent ($F(x) = G(x)$). Conclude that $J_{n+1}(-1) = E_n$, the nth Euler (tangent and secant) number ([175], Pansiot [261]).

HINT: See the preceding exercise.

11. Extend the results of the preceding exercise to R-enriched rooted trees. In particular, introduce the species $\mathcal{A}_{R,(q)}$ and $\mathcal{J}_{R,(q)}$ and the polynomial enumerators $A_{R,n}(q)$ and $J_{R,n}(q)$, of R-enriched rooted trees according to the number of inversions, and show that

a) $A_{R,n}(q) = n_q J_{R,n}(q),$

b) $(\mathcal{J}_{R,(q)})' = R(\mathcal{A}_{R,(q)})$ and $(\mathcal{J}_{R,(q)})'' = (\mathcal{A}_{R,(q)})' R'(\mathcal{A}_{R,(q)}),$ (96)

c) $\mathcal{A}_{R,(1)} = \mathcal{A}_R$ and $\mathcal{A}_{R,(0)} \equiv \mathcal{J}_{R,(0)} \equiv \mathcal{A}_R^{\uparrow},$

d) $\mathcal{A}_{R,(-1)} \equiv \mathcal{A}_{R_{\text{even}}}^{\uparrow} \equiv \mathcal{J}_{R,(-1),\text{odd}}$ and $(\mathcal{J}_{R,(-1),\text{even}})' \equiv R_{\text{odd}}(\mathcal{A}_{R_{\text{even}}}^{\uparrow}).$

Consider the particular cases where $R = L$ and $R = 1 + C$ and that of plane rooted trees (see Gessel, Sagan, and Yeh [130]).

12. HISTORY OF INCREASING ORDERED ROOTED TREES. Let $n \geq 1$ and $\mathbf{d} = (d_1, d_2, \ldots, d_n)$ be a sequence of integers ≥ 0 and let $\alpha \in \mathcal{A}_R^{\uparrow}[n]$ be an increasing ordered rooted tree on $[n]$ such that for $k = 1, 2, \ldots, n$, the indegree $d^-(k)$ of the vertex k is equal to d_k.

The *history of* α is, by definition, the sequence of increasing ordered rooted trees (with buds) $(\alpha_0, \alpha_1, \alpha_2, \ldots, \alpha_n)$ obtained by successively applying, starting from a simple bud ($= \alpha_0$), the sequence of n eclosions of type $XL(Z)(\partial/\partial Z)$ giving rise to the rooted tree α ($= \alpha_n$). Figure 20 represents the first five rooted trees of the history of the rooted tree α of Figure 6. For $k = 0, 1, \ldots, n$, let b_k be the number of buds of the rooted tree α_k. Evidently one has $b_0 = 1$ and $b_n = 0$.

a) Show that for $k = 1, \ldots, n$, $b_k = b_{k-1} + d_k - 1$ and that consequently

$$b_k = \sum_{i=1}^{k} d_i - k + 1. \qquad (97)$$

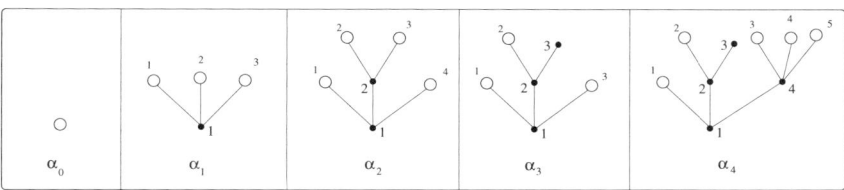

Fig. 20.

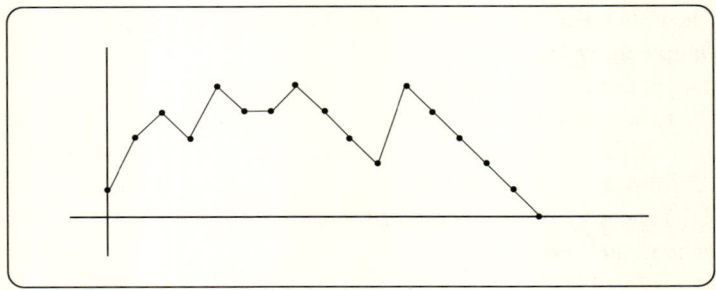

Fig. 21.

Conclude that there exists an increasing (ordered) rooted tree on $[n]$ such that for any $k \in [n]$, $d^-(k) = d_k$ if and only if

i) $\sum_{i=1}^{n} d_i = n - 1,$ ii) $\sum_{i=1}^{k} d_i > k - 1, \quad k = 1, \ldots, n - 1.$ (98)

b) If **d** satisfies condition (98), i), denote by $P(\mathbf{d})$ the polygonal line passing through the vertices $P_k = (k, b_k) \in \mathbb{N} \times \mathbb{Z}$, $k = 0, 1, \ldots, n$, where b_k is given by (97), going from $P_0 = (0, 1)$ to $P_n = (n, 0)$. If moreover condition (98), ii) is satisfied, $P(\mathbf{d})$ is called the *Lukasievich path* associated to the sequence **d** or to the rooted tree α. These paths are characterized by the conditions $b_0 = 1$, $b_k - b_{k-1} \geq -1$, for $k \in [n]$, $b_n = 0$ and $b_k > 0$ for $k \in [n-1]$. Figure 21 represents the Lukasievich path associated to the sequence $\mathbf{d} = (3, 2, 0, 3, 0, 1, 2, 0, 0, 0, 4, 0, 0, 0, 0, 0)$ and to the rooted tree in Figure 6.

Using the planar nature of an ordered rooted tree, we can number from 1 to b_k (left to right) the b_k buds of the rooted tree α_k. Set y_k equal to the number of the bud chosen for the $(k + 1)$th eclosion, $k = 0, 1, \ldots, n - 1$ (see Figure 20). One has $1 \leq y_k \leq b_k$.

Show that the history of α is entirely determined by the associated Lukasievich path and by the sequence $(y_0, y_1, \ldots, y_{n-1})$ of choices made with regard to the positions of the eclosions. Conclude that if **d** satisfies (98), i), then the number $a_{L,n}(\mathbf{d})$ of increasing ordered rooted trees on $[n]$ whose sequence of indegrees is **d** is equal to the product of the heights b_k ($k = 0, 1, \ldots, n - 1$) of the associated polygonal path:

$$a_{L,n}(\mathbf{d}) = d_1(d_1 + d_2 - 1) \cdots (d_1 + d_2 + \cdots + d_{n-1} - n + 2). \quad (99)$$

NOTE: This product will be zero if condition (98), ii) is not satisfied.

c) If **d** satisfies conditions (98), let $a_n(\mathbf{d})$ be the number of increasing rooted trees whose sequence of indegrees is equal to **d**. Show that (also see [122])

$$a_{L,n}(\mathbf{d}) = \mathbf{d}! \, a_n(\mathbf{d}), \quad (100)$$

where $\mathbf{d}! = d_1! d_2! \cdots d_n!$, and then that

$$a_n(\mathbf{d}) = \frac{d_1(d_1 + d_2 - 1) \cdots (d_1 + d_2 + \cdots + d_{n-1} - n + 2)}{\mathbf{d}!}. \tag{101}$$

d) Show that if $\mathbf{d}$ satisfies condition (98), i), then there exists a unique cyclic permutation $\hat{\mathbf{d}}$ of the sequence $\mathbf{d}$, such that the polygonal path $P(\hat{\mathbf{d}})$ is a Lukasievich path.

HINT: In the paths associated to the cyclic permutations of the sequence $\mathbf{d}$, the position i of the first vertex b_i of minimum height (= second coordinate) is cyclically permuted in the same manner as the sequence.

e) Let $\mathbf{m} = (m_0, m_1, m_2, \ldots)$ be a sequence of integers ≥ 0 such that $n = \sum_{i \geq 0} m_i$. We say that $\mathbf{m}$ is the indegree distribution of an increasing (or-dered) rooted tree α if for any i, m_i is the number of vertices of α of indegree equal to i. Show that $\mathbf{m}$ is a indegree distribution if and only if condition (30) is satisfied. Denote by Luk($\mathbf{m}$) the set of sequences $\mathbf{d} = (d_1, d_2, \ldots, d_n)$ of integers ≥ 0 such that for any i, $m_i =$ the number of occurrences of the in-teger i in the sequence $\mathbf{d}$ and such that $\mathbf{d}$ satisfies conditions (98). Note that conditions (30) and (98), i) are equivalent. Show that, under hypothesis (30),

$$|\text{Luk}(\mathbf{m})| = \frac{1}{n}\binom{n}{m_0, m_1, m_2, \ldots}. \tag{102}$$

f) Show that if the sequence of multiplicities $\mathbf{m}$ satisfies (30) and if $n = \sum_{i \geq 0} m_i$, then the coefficient $c(\mathbf{m})$ appearing in formula (29) is given by

$$c(\mathbf{m}) = \sum_{\mathbf{d} \in \text{Luk}(\mathbf{m})} \frac{d_1(d_1 + d_2 - 1) \cdots (d_1 + d_2 + \cdots + d_{n-1} - n + 2)}{d_1! d_2! \cdots d_n!}. \tag{103}$$

g) Show, by evaluating the heights of the Lukasievich path starting from the right, that under hypothesis (98), i), the conditions (98), ii) are equivalent to

$$d_{n-k} + \cdots + d_n < k + 1, \quad k = 0, \ldots, n-1, \tag{104}$$

and that formula (99) is equivalent to

$$a_{L,n}(\mathbf{d}) = (1 - d_n)(2 - d_n - d_{n-1}) \cdots (n - d_n - d_{n-1} - \cdots - d_1). \tag{105}$$

h) Let $k \geq 1$ and $\mathbf{d} = (d_1, \ldots, d_n)$ be a sequence of integers ≥ 0 such that $\sum_{i=1}^n d_i = n - k$. Let $a_n(k, \mathbf{d})$ (respectively $a_{L,n}(k, \mathbf{d})$) be the number of assemblies (respectively of lists) of k increasing rooted trees (respectively increasing ordered rooted trees) on $[n]$ whose sequence of indegrees is equal to $\mathbf{d}$. Show that these numbers are nonzero if and only if condition (104) is satisfied and that

i) $\quad a_{L,n}(k, \mathbf{d}) = (1 - d_n)(2 - d_n - d_{n-1}) \cdots$
$$(n - d_n - d_{n-1} - \cdots - d_1), \tag{106}$$

ii) $\quad a_{L,n}(k, \mathbf{d}) = k! d_1! \cdots d_n! a_n(k, \mathbf{d}).$

HINT: Add a new root numbered 0, of degree k.

i) Let $\mathbf{m} = (m_0, m_1, m_2, \ldots)$ be a sequence of integers ≥ 0 such that $\sum_i m_i = n$. Show that the coefficient $c(k, \mathbf{m})$ of $R^{m_0} R^{(1)m_1} R^{(2)m_2} \cdots$ in the polynomial expression $H_k(Z)$ defined by (31) is equal to the number of assemblies of k increasing rooted trees on $[n]$, with indegree distribution equal to $\mathbf{m}$. Show that $c(k, \mathbf{m})$ is nonzero if and only if $m_0 = k + \sum_{j \geq 1}(j - 1)m_j$ and give a formula for this coefficient, generalizing (103).

j) Verify that for $\mathcal{D} = R(Z)d/dZ$, writing $R_{(j)} = (\frac{d}{dZ})^j R(Z)$, for $j \geq 1$, and $d = d/dZ$, one has

$$\mathcal{D} = Rd,$$
$$\mathcal{D}^2 = RR_{(1)}d + R^2 d^2,$$
$$\mathcal{D}^3 = \left(RR_{(1)}^2 + R^2 R_{(2)}\right)d + 3R^2 R_{(1)} d^2 + R^3 d^3,$$
$$\mathcal{D}^4 = \left(RR_{(1)}^3 + 4R^2 R_{(1)} R_{(2)} + R^3 R_{(3)}\right)d + \left(7R^2 R_{(1)}^2 + 4R^3 R_{(2)}\right)d^2$$
$$+ 6R^3 R_{(1)} d^3 + R^4 d^4,$$
$$\mathcal{D}^5 = \left(RR_{(1)}^4 + 11R^2 R_{(1)}^2 R_{(2)} + 4R^3 R_{(2)}^2 + 7R^3 R_{(1)} R_{(3)} + R^4 R_{(4)}\right)d$$
$$+ \left(15R^2 R_{(1)}^3 + 30R^3 R_{(1)} R_{(2)} + 5R^4 R_{(3)}\right)d^2$$
$$+ \left(25R^3 R_{(1)}^2 + 10R^4 R_{(2)}\right)d^3 + 10R^4 R_{(1)} d^4 + R^5 d^5,$$
$$\mathcal{D}^6 = \left(RR_{(1)}^5 + 26R^2 R_{(1)}^3 R_{(2)} + 34R^3 R_{(1)} R_{(2)}^2 + 32R^3 R_{(1)}^2 R_{(3)}\right.$$
$$\left. + 15R^4 R_{(2)} R_{(3)} + 11R^4 R_{(1)} R_{(4)} + R^5 R_{(5)}\right)d$$
$$+ \left(31R^2 R_{(1)}^4 + 146R^3 R_{(1)}^2 R_{(2)} + 57R^4 R_{(1)} R_{(3)} + 34R^4 R_{(2)}^2\right.$$
$$\left. + 6R^5 R_{(4)}\right)d^2 + \left(90R^3 R_{(1)}^3 + 120R^4 R_{(1)} R_{(2)} + 15R^4 R_{(3)}\right)d^3$$
$$+ \left(65R^4 R_{(1)}^2 + 20R^5 R_{(2)}\right)d^4 + 15R^5 R_{(1)} d^5 + R^6 d^6.$$
$$\tag{107}$$

13. ITERATED INTEGRALS.

    a) Give a combinatorial interpretation and a pictorial representation of an iterated integral of the form
    $$\int_0^X F(T) \int_0^T G(U) \int_0^U H(V) \, dV \, dU \, dT, \tag{108}$$
    where F, G, and H are given $\mathbb{L}$-species.

    b) Let J be a finite set of indices and $\mathcal{F} = \{F_j\}_{j \in J}$ be a family of $\mathbb{L}$-species. One introduces the family $I_\mathcal{F}$ of the *iterated integrals* of $\mathcal{F}$, denoted more

5.2. Combinatorial Differential Equations

simply by I when the context is nonambiguous, which is indexed by the words $\mu \in J^*$ on "the alphabet" J. This family of $\mathbb{L}$-species $I = \{I_\mu\}_{\mu \in J^*}$ is defined recursively by

$$\begin{cases} I_\varepsilon = 1, & \text{if } \varepsilon \text{ is the empty word,} \\ I_{j\beta}(X) = \int_0^X F_j(T) I_\beta(T) \, dT, & \text{if } j \in J \text{ and } \beta \in J^*. \end{cases} \quad (109)$$

Show that

$$\sum_{\mu \in \sigma^*} I_\mu(X) = E\left(\sum_{j \in J} \int F_j\right), \quad (110)$$

where, for an arbitrary $\mathbb{L}$-species G, the notation $\int G = \int_0^X G(T) \, dT$ is used.

c) Let $J = \{1, 2\}$ and $\mathcal{F} = (F, G)$, that is, $F_1 = F$ and $F_2 = G$, where F and G are given $\mathbb{L}$-species, and denote by $I_{(F,G)}$ the family of associated iterated integrals. Show that the solution $Y = Y(X, Z)$ of the linear equation $Y' = F(X)Y + G(X)$, $Y(0) = Z$ (see Example 14), can be expressed in the form

$$Y = \sum_{k \geq 0} I_{(F,G), 1^k}(X) \, Z + \sum_{k \geq 0} I_{(F,G), 1^k 2}(X). \quad (111)$$

Show moreover that

$$I_{(F,G), 1^k} = E_k\left(\int F\right) \quad (112)$$

and that, if $F = 1$,

$$I_{(1,G), 1^k 2} = \underbrace{\int \cdots \int}_{k} G = E_k * G. \quad (113)$$

14. **GENERALIZED SEPARATION OF VARIABLES** (Leroux and Viennot [219]). Let J be a finite set of indices and $\mathcal{F} = \{F_j\}_{j \in J}$ and $\mathcal{R} = \{R_j\}_{j \in J}$ be two given families of $\mathbb{L}$-species. For $j \in J$, define the differential operator $\mathcal{D}_j = R_j(Z) d/dZ$ and extend this notation to words $\mu \in J^*$ in a recursive manner, by setting

$$\mathcal{D}_\varepsilon = \mathrm{Id}, \qquad \mathcal{D}_{j\beta} = \mathcal{D}_\beta \circ \mathcal{D}_j \quad (114)$$

(note the order reversion). Consider the solution $Y = A_H^\uparrow(X, Z)$ of the differential equation

$$Y' = \sum_{j \in J} F_j(X) R_j(Y), \qquad Y(0) = Z, \quad (115)$$

composed of $H(X, Y)$-enriched increasing rooted trees, with

$$H(X, Y) = \sum_j F_j(X) R_j(Y).$$

a) Show that a combinatorial separation of variables occurs, analogous to that of Proposition 13, and that we can write

$$\mathcal{A}_H^\uparrow(X, Z) = \sum_{\mu \in J^*} I_{\mathcal{F}, \mu}(X) \mathcal{D}_\mu(Z) \qquad (116)$$

and, more generally, for any $\mathbb{L}$-species G,

$$G\bigl(\mathcal{A}_H^\uparrow(X, Z)\bigr) = \sum_{\mu \in J^*} I_{\mathcal{F}, \mu}(X) \mathcal{D}_\mu(G(Z)). \qquad (117)$$

b) Show that formulas (116) and (117) are equally valid in the case of a system of differential equations of the form (115), where $Y = (Y_1, \ldots, Y_k)$ is a vector of $\mathbb{L}$-species, each F_j is an $\mathbb{L}$-species as before, each $R_j = (R_{j,1}, \ldots, R_{j,k})$ is a vector of k-sort $\mathbb{L}$-species, and $Z = (Z_1, \ldots, Z_k)$ is a vector of sorts, by setting, for $j \in J$, $\mathcal{D}_j = \sum_{i=1}^k R_{j,i}(Z_1, \ldots, Z_k) \partial/\partial Z_i$.

15. ELECTRIC CIRCUIT WITH A QUADRATIC RESISTANCE (Fliess [103], Leroux and Viennot [220]). Consider the differential equation

$$Y' = aY + bY^2 + U(X), \qquad Y(0) = 0, \qquad (118)$$

which models an electric circuit with a quadratic resistance. This equation is typical in control theory: $U(X)$ represents a variable input or *control* function (the current), depending on time, represented here by the variable X, and $Y = Y(X)$ is the output function (voltage). This equation is of form (115), with $J = \{0, 1\}$, $\mathcal{F} = (1, U)$ (i.e., $F_0 = 1$, $F_1 = U$), and $\mathcal{R} = (R_0, R_1)$, where $R_0(Y) = aY + bY^2$ and $R_1 = 1$.

a) Deduce the following formula for the solution $Y = V(X)$ of (118), where G is an arbitrary $\mathbb{L}$-species:

$$G(V(X)) = \sum_{\mu \in \{0,1\}^*} I_{(1,U),\mu}(X) \mathcal{D}_\mu(G(Z)) \bigg|_{Z=0}, \qquad (119)$$

where $\mathcal{D}_0 = (aZ + bZ^2) d/dZ$ and $\mathcal{D}_1 = d/dZ$.

b) For $\mu \in \{0, 1\}^*$, give a combinatorial interpretation of the iterated integral $I_{(1,U),\mu}(X)$ and of the weighted $\mathbb{L}$-species $\mathcal{D}_\mu(G(Z))|_{Z=0}$ with support equal to the empty set. By passing to the generating series we have, in particular, for $G(Z) = Z$,

$$V(X) = \sum_{\mu \in \{0,1\}^*} V_\mu I_{(1,U),\mu}(X), \qquad (120)$$

where

$$V_\mu = V_\mu(a,b) = \text{card } \mathcal{D}_\mu(Z)|_{Z=0}. \qquad (121)$$

The following formal series in noncommutative variables (0 and 1, or x_0 and x_1, if one prefers) is called a *Fliess series*:

$$S = \sum_{\mu \in \{0,1\}^*} V_\mu \, \mu. \qquad (122)$$

The goal is to compute explicitly the coefficients V_μ of the associated Fliess series. Denote by $W = W_v$, the weighted $\mathbb{L}$-species of increasing ordered rooted trees whose vertices are of indegree ≤ 2. For such a rooted tree $\alpha \in W[n]$, set $v(\alpha) = a^{m_1(\alpha)} b^{m_2(\alpha)}$, where $m_i(\alpha)$ denotes the number of vertices with indegree i in α. In addition to the list of indegrees $\mathbf{d}(\alpha) = (d^-(1), d^-(2), \ldots, d^-(n))$, one also associates to α a word $\mu = \mu(\alpha) = x_1 x_2 \cdots x_n \in \{0,1\}^n$, by the rule

$$x_k = \begin{cases} 0, & \text{if } d^-(k) = 1 \text{ or } 2, \\ 1, & \text{if } d^-(k) = 0. \end{cases} \qquad (123)$$

Show that for $\mu \in \{0,1\}^n$,

$$V_\mu(a,b) = \sum_{\substack{\alpha \in W[n] \\ \mu(\alpha) = \mu}} a^{m_1(\alpha)} b^{m_2(\alpha)}. \qquad (124)$$

c) Let $\mathbf{d} = (d_1, d_2, \ldots, d_n)$ be a sequence of integers with $0 \leq d_k \leq 2$, satisfying the conditions (98). In this case the associated Lukasievich path $P(\mathbf{d})$ is called a *Motzkin path*. These paths have also been considered in Exercise 3.4.10 with the difference that there they start at the point $(0, 0)$ and the last southeast step is suppressed. Use the results of Exercise 12 to give a more explicit description of the polynomial $V_\mu(a, b)$ in terms of Motzkin paths. Also consider the more general case of Equation (118) where the initial condition is $Y(0) = \gamma$, a formal variable.

16. Consider the case of the Duffing equation, corresponding to the cubic anharmonic oscillator (see work of Fliess, Lamnabhi–Lagarrigue, Leroux, and Viennot [104, 211, 217, 220])

$$Y'' = aY' + bY + cY^3 + U(X), \qquad Y(0) = \gamma, \qquad Y'(0) = \delta. \qquad (125)$$

In particular, starting from the combinatorial model, show how the coefficients of the Fliess series can be computed explicitly.

17. NEWTON–RAPHSON ITERATION FOR DIFFERENTIAL EQUATIONS (G. Labelle [183]). Let n be an integer ≥ 0. An increasing R-enriched rooted tree $a \in A_R^\uparrow[\ell]$

is called *light* if $|\ell| \leq n$ and *heavy* if $n+1 \leq |\ell| \leq 2n+2$. Let $\alpha = (A_R^\uparrow)_{\leq n}$ and $\beta = (A_R^\uparrow)_{\leq 2n+2} - \alpha$, the corresponding $\mathbb{L}$-species.

a) Show by a counting argument that $\beta =_{2n+2} \gamma$, where γ is the $\mathbb{L}$-species determined by the linear differential equation

$$\gamma' = G + F \cdot \gamma, \qquad \gamma(0) = 0, \tag{126}$$

where $G = R(\alpha) - \alpha'$ and $F = R'(\alpha)$. Taking into account the general solution to such an equation (see Example 14), deduce an iterative procedure with quadratic convergence $\alpha \mapsto \alpha^+ = \alpha + \beta$.

b) Adapt Algorithm 3.3.2 to the computation of the species β_{n+m+2}, $0 \leq m \leq n+1$, and of their generating series.

c) Extend this method to systems of differential equations.

18. a) Apply the iterative Newton–Raphson procedure described in Exercise 17 to the case of the differential equation

$$Y' = \mathrm{Alt}(Y), \qquad Y(0) = 0 \tag{127}$$

to compute the solution $Y = A_{\mathrm{Alt}}^\uparrow$ up to order $n = 14$.

b) Compare the result found in a) with the Taylor expansion of the explicit solution $A_{\mathrm{Alt}}^\uparrow(x) = \arcsin(e^x - 1)$.

c) Establish the asymptotic equivalence

$$|A_{\mathrm{Alt}}^\uparrow[n]| \sim \sqrt{2\log 2} \cdot \left(\frac{1}{e\log 2}\right)^n \cdot n^{n-1}, \quad n \longrightarrow \infty. \tag{128}$$

19. Let R be an $\mathbb{L}$-species such that $R(0) = 1$.

a) Solving for $A(x)$ the differential equation $A'(x) = R(A(x))$, $A(0) = 0$, show that the series $A(x)$ is the inverse, under substitution, of the series $\int_0^x dt/R(t)$. Deduce that $A(x)$ coincides with the unique solution y of either one of the functional equations

$$\text{i)} \quad y = x\hat{R} \diamond (y), \qquad \text{ii)} \quad y = x - \hat{G}(y), \tag{129}$$

where

$$\hat{R}(x) = \frac{x}{\int_0^x \frac{dt}{R(t)}}, \qquad \hat{G}(x) = x - \int_0^x \frac{dt}{R(t)}. \tag{130}$$

NOTE: The series $\hat{R}(x)$ (respectively $\hat{G}(x)$) corresponds to an $\mathbb{L}$-species if, for any $n \geq 0$, $n! \cdot [x^n]\hat{R}(x) \in \mathbb{N}$ (respectively $n! \cdot [x^n]\hat{G}(x) \in \mathbb{N}$). The example $R = 1 + X$ shows that the series $\hat{R}(x)$ and $\hat{G}(x)$ do not always correspond to $\mathbb{L}$-species.

b) Deduce from a), i) that the number a_n of increasing R-enriched rooted trees on n elements is given by

$$a_n = \hat{r}_{n-1}(n), \tag{131}$$

where $\hat{r}_k(\lambda)$ is the sequence of binomial type (see Definition 3.1.5) associated to the series $\hat{R}(x)$, that is,

$$(\hat{R}(x))^\lambda = \sum_{k\geq 0} \hat{r}_k(\lambda) \frac{x^k}{k!}. \tag{132}$$

c) Starting from a) and b), obtain explicit formulas enumerating the increasing R-enriched rooted trees in the following cases:

i) $R = (1+X)^2$,   ii) $R = E \cdot L$,   iii) $R = L(2X)$.   (133)

d) Show analytically that
   i) the solution $Y = \mathcal{A}_L^\uparrow$ of the differential equation

$$Y' = L(Y), \qquad Y(0) = 0 \tag{134}$$

is isomorphic to the solution of the functional equation

$$Y = X + E_2(Y), \qquad Y(0) = 0 \tag{135}$$

(also see Figure 6).
   ii) the solution $Y = \mathcal{A}_{L_{even}}^\uparrow$ of the differential equation (see Exercise 9)

$$Y' = L_{even}(Y), \qquad Y(0) = 0 \tag{136}$$

is isomorphic to the soluton of the functional equation

$$Y = X + L_3^c(Y), \qquad Y(0) = 0. \tag{137}$$

e) Show that one can express the series $R(x)$, starting from the series $\hat{R}(x)$ and $\hat{G}(x)$, by the formulas

$$R(x) = \frac{\hat{R}(x)}{1 - x\frac{\hat{R}'(x)}{\hat{R}(x)}}, \qquad R(x) = \frac{1}{1-\hat{G}'(x)}. \tag{138}$$

20. INCREASING MOBILES. Let $\mathcal{C}$ be the $\mathbb{L}$-species of cyclic permutations. The $\mathbb{L}$-species of *increasing mobiles* is characterized by the combinatorial differential equation $Y' = 1 + \mathcal{C}(Y)$, $Y(0) = 0$. In other words, an increasing mobile is an increasing $(1+\mathcal{C})$-enriched rooted tree. Denote by a_n the number of increasing mobiles on an ordered set with n elements.

a) Show that, for $n = 0$ to 12, the numbers a_n are given by

$$0, 1, 1, 2, 7, 36, 245, 2076, 21059, 248836, 3356609,$$
$$50896380, 856958911. \tag{139}$$

b) (Gessel, Sagan, and Yeh [130].) Show that

$$a_n = n! \cdot [x^n]\left(1 - e^{1 - E^{(-1)}(E(1) - x/e)}\right), \tag{140}$$

where $E(z) = \int_z^\infty du/ue^u$.

HINT: Use Exercise 5.2.19 a) and make a change of variable $1 - t = e^{1-u}$ in the integral $\int_0^x dt/R(t)$, with $R(t) = 1 + C(t) = 1 - \log(1 - t)$.

21. a) Let R be an $\mathbb{L}$-species such that $R(0) = 1$. Suppose that the hypotheses 3.4.71, i), ii) of Theorem 3.4.23 (Meir and Moon) are satisfied by the series $\hat{R}(x) = x / \int_0^x \frac{dt}{R(t)}$. Show that the number a_n of R-enriched increasing rooted trees on $[n]$ is given asymptotically by

$$a_n \sim \alpha \beta^n n^{n-1}, \tag{141}$$

where

$$\alpha = \sqrt{\frac{\hat{R}(\tau)}{\hat{R}''(\tau)}}, \qquad \beta = \frac{\hat{R}(\tau)}{\tau e}, \tag{142}$$

and τ is the smallest positive real satisfying the equation $R(\tau) = \infty$.

HINT: Show that $x\hat{R}'(x) - \hat{R}(x) = -x^2/(\int_0^x \frac{dt}{R(t)})^2 R(x)$.

b) Apply this result to the asymptotic enumeration of increasing R-enriched rooted trees in the case $R = (1 + X)^2$, $R = E \cdot L$, $R = L(2X)$.

c) Show why the asymptotic enumeration of increasing $(1 + C)$-enriched rooted trees (i.e., increasing mobiles) cannot be made by method a) of the present exercise even if, in this case, the coefficients of $\hat{R}(x) = x/\int_0^x \frac{dt}{1 - \log(1-t)}$ are all > 0.

NOTE: Gessel, Sagan, and Yeh [130] have nevertheless showed that the number of increasing mobiles on $[n+1]$ tends asymptotically to $(cn)^n \alpha_n \sqrt{2\pi n}$ where $c \approx 0.6169$, for certain constants $\alpha_k \geq 0$, satisfying $\sum_{k=0}^n \alpha_k \sim \log n$. In fact, they have showed that $c = 1/\int_1^\infty u^{-1} e^{-u} du$.

Appendix 1

Group Actions and Pólya Theory

This appendix is an elementary introduction to finite group actions on sets. It contains the fundamental concepts and results of this theory used in the book, such as: orbits, stabilizers, the Cauchy–Frobenius Theorem (alias Burnside's Lemma), cycle index polynomials, and the Pólya–Redfield Theorem. We place ourselves in the context of group actions on weighted sets.

Let G be a finite group and let $\mathbb{A}$ be a commutative ring with identity. We consider $\mathbb{A}$-weighted sets (Y, w) (i.e., functions $w : Y \longrightarrow \mathbb{A}$; see Section 2.3) which are finite, or more generally summable, when the ring $\mathbb{A}$ allows the introduction of such a concept (see Bourbaki, Anneaux topologiques [37] Lévesque [221]), for example the ring of formal series $\mathbb{A} = \mathbb{Q}[x, y, \ldots]$. Since morphisms of weighted sets are closed under composition, we obtain, for each weighted set (Y, w), the monoid End(Y, w) consisting of endomorphisms of (Y, w) and the group, Aut(Y, w), of automorphisms of (Y, w).

Group Actions

Definition 1. An *action of a (multiplicative) group G on an $\mathbb{A}$-weighted set (Y, w)* is a function

$$\alpha : G \times Y \longrightarrow Y \tag{1}$$

such that for any $g, h \in G$, $y \in Y$ (writing $g \cdot y = \alpha(g, y)$ to simplify notation),

$$\begin{aligned} &\text{i)} \quad g \cdot (h \cdot y) = (gh) \cdot y, \\ &\text{ii)} \quad 1 \cdot y = y, \quad \text{where 1 is the unit of } G, \\ &\text{iii)} \quad w(g \cdot y) = w(y). \end{aligned} \tag{2}$$

Remark 2. A group action α of G on (Y, w) is equivalent (see Exercise 1) to a group homomorphism

$$\varphi : G \longrightarrow \text{Aut}(Y, w), \tag{3}$$

where the relation between φ and α is given by the formula

$$\varphi(g)(y) = g \cdot y = \alpha(g, y). \tag{4}$$

One says that φ is a *permutation representation* of G, each element $g \in G$ being represented by a weight-preserving permutation $\varphi(g)$ of Y.

Definition 3. Let $\alpha : G \times Y \longrightarrow Y$ and $\beta : H \times Z \longrightarrow Z$ be two group actions on $\mathbb{A}$-weighted sets. A *morphism* of actions, $\theta : \alpha \longrightarrow \beta$, is a pair $\theta = (\psi, f)$ such that
  i) $\psi : G \longrightarrow H$ is a group homomorphism,
  ii) $f : Y \longrightarrow Z$ is a morphism of weighted sets,
  iii) (ψ, f) is *compatible* with the actions α and β, i.e., the following diagram commutes:

$$\begin{array}{ccc} G \times Y & \xrightarrow{\alpha} & Y \\ {\scriptstyle \psi \times f} \downarrow & & \downarrow {\scriptstyle f} \\ H \times Z & \xrightarrow{\beta} & Z \end{array} \tag{5}$$

Composition of morphisms of actions works componentwise; a group action *isomorphism* is an invertible morphism.

When considering two actions of the same group $G = H$, ψ is often the identity of G and its mention is omitted. In this case, a morphism is simply a weight-preserving function f which is compatible with the action, $f(g \cdot y) = g \cdot f(y)$, and an isomorphism is a bijective morphism. Moreover, if the function f is an inclusion, the action α is said to be a *sub-action* of the action β. Finally, the *product*, $\alpha \times \beta$, of two actions on the same group G is an action

$$\alpha \times \beta : G \times (Y \times Z) \longrightarrow (Y \times Z), \tag{6}$$

defined by $g \cdot (y, z) = (g \cdot y, g \cdot z)$.

Each action α of G on (Y, w) induces an equivalence relation $\sim_G$ on Y whose equivalence classes are the "orbits" of the action: For $x, y \in Y$, set

$$x \sim_G y \quad \text{if and only if there exists} \quad g \in G \quad \text{such that} \quad y = g \cdot x.$$

It is easily verified that $\sim_G$ is indeed an equivalence relation on Y. The equivalence class of an element $x \in Y$, called the *orbit* of x, is denoted by $\mathcal{O}(x)$, and we have

$$\mathcal{O}(x) = \{y \in Y \mid x \sim_G y\} = \{y \in Y \mid \exists g \in G : y = g \cdot x\}. \tag{7}$$

In the particular case where there is only one orbit, i.e., if $\mathcal{O}(x) = Y$, the action of G is said to be *transitive*.

Since Y is partitioned into a disjoint union of orbits associated to $\sim_G$, it follows that *every group action is isomorphic to a sum (disjoint union) of transitive actions*.

Given a subgroup H of a group G, consider the set

$$G/H = \{gH \mid g \in G\} \tag{8}$$

of (left) H-cosets. The group G acts naturally, by left multiplication, on this set G/H:

$$g_1 \cdot (g_2 H) = g_1 g_2 H. \tag{9}$$

This action is transitive since $g_2 H = (g_2 g_1^{-1}) \cdot (g_1 H)$. Two actions of this form associated to two subgroups H and K of G are isomorphic if and only if the subgroups H and K are conjugate, i.e., $K = gHg^{-1}$ for some $g \in G$ (see Exercise 3).

Proposition 4. *Let $\alpha : G \times Y \longrightarrow Y$ be a transitive action of G on a set Y and let $y \in Y$ and $H = \mathcal{G}_y$, the stabilizer subgroup of y (see Definition 5 below). Then α is isomorphic to the action of G on G/H by left multiplication.*

Proof. It is easily verified that the function $gy \mapsto gH$ defines a bijection $Y \xrightarrow{\sim} G/H$ which is compatible with the action of G. ∎

There follows a complete classification of the transitive actions of a (finite) group G, through the enumeration of the conjugacy classes of subgroups of G.

For a general group action, the quotient set $Y/\sim_G$, denoted more simply as Y/G, is the set of orbits of the action of G. It becomes an $\mathbb{A}$-weighted set if we put, for $\gamma = \mathcal{O}(x) \in Y/G$,

$$w(\gamma) = w(\mathcal{O}(x)) := w(x). \tag{10}$$

This definition is nonambiguous since, by assumption, $w(g \cdot x) = w(x)$, for any $g \in G$. Moreover, the weighted set $(Y/G, w)$ has the same finiteness and summability properties as (Y, w). The main problem considered here is that of finding the inventory of all orbits, i.e., to compute

$$|Y/G|_w = \sum_{\gamma \in Y/G} w(\gamma) \tag{11}$$

for each of the occurring cases. The basic tool is a general result due to Cauchy and Frobenius, and often known as Burnside's Lemma. Before stating and proving this result, we recall the definition of the stabilizer of an element.

Definition 5. Let $\alpha : G \times Y \longrightarrow Y$ be an action of a finite group G on an $\mathbb{A}$-weighted set (Y, w). If $x \in Y$, the *stabilizer* of x, denoted $\mathcal{G}_x$, is defined by

$$\mathcal{G}_x = \{g \in G \mid g \cdot x = x\}. \tag{12}$$

It is easily verified that $\mathcal{G}_x$ is a subgroup of G with the following properties.

Proposition 6. *Let* $\alpha : G \times Y \longrightarrow Y$ *be an action of a finite group G on an $\mathbb{A}$-weighted set (Y, w).*
  i) *If $y = h \cdot x$, then the stabilizer of y is obtained from the stabilizer of x by conjugation by h, i.e., $\mathcal{G}_y = h\mathcal{G}_x h^{-1}$.*
  ii) *Every orbit $\mathcal{O}(x)$ is finite and*

$$|\mathcal{O}(x)| = \frac{|G|}{|\mathcal{G}_x|}. \tag{13}$$

Proof.
  i) We have

$$g \in \mathcal{G}_y \iff g \cdot y = y \iff (gh) \cdot x = h \cdot x$$
$$\iff h^{-1}gh \cdot x = x$$
$$\iff h^{-1}gh \in \mathcal{G}_x$$
$$\iff g \in h\mathcal{G}_x h^{-1}.$$

ii) Since G is finite, it is clear that every orbit of G is finite. For $x \in Y$, let $f : G \longrightarrow \mathcal{O}(x)$ be the function defined by $f(g) = g \cdot x$. This function is clearly surjective and as G is finite, the orbit $\mathcal{O}(x)$ is also finite. Moreover, each fiber $f^{-1}(y)$ is of the form $h\mathcal{G}_x$, if $y = h \cdot x$. Indeed, we have

$$f(g) = y \iff g \cdot x = h \cdot x$$
$$\iff h^{-1}g \in \mathcal{G}_x$$
$$\iff g \in h\mathcal{G}_x.$$

Thus all the fibers have the same cardinality $|\mathcal{G}_x|$ and, consequently, $|\mathcal{O}(x)| = |G|/|\mathcal{G}_x|$. ∎

Proposition 7. CAUCHY–FROBENIUS THEOREM, ALIAS BURNSIDE'S LEMMA. *The inventory of the orbits of the action of a finite group G on a finite or summable*

$\mathbb{A}$-weighted set (Y, w) is given by

$$|Y/G|_w = \frac{1}{|G|} \sum_{g \in G} |\text{Fix}(\varphi(g))|_w, \qquad (14)$$

where for each $g \in G$, $\text{Fix}(\varphi(g))$ denotes the subset of Y of fixed points of the permutation $\varphi(g)$:

$$\text{Fix}(\varphi(g)) = \{x \in Y \mid g \cdot x = x\}. \qquad (15)$$

Proof. We have

$$|Y/G|_w = \sum_{\gamma \in Y/G} w(\gamma)$$

$$= \sum_{\gamma \in Y/G} \frac{1}{|\gamma|} \sum_{x \in \gamma} w(x)$$

$$= \sum_{x \in Y} \frac{w(x)}{|\mathcal{O}(x)|}$$

$$= \sum_{x \in Y} \frac{|\mathcal{G}_x|}{|G|} w(x)$$

$$= \frac{1}{|G|} \sum_{x \in Y} \sum_{g \in G} w(x) \chi(g \cdot x = x)$$

$$= \frac{1}{|G|} \sum_{g \in G} |\text{Fix}(\varphi(g))|_w. \qquad \blacksquare$$

Pólya Theory

Definition 8. The *cycle index polynomial* of an action of a finite group G on a finite set Y (weighted trivially by the constant function 1) is the polynomial $P_{G:Y}(y_1, y_2, y_3, \ldots)$ in the formal variables $y_1, y_2, y_3, \ldots$ defined by

$$P_{G:Y}(y_1, y_2, y_3, \ldots) = \frac{1}{|G|} \sum_{g \in G} y_1^{\varphi(g)_1} y_2^{\varphi(g)_2} y_3^{\varphi(g)_3} \cdots, \qquad (16)$$

where $\varphi : G \longrightarrow \mathcal{S}[Y]$ is the homomorphism associated to this action of G on Y and where, for any permutation σ of Y and any $k \geq 1$, σ_k denotes the number of cycles of σ of length k.

Example 9. THE GROUP OF ROTATIONS OF A CUBE. Consider a cube whose vertices are labeled by the set $\mathcal{V} = \{1, 2, \ldots, 8\}$. Denote by G the group of rotations in

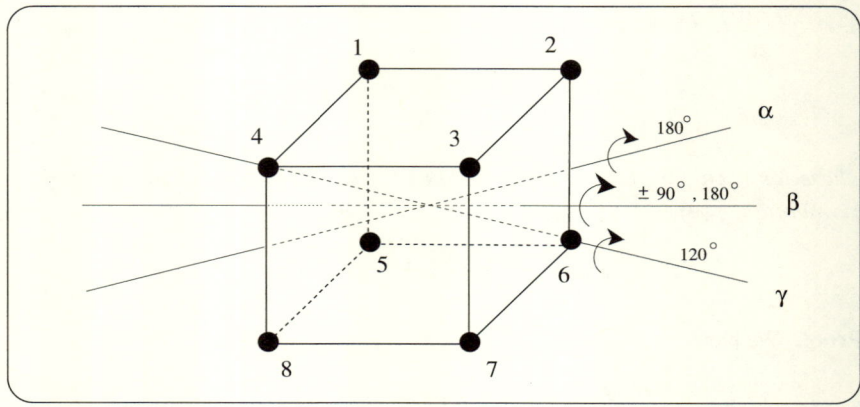

Fig. 1.

3-space which leaves this cube globally fixed. This group acts on $\mathcal{V}$ by permuting the vertices, and we obtain the cycle index polynomial

$$P_{G:\mathcal{V}}(y_1, y_2, y_3, y_4) = \frac{1}{24}\left(y_1^8 + 8y_1^2 y_3^2 + 9y_2^4 + 6y_4^2\right) \qquad (17)$$

by the following computation:

- There are 6 rotation axes of type α formed by joining the midpoints of opposite edges (see Figure 1). Each gives rise to a 180° rotation of the cube, producing a permutation of the vertices of type y_2^4, for example the permutation $\alpha_1 = (1, 7)(2, 6)(3, 5)(4, 8)$. In this case, the contribution to the polynomial will be $6\,y_2^4$.
- There are 3 rotation axes of type β formed by joining the midpoints of opposite faces. Rotating through an angle of $\pm 90°$ gives permutations of cycle type y_4^2, for example the permutation $\beta_1 = (1, 5, 8, 4)(2, 6, 7, 3)$. Rotating through an angle of 180° gives permutations of cycle type y_2^4, for example the permutation $\beta_2 = (1, 8)(2, 7)(3, 6), (4, 5)$. Thus the corresponding contribution for these axes is $6\,y_4^2 + 3\,y_2^4$.
- There are 4 axes of symmetry of type γ formed by joining opposite vertices and rotating the cube through an angle of $\pm 120°$. This gives permutations of the vertices of cycle type $y_1^2 y_3^2$, for example the permutation $\gamma_1 = (1, 8, 3)\,(2, 5, 7)\,(4)\,(6)$. The contribution to the cycle index polynomial is in this case $8\,y_1^2\,y_3^2$.
- Finally there is the identity rotation giving rise to the identity permutation of cycle type y_1^8.

We conclude that $|G| = 6 + 9 + 8 + 1 = 24$ and the computation of the cycle index polynomial (16) is then immediate.

Additionally, the group G acts on the set $\mathcal{E}$ of the edges of the cube as well as on the set $\mathcal{F}$ of the faces of the cube (see Exercise 4). In this last case, the cycle index polynomial is

$$P_{G:\mathcal{F}} = \frac{1}{24}\left(y_1^6 + 3\,y_1^2\,y_2^2 + 6\,y_1^2\,y_4 + 6\,y_2^3 + 8\,y_3^2\right). \quad (18)$$

Example 10. The symmetric group $\mathcal{S}_n$ acts by permutation on $[n] = \{1, 2, \ldots, n\}$, $n \geq 0$. The corresponding cycle index polynomial is usually denoted more simply as: $P_{\mathcal{S}_n} = P_{\mathcal{S}_n : [n]}$. For example,

$$P_{\mathcal{S}_4}(y_1, y_2, y_3, y_4) = \frac{1}{24}\left(y_1^4 + 6\,y_1^2 y_2 + 8\,y_1 y_3 + 3\,y_2^2 + 6\,y_4\right). \quad (19)$$

Example 11. The symmetric group $\mathcal{S}_n$ also acts on the set $\wp^{[2]}[n]$ of unordered pairs of elements of $[n]$ by setting, for $\sigma \in \mathcal{S}_n, \sigma \cdot \{a, b\} = \{\sigma(a), \sigma(b)\}$. The cycle index polynomial of this action is denoted by $P_{\mathcal{S}_n^{(2)}} = P_{\mathcal{S}_n : \wp^{[2]}[n]}$. For example we have

$$P_{\mathcal{S}_4^{(2)}}(y_1, y_2, y_3, y_4) = \frac{1}{24}\left(y_1^6 + 9\,y_1^2 y_2^2 + 8\,y_3^2 + 6\,y_2 y_4\right). \quad (20)$$

The general formula for $P_{\mathcal{S}_n^{(2)}}$ is easily deduced from relation (2.2.20) (see also Harary and Palmer [144]):

$$P_{\mathcal{S}_n^{(2)}}(y_1, y_2, \ldots) = \frac{1}{n!} \sum_{\sigma \in \mathcal{S}_n} \prod_{1 \leq i < j \leq n} y_{[i,j]}^{(i,j)\sigma_i \sigma_j} \prod_{k=1}^{n} y_k^{k\binom{\sigma_k}{2} + \frac{1}{2}(k-2+(k \bmod 2)) + \sigma_{2k}}. \quad (21)$$

Let $\alpha : G \times D \longrightarrow D$ be a group action, where G and D are finite, and $\varphi : G \longrightarrow \mathcal{S}[D]$, the associated group homomorphism. For arbitrary sets W and U, we introduce the following sets:

- $\Phi[D, W] := W^D = $ the set of functions $f : D \longrightarrow W$, called the *colorings* of D by W.
- $\mathrm{Bij}[D, U] := $ the set of bijections $\beta : D \widetilde{\longrightarrow} U$, called the *labelings* of D by U.

Hence a coloring is an arbitrary function, while a labeling is necessarily bijective. There exists an action induced by G on each of these sets defined, for $g \in G$, $f \in \Phi[D, W], \beta \in \mathrm{Bij}[D, U], d \in D$, by

$$(g \cdot f)(d) = f(g^{-1} \cdot d), \qquad (g \cdot \beta)(d) = \beta(g^{-1} \cdot d). \quad (22)$$

One shows easily that these formulas define group actions of G. The orbits of these actions are respectively called G-*reduced colorings* and G-*reduced labelings*. We wish to identify two colorings or two labelings when one is obtained from the other by the action of an element $g \in G$. We will thus focus our attention on the quotient sets

$$\Phi[D, W]/G \quad \text{and} \quad \mathrm{Bij}[D, U]/G. \quad (23)$$

Example 12. We have seen that the rotation group G of the cube acts on the set $V = \{1, 2, \ldots, 8\}$ of its vertices (see Example 9). Let us take the set of *colors*

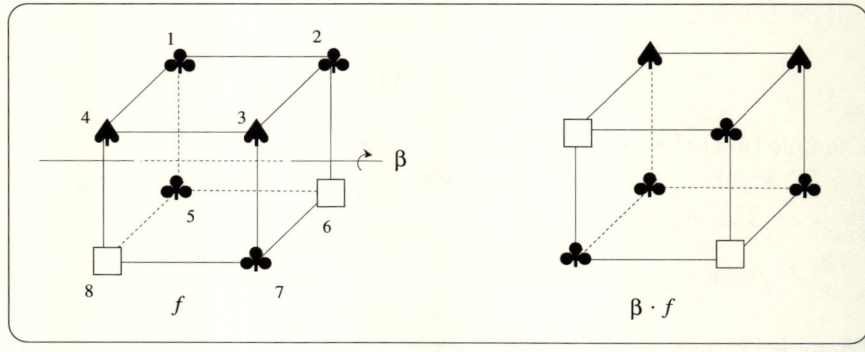

Fig. 2.

to be $W = \{\clubsuit, \square, \spadesuit\}$, and consider the coloring of the vertices given by the function

$$f = \begin{pmatrix} 1 & 2 & 3 & 4 & 5 & 6 & 7 & 8 \\ \clubsuit & \clubsuit & \spadesuit & \spadesuit & \clubsuit & \square & \clubsuit & \square \end{pmatrix}. \tag{24}$$

Let β be the 90° rotation about the axis passing through the midpoints of the opposite faces $(1, 5, 8, 4)$ and $(2, 6, 7, 3)$. The colorings f and $\beta \cdot f = f \circ \varphi(\beta^{-1})$, where $\varphi(\beta)$ is the permutation $(1, 5, 8, 4)(2, 6, 7, 3)$, are represented in Figure 2. The reader can check that the formula $\beta \cdot f = f \circ \varphi(\beta^{-1})$ corresponds to turning the coloring f according to the rotation β. It becomes clear that the orbit of f under the action of G, that is, G-reduced colorings, corresponds to the intuitive notion of coloration of the vertices of a cube which is allowed to turn around freely. We agree to identify the set $\Phi[S, W]/G$ of orbits of colorings under the action of the rotation group G of the cube with the set of *vertex colorations of the cube*. One obtains in the same fashion the set $\Phi[\mathcal{E}, W]/G$ of edge colorations and the set $\Phi[\mathcal{F}, W]/G$ of face colorations of the cube.

In an analogous manner, we can identify G-reduced labelings of the vertices of a cube by a set U, say $U = \{a, b, c, \ldots, h\}$, with labelings of the vertices of a cube which can be freely rotated.

Remark 13. The enumeration of G-reduced labelings is easy. Let $\varphi : G \longrightarrow S[D]$ be the group homomorphism associated to the action $G \times D \longrightarrow D$ and denote by $\text{Ker}\,\varphi$ and $\varphi(G)$, respectively, the kernel and the image of φ. For $\beta \in \text{Bij}[D, U]$ and $g \in G$, we have

$$g \cdot \beta = \beta \iff \beta \circ \varphi(g^{-1}) = \beta$$

$$\iff g \in \text{Ker}\,\varphi.$$

A simple application of the Cauchy–Frobenius Theorem gives

$$|\text{Bij}[D, U]/G| = \frac{1}{|G|} |\text{Ker}\,\varphi|\, n! \qquad (25)$$
$$= \frac{n!}{|\varphi(G)|}.$$

Suppose now that the set W of colors is $\mathbb{A}$-weighted by a weight function $w : W \longrightarrow \mathbb{A}$. The set W can be finite or summable and $|W|_w = \sum_{y \in W} w(y)$ denotes *the inventory of colors*. In this case, the set $\Phi[D, W]$ of colorings is itself weighted by a function w defined by

$$w(f) = \prod_{x \in D} w(f(x)), \qquad (26)$$

corresponding to a multiplicative census of the colors used for the coloring f. The inventory of all colorings is easily produced when D is finite since we have the following formula, generalizing the classical equality $|W^D| = |W|^{|D|}$.

Proposition 14. *Let D be a finite set and (W, w) be a finite or summable $\mathbb{A}$-weighted set. One has*

$$|\Phi[D, W]|_w = |W|_w^{|D|}. \qquad (27)$$

Proof. By hypothesis D is finite, thus

$$|\Phi[D, W]|_w = \sum_{f \in \Phi[D,W]} w(f)$$
$$= \sum_{f \in \Phi[D,W]} \prod_{x \in D} w(f(x))$$
$$= \prod_{x \in D} \sum_{y \in W} w(y)$$
$$= |W|_w^{|D|}. \qquad \blacksquare$$

Example 15. a) Let $W = \{\text{green, red, black}\}$ and set $w(\text{green}) = g$, $w(\text{red}) = r$, $w(\text{black}) = b$, where g, r, and b are indeterminates, with $\mathbb{A} = \mathbb{Q}[g, r, b]$. We have $|W|_w = g + r + b$ and, for the colorings f and $\beta \cdot f$ (see Figure 2) of the cube vertices, $w(f) = g^2 r^2 b^4 = w(\beta \cdot f)$. Moreover, we have

$$|\Phi[S, W]|_w = (g + r + b)^8. \qquad (28)$$

b) Let $W = \{0, 1\}$, $\mathbb{A} = \mathbb{Z}[y]$, $w(0) = 1$, and $w(1) = y$. Then a coloring $f : D \longrightarrow \{0, 1\}$ can be considered as the characteristic function $f = \chi_A$ of a subset A of D, and we have $w(\chi_A) = y^{|A|}$. It follows that $\Phi[D, W] \cong \wp[D]$ and

$$|\wp[D]|_w = (1 + y)^{|D|}. \tag{29}$$

c) More generally, let $W = \mathbb{N} = \{0, 1, 2, \ldots\}$ and $\mathbb{R} = \mathbb{Z}[\![y]\!]$ and set $w(i) = y^i$. We have

$$\begin{aligned}|\mathbb{N}|_w &= 1 + y + y^2 + y^3 + \cdots \\ &= \frac{1}{1-y},\end{aligned} \tag{30}$$

and, in this case, a coloring $f : D \longrightarrow \mathbb{N}$ can be identified with a subset of D allowing repetitions where, for $x \in D$, $f(x)$ indicates the multiplicity of the element x in the multiset. It is natural to set $|f| = \sum_{x \in D} f(x)$, the weight of f being given by

$$w(f) = \prod_{x \in D} w(f(x)) = \prod_{x \in D} y^{f(x)} = y^{|f|}. \tag{31}$$

Finally we have

$$|\Phi[D, \mathbb{N}]|_w = \left(\frac{1}{1-y}\right)^{|D|}. \tag{32}$$

d) Let $W = [m] = \{1, 2, \ldots, m\}$ and $w(i) = t_i$, where the $t_1, t_2, \ldots$, are formal variables. We have $|[m]|_w = t_1 + t_2 + \cdots + t_m$. It is possible to take $m = \infty$ in which case $[\infty] = \mathbb{N} \setminus \{0\}$ and $|[\infty]|_w = t_1 + t_2 + \cdots = : p_1(t_1, t_2, \ldots)$. This situation is very general since we may substitute arbitrary values for the indeterminates t_i subject to a condition of summability.

e) Let $F = F_v(X)$ be an $\mathbb{A}$-weighted species of structures. Denote by W the set of types of F-structures and, for $\tau \in W$, set $w(\tau) = v(\tau) x^{n(\tau)}$, where $n(\tau)$ is the underlying cardinality of an F-structure of type τ. We evidently have $|W|_w = \tilde{F}_v(x)$. In this case, a function $f \in \Phi[D, W]$ can be considered as a family $f = \{f_i\}_{i \in D}$ of types of F-structures with $w(f) = x^n \prod_{i \in D} v(f_i)$, where $n = \sum_{i \in D} n(f_i)$ is the total underlying cardinality. We conclude that $|\Phi(D, W)|_w = (\tilde{F}_v(x))^{|D|}$.

Observe that the extension of the action $\alpha : G \times D \longrightarrow D$ of a finite group G to the colorings $f \in \Phi[D, W]$ is compatible with the weighting $w : \Phi[D, W] \longrightarrow \mathbb{A}$ induced from a weighting $w : W \longrightarrow \mathbb{A}$. Indeed, we have, for $g \in G$ and $f \in \Phi[D, W]$,

$$w(g \cdot f) = \prod_{x \in D} w((g \cdot f)(x)) = \prod_{x \in D} w(f(\varphi(g^{-1})(x))) = \prod_{x \in D} w(f(x)) = w(f), \tag{33}$$

since $\varphi(g^{-1})$ is a permutation of D. We can thus speak of the weight of a G-reduced coloring and try to compute the inventory of $\Phi[D, W]/G$. The following theorem enables this computation to be made in general. It is the fundamental result of Pólya theory. This formula brings in the weightings $w^2, w^3, \ldots$ of a set W, obtained from a given weighting $w : W \longrightarrow \mathbb{A}$ in the following manner: For $y \in W$ and $k \geq 0$,

$$w^k(y) = (w(y))^k \quad (\in \mathbb{A}). \tag{34}$$

Example 16. For each of the preceding examples (see Example 15), we have

a) $|W|_{w^k} = g^k + r^k + b^k,$

b) $|\{0, 1\}|_{w^k} = 1 + y^k,$

c) $|\mathbb{N}|_{w^k} = (1 - y^k)^{-1},$ \hfill (35)

d) $|[\infty]|_{w^k} = t_1^k + t_2^k + \cdots = p_k(t_1, t_2, \ldots),$

e) $|W|_{w^k} = \widetilde{F}_{v^k}(x^k).$

Theorem 17. PÓLYA–REDFIELD. (See Pólya [263], [264] and Redfield [275].) *Let $\alpha : G \times D \longrightarrow D$ be an action of a finite group G on a finite set D and (W, w) an $\mathbb{A}$-weighted set (finite or summable). Then the inventory of G-reduced colorings of D in W is given by*

$$\begin{aligned} |\Phi[D, W]/G|_w &= P_{G:D}(y_1, y_2, y_3, \ldots)|_{y_k := |W|_{w^k}} \\ &= P_{G:D}(|W|_w, |W|_{w^2}, |W|_{w^3}, \ldots). \end{aligned} \tag{36}$$

Proof. Denote by $\overline{\varphi} : G \longrightarrow \text{Aut}(\Phi[D, W], w)$ the group homomorphism associated to the action induced by that of G on D. Recall that for $g \in G$ and $f \in \Phi[D, W]$, we have

$$\overline{\varphi}(g)(f) = f \circ \varphi(g^{-1}). \tag{37}$$

It follows from Proposition 7 (Cauchy–Frobenius Theorem, alias Burnside's Lemma) that

$$|\Phi[D, W]/G|_w = \frac{1}{|G|} \sum_{g \in G} |\text{Fix}\overline{\varphi}(g)|_w. \tag{38}$$

But for a function $f \in \Phi[D, W]$, we have

$$f \in \text{Fix}\overline{\varphi}(g) \iff \overline{\varphi}(g)(f) = f$$
$$\iff f \circ \varphi(g^{-1}) = f$$
$$\iff f \text{ is constant on each of the cycles of } \varphi(g^{-1})$$
$$\iff f \text{ is constant on each of the cycles of } \varphi(g),$$

and in this case, if $C(\varphi(g))$ denotes the set of cycles of $\varphi(g)$, and if for $c \in C(\varphi(g))$, $\ell(c)$ denotes the length of the cycle c and $f(x) = f(c)$ for all $x \in c$, then

$$w(f) = \prod_{x \in D} w(f(x))$$
$$= \prod_{c \in C(\varphi(g))} \prod_{x \in C} w(f(x))$$
$$= \prod_{c \in C(\varphi(g))} w^{\ell(c)}(f(c))$$

so that

$$\frac{1}{|G|} \sum_{g \in G} |\text{Fix}\overline{\varphi}(g)|_w = \frac{1}{|G|} \sum_{g \in G} \sum_{f \in \text{Fix}\overline{\varphi}(g)} w(f)$$
$$= \frac{1}{|G|} \sum_{g \in G} \sum_{f \in \text{Fix}\overline{\varphi}(g)} \prod_{c \in C(\varphi(g))} w^{\ell(c)}(f(c))$$
$$= \frac{1}{|G|} \sum_{g \in G} \sum_{f : C(\varphi(g)) \to W} \prod_{c \in C(\varphi(g))} w^{\ell(c)}(f(c))$$
$$= \frac{1}{|G|} \sum_{g \in G} \prod_{c \in C(\varphi(g))} \sum_{y \in W} w^{\ell(c)}(y)$$
$$= \frac{1}{|G|} \sum_{g \in G} \prod_{c \in C(\varphi(g))} |W|_{w^{\ell(c)}}$$
$$= \frac{1}{|G|} \sum_{g \in G} |W|_w^{\varphi(g)_1} |W|_{w^2}^{\varphi(g)_2} \cdots$$
$$= P_{G:D}(|W|_w, |W|_{w^2}, \ldots),$$

which completes the proof. ∎

Corollary 18. *If the set W of colors is finite then the number of G-reduced colorings of D in W is given by*

$$|\Phi[D, W]/G| = P_{G:D}(|W|, |W|, \ldots). \tag{39}$$

Proof. Choose the weighting $w : W \longrightarrow \mathbb{Z}$ to be the constant 1. ∎

Example 19. Let us call *colorations* the G-reduced colorings of the vertices, edges, or faces of the cube, where G is the rotation group of the cube, with an inventory of colors $|W|_w = g + b + r$ (see Examples 9, 12, and 16). The inventory of colorations is then obtained by applying Pólya–Redfield Theorem. For example, for the face colorations of the cube, we find, using (18) and (35), a),

$$\begin{aligned}
c_{\mathcal{F}}(g,b,r) &:= |\Phi[\mathcal{F},W]/G|_w \\
&= P_{G:\mathcal{F}}(|W|_w, |W|_{w^2}, |W|_{w^3}, |W|_{w^4}) \\
&= (g^6 + b^6 + r^6) + (g^5 b + g^5 r + b^5 g + r^5 g + b^5 r + r^5 b) \\
&\quad + 2(g^4 b r + b^4 g r + r^4 g b) + 2(g^3 b^3 + g^3 r^3 + b^3 r^3) \\
&\quad + 2(g^4 b^2 + g^4 r^2 + b^4 g^2 + r^4 g^2 + b^4 r^2 + r^4 b^2) + 6g^2 b^2 r^2 \\
&\quad + 3(g^3 b^2 r + g^3 r^2 b + b^3 g^2 r + b^3 r^2 g + r^3 g^2 b + r^3 b^2 g).
\end{aligned} \quad (40)$$

For instance, the term $3g^3 b^2 r$ indicates that there are 3 distinct face colorations of the cube using the color green thrice, the color blue twice, and the color red once.

It is clear that the polynomial $c_{\mathcal{F}}(g,b,r)$ is symmetric in g, b, and r and that the total number of colorations of the faces of the cube is given by

$$\begin{aligned}
c_{\mathcal{F}}(1,1,1) &= P_{G:\mathcal{F}}(3,3,3,3) \\
&= 57.
\end{aligned} \quad (41)$$

Example 20. A simple graph on a set of vertices $[n]$ can be considered as a choice of a set of edges, i.e., of a subset A of $\wp^{[2]}[n]$, or else as a characteristic function

$$\chi_A : \wp^{[2]}[n] \longrightarrow \{0, 1\},$$

weighted by $w(\chi_A) = y^{|A|}$. The variable y is said to be an *edge counter*. Moreover the action of the symmetric group $\mathcal{S}_n$ on $\wp^{[2]}[n]$ (see Example 11) induces on $\Phi[\wp^{[2]}[n], \{0,1\}]$ an $\mathcal{S}_n$ action given by (22). It is easily verified that for a permutation σ of $[n]$, and $A \subseteq \wp^{[2]}[n]$,

$$\sigma \cdot \chi_A = \chi_{\wp^{[2]}[\sigma](A)}, \quad (42)$$

i.e., that the edges of the image graph are the images under σ of the edges of the original graph. This means that these two graphs are isomorphic and the (quotient) set of $\mathcal{S}_n$-reduced colorings is in bijection with the set $T_n(\mathcal{G})$ of unlabeled graphs of order n, weighted by an edge counter. Applying the Pólya–Redfield Theorem gives the generating function

$$\begin{aligned}
T_n(\mathcal{G})(y) &= \left|\Phi[\wp^{[2]}[n], \{0,1\}]/\mathcal{S}_n\right|_w \\
&= P_{\mathcal{S}_n^{(2)}}(1+y, 1+y^2, \ldots).
\end{aligned} \quad (43)$$

For example, using (20), we see that the generating polynomial of unlabeled graphs of order 4 according to the number of edges is

$$\begin{aligned}
T_4(\mathcal{G})(y) &= P_{\mathcal{S}_4^{(2)}}(1+y, 1+y^2, 1+y^3, 1+y^4) \\
&= 1 + y + 2y^2 + 3y^3 + 2y^4 + y^5 + y^6.
\end{aligned} \quad (44)$$

Exercises for Appendix 1

1. Show that formula (4) establishes an equivalence between actions $\alpha : G \times Y \longrightarrow Y$ on a weighted set $Y = (Y, w)$, and permutation representations of G, that is, group homomorphisms $\varphi : G \longrightarrow \text{Aut}(Y, w)$.

2. Given an action $G \times Y \longrightarrow Y$ of a group G on a weighted set (Y, w), show that the relation $\sim_G$ on Y is an equivalence relation, whose equivalence classes are the orbits (7), and is compatible with the weighting, that is

$$x \sim_G y \implies w(x) = w(y). \qquad (45)$$

3. a) Show that formula (9) defines a transitive group action $G \times G/H \longrightarrow G/H$ and prove Proposition 4.

    b) Let H and K be two subgroups of a finite group G. Show that the associated actions $G \times G/H \longrightarrow G/H$ and $G \times G/K \longrightarrow G/K$ are isomorphic if and only if there exists an element $g \in G$ such that $K = g^{-1}Hg$.

4. The rotation group G of a cube acts on the set $\mathcal{V}$ of its 8 vertices, the set $\mathcal{E}$ of its 12 edges, and the set $\mathcal{F}$ of its 6 faces (see Examples 9, 12, and 19).

    a) Show that the cycle index polynomials $P_{G:\mathcal{F}}$ and $P_{G:\mathcal{E}}$ are respectively given by (18) and by

    $$P_{G:\mathcal{E}} = \frac{1}{24}(y_1^{12} + 6y_1^2 y_2^5 + 3y_2^6 + 8y_3^4 + 6y_4^3). \qquad (46)$$

    b) Compute the inventory of vertex colorations of a cube using three colors, as well as the total number of such colorations.

    c) Answer the same question for the case of edge colorations of the cube.

5. Enumerate (algebraically and combinatorially) unlabeled graphs of order 5 according to the number of edges.

6. The cyclic group C_n and the dihedral group D_n can be considered respectively as the plane rotation group and the isometry group of a regular polygon with n sides. One has

$$C_n = \{1, \rho, \rho^2, \ldots \rho^{n-1}\} \quad D_n = \{1, \rho, \rho^2, \ldots \rho^{n-1}, \tau, \tau\rho, \tau\rho^2, \ldots \tau\rho^{n-1}\},$$

where ρ is the plane rotation through an angle of $2\pi/n$ about the center of the polygon and τ is the reflection by an axis passing through the center of the polygon and the center of one of its sides.

a) Show that the cycle index polynomial of the action of C_n on the set V of vertices of the polygon is given by

$$P_{C_n : V} = \frac{1}{n} \sum_{d | n} \phi(d) y_d^{n/d}, \qquad (47)$$

where ϕ is the Euler ϕ-function. Deduce a formula for the number of colorations of the vertices of this polygon with c colors, up to (plane) rotations.

b) Show that the cycle index polynomial of the action of D_n on V is given by

$$P_{D_n:V} = \frac{1}{2}P_{C_n:V} + \begin{cases} \frac{1}{2}y_1 y_2^{(n-1)/2}, & \text{if } n \text{ is odd,} \\ \frac{1}{4}\left(y_2^{n/2} + y_1^2 y_2^{(n-2)/2}\right), & \text{if } n \text{ is even.} \end{cases} \quad (48)$$

Deduce expressions for the number of colorations of this polygon with c colors, up to isometry.

NOTE: The number of colorations of n sectors of a casino roulette, with c colors, corresponds to the number of colorations found in a). Likewise, the number of single strand necklaces with n colored beads corresponds to the number found in b).

7. a) Show that two isomorphic actions have the same cycle index polynomial.
   b) Let $\varphi : G \longrightarrow S[D]$ be the group homomorphism associated to a set action $\alpha : G \times D \longrightarrow D$. Show that

$$P_{G:D} = P_{\varphi(G):D}. \quad (49)$$

8. a) Give a formula, as a function of the number c of colors, for the number of distinct colorings of a regular tetrahedron, up to rotation, where each vertex, each edge, and each face is colored.
   b) Answer the same question for a regular n-dimensional simplex in n-dimensional space, up to orientation preserving isometry.

9. Let G and H be two finite groups and $\alpha : (G \times H) \times Y \longrightarrow Y$ be an action of the product group $G \times H$ on a finite set Y.
   a) Show that α corresponds to two group actions $\alpha_G : G \times Y \longrightarrow Y$ and $\alpha_H : H \times Y \longrightarrow Y$ which commute with each other in the sense that for any $g \in G, h \in H$, and $y \in Y$,

$$g \cdot (h \cdot y) = h \cdot (g \cdot y), \quad (50)$$

   and such that $(g, h) \cdot y = g \cdot (h \cdot y)$.
   b) Show that the group H acts on the set Y/G of orbits of the action of G and that the number of orbits fixed by an element $h \in H$ is

$$|\text{Fix}_{Y/G}(h)| = \frac{1}{|G|} \sum_{g \in G} |\text{Fix}_Y(g, h)|. \quad (51)$$

   c) Generalize these results to the case of actions on a summable weighted set (Y, w).
   HINT: Follow the proof of the Cauchy–Frobenius Theorem, using the fact that for $x \in Y$, if $h \cdot x = g_0 \cdot x$ then $|\mathcal{G}_x| = |\mathcal{G}_x \cdot g_0^{-1}|$ and $g \in \mathcal{G}_x \cdot g_0^{-1} \iff g \cdot (h \cdot x) = x$.

Appendix 2

Miscellaneous Tables

This apppendix contains various tables serving as references or complementing the text. The references given correspond either to formula or exercise numbers. Tables 1 and 2 give the (exponential) generating series of a certain number of basic species. Tables 3 and 4 describe the unlabeled and the asymmetry types generating series and the cycle and asymmetry index series of some elementary species. The molecular species on $n \leq 5$ points are analyzed in Table 5. One finds there, for each molecular species M, the number of structures $|M[n]|$ on $[n]$, the derivative M', and the values of the coefficients fix $M[\sigma]$ of the cycle index series Z_M. Table 6 contains the molecular decomposition and the cycle and asymmetry index series of the species $\mathcal{G}$ of simple graphs up to order 6. The molecular decompositions of the species of trees and rooted trees, up to order $n = 10$, are described in Tables 7 and 8. Tables 9 through 13 contain the generic series, up to order 6, of the species $\mathfrak{a}_R$ of R-enriched trees and the species $\mathcal{A}_R$ of R-enriched rooted trees. Table 14 gives the complete molecular decomposition of $F(X + Y)$ for each atomic species $F(X)$ on $n \leq 5$ points. As a complement to Table 4 in Section 4.4, Table 15 gives the asymmetry index series of molecular species of order $n = 5$ (see Pineau [262] where atomic species of order 6 are also covered). Tables 7 through 15 have been computed with the help of Maple [47] and Darwin [21, 44].

Table 1. *Labeled enumeration.*
1.1. *(Exponential) generating series of particular species.*

Species	F	F(x)	Ref.
Empty	0	0	(1.2.8)
Empty set	1	1	(1.2.8)
Singletons	X	x	(1.2.8)
Linear orders	L	$1/(1-x)$	(1.2.8)
Permutations	S	$1/(1-x)$	(1.2.8)
Even permutations	A	$(2-x^2)/(2(1-x))$	(4.4.69)
Chains	Cha	$(2-x^2)/(2(1-x))$	(2.6.79)
Sets	E	e^x	(1.2.8)
Subsets	$\wp$	e^{2x}	(1.2.8)
Elements	ε	xe^x	(1.2.8)
Oriented sets	$E^\pm$	$2e^x - (1+x)$	(2.6.44)
Involutions	Inv	$e^{x+x^2/2}$	
Even sets	E_{even}	$\cosh(x)$	(1.3.12)
Odd sets	E_{odd}	$\sinh(x)$	(1.3.12)
Cyclic permutations	$\mathcal{C}$	$-\log(1-x)$	(1.4.16)
Derangements	Der	$e^{-x}/(1-x)$	(1.3.29)
Set partitions	Par	$\exp(e^x - 1)$	(1.4.13)
Binary rooted trees	$\mathcal{B}$	$(1 - \sqrt{1-4x})/(2x)$	(3.0.3)
Simple graphs	$\mathcal{G}$	$\sum_{n\geq 0} 2^{\binom{n}{2}} x^n/n!$	(1.2.8)
Directed graphs	$\mathcal{D}$	$\sum_{n\geq 0} 2^{n^2} x^n/n!$	(1.2.8)
Endofunctions	End	$\sum_{n\geq 0} n^n x^n/n!$	(1.2.8)
Rooted trees	$\mathcal{A}$	$\sum_{n\geq 1} n^{n-1} x^n/n!$	(2.1.13)
Octopuses	Oct	$\ln(1-x) - \ln(1-2x)$	(1.4.63)
Ballots	Bal	$1/(2 - e^x)$	(1.3.46)

Table 1.1. (*cont.*)

Species	F	F(x)	Ref.
Commutative parenthesizations	$\mathcal{P}$	$1 - \sqrt{1-2x}$	(3.3.88)
Trees	$\mathfrak{a}$	$\sum_{n\geq 1} n^{n-2} x^n/n!$	(2.1.13)

1.2. *(Exponential) generating series of general species.*

Species	F	F(x)	Ref.
Even part	F_{even}	$(F(x) + F(-x))/2$	(1.3.54)
Odd part	F_{odd}	$(F(x) - F(-x))/2$	(1.3.54)
Connected structures	F^c	$\log F(x)$	(1.4.60)
Derivative	F'	$\dfrac{d}{dx} F(x)$	(1.3.4)
Pointing	$F^{\bullet}$	$x \dfrac{d}{dx} F(x)$	(2.1.7)
Sum	$F + G$	$F(x) + G(x)$	(1.3.4)
Product	$F \cdot G$	$F(x) \cdot G(x)$	(1.3.4)
Cartesian product	$F \times G$	$F(x) \times G(x)$	(2.1.26)
Substitution	$F \circ G$	$F(G(x))$	(1.3.4)
Functorial composition	$F \square G$	$F(x) \square G(x)$	(2.2.13)
R-enriched rooted trees	$\mathcal{A}_R = XR(\mathcal{A}_R)$	$\sum_{n\geq 1} r_{n-1}(n) x^n/n!$, $(R(x))^\lambda = \sum_{k\geq 0} r_k(\lambda) x^k/k!$	(3.1.46)
R-enriched trees	$\mathfrak{a}_R$	$r_0 x + \sum_{n\geq 2} \rho_{n-2}(n) x^n/n!$, $(R'(x))^\lambda = \sum_{k\geq 0} \rho_k(\lambda) x^k/k!$	(3.1.54)
R-enriched endofunctions	End_R	$\sum_{n\geq 0} r_n(n) x^n/n!$, $(R(x))^\lambda = \sum_{k\geq 0} r_k(\lambda) x^k/k!$	(3.1.48)

Table 2. *(Exponential) generating series for weighted species.*

Species	F	Parameter, counter	$F_w(x)$	Ref.
Subsets	$\wp_w$	elements, y	$e^{(1+y)x}$	(2.3.88)
Permutations	$\mathcal{S}_w$	cycles, α	$(1-x)^{-\alpha}$	(2.3.23)
Involutions	Inv_w	fixed points, t transpositions, -1	$e^{tx-x^2/2}$	(2.3.41)
Set partitions	Par_w	parts, t	$\exp(t(e^x-1))$	(2.3.24)
Simple graphs	$\mathcal{G}_w$	edges, y	$\sum_{n\geq 0}(1+y)^{\binom{n}{2}}x^n/n!$	(2.4.44)
Directed graphs	$\mathcal{D}_w$	arcs, y	$\sum_{n\geq 0}(1+y)^{n^2}x^n/n!$	
F_w	$F_{w^{(\alpha)}}$	connected components, α	$(F_w(x))^\alpha$	(2.3.73)

Table 3. *Unlabeled enumeration.*
3.1. Type (ordinary) generating series of particular species.

Species	F	$\widetilde{F}(x)$	Ref.
Sets	E	$(1-x)^{-1}$	(1.2.11)
Linear orders	L	$(1-x)^{-1}$	(1.2.11)
Even sets	E_{even}	$1/(1-x^2)$	(3.1.12)
Odd sets	E_{odd}	$x/(1-x^2)$	(3.1.12)
Subsets	$\wp$	$(1-x)^{-2}$	(1.2.11)
Permutations	$\mathcal{S}$	$\prod_{n\geq 1}(1-x^k)^{-1}$	(1.2.11)
Even permutations	$\mathcal{A}$	$\frac{1}{2}\prod_{n\geq 1}(1-x^n)^{-1}+\frac{1}{2}\prod_{n\geq 1}(1+x^{2n-1})$	(4.4.69)
Set partitions	Par	$\prod_{n\geq 1}(1-x^k)^{-1}$	(1.4.13)
Involutions	Inv	$(1-x)^{-1}(1-x^2)^{-1}$	
Cyclic permutations	$\mathcal{C}$	$x(1-x)^{-1}$	(1.2.11)

Table 3.1. (*cont.*)

Species	F	$\widetilde{F}(x)$	Ref.
Elements	ε	$x(1-x)^{-1}$	(1.2.11)
Derangements	Der	$\prod_{k \geq 2}(1-x^k)^{-1}$	(1.3.29)
Oriented sets	$E^{\pm}$	$1/(1-x)$	(2.6.44)
Ballots	Bal	$(1-x)/(1-2x)$	(1.3.46)
Octopuses	Oct	$-x/(1-x) - \sum_{n \geq 1} \frac{\phi(n)}{n} \log(1-2x^n)$	(1.4.63)
Chains	Cha	$1/(1-x)$	(2.6.79)
Rooted trees	$\mathcal{A}$	$\widetilde{\mathcal{A}}(x) = x \exp\left(\sum_{n \geq 1} \widetilde{\mathcal{A}}(x^n)/n\right)$	(4.1.44)
Trees	$\mathfrak{a}$	$\widetilde{\mathcal{A}}(x) + \frac{1}{2}\widetilde{\mathcal{A}}(x^2) - \frac{1}{2}(\widetilde{\mathcal{A}}(x))^2$	(4.1.5)
Endofunctions	End	$\prod_{n \geq 1} 1/(1-\widetilde{\mathcal{A}}(x^n))$	(1.4.18)
Commutative parenthesizations	$\mathcal{P}$	$\lim_{n \to \infty} 1 - \sqrt{-2x + \sqrt{-2x^2 + \cdots + \sqrt{1-2x^{2^n}}}}$	(3.3.90)

3.2. Type (ordinary) generating series of general species.

Species	F	$\widetilde{F}(x)$	Ref.
Even part	F_{even}	$(\widetilde{F}(x) + \widetilde{F}(-x))/2$	Ex. 1.3.3
Odd part	F_{odd}	$(\widetilde{F}(x) - \widetilde{F}(-x))/2$	Ex. 1.3.3
Connected structures	F^c	$\sum_{k \geq 1} \frac{\mu(k)}{k} \log \widetilde{F}(x^k)$	(2.3.71)
Derivative	F'	$\left(\frac{\partial}{\partial x_1} Z_F\right)(x, x^2, x^3, \ldots)$	(1.4.31)
Pointing	$F^{\bullet}$	$x\left(\frac{\partial}{\partial x_1} Z_F\right)(x, x^2, x^3, \ldots)$	(2.1.7)
Sum	$F + G$	$\widetilde{F}(x) + \widetilde{G}(x)$	(1.3.11)

Table 3.2. (cont.)

Species	F	$\widetilde{F}(x)$	Ref.
Product	$F \cdot G$	$\widetilde{F}(x)\widetilde{G}(x)$	(1.3.27)
Cartesian product	$F \times G$	$(Z_F \times Z_G)(x, x^2, x^3, \ldots)$	(2.1.26)
Substitution	$F \circ G$	$Z_F(\widetilde{G}(x), \widetilde{G}(x^2), \widetilde{G}(x^3), \ldots)$	(1.4.5)
Functorial composition	$F \square G$	$(Z_F \square Z_G)(x, x^2, x^3, \ldots)$	(2.2.13)
F_w, with connected component counter α	$F_{w^{(\alpha)}}$	$\prod_{k\geq 1}\left(\widetilde{F_{w^k}}(x^k)\right)^{\lambda_k(\alpha)}$	(2.3.73)

Note: Ex. i.j.k means Exercise i.j.k.

3.3. *Asymmetry generating series of particular species.*

Species	F	$\overline{F}(x)$	Ref.
Sets	E	$1+x$	(4.4.27)
Linear orders	L	$(1-x)^{-1}$	(4.4.27)
Even sets	E_{even}	1	(1.3.12)
Odd sets	E_{odd}	x	(1.3.12)
Subsets	$\wp$	$(1+x)^2$	(1.3.33)
Permutations	$\mathcal{S}$	$1+x$	(4.4.27)
Even permutations	$\mathcal{A}$	$1+x$	(4.4.70)
Partitions	Par	$1+x$	(1.4.12)
Involutions	Inv	$1+x$	
Cyclic permutations	$\mathcal{C}$	x	
Elements	ε	$x(1+x)$	
Derangements	Der	1	
Oriented sets	$E^{\pm}$	$1+x+x^2$	(4.4.56)
Ballots	Bal	$1/(1-x)$	(1.4.11)
Octopuses	Oct	$-x - \sum_{n\geq 1}\dfrac{\mu(n)}{n}\log(1-2x^n)$	(1.4.62)

Table 3.3. (cont.)

Species	F	$\overline{F}(x)$	Ref.
Chains	Cha	$1 + x$	Ex. 1.1.5
Rooted trees	$\mathcal{A}$	$\overline{\mathcal{A}}(x) = \exp\left(\sum_{n\geq 1}(-1)^{n-1}\overline{\mathcal{A}}(x^n)/n\right)$	(4.4.36)
Trees	$\mathfrak{a}$	$\overline{\mathcal{A}}(x) - \frac{1}{2}\overline{\mathcal{A}}(x^2) - \frac{1}{2}\left(\overline{\mathcal{A}}(x)\right)^2$	(4.4.37)
Endofunctions	End	$(1 - \overline{\mathcal{A}}(x^2))/(1 - \overline{\mathcal{A}}(x))$	(4.4.31)
Commutative parenthesizations	$\mathcal{P}$	$\lim_{n\to\infty} 1 - \sqrt{2 - 2x - \sqrt{2 - 2x^2 - \cdots - \sqrt{1 - 2x^{2^n}}}}$	Ex. 3.3.5

3.4. *Asymmetry generating series of general species.*

Species	F	$\overline{F}(x)$	Ref.
Even part	F_{even}	$(\overline{F}(x) + \overline{F}(-x))/2$	Ex. 1.3.3
Odd part	F_{odd}	$(\overline{F}(x) - \overline{F}(-x))/2$	Ex. 1.3.3
Connected structures	F^c	$\Lambda(x) + \Lambda(x^2) + \Lambda(x^4) + \Lambda(x^8) + \cdots$ $\Lambda(x) = \sum_{k\geq 1} \frac{\mu(k)}{k} \log \overline{F}(x^k)$	(4.4.28)
Derivative	F'	$\left(\frac{\partial}{\partial x_1}\Gamma_F\right)(x, x^2, x^3, \ldots)$	(4.4.24)
Pointing	$F^{\bullet}$	$x\left(\frac{\partial}{\partial x_1}\Gamma_F\right)(x, x^2, x^3, \ldots)$	(4.4.24)
Sum	$F + G$	$\overline{F}(x) + \overline{G}(x)$	(4.4.24)
Product	$F \cdot G$	$\overline{F}(x)\overline{G}(x)$	(4.4.24)
Cartesian product	$F \times G$	$\Gamma_{F\times G}(x, x^2, x^3, \ldots)$	
Substitution	$F \circ G$	$\Gamma_F(\overline{G}(x), \overline{G}(x^2), \overline{G}(x^3), \ldots)$	(4.4.23)
Functorial composition	$F \square G$	$\Gamma_{F \square G}(x, x^2, x^3, \ldots)$	
F_w, with connected component counter α	$F_{w^{(\alpha)}}$	$\prod_{k\geq 1}\left(\overline{F_{w^k}}(x^k)\right)^{\gamma_k(\alpha)}$	Ex. A.2.8

Table 4. *Index series.*
4.1. Cycle index series of particular species.

Species	F	Z_F	Ref.
Sets	E	$\exp\left(\dfrac{x_1}{1} + \dfrac{x_2}{2} + \dfrac{x_3}{3} + \cdots\right)$	(1.2.16)
Oriented sets	$E^{\pm}$	$1 + x_1 + \displaystyle\sum_{n\geq 2}\dfrac{1}{n!}\sum_{\sigma\in S_n}(1+\mathrm{sgn}(\sigma))x_1^{\sigma_1}x_2^{\sigma_2}\cdots$	(2.6.44)
Even sets	E_{even}	$\exp\left(\dfrac{x_2}{2} + \dfrac{x_4}{4} + \cdots\right)\cosh\left(\dfrac{x_1}{1} + \dfrac{x_3}{3} + \cdots\right)$	(1.3.12)
Odd sets	E_{odd}	$\exp\left(\dfrac{x_2}{2} + \dfrac{x_4}{4} + \cdots\right)\sinh\left(\dfrac{x_1}{1} + \dfrac{x_3}{3} + \cdots\right)$	(1.3.12)
Linear orders	L	$(1-x_1)^{-1}$	(1.2.15)
Chains	Cha	$\dfrac{2-x_1^2}{2(1-x_1)} + \dfrac{(1+x_1)x_2}{2(1-x_2)}$	(2.6.79)
Subsets	$\wp$	Z_E^2	(1.3.35)
Permutations	S	$\displaystyle\prod_{k\geq 1}(1-x_k)^{-1}$	(1.4.17)
Set partitions	Par	$Z_E \circ (Z_E - 1)$	(1.4.13)
Ballots	Bal	$1/(2 - Z_E)$	(1.3.46)
Cyclic permutations	C	$-\displaystyle\sum_{k=1}^{\infty}\dfrac{\phi(k)}{k}\log(1-x_k)$	(1.4.18)
Derangements	Der	$Z_E^{-1} Z_S$	(1.3.29)
Elements	ε	$x_1 Z_E$	(1.2.15)
Octopuses	Oct	$\displaystyle\sum_{n\geq 1}\dfrac{\phi(n)}{n}\log\left(\dfrac{1-x_n}{1-2x_n}\right)$	(1.4.63)
Simple graphs	$\mathcal{G}$	$\mathrm{coeff}_\sigma Z_\mathcal{G} = 2^{\frac{1}{2}\sum_{i,j\geq 1}(i,j)\sigma_i\sigma_j - \frac{1}{2}\sum_{k\geq 1}(k\bmod 2)\sigma_k}$	(A.2.6)
Directed graphs	$\mathcal{D}$	$\mathrm{coeff}_\sigma Z_\mathcal{D} = 2^{\sum_{i,j\geq 1}(i,j)\sigma_i\sigma_j}$	(2.2.30)
Rooted trees	$\mathcal{A}$	$\mathrm{coeff}_\sigma(Z_\mathcal{A}) = \sigma_1^{\sigma_1-1}\displaystyle\prod_{k\geq 2}\left(\phi_k^{\sigma_k} - k\sigma_k\phi_k^{\sigma_k-1}\right)$ where $\phi_k = \displaystyle\sum_{d\mid k} d\sigma_d$	(3.2.61)

Note: For the cycle index series of the species $\mathfrak{a}$, of trees, End, of endofunctions, $\mathcal{P}$, of commutative parenthesizations, and A, of even permutations, see formulas (4.1.47), (3.2.63), (3.3.90), and (4.4.69), respectively.

4.2. Cycle index series of general species.

Species	F	Z_F	Ref.
Even part	F_{even}	$(Z_F(x_1, x_2, x_3, \ldots) + Z_F(-x_1, x_2, -x_3, \ldots))/2$	(1.3.54)
Odd part	F_{odd}	$(Z_F(x_1, x_2, x_3, \ldots) - Z_F(-x_1, x_2, -x_3, \ldots))/2$	(1.3.54)
Connected structures	F^c	$\sum_{k \geq 1} \frac{\mu(k)}{k} \log Z_F \circ x_k$	(2.3.71)
Derivative	F'	$\left(\frac{\partial}{\partial x_1} Z_F\right)(x_1, x_2, x_3, \ldots)$	(1.4.31)
Pointing	$F^{\bullet}$	$x\left(\frac{\partial}{\partial x_1} Z_F\right)(x_1, x_2, x_3, \ldots)$	(2.1.7)
Sum	$F + G$	$Z_F(x_1, x_2, x_3, \ldots) + Z_G(x_1, x_2, x_3, \ldots)$	(1.3.11)
Product	$F \cdot G$	$Z_F(x_1, x_2, x_3, \ldots) Z_G(x_1, x_2, x_3, \ldots)$	(1.3.27)
Cartesian product	$F \times G$	$Z_F(x_1, x_2, x_3, \ldots) \times Z_G(x_1, x_2, x_3, \ldots)$	(2.1.26)
Substitution	$F \circ G$	$Z_F(Z_G \circ x_1, Z_G \circ x_2, Z_G \circ x_3, \ldots)$	(1.4.5)
Functorial composition	$F \square G$	$Z_F(x_1, x_2, x_3, \ldots) \square Z_G(x_1, x_2, x_3, \ldots)$	(2.2.12)
F_w, with connected component counter α	$F_{w^{(\alpha)}}$	$\prod_{k \geq 1} \left(Z_{F_{w^k}} \circ (x_k)\right)^{\lambda_k(\alpha)}$	(2.3.73)

Note: For the cycle index series of the species $\mathcal{A}_R$, of R-enriched rooted trees, and $\mathfrak{a}_R$, of R-enriched trees, see Proposition 3.2.14, and (4.1.51).

4.3. Asymmetry index series of particular species.

Species	F	Γ_F	Ref.
Sets	E	$\exp\left(\frac{x_1}{1} - \frac{x_2}{2} + \frac{x_3}{3} - \cdots\right)$	(4.4.27)
Oriented sets	$E^{\pm}$	$\left(2 + \sum_{k \geq 2}(-1)^k x_k\right)\Gamma_E - 1 - x_1$	(4.4.56)
Even sets	E_{even}	$\exp\left(-\frac{x_2}{2} - \frac{x_4}{4} - \cdots\right) \cosh\left(\frac{x_1}{1} + \frac{x_3}{3} + \cdots\right)$	Example 1.3.4

Table 4.3. (*cont.*)

Species	F	Γ_F	Ref.
Odd sets	E_{odd}	$\exp\left(-\dfrac{x_2}{2} - \dfrac{x_4}{4} - \cdots\right) \sinh\left(\dfrac{x_1}{1} + \dfrac{x_3}{3} + \cdots\right)$	Example 1.3.4
Linear orders	L	$(1 - x_1)^{-1}$	(4.4.27)
Chains	Cha	$\dfrac{2 - x_1^2}{2(1 - x_1)} - \dfrac{(1 + x_1)x_2}{2(1 - x_2)}$	(1.4.68)
Subsets	$\wp$	Γ_E^2	(1.3.33)
Permutations	$\mathcal{S}$	$(1 - x_2)(1 - x_1)^{-1}$	(4.4.27)
Set partitions	Par	$\Gamma_E \circ (\Gamma_E - 1)$	
Ballots	Bal	$1/(2 - \Gamma_E)$	
Cyclic permutations	$\mathcal{C}$	$-\displaystyle\sum_{k=1}^{\infty} \dfrac{\mu(k)}{k} \log(1 - x_k)$	(4.4.27)
Derangements	Der	$\Gamma_E^{-1} \Gamma_{\mathcal{S}}$	
Elements	ε	$x_1 \Gamma_E$	
Octopuses	Oct	$\displaystyle\sum_{n \geq 1} \dfrac{\mu(n)}{n} \log\left(\dfrac{1 - x_n}{1 - 2x_n}\right)$	(1.4.63)
Rooted trees	$\mathcal{A}$	$\operatorname{coeff}_\sigma(\Gamma_\mathcal{A}) = \sigma_1^{\sigma_1 - 1} \displaystyle\prod_{k \geq 2}\left(\theta_k^{\sigma_k} - k\sigma_k \theta_k^{\sigma_k - 1}\right)$ where $\theta_k = \displaystyle\sum_{d \mid k}(-1)^{(k/d)-1} d\sigma_d$	(4.4.40)

Note: For the asymmetry index series of the species $\mathfrak{a}$, of trees, End, of endofunctions, $\mathcal{P}$, of commutative parenthesizations, and A, of even permutations, see Exercise 4.4.13, (4.4.63), Exercise 3.3.5, and (4.4.70), respectively.

4.4. *Asymmetry index series of general species.*

Species	F	Γ_F	Ref.
Even part	F_{even}	$(\Gamma_F(x_1, x_2, x_3, \ldots)$ $+ \Gamma_F(-x_1, x_2, -x_3, \ldots))/2$	(1.3.54)
Odd part	F_{odd}	$(\Gamma_F(x_1, x_2, x_3, \ldots)$ $- \Gamma_F(-x_1, x_2, -x_3, \ldots))/2$	(1.3.54)

418 Appendix 2

Table 4.4. (cont.)

Species	F	Γ_F	Ref.
Connected structures	F^c	$(x_1 + x_2 + x_4 + x_8 + \cdots) \circ \sum_{k \geq 1} \frac{\mu(k)}{k} \log \Gamma_F \circ x_k$	(2.3.71)
Derivative	F'	$\left(\frac{\partial}{\partial x_1} \Gamma_F\right)(x_1, x_2, x_3, \ldots)$	(4.4.24)
Pointing	$F^\bullet$	$x \left(\frac{\partial}{\partial x_1} \Gamma_F\right)(x_1, x_2, x_3, \ldots)$	(4.4.24)
Sum	$F + G$	$\Gamma_F(x_1, x_2, x_3, \ldots) + \Gamma_G(x_1, x_2, x_3, \ldots)$	(4.4.24)
Product	$F \cdot G$	$\Gamma_F(x_1, x_2, x_3, \ldots) \Gamma_G(x_1, x_2, x_3, \ldots)$	(4.4.24)
Cartesian product	$F \times G$	$\Gamma_{F \times G}(x_1, x_2, x_3, \ldots)$	
Substitution	$F \circ G$	$\Gamma_F(\Gamma_G \circ x_1, \Gamma_G \circ x_2, \Gamma_G \circ x_3, \ldots)$	(4.4.23)
Functorial composition	$F \square G$	$\Gamma_{F \square G}(x_1, x_2, x_3, \ldots)$	
F_w, with connected component counter α	$F_{w^{(\alpha)}}$	$\prod_{k \geq 1} \left(\Gamma_{F_{w^k}} \circ (x_k)\right)^{\gamma_k(\alpha)}$	Ex. A.2.8

Note: For the asymmetry index series of the species $\mathcal{A}_R$, of R-enriched rooted trees, and $\mathfrak{a}_R$, of R-enriched trees, see (4.4.34) and Exercise 4.4.13.

Table 5. *Molecular species, derivatives, and cycle index series coefficients* ($n \leq 4$).

				Cycle type of σ	
n	M	$\|M[n]\|$	M'		fix $M[\sigma]$
				0	
0	1	1	0		1
				1	
1	X	1	1	1	1

Table 5. (*cont.*)

n	M	$|M[n]|$	M'	Cycle type of σ — fix $M[\sigma]$				
				1^2	2^1			
2	E_2	1	X	1	1			
	X^2	2	$2X$	2	0			
				1^3	$1^1 2^1$	3^1		
3	E_3	1	E_2	1	1	1		
	$\mathcal{C}_3$	2	X^2	2	0	2		
	XE_2	3	$E_2 + X^2$	3	1	0		
	X^3	6	$3X^2$	6	0	0		
				1^4	$1^2 2^1$	$1^1 3^1$	2^2	4
4	E_4	1	E_3	1	1	1	1	1
	$E_4^\pm$	2	$\mathcal{C}_3$	2	0	2	2	0
	$E_2 \circ E_2$	3	XE_2	3	1	0	3	1
	XE_3	4	$E_3 + XE_2$	4	2	1	0	0
	E_2^2	6	$2XE_2$	6	2	0	2	0
	P_4^{bic}	6	X^3	6	0	0	6	0
	$\mathcal{C}_4$	6	X^3	6	0	0	2	2
	$X\mathcal{C}_3$	8	$\mathcal{C}_3 + X^3$	8	0	2	0	0
	$X^2 E_2$	12	$2XE_2 + X^3$	12	2	0	0	0
	$E_2(X^2)$	12	$2X^3$	12	0	0	4	0
	X^4	24	$4X^3$	24	0	0	0	0

Table 5. (cont.) Molecular species, derivatives and cycle index series coefficients ($n = 5$) (J. Labelle [200]).

M	$\|M[n]\|$	M'	\multicolumn{7}{c}{Cycle type of σ — fix $M[\sigma]$}						
			1^5	$1^3 2^1$	$1^2 3^1$	$1^1 4^1$	$1^1 2^2$	$2^1 3^1$	5
E_5	1	E_4	1	1	1	1	1	1	1
$E_5^{\pm}$	2	$E_4^{\pm}$	2	0	2	0	2	0	2
XE_4	5	$E_4 + XE_3$	5	3	2	1	1	0	0
$P_5/\mathbb{Z}_2$	6	C_4	6	0	0	2	2	0	1
$XE_4^{\pm}$	10	$E_4^{\pm} + XC_3$	10	0	4	0	2	0	0
$E_2 E_3$	10	$XE_3 + E_2^2$	10	4	1	0	2	1	0
P_5	12	$E_2(X^2)$	12	0	0	0	4	0	2
$XE_2(E_2)$	15	$E_2(E_2) + X^2 E_2$	15	3	0	1	3	0	0
$X^2 E_3$	20	$2XE_3 + X^2 E_2$	20	6	2	0	0	0	0
$E_2 C_3$	20	$XC_3 + X^2 E_2$	20	2	2	0	0	2	0
$(X^2 C_3)/\mathbb{Z}_2$	20	$XC_3 + E_2(X^2)$	20	0	2	0	4	0	0
C_5	24	X^4	24	0	0	0	0	0	4
XE_2^2	30	$E_2^2 + 2X^2 E_2$	30	6	0	0	2	0	0
XP_4^{bic}	30	$X^4 + P_4^{bic}$	30	0	0	0	6	0	0
XC_4	30	$X^4 + C_4$	30	0	0	2	2	0	0
$X^2 C_3$	40	$X^4 + 2XC_3$	40	0	4	0	0	0	0
$X^3 E_2$	60	$X^4 + 3X^2 E_2$	60	6	0	0	0	0	0
$XE_2(X^2)$	60	$2X^4 + E_2(X^2)$	60	0	0	0	4	0	0
X^5	120	$5X^4$	120	0	0	0	0	0	0

Note: See Chiricota [53] for a table of cycle index series of molecular species of order $n \leq 7$.

Table 6. *Molecular decomposition and index series of the species $\mathcal{G}$ of simple graphs up to order 6 (Ikollo [154]).*

$$\begin{aligned}
\mathcal{G} =\ & 1 + X + 2E_2 + 2E_3 + 2X \cdot E_2 + 2E_4 + 2E_2 \circ E_2 + 2X \cdot E_3 \\
& + 2E_2^2 + 2X^2 \cdot E_2 + E_2 \circ X^2 + 2E_5 + 6E_2 \cdot E_3 + 6X \cdot E_2^2 \\
& + 5X \cdot (E_2 \circ X^2) + 4X \cdot (E_2 \circ E_2) + 2X \cdot E_4 + 2X^2 \cdot E_3 + 6X^3 \cdot E_2 + P_5 \\
& + 2E_6 + 6E_2 \cdot E_4 + 16X \cdot E_2 \cdot E_3 + 6E_2 \cdot (E_2 \circ E_2) + 12E_2 \cdot (E_2 \circ X^2) \\
& + 2E_3^2 + 22X^2 \cdot E_2^2 + 2P_6 + 2E_3 \circ E_2 + 2E_2 \circ E_3 + 2X^2 \cdot E_4 \\
& + 6E_2 \circ (X \cdot E_2) + 26X^4 \cdot E_2 + 16X^2 \cdot (E_2 \circ X^2) + 4E_2^3 + 6X^3 \cdot E_3 \\
& + 2X \cdot P_5 + 2X \cdot E_5 + 4X^2 \cdot (E_2 \circ E_2) + 8X^6 + 4E_2 \circ X^3 + 2P_6^{\text{opp}} \\
& + 2E_3 \circ X^2 + \cdots
\end{aligned}$$

$$\begin{aligned}
Z_\mathcal{G} =\ & 1 + x_1 + x_2 + x_1^2 + \frac{4}{3}x_1^3 + 2x_1 x_2 + \frac{2}{3}x_3 + \frac{8}{3}x_1^4 + 4x_1^2 x_2 + \frac{4}{3}x_1 x_3 + 2x_2^2 + x_4 \\
& + \frac{128}{15}x_1^5 + \frac{32}{3}x_1^3 x_2 + \frac{8}{3}x_1^2 x_3 + 2x_1 x_4 + 8x_1 x_2^2 + \frac{4}{3}x_2 x_3 + \frac{4}{5}x_5 + \frac{2048}{45}x_1^6 \\
& + \frac{128}{3}x_1^4 x_2 + \frac{64}{9}x_1^3 x_2 + 4x_1^2 x_4 + \frac{8}{5}x_1 x_5 + \frac{32}{3}x_2^3 + 32 x_1^2 x_2^2 + \frac{16}{9}x_3^2 \\
& + \frac{16}{3}x_1 x_2 x_3 + 4x_2 x_4 + \frac{4}{3}x_6 + \cdots
\end{aligned}$$

$$\begin{aligned}
\Gamma_\mathcal{G} =\ & 1 + x_1 - x_2 + x_1^2 + \frac{4}{3}x_1^3 - 2x_1 x_2 + \frac{2}{3}x_3 + \frac{8}{3}x_1^4 - 4x_1^2 x_2 + \frac{4}{3}x_1 x_3 + \frac{128}{15}x_1^5 \\
& - \frac{32}{3}x_1^3 x_2 + \frac{8}{3}x_1^2 x_3 - \frac{4}{3}x_2 x_3 + \frac{4}{5}x_5 + \frac{2048}{45}x_1^6 - \frac{128}{3}x_1^4 x_2 + \frac{64}{9}x_1^3 x_2 - \frac{5}{2}x_2^3 \\
& + \frac{29}{18}x_3^2 + \frac{8}{5}x_1 x_5 + 4x_2 x_4 - \frac{16}{3}x_1 x_2 x_3 - \frac{4}{3}x_6 + \cdots
\end{aligned}$$

Note: The species P_6^{opp} is described in Exercise 3.

Table 7. *Molecular decomposition of the species $\mathfrak{a}$ of trees on $n \leq 10$ points (Chiricota [53]).*

n	$\mathfrak{a}_n$
1	X
2	$E_2(X)$
3	$X E_2(X)$
4	$X E_3(X) + E_2(X^2)$

Table 7. (cont.)

n	a_n
5	$X^3 E_2(X) + X E_4(X) + X E_2(X^2)$
6	$X^4 E_2(X) + X^3 E_3(X) + E_2(X^2)X^2 + X E_5(X) + E_2(X E_2(X)) + E_2(X^3)$
7	$X^7 + 2X^5 E_2(X) + X^4 E_3(X) + X^3 E_4(X) + X^2 E_2(X) E_3(X)$ $+ E_2(X^2) X E_2(X) + X E_6(X) + E_3(X^2) X + E_2(X^3) X + E_2(X E_2(X)) X$
8	$X^8 + 4X^6 E_2(X) + 2X^5 E_3(X) + X^4 E_2(X)^2 + X^4 E_4(X) + X^4 E_2(X^2)$ $+ X^3 E_2(X) E_3(X) + X^3 E_5(X) + X^2 E_2(X) E_4(X) + E_2(X^2) X^2 E_2(X)$ $+ E_2(X E_2(X)) X^2 + E_3(X^2) X^2 + E_2(X^3) X^2 + E_2(X^2) X E_3(X)$ $+ X E_7(X) + E_2(X E_3(X)) + 2 E_2(X^4) + E_2\big(X^2 E_2(X)\big)$
9	$3X^9 + 8X^7 E_2(X) + 4X^6 E_3(X) + 3X^5 E_2(X)^2 + 2X^5 E_4(X) + 3X^5 E_2(X^2)$ $+ 4X^4 E_2(X) E_3(X) + X^4 E_5(X) + X^3 E_2(X) E_4(X) + 2X^3 E_2(X^2) E_2(X)$ $+ X^3 E_6(X) + X^3 E_2(X E_2(X)) + X^3 E_2(X^3) + X^2 E_2(X) E_5(X)$ $+ X^2 E_3(X) E_4(X) + X^2 E_2(X^2) E_3(X) + X E_2(X^3) E_2(X)$ $+ X E_2(X E_2(X)) E_2(X) + X E_3(X^2) E_2(X) + X E_2(X^2) E_4(X)$ $+ X E_8(X) + X E_4(X^2) + X E_2\big(X^2 E_2(X)\big) + 2 X E_2(X^4) + X E_2(X E_3(X))$
10	$6X^{10} + 17X^8 E_2(X) + 8X^7 E_3(X) + 7X^6 E_2(X)^2 + 4X^6 E_4(X) + 5X^6 E_2(X^2)$ $+ 8X^5 E_2(X) E_3(X) + 2X^5 E_5(X) + X^4 E_2(X)^3 + 4X^4 E_2(X) E_4(X)$ $+ X^4 E_3(X)^2 + 5X^4 E_2(X^2) E_2(X) + X^4 E_6(X) + X^4 E_3(X^2) + X^4 E_2(X^3)$ $+ 2X^4 E_2(X E_2(X)) + X^3 E_2(X)^2 E_3(X) + X^3 E_2(X) E_5(X) + X^3 E_3(X) E_4(X)$ $+ 2X^3 E_2(X^2) E_3(X) + X^3 E_7(X) + X^2 E_2(X^2) E_2(X)^2 + X^2 E_2(X) E_6(X)$ $+ X^2 E_3(X) E_5(X) + X^2 E_3(X^2) E_2(X) + X^2 E_2(X^3) E_2(X) + X^2 E_2(X^2) E_4(X)$ $+ X^2 E_2\big(X^2 E_2(X)\big) + 2X^2 E_2(X^4) + X^2 E_2(X E_3(X)) + X^2 E_4(X^2)$ $+ X E_3(X^2) E_3(X) + X E_2(X E_2(X)) E_3(X) + X E_2(X^3) E_3(X)$ $+ X E_2(X^2) E_5(X) + X E_3(X E_2(X)) + X E_3(X^3) + X E_9(X) + E_2\big(X E_2(X^2)\big)$ $+ E_2\big(X^2 E_3(X)\big) + E_2(X E_4(X)) + 3 E_2(X^5) + 3 E_2\big(X^3 E_2(X)\big)$

Table 8. *Molecular decomposition of the species $\mathcal{A}$ of rooted trees on $n \leq 10$ points (Chiricota [53]).*

n	$\mathcal{A}_n$
1	X
2	X^2
3	$X^3 + XE_2(X)$
4	$2X^4 + X^2 E_2(X) + X E_3(X)$
5	$3X^5 + 3X^3 E_2(X) + X^2 E_3(X) + X E_4(X) + X E_2(X^2)$
6	$6X^6 + 6X^4 E_2(X) + 3X^3 E_3(X) + X^2 E_2(X)^2 + X^2 E_4(X) + 2X^2 E_2(X^2)$ $+ X E_5(X)$
7	$12X^7 + 14X^5 E_2(X) + 6X^4 E_3(X) + 2X^3 E_2(X)^2 + 3X^3 E_4(X) + 3X^3 E_2(X^2)$ $+ 2X^2 E_2(X) E_3(X) + X^2 E_5(X) + X E_2(X) E_2(X^2) + X E_2(X E_2(X))$ $+ X E_6(X) + X E_3(X^2) + X E_2(X^3)$
8	$25X^8 + 32X^6 E_2(X) + 14X^5 E_3(X) + 8X^4 E_2(X)^2 + 6X^4 E_4(X) + 7X^4 E_2(X^2)$ $+ 5X^3 E_2(X) E_3(X) + 3X^3 E_5(X) + 2X^2 E_2(X) E_4(X) + 3X^2 E_2(X) E_2(X^2)$ $+ X^2 E_3(X)^2 + 2X^2 E_2(X E_2(X)) + X^2 E_6(X) + 2X^2 E_3(X^2) + 2X^2 E_2(X^3)$ $+ X E_3(X) E_2(X^2) + X E_7(X)$
9	$52X^9 + 75X^7 E_2(X) + 32X^6 E_3(X) + 23X^5 E_2(X)^2 + 14X^5 E_4(X) + 17X^5 E_2(X^2)$ $+ 18X^4 E_2(X) E_3(X) + 6X^4 E_5(X) + X^3 E_2(X)^3 + 5X^3 E_2(X) E_4(X)$ $+ 9X^3 E_2(X) E_2(X^2) + 2X^3 E_3(X)^2 + 4X^3 E_2(X E_2(X)) + 3X^3 E_6(X)$ $+ 3X^3 E_3(X^2) + 4X^3 E_2(X^3) + 2X^2 E_2(X) E_5(X) + 2X^2 E_3(X) E_4(X)$ $+ 3X^2 E_3(X) E_2(X^2) + X^2 E_7(X) + X^2 E_7(X)1 + X E_2(X) E_2(X E_2(X))$ $+ X E_2(X) E_3(X^2) + X E_2(X) E_2(X^3) + X E_4(X) E_2(X^2)$ $+ X E_2(X E_3(X)) + X E_2(X^2 E_2(X)) + X E_8(X) + 2X E_2(X^4) + X E_4(X^2)$

Table 8. (cont.)

n	$\mathcal{A}_n$
10	$113X^{10} + 177X^8 E_2(X) + 75X^7 E_3(X) + 68X^6 E_2(X)^2 + 32X^6 E_4(X)$
	$\quad + 39X^6 E_2(X^2) + 51X^5 E_2(X) E_3(X) + 14X^5 E_5(X) + 5X^4 E_2(X)^3$
	$\quad + 18X^4 E_2(X) E_4(X) + 26X^4 E_2(X) E_2(X^2) + 8X^4 E_3(X)^2 + 9X^4 E_2(X E_2(X))$
	$\quad + 6X^4 E_6(X) + 7X^4 E_3(X^2) + 8X^4 E_2(X^3) + 4X^3 E_2(X)^2 E_3(X)$
	$\quad + 5X^3 E_2(X) E_5(X) + 5X^3 E_3(X) E_4(X) + 9X^3 E_3(X) E_2(X^2)$
	$\quad + 2X^2 E_2(X)^2 E_2(X^2) + 3X^3 E_7(X) + 2X^2 E_2(X) E_2(X E_2(X))$
	$\quad + 2X^2 E_2(X) E_6(X) + 3X^2 E_2(X) E_3(X^2) + 3X^2 E_2(X) E_2(X^3)$
	$\quad + 2X^2 E_3(X) E_5(X) + X^2 E_4(X)^2 + 3X^2 E_4(X) E_2(X^2) + X^2 E_2(X^2)^2$
	$\quad + 2X^2 E_2(X E_3(X)) + 2X^2 E_2(X^2 E_2(X)) + X^2 E_8(X) + 4X^2 E_2(X^4)$
	$\quad + 2X^2 E_4(X^2) + X E_3(X) E_2(X E_2(X)) + X E_3(X) E_3(X^2) + X E_3(X) E_2(X^3)$
	$\quad + X E_2(X^2) E_5(X) + X E_3(X^3) + X E_3(X E_2(X)) + X E_9(X)$

Table 9. *Coefficients of the generic (exponential) generating series of R-enriched trees and R-enriched rooted trees ($n \leq 6$) when*

$$R(x) = \sum_{n \geq 0} r_n \frac{x^n}{n!}.$$

n	$n! [x^n] \mathfrak{a}_R(x)$	$n! [x^n] \mathcal{A}_R(x)$
0	0	0
1	r_0	r_0
2	r_1^2	$2 r_0 r_1$
3	$3 r_1^2 r_2$	$3 r_0^2 r_2 + 6 r_0 r_1^2$
4	$4 r_1^3 r_3 + 12 r_1^2 r_2^2$	$4 r_0^3 r_3 + 36 r_0^2 r_1 r_2 + 24 r_0 r_1^3$
5	$5 r_1^4 r_4 + 60 r_1^3 r_2 r_3 + 60 r_1^2 r_2^3$	$5 r_0^4 r_4 + 80 r_0^3 r_1 r_3 + 60 r_0^3 r_2^2 + 360 r_0^2 r_1^2 r_2 + 120 r_0 r_1^4$
6	$6 r_1^5 r_5 + 120 r_1^4 r_2 r_4 + 90 r_1^4 r_3^2$ $\quad + 720 r_1^3 r_2^2 r_3 + 360 r_1^2 r_2^4$	$6 r_0^5 r_5 + 150 r_0^4 r_1 r_4 + 300 r_0^4 r_2 r_3 + 1200 r_0^3 r_1^2 r_3$ $\quad + 1800 r_0^3 r_1 r_2^2 + 3600 r_0^2 r_1^3 r_2 + 720 r_0 r_1^5$

Table 10. *Coefficients of the generic generating series of unlabeled R-enriched trees and R-enriched rooted trees ($n \leq 6$) when*

$$Z_R(x_1, x_2, \ldots) = \sum_{n_1, n_2, \ldots} r_{n_1, n_2, \ldots} \frac{x_1^{n_1} x_2^{n_2} \cdots}{1^{n_1} n_1! 2^{n_2} n_2! \cdots}.$$

n	$[x^n] \widetilde{\mathfrak{a}}_R(x)$	$[x^n] \mathcal{A}_R\widetilde{\ }(x)$
0	0	0
1	r_0	r_0
2	$\frac{1}{2}(r_1^2 + r_1)$	$\frac{1}{2} r_0 r_1$
3	$\frac{1}{2}(r_1^2 r_2 + r_1 r_{01})$	$\frac{1}{2}(r_0^2 r_2 + 2 r_0 r_1^2 + r_0 r_{01})$
4	$\frac{1}{6}(r_1^3 r_3 + 3 r_1^2 r_2^2 + 3 r_1^2 r_{11} + 3 r_1 r_2$ $+ 2 r_1 r_{001})$	$\frac{1}{6}(r_0^3 r_3 + 9 r_0^2 r_1 r_2 + 3 r_0^2 r_{11} + r_0 r_1^3$ $+ 3 r_0 r_1 r_{01} + 2 r_0 r_{001})$
5	$\frac{1}{24}(r_1^4 r_4 + 12 r_1^3 r_2 r_3 + 6 r_1^3 r_{21} + 12 r_1^2 r_2^3$ $+ 12 r_1^2 r_2 r_{11} + 3 r_1^2 r_{02} + 8 r_1^2 r_{101}$ $+ 12 r_1 r_2 r_{01} + 6 r_1 r_{0001})$	$\frac{1}{24}(r_0^4 r_4 + 16 r_0^3 r_1 r_3 + 12 r_0^3 r_2^2 + 6 r_0^3 r_{21}$ $+ 72 r_0^2 r_1^2 r_2 + 24 r_0^2 r_1 r_{11} + 12 r_0^2 r_2 r_{01}$ $+ 3 r_0^2 r_{02} + 8 r_0^2 r_{101} + 24 r_0 r_1^4$ $+ 12 r_0 r_1^2 r_{01} + 12 r_0 r_1 r_{01} + 8 r_0 r_1 r_{001}$ $+ 6 r_0 r_{0001})$
6	$\frac{1}{120}(r_1^5 r_5 + 20 r_1^4 r_2 r_4 + 15 r_1^4 r_3^2 + 10 r_1^4 r_{31}$ $+ 120 r_1^3 r_2^2 r_3 + 60 r_1^3 r_2 r_{21} + 30 r_1^3 r_3 r_{11}$ $+ 15 r_1^3 r_{12} + 20 r_1^3 r_{201} + 60 r_1^2 r_2^4$ $+ 60 r_1^2 r_2^2 r_{11} + 60 r_1^2 r_2 r_{11}$ $+ 40 r_1^2 r_2 r_{101} + 30 r_1^2 r_3 + 15 r_1^2 r_{11}^2$ $+ 20 r_1^2 r_{011} + 30 r_1^2 r_{1001} + 60 r_1 r_2^2$ $+ 30 r_1 r_{11} + 24 r_1 r_{00001})$	$\frac{1}{120}(r_0^5 r_5 + 25 r_0^4 r_1 r_4 + 50 r_0^4 r_2 r_3$ $+ 10 r_0^4 r_{31} + 200 r_0^3 r_1^2 r_3 + 300 r_0^3 r_1 r_2^2$ $+ 90 r_0^3 r_1 r_{21} + 90 r_0^3 r_2 r_{11} + 30 r_0^3 r_{01} r_3$ $+ 15 r_0^3 r_{12} + 20 r_0^3 r_{201} + 600 r_0^2 r_1^3 r_2$ $+ 180 r_0^2 r_1^2 r_{11} + 180 r_0^2 r_1 r_2 r_{01}$ $+ 60 r_0^2 r_1 r_{11} + 15 r_0^2 r_1 r_{02}$ $+ 80 r_0^2 r_1 r_{101} + 40 r_0^2 r_2 r_{001}$ $+ 30 r_0^2 r_{01} r_{11} + 20 r_0^2 r_{011} + 30 r_0^2 r_{1001}$ $+ 120 r_0 r_1^5 + 60 r_0 r_1^3 r_{01} + 60 r_0 r_1^2 r_{01}$ $+ 40 r_0 r_1^2 r_{001} + 30 r_0 r_1 r_{0001}$ $+ 24 r_0 r_{00001})$

Table 11. *(G. Labelle [187].)* Coefficients of the generic asymmetry types generating series of R-enriched trees and R-enriched rooted trees ($n \leq 6$) when

$$\Gamma_R(x_1, x_2, \ldots) = \sum_{n_1, n_2, \ldots} r_{n_1, n_2, \ldots} \frac{x_1^{n_1} x_2^{n_2} \cdots}{1^{n_1} n_1! 2^{n_2} n_2! \cdots}.$$

n	$[x^n]\,\overline{\mathfrak{a}}_R(x)$	$[x^n]\,\overline{\mathcal{A}}_R(x)$
0	0	0
1	r_0	r_0
2	$\frac{1}{2}(r_1^2 - r_1)$	$r_0 r_1$
3	$\frac{1}{2}(r_1^2 r_2 + r_1 r_{01})$	$\frac{1}{2}(r_0^2 r_2 + 2r_0 r_1^2 + r_0 r_{01})$
4	$\frac{1}{6}(r_1^3 r_3 + 3r_1^2 r_2^2 + 3r_1^2 r_{11} - 3r_1 r_2 + 2r_1 r_{001})$	$\frac{1}{6}(r_0^3 r_3 + 9r_0^2 r_1 r_2 + 3r_0^2 r_{11} + r_0 r_1^3 + 3r_0 r_1 r_{01} + 2r_0 r_{001})$
5	$\frac{1}{24}(r_1^4 r_4 + 12r_1^3 r_2 r_3 + 6r_1^3 r_{21} + 12r_1^2 r_2^3 + 12r_1^2 r_2 r_{11} + 3r_1^2 r_{02} + 8r_1^2 r_{101} + 12r_1 r_2 r_{01} + 6r_1 r_{0001})$	$\frac{1}{24}(r_0^4 r_4 + 16r_0^3 r_1 r_3 + 12r_0^3 r_2^2 + 6r_0^3 r_{21} + 72r_0^2 r_1^2 r_2 + 24r_0^2 r_1 r_{11} + 12r_0^2 r_2 r_{01} + 3r_0^2 r_{02} + 8r_0^2 r_{101} + 24r_0 r_1^4 + 12r_0 r_1^2 r_{01} + 12r_0 r_1 r_{01} + 8r_0 r_1 r_{001} + 6r_0 r_{0001})$
6	$\frac{1}{120}(r_1^5 r_5 + 20r_1^4 r_2 r_4 + 15r_1^4 r_3^2 + 10r_1^4 r_{31} + 120r_1^3 r_2^2 r_3 + 60r_1^3 r_2 r_{21} + 30r_1^3 r_3 r_{11} + 15r_1^3 r_{12} + 20r_1^3 r_{201} + 60r_1^2 r_2^4 + 60r_1^2 r_2^2 r_{11} + 60r_1^2 r_2 r_{11} + 40r_1^2 r_2 r_{101} - 30r_1^2 r_3 + 15r_1^2 r_{11}^2 + 20r_1^2 r_{011} + 30r_1^2 r_{1001} - 60r_1 r_2^2 - 30r_1 r_{11} + 24r_1 r_{00001})$	$\frac{1}{120}(r_0^5 r_5 + 25r_0^4 r_1 r_4 + 50r_0^4 r_2 r_3 + 10r_0^4 r_{31} + 200r_0^3 r_1^2 r_3 + 300r_0^3 r_1 r_2^2 + 90r_0^3 r_1 r_{21} + 90r_0^3 r_2 r_{11} + 30r_0^3 r_{01} r_3 + 15r_0^3 r_{12} + 20r_0^3 r_{201} + 600r_0^2 r_1^3 r_2 + 180r_0^2 r_1^2 r_{11} + 180r_0^2 r_1 r_2 r_{01} + 60r_0^2 r_1 r_{11} + 15r_0^2 r_1 r_{02} + 80r_0^2 r_1 r_{101} + 40r_0^2 r_2 r_{001} + 30r_0^2 r_{01} r_{11} + 20r_0^2 r_{011} + 30r_0^2 r_{1001} + 120r_0 r_1^5 + 60r_0 r_1^3 r_{01} + 60r_0 r_1^2 r_{01} + 40r_0 r_1^2 r_{001} + 30r_0 r_1 r_{0001} + 24r_0 r_{00001})$

Table 12. *Coefficients of the generic cycle index series of R-enriched trees and R-enriched rooted trees ($n \leq 6$) when*

$$Z_R(x_1, x_2, \ldots) = \sum_{n_1, n_2, \ldots} r_{n_1, n_2, \ldots} \frac{x_1^{n_1} x_2^{n_2} \cdots}{1^{n_1} n_1! 2^{n_2} n_2! \cdots}.$$

$(n_1, n_2, \ldots)$	$1^{n_1} n_1! 2^{n_2} n_2! \cdots [x_1^{n_1} x_2^{n_2} \cdots] Z_{\mathfrak{a}_R}$	$1^{n_1} n_1! 2^{n_2} n_2! \cdots [x_1^{n_1} x_2^{n_2} \cdots] Z_{\mathcal{A}_R}$
( )	0	0
(1)	r_0	r_0
(2)	r_1^2	$2r_0 r_1$
(01)	r_1	0
(3)	$3r_1^2 r_2$	$3r_0^2 r_2 + 6r_0 r_1^2$
(11)	$r_1 r_{01}$	$r_0 r_{01}$
(001)	0	0
(4)	$4r_1^3 r_3 + 12r_1^2 r_2^2$	$4r_0^3 r_3 + 36r_0^2 r_1 r_2 + 24r_0 r_1^3$
(21)	$2r_1^2 r_{11}$	$2r_0^2 r_{11} + 2r_0 r_1 r_{01}$
(101)	$r_1 r_{001}$	$r_0 r_{001}$
(02)	$4r_1 r_2$	0
(0001)	0	0
(5)	$5r_1^4 r_4 + 60 r_1^3 r_2 r_3 + 60 r_1^2 r_2^3$	$5r_0^4 r_4 + 80r_0^3 r_1 r_3 + 60r_0^3 r_2^2$ $+ 360 r_0^2 r_1^2 r_2 + 120 r_0 r_1^4$
(31)	$3r_1^3 r_{21} + 6r_1^2 r_2 r_{11}$	$3r_0^3 r_{21} + 12r_0^2 r_1 r_{11} + 6r_0^2 r_2 r_{01}$ $+ 6r_0 r_1^2 r_{01}$
(201)	$2r_1^2 r_{101}$	$2r_0^2 r_{101} + 2r_0 r_1 r_{001}$
(12)	$r_1^2 r_{02} + 4r_1 r_2 r_{01}$	$r_0^2 r_{02} + 4r_0 r_1 r_{01}$
(1001)	$r_1 r_{0001}$	$r_0 r_{0001}$
(011)	0	0
(00001)	0	0
(6)	$6r_1^5 r_5 + 120 r_1^4 r_2 r_4 + 90 r_1^4 r_3^2$ $+ 720 r_1^3 r_2^2 r_3 + 360 r_1^2 r_2^4$	$6r_0^5 r_5 + 150 r_0^4 r_1 r_4 + 300 r_0^4 r_2 r_3$ $+ 1200 r_0^3 r_1^2 r_3 + 1800 r_0^3 r_1 r_2^2$ $+ 3600 r_0^2 r_1^3 r_2 + 720 r_0 r_1^5$

Table 12. (cont.)

$(n_1, n_2, \ldots)$	$1^{n_1} n_1! 2^{n_2} n_2! \cdots [x_1^{n_1} x_2^{n_2} \cdots] Z_{a_R}$	$1^{n_1} n_1! 2^{n_2} n_2! \cdots [x_1^{n_1} x_2^{n_2} \cdots] Z_{\mathcal{A}_R}$
(41)	$4r_1^4 r_{31} + 24 r_1^3 r_2 r_{21} + 12 r_1^3 r_3 r_{11} + 24 r_1^2 r_2^2 r_{11}$	$4r_0^4 r_{31} + 36 r_0^3 r_1 r_{21} + 36 r_0^3 r_2 r_{11} + 12 r_0^3 r_{01} r_3 + 72 r_0^2 r_1^2 r_{11} + 72 r_0^2 r_1 r_2 r_{01} + 24 r_0 r_1^3 r_{01}$
(301)	$3r_1^3 r_{201} + 6 r_1^2 r_2 r_{101}$	$3r_0^3 r_{201} + 12 r_0^2 r_1 r_{101} + 6 r_0^2 r_2 r_{001} + 6 r_0 r_1^2 r_{001}$
(22)	$2r_1^3 r_{12} + 8 r_1^2 r_2 r_{11} + 2 r_1^2 r_{11}^2$	$2r_0^3 r_{12} + 8 r_0^2 r_1 r_{11} + 2 r_0^2 r_1 r_{02} + 4 r_0^2 r_{01} r_{11} + 8 r_0 r_1^2 r_{01}$
(2001)	$2r_1^2 r_{1001}$	$2r_0^2 r_{1001} + 2 r_0 r_1 r_{0001}$
(111)	$r_1^2 r_{011}$	$r_0^2 r_{011}$
(10001)	$r_1 r_{00001}$	$r_0 r_{00001}$
(03)	$12 r_1^2 r_3 + 24 r_1 r_2^2$	0
(0101)	$2 r_1 r_{11}$	0
(002)	0	0
(000001)	0	0

Table 13. *(G. Labelle [187].) Coefficients of the generic asymmetry index series of R-enriched trees and R-enriched rooted trees ($n \leq 6$) when*

$$\Gamma_R(x_1, x_2, \ldots) = \sum_{n_1, n_2, \ldots} r_{n_1, n_2, \ldots} \frac{x_1^{n_1} x_2^{n_2} \cdots}{1^{n_1} n_1! 2^{n_2} n_2! \cdots}.$$

$(n_1, n_2, \ldots)$	$1^{n_1} n_1! 2^{n_2} n_2! \cdots [x_1^{n_1} x_2^{n_2} \cdots] \Gamma_{a_R}$	$1^{n_1} n_1! 2^{n_2} n_2! \cdots [x_1^{n_1} x_2^{n_2} \cdots] \Gamma_{\mathcal{A}_R}$
( )	0	0
(1)	r_0	r_0
(2)	r_1^2	$2 r_0 r_1$
(01)	$-r_1$	0
(3)	$3 r_1^2 r_2$	$3 r_0^2 r_2 + 6 r_0 r_1^2$

Miscellaneous Tables 429

Table 13. (cont.)

$(n_1, n_2, \ldots)$	$1^{n_1} n_1! 2^{n_2} n_2! \cdots [x_1^{n_1} x_2^{n_2} \cdots] \Gamma_{a_R}$	$1^{n_1} n_1! 2^{n_2} n_2! \cdots [x_1^{n_1} x_2^{n_2} \cdots] \Gamma_{\mathcal{A}_R}$
(11)	$r_1 r_{01}$	$r_0 r_{01}$
(001)	0	0
(4)	$4r_1^3 r_3 + 12 r_1^2 r_2^2$	$4 r_0^3 r_3 + 36 r_0^2 r_1 r_2 + 24 r_0 r_1^3$
(21)	$2 r_1^2 r_{11}$	$2 r_0^2 r_{11} + 2 r_0 r_1 r_{01}$
(101)	$r_1 r_{001}$	$r_0 r_{001}$
(02)	$-4 r_1 r_2$	0
(0001)	0	0
(5)	$5 r_1^4 r_4 + 60 r_1^3 r_2 r_3 + 60 r_1^2 r_2^3$	$5 r_0^4 r_4 + 80 r_0^3 r_1 r_3 + 60 r_0^3 r_2^2$ $+ 360 r_0^2 r_1^2 r_2 + 120 r_0 r_1^4$
(31)	$3 r_1^3 r_{21} + 6 r_1^2 r_2 r_{11}$	$3 r_0^3 r_{21} + 12 r_0^2 r_1 r_{11}$ $+ 6 r_0^2 r_2 r_{01} + 6 r_0 r_1^2 r_{01}$
(201)	$2 r_1^2 r_{101}$	$2 r_0^2 r_{101} + 2 r_0 r_1 r_{001}$
(12)	$r_1^2 r_{02} + 4 r_1 r_2 r_{01}$	$r_0^2 r_{02} + 4 r_0 r_1 r_{01}$
(1001)	$r_1 r_{0001}$	$r_0 r_{0001}$
(011)	0	0
(00001)	0	0
(6)	$6 r_1^5 r_5 + 120 r_1^4 r_2 r_4 + 90 r_1^4 r_3^2$ $+ 720 r_1^3 r_2^2 r_3 + 360 r_1^2 r_2^4$	$6 r_0^5 r_5 + 150 r_0^4 r_1 r_4 + 300 r_0^4 r_2 r_3$ $+ 1200 r_0^3 r_1^2 r_3 + 1800 r_0^3 r_1 r_2^2$ $+ 3600 r_0^2 r_1^3 r_2 + 720 r_0 r_1^5$
(41)	$4 r_1^4 r_{31} + 24 r_1^3 r_2 r_{21}$ $+ 12 r_1^3 r_3 r_{11} + 24 r_1^2 r_2^2 r_{11}$	$4 r_0^4 r_{31} + 36 r_0^3 r_1 r_{21} + 36 r_0^3 r_2 r_{11}$ $+ 12 r_0^3 r_{01} r_3 + 72 r_0^2 r_1^2 r_{11}$ $+ 72 r_0^2 r_1 r_2 r_{01} + 24 r_0 r_1^3 r_{01}$
(301)	$3 r_1^3 r_{201} + 6 r_1^2 r_2 r_{101}$	$3 r_0^3 r_{201} + 12 r_0^2 r_1 r_{101}$ $+ 6 r_0^2 r_2 r_{001} + 6 r_0 r_1^2 r_{001}$

Table 13. (cont.)

$(n_1, n_2, \ldots)$	$1^{n_1} n_1! 2^{n_2} n_2! \cdots [x_1^{n_1} x_2^{n_2} \cdots] \Gamma_{a_R}$	$1^{n_1} n_1! 2^{n_2} n_2! \cdots [x_1^{n_1} x_2^{n_2} \cdots] \Gamma_{\mathcal{A}_R}$
(22)	$2r_1^3 r_{12} + 8r_1^2 r_2 r_{11} + 2r_1^2 r_{11}^2$	$2r_0^3 r_{12} + 8r_0^2 r_1 r_{11} + 2r_0^2 r_1 r_{02}$ $+ 4r_0^2 r_{01} r_{11} + 8r_0 r_1^2 r_{01}$
(2001)	$2r_1^2 r_{1001}$	$2r_0^2 r_{1001} + 2r_0 r_1 r_{0001}$
(111)	$r_1^2 r_{011}$	$r_0^2 r_{011}$
(10001)	$r_1 r_{00001}$	$r_0 r_{00001}$
(03)	$-12r_1^2 r_3 - 24r_1 r_2^2$	0
(0101)	$-2r_1 r_{11}$	0
(002)	0	0
(000001)	0	0

Table 14. *Molecular decomposition of $F(X + Y)$ for the atomic species F on $n \leq 5$ points (G. Labelle [184]).*

n	$F(X)$	$F(X + Y)$
1	X	$X + Y$
2	$E_2(X)$	$E_2(X) + XY + E_2(Y)$
3	$E_3(X)$	$E_3(X) + E_2(X)Y + XE_2(Y) + E_3(Y)$
	$\mathcal{C}_3(X)$	$\mathcal{C}_3(X) + X^2 Y + XY^2 + \mathcal{C}_3(Y)$
4	$E_4(X)$	$E_4(X) + E_3(X)Y + E_2(X)E_2(Y) + XE_3(Y) + E_4(Y)$
	$E_4^{\pm}(X)$	$E_4^{\pm}(X) + \mathcal{C}_3(X)Y + E_2(XY) + X\mathcal{C}_3(Y) + E_4^{\pm}(Y)$
	$E_2 \circ E_2(X)$	$E_2 \circ E_2(X) + XE_2(X)Y + E_2(X)E_2(Y) + E_2(XY)$ $+ XYE_2(Y) + E_2 \circ E_2(Y)$
	$P_4^{bic}(X)$	$P_4^{bic}(X) + X^3 Y + 3E_2(XY) + XY^3 + P_4^{bic}(Y)$
	$\mathcal{C}_4(X)$	$\mathcal{C}_4(X) + X^3 Y + X^2 Y^2 + E_2(XY) + XY^3 + \mathcal{C}_4(Y)$
	$E_2(X^2)$	$E_2(X^2) + 2X^3 Y + 2X^2 Y^2 + 2E_2(XY) + 2XY^3 + E_2(Y^2)$

Table 14. (cont.)

n	$F(X)$	$F(X+Y)$
5	$E_5(X)$	$E_5(X) + E_4(X)Y + E_3(X)E_2(Y) + E_2(X)E_3(Y)$
		$+ XE_4(Y) + E_5(Y)$
	$E_5^\pm(X)$	$E_5^\pm(X) + E_4^\pm(X)Y + \bigl(\mathcal{C}_3(X)Y^2\bigr)/\mathbb{Z}_2 + \bigl(X^2\mathcal{C}_3(Y)\bigr)/\mathbb{Z}_2$
		$+ XE_4^\pm(Y) + E_5^\pm(Y)$
	$P_5(X)/\mathbb{Z}_2$	$P_5(X)/\mathbb{Z}_2 + \mathcal{C}_4(X)Y + XE_2(XY) + E_2(XY)Y$
		$+ X\mathcal{C}_4(Y) + P_5(Y)/\mathbb{Z}_2$
	$P_5(X)$	$P_5(X) + E_2(X^2)Y + 2XE_2(XY) + 2E_2(XY)Y$
		$+ XE_2(Y^2) + P_5(Y)$
	$(X^2\mathcal{C}_3(X))/\mathbb{Z}_2$	$\bigl(X^2\mathcal{C}_3(X)\bigr)/\mathbb{Z}_2 + X\mathcal{C}_3(X)Y + E_2(X^2)Y + \bigl(\mathcal{C}_3(X)Y^2\bigr)/\mathbb{Z}_2$
		$+ X^3Y^2 + XE_2(XY) + E_2(XY)Y + X^2Y^3$
		$+ \bigl(X^2\mathcal{C}_3(Y)\bigr)/\mathbb{Z}_2 + XE_2(Y^2) + X\mathcal{C}_3(Y)Y + \bigl(Y^2\mathcal{C}_3(Y)\bigr)/\mathbb{Z}_2$
	$\mathcal{C}_5(X)$	$\mathcal{C}_5(X) + X^4Y + 2X^3Y^2 + 2X^2Y^3 + XY^4 + \mathcal{C}_5(Y)$

Table 15. *Asymmetry index series of molecular species ($n = 5$) (Pineau [262])*.

M	$\Gamma_M(x_1, x_2, \ldots)$
E_5	$\frac{1}{120}x_1^5 - \frac{1}{12}x_1^3 x_2 + \frac{1}{6}x_1^2 x_3 + \frac{1}{8}x_1 x_2^2 - \frac{1}{4}x_1 x_4 - \frac{1}{6}x_2 x_3 + \frac{1}{5}x_5$
$E_5^\pm$	$\frac{1}{60}x_1^5 - \frac{1}{6}x_1^2 x_3 - \frac{1}{4}x_1 x_2^2 + \frac{1}{2}x_1 x_4 + \frac{1}{2}x_2 x_3 - \frac{3}{5}x_5$
$X \cdot E_4$	$\frac{1}{24}x_1^5 - \frac{1}{4}x_1^3 x_2 + \frac{1}{3}x_1^2 x_3 + \frac{1}{8}x_1 x_2^2 - \frac{1}{4}x_1 x_4$
$P_5/\mathbb{Z}_2$	$\frac{1}{20}x_1^5 - \frac{1}{4}x_1 x_2^2 + \frac{1}{5}x_5$
$X \cdot E_4^\pm$	$\frac{1}{12}x_1^5 - \frac{1}{3}x_1^2 x_3 - \frac{1}{4}x_1 x_2^2 + \frac{1}{2}x_1 x_4$
$E_2 \cdot E_3$	$\frac{1}{12}x_1^5 - \frac{1}{3}x_1^3 x_2 + \frac{1}{6}x_1^2 x_3 + \frac{1}{4}x_1 x_2^2 - \frac{1}{6}x_2 x_3$
P_5	$\frac{1}{10}x_1^5 - \frac{1}{2}x_1 x_2^2 + \frac{2}{5}x_5$
$X \cdot P_4$	$\frac{1}{8}x_1^5 - \frac{1}{4}x_1^3 x_2 - \frac{1}{8}x_1 x_2^2 + \frac{1}{4}x_1 x_4$

Table 15. (cont.)

M	$\Gamma_M(x_1, x_2, \ldots)$
$X^2 \cdot E_3$	$\frac{1}{6}x_1^5 - \frac{1}{2}x_1^3 x_2 + \frac{1}{3}x_1^2 x_3$
$E_2 \cdot C_3$	$\frac{1}{6}x_1^5 - \frac{1}{6}x_1^3 x_2 - \frac{1}{6}x_1^2 x_3 + \frac{1}{6}x_2 x_3$
$(X^2 \cdot C_3)/\mathbb{Z}_2$	$\frac{1}{6}x_1^5 - \frac{1}{6}x_1^2 x_3 - \frac{1}{2}x_1 x_2^2 + \frac{1}{2}x_2 x_3$
C_5	$\frac{1}{5}x_1^5 - \frac{1}{5}x_5$
$X \cdot E_2^2$	$\frac{1}{4}x_1^5 - \frac{1}{2}x_1^3 x_2 + \frac{1}{4}x_1 x_2^2$
$X \cdot P_4^{bic}$	$\frac{1}{4}x_1^5 - \frac{3}{4}x_1 x_2^2 + \frac{1}{2}x_1 x_4$
$X \cdot C_4$	$\frac{1}{4}x_1^5 - \frac{1}{4}x_1 x_2^2$
$X^2 \cdot C_3$	$\frac{1}{3}x_1^5 - \frac{1}{3}x_1^2 x_3$
$X^3 \cdot E_2$	$\frac{1}{2}x_1^5 - \frac{1}{2}x_1^3 x_2$
$X \cdot (E_2 \circ X^2)$	$\frac{1}{2}x_1^5 - \frac{1}{2}x_1 x_2^2$
X^5	x_1^5

Note: See Table 4 in Section 4.4 for orders $n \leq 4$.

Exercises for Appendix 2

1. Verify, by geometric arguments or by combinatorial differential calculus, the formulas in Table 5 for the derivative of the given molecular species.
2. Starting from the data in Table 6, show that the asymmetry generating series $\overline{\mathcal{G}}(x)$ of the species of simple graphs begins with the terms

$$\overline{\mathcal{G}}(x) = 1 + x + 8x^6 + \cdots. \tag{1}$$

3. The molecular species P_6^{opp} is defined as the species of polygons of order 6 with two distinguished opposite vertices.

a) Identify the two simple graphs of order 6 explaining the occurrence of $2P_6^{\text{opp}}$ in the molecular decomposition of the species of graphs of order ≤ 6 (see Table 6).
b) Show that $P_6^{\text{opp}} = X^6/K$, where K is the subgroup of $\mathcal{S}_6$ generated by $\{\sigma, \tau\}$, where $\sigma = (1, 3)(4, 6)$ and $\tau = (1, 6)(2, 5)(3, 4)$.
c) Show that P_6^{opp} is atomic.

4. Verify that Tables 7 and 8 are compatible with the dissymmetry formula for trees,

$$\mathfrak{a} + \mathcal{A}^2 = \mathcal{A} + E_2(\mathcal{A}), \tag{2}$$

at least up to order $n = 6$.

5. With the aid of suitable geometric drawings, give a combinatorial explanation for each of the coefficients of the generic series in Table 9.

6. Observe that upon replacing all negative signs ("$-$") in Table 11 by plus signs ("$+$"), Table 10 is obtained. The same procedure allows one to pass from Table 13 to Table 12. Show that what is happening here is a general rule.

7. a) Verify the molecular decompositions of the species $F(X + Y)$ given in Table 14.
b) By considering the molecular decomposition of $M(X+Y)$ for each molecular species $M(X)$, generalizations of Newton's binomial formula are obtained. Indeed, the case $M(X) = X^n$ corresponds to the usual binomial formula

$$(X + Y)^n = \sum_{i+j=n} \frac{n!}{i!j!} X^i Y^j. \tag{3}$$

Explicitly describe the molecular decomposition of $E_n(X + Y)$ and of $C_n(X + Y)$.

8. Verify the asymmetry index series of molecular species of degree 5 in Table 15.
HINT: Use Table 4 in Section 4.4 in the case of nonatomic molecular species. For atomic species, use Proposition 4.4.11 or the explicit formula (4.4.51) of Exercise 4.4.7.

Bibliography

[1] A.V. Aho, J.E. Hopcroft, and J.D. Ullman, *Data Structures and Algorithms*, Addison–Wesley, Reading, MA, **1983**.
[2] D. André, Sur les permutations alternées, *Journal de Mathématiques Pures et Appliquées*, 5, **1879**, 31–46.
[3] T.M. Apostol, *Introduction to Analytic Number Theory*, Springer–Verlag, Berlin, Heidelberg, and New York, **1976**.
[4] P. Auger, *Étude exacte et asymptotique de divers paramètres de structures arborescentes*, M.Sc. Thesis, Université du Québec à Montréal, **1993**.
[5] R. Baeza-Yates and G.H. Gonnet, *Handbook of Algorithms and Data Structures: in Pascal and C*, 2nd, Addison–Wesley, Reading, MA, **1991**.
[6] M. Bajraktarević, Sur une équation fonctionnelle, *Glasnik Matematičko-Fizički I Astronomski*, 12, **1957**, 201–205.
[7] F. Bédard and A. Goupil, The Lattice of Conjugacy Classes of the Symmetric Group, *Canadian Bulletin of Mathematics*, 35, **1992**, 152–160.
[8] E.A. Bender, Central and Local Limit Theorems Applied to Asymptotic Enumeration, *Journal of Combinatorial Theory, Series A*, 15, **1973**, 91–111.
[9] E.A. Bender, Asymptotic Methods in Enumeration, *SIAM Review*, 16, **1974**, 485–515.
[10] E.A. Bender and E.R. Canfield, The Asymptotic Number of Labeled Connected Graphs With a Given Number of Vertices and Edges, *Random Structures and Algorithms*, 1, **1990**, 127–169.
[11] E.A. Bender and J.R. Goldman, Enumerative Uses of Generating Functions, *Indiana University Mathematics Journal*, 20, **1971**, 753–765.
[12] E.A. Bender and S.G. Williamson, *Foundations of Applied Combinatorics*, Addison–Wesley, Reading, MA, **1991**.
[13] C. Berge, *Principles of Combinatorics*, Academic, New York, **1970**.
[14] F. Bergeron, *Une systématique de la combinatoire énumérative*, Ph.D. Dissertation, Université de Montréal, Canada, **1985**.
[15] F. Bergeron, Une approche combinatoire de la méthode de Weisner, in: *Polynômes Orthogonaux et Applications*, editors: C. Brezinski, A. Draux, A. P. Magnus, P. Maroni, and A. Ronveaux, Lecture Notes in Mathematics 1171, Springer–Verlag, Berlin, Heidelberg, and New York, **1985**, 111–119.
[16] F. Bergeron, Combinatorial Representations of Lie Groups and Algebras, in: [196], **1986**, 34–47.
[17] F. Bergeron, Une combinatoire du pléthysme, *Journal of Combinatorial Theory, Series A*, 46, **1987**, 291–305.
[18] F. Bergeron, A Combinatorial Outlook on Symmetric Functions, *Journal of Combinatorial Theory, Series A*, 50, **1989**, 226–234.
[19] F. Bergeron, Algorithms for the Sequential Generation of Combinatorial Structures, *Discrete Applied Mathematics*, 24, **1989**, 29–35.

[20] F. Bergeron, Combinatoire des polynômes orthogonaux classiques: une approche unifiée, *European Journal of Combinatorics*, 11, **1990**, 393–401.
[21] F. Bergeron and G. Cartier, Darwin: Computer Algebra and Enumerative Combinatorics, in: *Proceedings of STACS-88*, editors: R. Cori and M. Wirsing, Lecture Notes in Computer Sciences 294, Springer–Verlag, Berlin, Heidelberg, and New York, **1988**, 393–394.
[22] F. Bergeron, P. Flajolet, and B. Salvy, Varieties of Increasing Trees, in: *Proceedings of the 17th Colloquium on Trees in Algebra and Programming*, editor: J.-C. Raoult, Lecture Notes in Computer Sciences 581, Springer–Verlag, Berlin, Heidelberg, and New York, **1991**, 24–48.
[23] F. Bergeron, G. Labelle, and P. Leroux, Functional Equations for Data Structures, in: *Proceedings of STACS-88*, editors: R. Cori and M. Wirsing, Lecture Notes in Computer Sciences 294, Springer–Verlag, Berlin, Heidelberg, and New York, **1988**, 73–80.
[24] F. Bergeron, G. Labelle, and P. Leroux, Computation of the Expected Number of Leaves in a Tree Having a Given Automorphism, *Discrete Applied Mathematics*, 34, **1991**, 49–66.
[25] F. Bergeron and S. Plouffe, Computing the Generating Function of a Series Given its First Few Terms, *Experimental Mathematics*, 1, **1992**, 307–312.
[26] F. Bergeron and C. Reutenauer, Interprétation combinatoire des puissances d'un opérateur différentiel linéaire, *Annales des Sciences Mathématiques du Québec*, 11, **1987**, 269–278.
[27] F. Bergeron and C. Reutenauer, Combinatorial Resolution of Systems of Differential Equations III: A Special Class of Differentiably Algebraic Series, *European Journal of Combinatorics*, 11, **1990**, 501–512.
[28] F. Bergeron and U. Sattler, Constructibly Differentially Algebraic Series in Several Variables, *Theoretical Computer Science*, 144, **1995**, 59–65.
[29] F. Bergeron and Y.N. Yeh, The Factoriality of the Ring of s-Species, *Journal of Combinatorial Theory, Series A*, 55, **1990**, 194–203.
[30] J. Berstel and C. Reutenauer, Zeta Functions of Formal Languages, *Transactions of the American Mathematical Society*, 321, **1990**, 533–546.
[31] J. Bétréma and A. Zvonkin, Planes Trees and Shabat Polynomials, in: *Discrete Mathematics*, 153, **1996**, 47–58.
[32] F. Bonetti, G.C. Rota, D. Senato, and A.M. Venezia, Symmetric Functions and Symmetric Species, *Combinatorics'84* (Bari, 1984), North–Holland Mathematical Studies 123, Amsterdam, **1986**, 107–113.
[33] F. Bonetti, G.C. Rota, D. Senato, and A.M. Venezia, On the Foundation of Combinatorial Theory X, A Categorical Setting for Symmetric Functions, *Studies in Applied Mathematics*, 86, **1992**, 1–29.
[34] K.S. Booth and G.S. Lueker, Testing for the Consecutive Ones Property, Interval Graphs, and Graph Planarity using PQ-Tree Algorithms, *Journal of Computer and System Sciences*, 13, **1976**, 335–379.
[35] P. Bouchard, Y. Chiricota, and G. Labelle, Arbres, arborescences et racines carrées symétriques, *Discrete Mathematics*, 139, **1995**, 49–56.
[36] P. Bouchard and M. Ouellette, Décomposition arborescente de Mario Ouellette pour les espèces de structures, *Actes du séminaire lotharingien de combinatoire*, Publications de l'Institut de recherche mathématiques, Strasbourg, France, **1990**, 5–13.
[37] N. Bourbaki, *Anneaux topologiques*, Herman, Paris, **1971**.
[38] G. Brassard and P. Bratley, *Algorithmique, conception et analyse*, Masson/Les Presses de Université de Montréal, Canada, **1987**.
[39] R.P. Brent and H.T. Kung, Fast Algorithms for Manipulating Formal Power Series, *Journal of the ACM*, 25, **1978**, 581–595.
[40] T.J. L'A Bromwich, *Introduction to the Theory of Infinite Series*, Macmillan, New York, **1965**.
[41] W.H. Burge, An Analysis of a Tree Sorting Method and Some Properties of a Set of Trees, in: *Proceedings of the First USA-Japan Computer Conference*, **1972**, 372–379.
[42] W. Burnside, *Theory of Groups of Finite Order*, Dover, New York, **1955**.
[43] E.R. Canfield, Remarks on an Asymptotic Method in Combinatorics, *Journal of Combinatorial Theory, Series A*, 37, **1984**, 348–352.

[44] G. Cartier, *Darwin, Manuel d'Utilisateur*, Publications du Laboratoire de combinatoire et d'informatique mathématique, Université du Québec à Montréal, **1990**.
[45] P. Cartier and D. Foata, *Problèmes Combinatoires de Commutation et de Réarrangements*, Lecture Notes in Mathematics 85, Springer–Verlag, Berlin, Heidelberg, and New York, **1969**.
[46] A. Cayley, A Theorem on Trees, *Quarterly Journal of Mathematics*, Oxford Series 23, **1889**, 376–378; Collected papers, Cambridge 13, **1897**, 26–28.
[47] B.W. Char, K.O. Geddes, G.H. Gonnet, M.B. Monagan, and S.M. Watt, *Maple Reference Manual, 5th edition*, Watcom, Waterloo, Ontario, **1988**.
[48] W.Y.C. Chen, A General Bijective Algorithm for Trees, *Proceedings of the National Academy of Sciences USA*, 87, **1990**, 9635–9639.
[49] W.Y.C. Chen, The Theory of Compositionals, *Discrete Mathematics*, 22, **1993**, 59–87.
[50] W.Y.C. Chen, Compositional Calculus, *Journal of Combinatorial Theory, Series A*, 64, **1993**, 149–188.
[51] W.Y.C. Chen, A Bijection for Enriched Trees, *European Journal of Combinatorics*, 15, **1994**, 337–343.
[52] N. Chiba, T. Nishizeki, S. Abe, and T. Ozawa, A Linear Algorithm for Embedding Planar Graphs Using PQ-Trees, *Journal of Computer and System Sciences*, 30, **1985**, 54–76.
[53] I. Chiricota, *Structures combinatoires et calcul symbolique*, Ph.D. Dissertation, Université du Québec à Montréal, Publications du Laboratoire de combinatoire et d'informatique mathématique Vol. 12, **1993**.
[54] Y. Chiricota, Classification des espèces moléculaires de degré 6 et 7, *Annales des Sciences Mathématiques du Québec*, 17, **1993**, 11–37.
[55] Y. Chiricota and G. Labelle, Familles de solutions combinatoires de l'équation différentielle $Y' = 1 + Y^2$ et des équations différentielles autonomes, *Discrete Mathematics*, 115, **1993**, 77–93.
[56] Y. Chiricota and V. Strehl, Inversion of Polynomial Species and Jacobian Conjecture, *Bayreuther Mathematische Schriften*, 44, **1993**, 53–61.
[57] L. Chottin and R. Cori, Enumération d'arbres et formules d'inversion de séries formelles, *Journal of Combinatorial Theory, Series B*, 31, **1981**, 23–45.
[58] L. Comtet, *Advanced Combinatorics*, Reidel, Dordrecht, The Netherlands, **1974**.
[59] I. Constantineau, *Calcul combinatoire de séries indicatrices de cycles*, Ph.D. Dissertation, Université du Québec à Montréal, Publications du Laboratoire de combinatoire et d'informatique mathématique Vol. 5, **1991**.
[60] I. Constantineau, Auto-similarité dans la combinatoire des polynômes orthogonaux, *Theoretical Computer Science*, 117, **1993**, 153–167.
[61] I. Constantineau, Le nombre d'arbres m-Husimi invariants sous une permutation des sommets, *Discrete Mathematics*, 139, **1995**, 89–103.
[62] I. Constantineau and G. Labelle, Une généralisation automorphe des nombres de Stirling, *Discrete Mathematics*, 157, **1996**, 53–64.
[63] I. Constantineau and J. Labelle, Le nombres d'endofonctions et d'arborescences laissées fixes par l'action d'une permutation, *Annales des Sciences Mathématiques du Québec*, 13, **1989**, 33–38.
[64] I. Constantineau and J. Labelle, On combinatorial structures kept fixed by the action of a given permutation, *Studies in Applied Mathematics*, 84, **1991**, 105–118.
[65] I. Constantineau and J. Labelle, On the Construction of Permutations of a Given Type Kept Fixed by Conjugation, *Journal of Combinatorial Theory, Series A*, 62, **1993**, 199–209.
[66] M. Content, F. Lemay, and P. Leroux, Catégories de Möbius et fonctorialité, un cadre pour l'inversion de Möbius, *Journal of Combinatorial Theory, Series A*, 28, **1980**, 169–190.
[67] R. Cori and B. Vauquelin, Planar Maps are Well Labelled Trees, *Canadian Journal of Mathematics*, 33, **1981**, 1023–1042.
[68] N.G. de Bruijn, Enumerative Combinatorial Problems Concerning Structures, *Nieuw Archief voor Wiskunde* (3), XI, **1963**, 142–161.
[69] N.G. de Bruijn, *Asymptotic Methods in Analysis*, Dover, New York, **1981**.
[70] H. Décoste, *Séries indicatrices d'espèces pondérées et q-analogues*, Ph.D. Dissertation, Université de Montréal, Canada, **1989**. Publications du Laboratoire de combinatoire et d'informatique mathématique Vol. 2, Université du Québec à Montréal, **1990**.

[71] H. Décoste, Séries indicatrices et q-series, Theoretical Computer Science, 117, **1993**, 169–186.
[72] H. Décoste and G. Labelle, Le q-dénombrement générique d'une espèce: existence et méthode de calcul, *Discrete Mathematics*, 153, **1996**, 59–67.
[73] H. Décoste, G. Labelle, and P. Leroux, Une approche combinatoire pour l'itération de Newton-Raphson, *Advances in Applied Mathematics*, 3, **1982**, 407–416.
[74] H. Décoste, G. Labelle, and P. Leroux, The Functorial Composition of Species, a Forgotten Operation, *Discrete Mathematics*, 99, **1992**, 31–48.
[75] M. Delest and G.X. Viennot, Algebraic Languages and Polyominal Enumeration, *Theoretical Computer Science*, 34, **1984**, 169–206.
[76] L. Devroye, Applications of the Theory of Records in the Study of Random Trees, *Acta Informatica*, 26, **1988**, 123–130.
[77] L. Devroye and L. Laforest, An Analysis of Random d-Dimensional Quad Trees, *SIAM Journal of Computing*, 19, **1990**, 821–832.
[78] V. Domocos and W.R. Schmitt, An Application of Linear Species, *Discrete Mathematics*, 132, **1994**, 377–381.
[79] W.F. Doran IV, The Lattice of Periods of a Group Action, *Advances in Mathematics*, 110, **1995**, 88–108.
[80] P. Doubilet, G.-C. Rota, and R.P. Stanley, The Idea of Generating Function, in: *Finite Operator Calculus*, editor: G.-C. Rota, Academic, New York, **1975**, 83–134.
[81] S. Dubuc, Une équation fonctionnelle pour diverses constructions géométriques, *Annales des Sciences Mathématiques du Québec*, 9, **1985**, 407–416.
[82] S. Dubuc and A. Elqortobi, Enveloppe convexe d'une courbe de Mandelbrot, *Annales des Sciences Mathématiques du Québec*, 11, **1987**, 45–64.
[83] D. Dumont, Une approche combinatoire des fonctions elliptiques de Jacobi, *Advances in Mathematics*, 41, **1981**, 1–39.
[84] A. Dür, *Mobius Functions, Incidence Algebras and Power Series Representations*, Lecture Notes in Mathematics 1202, Springer–Verlag, Berlin, Heidelberg, and New York, **1986**.
[85] P.H. Edelman, R. Simion, and D. White, Partition Statistics on Permutations, *Discrete Mathematics*, 99, **1992**, 63–68.
[86] R. Ehrenborg and M. Méndez, A Bijective Proof of Infinite Variated Good's Inversion, *Advances in Mathematics*, 103, **1994**, 221–259.
[87] C. Ehresmann, *Catégories et Structures*, Dunod, Paris, **1965**.
[88] S. Eilenberg, *Automata, Languages and Machines*, Academic, New York, Vol. A, **1974**, Vol. B, **1976**.
[89] P. Fatou, Sur les équations fonctionnelles, *Bulletin de la Société Mathématique de France*, 48, **1920**, 208–314.
[90] G. Fayole, P. Flajolet, and M. Hofri, On a Functional Equation Arising in the Analysis of a Protocol for Multi-Access Broadcast Channel, *Advances in Applied Probability*, 18, **1986**, 441–472.
[91] P. Flajolet, Elements of a General Theory of Combinatorial Structures, in: *Fundamentals of Computation Theory*, editor: L. Budach, Lecture Notes in Computer Sciences 199, Springer–Verlag, Berlin, Heidelberg, and New York, **1985**, 112–127.
[92] P. Flajolet, Random Tree Models in the Analysis of Algorithms, in: *Performance'87*, editors: P.-J. Courtois and G. Latouche, Elsevier Science Publishers, New York, **1988**, 171–187.
[93] P. Flajolet and J. Françon, Elliptic Functions, Continued Fractions and Doubled Permutations, *European Journal of Combinatorics*, 10, **1989**, 235–241.
[94] P. Flajolet, G. Gonnet, C. Puech, and J.M. Robson, Analytic Variations on Quad Trees, *Algorithmica*, 10, **1993**, 473–500.
[95] P. Flajolet, G. Labelle, and L. Laforest, Hypergeometrics and the Cost Structure of Quadtrees, *Random Structures and Algorithms*, 7, **1995**, 117–144.
[96] P. Flajolet and A.M. Odlyzko, Limit Distributions for Coefficients of Iterates of Polynomials with Applications to Combinatorial Enumerations, *Proceedings of the Cambridge Philosophical Society*, 96, **1984**, 237–253.
[97] P. Flajolet and A.M. Odlyzko, Singularity Analysis of Generating Functions, *SIAM Journal on Discrete Mathematics*, 3, **1990**, 216–240.

[98] P. Flajolet and H. Prodinger, Level Number Sequences for Trees, *Discrete Mathematics*, 65, **1987**, 149–156.
[99] P. Flajolet, B. Salvy, and P. Zimmermann, Automatic Average-case Analysis of Algorithms, *Theoretical Computer Science*, 79, **1991**, 37–109.
[100] P. Flajolet and M. Soria, Gaussian Limiting Distributions for the Number of Components in Combinatorial Structures, *Journal of Combinatorial Theory, Series A*, 53, **1990**, 165–182.
[101] P. Flajolet and J.M. Steyaert, Patterns and Pattern Matching in Trees, an Analysis, *Information Control*, 58, **1988**, 19–58.
[102] P. Flajolet and J.S. Vitter, Analysis of Algorithms and Data Structures, in: *Handbook of Theoretical Computer Science*, editor: J. van Leeuwen, Vol. A: *Algorithms and Complexity*, North Holland, Amsterdam, **1990**, 431–524.
[103] M. Fliess, Fonctionnelles causales non linéaires et indéterminées non commutatives, *Bulletin de la société mathématique de France*, 109, **1981**, 3–40.
[104] M. Fliess and F. Lamnabhi–Lagarrigue, Application of a New Functional Expansion to the Cubic Anharmonic Oscillator, *Journal of Mathematical Physics*, 23, **1982**, 495–502.
[105] D. Flores-de-Chela and M.A. Mendez, Equivariant Representations of the Symmetric Group, *Advances in Mathematics*, 97, **1993**, 191–230.
[106] D. Foata, *La Série Génératrice Exponentielle dans les Problèmes d'Énumération*, Presses de l'Université de Montréal, Montréal, Québec, **1974**.
[107] D. Foata, A Combinatorial Proof of the Mehler Formula, *Journal of Combinatorial Theory, Series A*, 24, **1978**, 367–376.
[108] D. Foata, A Combinatorial Proof of Jacobi's Identity, *Annals of Discrete Mathematics*, 6, **1980**, 125–135.
[109] D. Foata, Combinatoire des identités sur les polynômes orthogonaux, in: *Proceedings of the International Congress of Mathematicians* Vol. 2, Warsaw 1983, PWN–Polish Scientific Publications, Warsawa, and North Holland, Amsterdam, **1984**, 1541–1553.
[110] D. Foata and J. Labelle, Modèles combinatoires pour les polynômes de Meixner, *European Journal of Combinatorics*, 4, **1983**, 305–311.
[111] D. Foata and P. Leroux, Polynômes de Jacobi, interprétation combinatoire et fonction génératrice, *Proceedings of the American Mathematical Society*, 87, **1983**, 47–53.
[112] D. Foata and M.P. Schutzenberger, *Théorie géométrique des polynômes Eulériens*, Lecture Notes in Mathematics 138, Springer–Verlag, Berlin, Heidelberg, and New York, **1970**.
[113] D. Foata and V. Strehl, Rearrangement of the Symmetric Group and Enumeration Properties of the Tangent and Secant Numbers, *Mathematische Zeitschrift*, 137, **1974**, 257–264.
[114] D. Foata and V. Strehl, Combinatorics of Laguerre Polynomials, *Enumeration and Design*, editors: D.M. Jackson and S.A. Vanstone, Academic, New York, **1984**, 123–140.
[115] D. Foata and D. Zeilberger, Laguerre Polynomials, Weighted Derangements, and Positivity, *SIAM Journal on Discrete Mathematics*, 1, **1988**, 425–433.
[116] G.W. Ford and G.E. Uhlenbeck, Combinatorial Problems in the Theory of Graphs III, *Proceedings of the National Academy of Sciences USA* 42, **1956**, 529–535.
[117] J. Françon, On the Analysis of Algorithms for Trees, *Theoretical Computer Science*, 12, **1976**, 35–50.
[118] J. Françon, Arbres binaires de recherche: Propriétés combinatoires et applications, *RAIRO Informatique Théorique*, 4, **1977**, 155–169.
[119] J. Françon, Sur le nombre de registres nécessaires à l'évaluation d'une expression arithmétique, *RAIRO Informatique Théorique*, 18, **1984**, 355–364.
[120] J. Françon, Fractions continues q-analogiques pour certaines familles de permutations et de partitions, *Annales des Sciences Mathématiques du Québec*, 16, **1992**, 175–182.
[121] J. Françon and G. Viennot, Permutations selon les pics, creux, doubles descentes, nombres d'Euler et nombres de Genocchi, *Discrete Mathematics*, 28, **1979**, 21–35.
[122] J.F. Gagné, *Rapport existant entre la théorie des espèces et les équations différentielles*, M.Sc. Thesis, Université du Québec à Montréal, **1985**.
[123] D. Gardy, Some Results on the Asymptotic Behaviour of Coefficients of Large Powers of Functions, *Discrete Mathematics*, 139, **1995**, 189–217.

[124] A. Garsia and S.A. Joni, *Lecture Notes in Combinatorics*, Department of Mathematics, University of California at San Diego, La Jolla, CA, **1976**.
[125] A.M. Garsia and J. Remmel, A Novel Form of q-Lagrange Inversion, *Houston Journal of Mathematics*, 12, **1986**, 503–523.
[126] I. Gessel, *Generating Functions and Enumeration of Sequence*, Ph.D. Dissertation, Massachusetts Institute of Technology, Cambridge, MA, **1977**.
[127] I. Gessel, A Combinatorial Proof of the Multivariable Lagrange Inversion Formula, *Journal of Combinatorial Theory, Series A*, 45, **1987**, 178–195.
[128] I. Gessel, Enumerative Applications of a Decomposition for Graphs and Digraphs, *Discrete Mathematics*, 139, **1995**, 257–271.
[129] I. Gessel and G. Labelle, Lagrange Inversion for Species, *Journal of Combinatorial Theory, Series A*, 72, **1995**, 95–117.
[130] I. Gessel, B. Sagan, and Y.N. Yeh, Enumeration of Trees by Inversions, *Journal of Graph Theory*, 19, **1995**, 435–459.
[131] I. Gessel and D.L. Wang, Depth-First Search as a Combinatorial Correspondence, *Journal of Combinatorial Theory, Series A*, 26, **1979**, 308–313.
[132] I.J. Good, Generalization to Several Variables of Lagrange's Expansion, with Applications to Stochastic Process, *Proceedings of the Cambridge Philosophical Society*, 56, **1960**, 367–380.
[133] I.P. Goulden and D.M. Jackson, *Combinatorial Enumeration*, Wiley, New York, **1983**.
[134] I.P. Goulden and D.M. Jackson, The Combinatorial Relationship Between Trees, Cacti and Certain Connection Coefficients for the Symmetric Group, *European Journal of Combinatorics*, 13, **1992**, 357–365.
[135] A.O. Guelfond, *Calcul des Différences Finies*, Dunod, Paris, **1963**.
[136] R.L. Graham, D.E. Knuth, and O. Patashnik, *Concrete Mathematics*, Addison–Wesley, Reading, MA, **1989**.
[137] D.H. Greene, *Labelled Formal Languages and Their Uses*, Ph.D. Dissertation, Department of Computer Science, Stanford University, Stanford CA, **1983**.
[138] W. Gröbner and H. Knapp, *Contributions to the Theory of Lie Series*, Bibliographisches Institut–Mannheim, Hochschultaschenbücher Verlag, Mannheim, Germany, **1967**.
[139] R. Hager, A. Kerber, R. Laue, D. Moser, and W. Weber, Construction of Orbit Representatives, *Bayreuther Mathematische Schriften*, 35, **1991**, 157–169.
[140] P. Hanlon, A Cycle Sum Inversion Theorem, *Journal of Combinatorial Theory, Series A*, 30, **1981**, 248–269.
[141] P. Hanlon and R.W. Robinson, Counting Bridgeless Graphs, *Journal of Combinatorial Theory, Series B*, 33, **1982**, 276–305.
[142] F. Harary, *Graph Theory*, Addison–Wesley, Reading, MA, **1969**.
[143] F. Harary and R.Z. Norman, The Dissimilarity Characteristic of Husimi Trees, *Annals of Mathematics*, 58, **1953**, 134–141.
[144] F. Harary and E. Palmer, *Graphical Enumeration*, Academic, New York, **1973**.
[145] F. Harary, E.M. Palmer, and R.W. Robinson, Counting Free Binary Trees Admitting a Given Height, *Journal of Combinatorics, Information, and System Sciences*, 17, **1992**, 175–181.
[146] F. Harary and G. Prins, The Number of Homeomorphically Irreducible Trees and Other Species, *Acta Mathematica*, 101, **1959**, 141–162.
[147] F. Harary, G. Prins, and W.T. Tutte, The Number of Plane Trees, *Indagaciones Math*, 26, **1964**, 319–329.
[148] F. Harary, R.W. Robinson, and A.J. Schwenk, Twenty-step Algorithm for Determining the Asymptotic Number of Trees of Various Species, *Journal of the Australian Mathematical Society (Series A)*, 20, **1975**, 483–503.
[149] F. Harary and G.E. Uhlenbeck, On the Number of Husimi Trees, *Proceedings of the National Academy of Sciences USA* 39, **1953**, 315–322.
[150] W.K. Hayman, A Generalization of Stirling's Formula, *Journal für die reine und angewandte Mathematik*, 196, **1956**, 67–95.
[151] P. Hennequin, Combinatorial Analysis of Quicksort Algorithm, *RAIRO Informatique Théorique*, 23, **1989**, 317–333.

[152] J. Hornegger and R. Pirastu, Die Speziestheorie und die Implementation iher Zykelindexreihen in Scratchpad II, *Bayreuther Mathematische Schriften*, 43, **1993**, 11–34.

[153] A. Hurwitz, Über einige Verallgemeinerungen der Lebniz'schen Differentiationsformel und des polynomischen Lehrsatzes, *Zeitschrift für Mathematik und Physik*, 35, **1890**, 56–58.

[154] M. Ikollo Ndoumbe, *Espèces asymétriques et séries indicatrices d'asymétrie*, M.Sc. Thesis, Université du Québec à Montréal, **1992**.

[155] S. Janson, D.E. Knuth, T. Luczak, and B. Pittel, The Birth of the Giant Component, *Random Structures and Algorithms*, 4, **1993**, 233–358.

[156] S.A. Joni, Lagrange Inversion in Higher Dimensions and Umbral Operators, *Linear and Multilinear Algebra*, 6, **1978**, 111–121.

[157] S.A. Joni and G.C. Rota, Coalgebras and Bialgebras in Combinatorics, *Studies in Applied Mathematics*, 61, **1979**, 93–139.

[158] A. Joyal, Une théorie combinatoire des séries formelles, *Advances in Mathematics*, 42, **1981**, 1–82.

[159] A. Joyal, β-anneaux et vecteurs de Witt, *Comptes Rendus Mathématiques de l'Académie des Sciences*, La société royale du Canada, VII, **1985**, 177–182.

[160] A. Joyal, β-anneaux et λ-anneaux, *Comptes Rendus Mathématiques de l'Académie des Sciences*, La société royale du Canada, VII, **1985**, 227–232.

[161] A. Joyal, Règle des signes en algèbre combinatoire, *Comptes Rendus Mathématiques de l'Académie des Sciences*, La société royale du Canada, VII, **1985**, 285–290.

[162] A. Joyal, Calcul intégral combinatoire et homologie du groupe symétrique, *Comptes Rendus Mathématiques de l'Académie des Sciences*, La société royale du Canada, VII, **1985**, 337–342.

[163] A. Joyal, Foncteurs analytiques et espèces de structures, in: [196], **1986**, 126–159.

[164] A. Joyal and R. Street, The Category of Representations of the General Linear Group over a Finite Field, *Journal of Algebra*, 176, **1995**, 908–946.

[165] A.C. Karabeg, *PQ-tree Data Structure and some Graph Embedding Problems*, Ph.D. Dissertation, University of California at San Diego, La Jolla, CA, **1988**.

[166] G.M. Kelly, On Clubs and Doctrines, in: *Category Seminar*, editor: G.M. Kelly, Lecture Notes in Mathematics 420, Springer–Verlag, Berlin, Heidelberg, and New York, **1974**, 181–256.

[167] R. Kemp, A Note on the Number of Leftist Trees, *Information Processing Letters*, 25, **1987**, 227–232.

[168] A. Kerber, Enumeration under Finite Group Action: Symmetry Classes of Mappings, in: [196], **1986**, 160–176.

[169] A. Kerber, *Algebraic Combinatorics via Finite Group Actions*, Bibliographisches Institut-Mannheim, Hochschultaschenbücher–Verlag, Mannheim, Germany, **1991**.

[170] A. Kerber, La théorie combinatoire sous-tendant la théorie des représentations linéaires des groupes symétriques finis, in: *Mots*, editor: M. Lothaire, Hermés, Paris, **1990**, 246–253.

[171] A. Kerber and K.-J. Thürlings, *Counting Symmetry Classes of Functions by Weight and Automorphism Group*, Lecture Notes in Mathematics 969, Springer–Verlag, Berlin, Heidelberg, and New York, **1992**, 195–203.

[172] D.E. Knuth, *The Art of Computer Programming*, Addison–Wesley, Reading, MA, **1981**.

[173] D.E. Knuth and H.S. Wilf, A Short Proof of Darboux's Lemma, *Applied Mathematical Letters*, 2, **1989**, 139–140.

[174] D. Knutson, λ-*Rings and the Representations of the Symmetric Group*, Lecture Notes in Mathematics 308, Springer–Verlag, Berlin, Heidelberg, and New York, **1973**.

[175] G. Kreweras, Une famille de polynômes ayant plusieurs propriétés énumératives, *Periodica Math. Hungarica*, 11, **1980**, 309–320.

[176] M. Kuczma, B. Choczewski, and R. Ger, *Iterative Functional Equations, Encyclopedia of Mathematics and its Applications* 32, Cambridge University Press, New York, **1990**.

[177] G. Labelle, Sur l'inversion et l'itération continue des séries formelles, *European Journal of Combinatorics*, 1, **1980**, 113–138.

[178] G. Labelle, Une nouvelle démonstration combinatoire des formules d'inversion de Lagrange, *Advances in Mathematics*, 42, **1981**, 217–247.

[179] G. Labelle, Une combinatoire sous-jacente au théorème des fonctions implicites, *Journal of Combinatorial Theory, Series A*, 40, **1985**, 377–393.
[180] G. Labelle, Eclosions combinatoires appliquées à l'inversion multidimensionnelle des séries formelles, *Journal of Combinatorial Theory, Series A*, 39, **1985**, 52–82.
[181] G. Labelle, Some New Computational Methods in the Theory of Species, in: [196], **1986**, 192–209.
[182] G. Labelle, Interpolation dans les K-espèces, in: *Actes du séminaire lotharingien de combinatoire, 14e session*, editor: V. Strehl, Publications de l'Institut de recherche mathématiques, Strasbourg, France, **1986**, 60–70.
[183] G. Labelle, On Combinatorial Differential Equations, *Journal of Mathematical Analysis Applications*, 113, **1986**, 344–381.
[184] G. Labelle, Dérivées directionnelles et développements de Taylor combinatoires, *Discrete Mathematics*, 79, **1989**, 279–297.
[185] G. Labelle, On the Generalized Iterates of Yeh's Combinatorial K-Species, *Journal of Combinatorial Theory, Series A*, 50, **1989**, 235–258.
[186] G. Labelle, On Asymmetric Structures, *Discrete Mathematics*, 99, **1992**, 141–164.
[187] G. Labelle, Counting Asymmetric Enriched Trees, *Journal of Symbolic Computation*, 14, **1992**, 211–242.
[188] G. Labelle, Sur la symétrie et l'asymétrie des structures combinatoires, *Theoretical Computer Science*, 117, **1993**, 3–22.
[189] G. Labelle, Some Combinatorial Results First Found Using Computer Algebra, *Journal of Symbolic Computation*, 20, **1995**, 567–594.
[190] G. Labelle, J. Labelle, and K. Pineau, Sur une généralisation des séries indicatrices d'espèces, *Journal of Combinatorial Theory, Series A*, 19, **1995**, 17–35.
[191] G. Labelle and L. Laforest, Sur la distribution de l'arité de la racine d'une arborescence hyperquaternaire à d dimensions, *Discrete Mathematics*, 139, **1995**, 287–302.
[192] G. Labelle and L. Laforest, Combinatorial Variations on Multidimensional Quadtrees, *Journal of Combinatorial Theory, Series A*, 69, **1995**, 1–16.
[193] G. Labelle and L. Laforest, Etude de constantes universelles pour les arborescences hyperquaternaires de recherche, *Discrete Mathematics*, 153, **1996**, 199–211.
[194] G. Labelle and P. Leroux, Enumeration of (Uni- or Bicolored) Plane Trees According to their Degree Distribution, *Discrete Mathematics*, 157, **1996**, 227–240.
[195] G. Labelle and P. Leroux, Extension of the Exponential Formula in Enumerative Combinatorics, *Electronic Journal of Combinatorics*, 3 (2), **1996**, # R12, 14 p.
[196] G. Labelle and P. Leroux, editors, *Combinatoire Énumérative*, Lecture Notes in Mathematics 1234, Springer–Verlag, Berlin, Heidelberg, and New York, **1986**.
[197] G. Labelle and K. Pineau, Ensembles orientés et permutations paires: séries indicatrices d'asymétrie et q-séries, *Advances in Applied Mathematics*, 15, **1994**, 452–475.
[198] J. Labelle, *Théorie des Graphes*, Modulo, Montréal, **1981**.
[199] J. Labelle, Applications diverses de la théorie combinatoire des espèces de structures, *Annales des Sciences Mathématiques du Québec*, 7, **1983**, 59–94.
[200] J. Labelle, Quelques espèces sur les ensembles de petite cardinalité, *Annales des Sciences Mathématiques du Québec*, 11, **1985**, 31–58.
[201] J. Labelle, Langages de Dyck généralisés, *Annales des Sciences Mathématiques du Québec*, 17, **1993**, 1–13.
[202] J. Labelle and Y.N. Yeh, A Combinatorial Model for Hahn Polynomials, in: *Actes du séminaire lotharingien de combinatoire, 16e session*, editor: D. Foata, Publications de l'Institut de recherche mathématiques, Strasbourg, France, **1988**, 99–107.
[203] J. Labelle and Y.N. Yeh, Some Combinatorics of the Hypergeometric Series, *European Journal of Combinatorics*, 9, **1988**, 593–605.
[204] J. Labelle and Y.N. Yeh, The Combinatorics of Laguerre, Charlier and Hermite Polynomials, *Studies in Applied Mathematics*, 80, **1989**, 25–36.
[205] J. Labelle and Y.N. Yeh, The Relation Between Burnside Rings and Combinatorial Species, *Journal of Combinatorial Theory, Series A*, 50, **1989**, 269–284.

[206] J. Labelle and Y.N. Yeh, Combinatorial Proofs of Symmetry Formulas for the Generalized Hypergeometric Series, *Journal of Mathematical Analysis Applications*, 139, **1989**, 36–48.
[207] J. Labelle and Y.N. Yeh, Dyck Paths of Knight Moves, *Discrete Applied Mathematics*, 24, **1989**, 213–221.
[208] J. Labelle and Y.N. Yeh, Combinatorial Proofs of some Limit Formulas Involving Orthogonal Polynomials, *Discrete Mathematics*, 79, **1989–90**, 77–93.
[209] J. Labelle and Y.N. Yeh, Generalized Dyck Paths, *Discrete Mathematics*, 82, **1990**, 1–6.
[210] L. Laforest, *Etude des arbres Hyperquaternaires*, Ph.D. Dissertation, Université McGill, Publications du Laboratoire de combinatoire et d'informatique mathématique Vol. 3, Université du Québec à Montréal, **1990**.
[211] F. Lamnabhi–Lagarrigue, P. Leroux, and X.G. Viennot, Combinatorial Approximations of Volterra Series by Bilinear Systems, in: *Analysis of Controlled Dynamical Systems*, editors: B. Bonnard, B. Bride, J.-P. Gauthier, and I. Gupka, Progress in Systems and Control Theory 8, Birkhäusser, Boston, **1991**, 304–315.
[212] P. Leroux, The Isomorphism Problem for Incidence Algebras of Möbius Categories, *Illinois Journal of Mathematics*, 26, **1982**, 52–61.
[213] P. Leroux, Methoden der Anzahlbestimmung für Einige Klassen von Graphen, *Bayreuther Mathematische Schriften*, 26, **1988**, 1–36.
[214] P. Leroux, Reduced Matrices and $q - \log$ Concavity Properties of q-Stirling Numbers, *Journal of Combinatorial Theory, Series A*, 54, **1990**, 64–84.
[215] P. Leroux and B. Miloudi, Généralisations de la formule d'Otter, *Annales des Sciences Mathématiques du Québec*, 16, **1992**, 53–80.
[216] P. Leroux and V. Strehl, Jacobi Polynomials: Combinatorics of the Basic Identities, *Discrete Mathematics*, 57, **1985**, 167–187.
[217] P. Leroux and G.X. Viennot, Combinatorial Resolution of Systems of Differential Equations I: Ordinary Differential Equations, in: [196], **1986**, 233–253.
[218] P. Leroux and G.X. Viennot, Résolution combinatoire des équations différentielles II: Calcul intégral combinatoire, *Annales des Sciences Mathématiques du Québec*, 12, **1988**, 210–245.
[219] P. Leroux and G.X. Viennot, Combinatorial Resolution of Systems of Differential Equations IV: Separation of Variables, *Discrete Mathematics*, 72, **1988**, 237–250.
[220] P. Leroux and G.X. Viennot, A Combinatorial Approach to Non Linear Functional Expansions: An Introduction With an Example, *Theoretical Computer Science*, 79, **1991**, 179–193.
[221] J.-B. Lévesque, *Sur la pondération en combinatoire énumérative*, M.Sc. Thesis, Université du Québec à Montréal, **1989**.
[222] C.I. Liu, *Introduction to Combinatorial Mathematics*, McGraw–Hill, New York, **1968**.
[223] V.A. Liskovets and T.R.Walsh, Ten Steps to Counting Planar Graphs, *Congressus Numerantium*, 60, **1987**, 269–277.
[224] J.-L. Loday, La renaissance des opérades, *Séminaire Bourbaki*, 792, **1994–95**.
[225] D.E. Loeb, Sequences of Symmetric Functions of Binomial Type, *Studies in Applied Mathematics*, 83, **1990**, 1–30.
[226] A. Longtin, Une combinatoire non-commutative pour l'étude des nombres sécants, in: [196], **1986**, 246–266.
[227] M. Lothaire, *Combinatorics on Words*, Encyclopedia of Mathematics Vol. 17, Addison–Wesley, Reading, MA, **1983**.
[228] I.G. Macdonald, *Symmetric Functions and Hall Polynomials*, 2nd edition, Clarendon, Oxford, **1995**.
[229] S. MacLane, *Categories for the Working Mathematician*, Springer–Verlag, Berlin, Heidelberg, and New York, **1971**.
[230] S. MacLane, *Coherence in Categories*, Lecture Notes in Mathematics 281, Springer–Verlag, Berlin, Heidelberg, and New York, **1972**.
[231] P.A. Macmahon, *Combinatorial Analysis*, 2 volumes, Cambridge University Press, London, 1915–1916. Reprinted: Chelsea, New York, **1960**.
[232] C.L. Mallows and J. Riordan, The Inversion Enumerator for Labeled Trees, *Bulletin of the American Mathematical Society*, 74, **1968**, 92–94.

[233] J.P. May, *The Geometry of Iterated Loops*, Lecture Notes in Mathematics 271, Springer–Verlag, Berlin, Heidelberg, and New York, **1972**.
[234] A. Meir and J.W. Moon, On the Altitude of Nodes in Random Trees, *Canadian Journal of Mathematics*, 30, **1978**, 997–1015.
[235] A. Meir and J.W. Moon, On the Branch Sizes of Rooted Unlabeled Trees, in: Graph Theory and its Applications: East and West, Jinan 1986, *Annals of the New York Academy of Sciences*, 576, **1989**, 399–407.
[236] A. Meir and J.W. Moon, Some Asymptotic Results Useful in Enumerative Problems, *Aequationes Math.*, 33, **1987**, 260–268.
[237] A. Meir and J.W. Moon, On Nodes of Out-Degree One in Random Trees, *Combinatorics*, 52, **1987**, 405–416.
[238] A. Meir and J.W. Moon, Recursive Trees with no Nodes of Out-Degree One, *Congressus Numerantium*, 66, **1988**, 49–62.
[239] A. Meir and J.W. Moon, On an Asymptotic Method in Enumeration, *Journal of Combinatorial Theory, Series A*, 51, **1989**, 77–89.
[240] A. Meir and J.W. Moon, On the Maximum Out-Degree in Random Trees, *Australian Journal of Combinatorics*, 2, **1990**, 147–156.
[241] A. Meir and J.W. Moon, The Asymptotic Behavior of Coefficients of Powers of Certain Generating Functions, *European Journal of Combinatorics*, 11, **1990**, 581–587.
[242] A. Meir and J.W. Moon, On Nodes of Large Out-Degree in Random Trees, *Congressus Numerantium*, 82, **1991**, 3–13.
[243] A. Meir, J.W. Moon, and J. Mycielsky, Hereditary Finite Sets and Identity Trees, *Journal of Combinatorial Theory, Series B*, 35, **1983**, 142–155.
[244] K. Mehlhorn, *Data Structures and Algorithms, I: Sorting and Searching*, Springer–Verlag, Berlin, Heidelberg, and New York, **1985**.
[245] M. Méndez, Tensor Species and Symmetric Functions, *Proceedings of the National Academy of Sciences USA*, 88, **1991**, 9892–9894.
[246] M. Méndez, Multisets and the Combinatorics of Symmetric Functions, *Advances in Mathematics*, 102, **1993**, 95–125.
[247] M. Méndez and O. Nava, Colored Species, C-Monoids and Plethysm, *Journal of Combinatorial Theory, Series A*, 64, **1993**, 102–129.
[248] M. Méndez and J. Yang, Möbius Species, *Advances in Mathematics*, 85, **1991**, 83–128.
[249] N.C. Metropolis and G.C. Rota, The Cyclotomic Identity, in: *Combinatorics and Algebra*, Contemporary Mathematics, American Mathematical Society, 34, **1984**, 19–28.
[250] N.C. Metropolis and G.C. Rota, Witt Vectors and the Algebra of Necklaces, *Advances in Mathematics*, 50, **1983**, 95–125.
[251] J.W. Moon, *Counting Labelled Trees*, Canadian Mathematical Monographs 1, Canadian Mathematical Society, **1970**.
[252] J.W. Moon, Some Enumerative Results on Series Parallel Networks, *Annals of Discrete Mathematics*, 33, **1987**, 199–226.
[253] R. Mullin and G.C. Rota, On the Foundation of Combinatorial Theory III: Theory of Binomial Enumeration, in: *Graph Theory and its Applications*, editor: B. Harris, Academic, New York, **1970**, 167–213.
[254] A. Nickel, *Abzählung und Konstruction diskreter und die Theorie des Spezies*, Diplomarbeit, Universität Bayreuth, Bayreuth, Germany, **1989**.
[255] O. Nava, On the Combinatorics of Plethysm, *Journal of Combinatorial Theory, Series A*, 46, **1987**, 215–251.
[256] O. Nava and G.C. Rota, Plethysm, Categories and Combinatorics, *Advances in Mathematics*, 58, **1985**, 61–68.
[257] R.Z. Norman, *On the Number of Linear Graphs With Given Blocks*, Ph.D. Dissertation, University of Michigan, Ann Arbor, **1954**.
[258] A. Odlyzko, Periodic Oscillations of Power Series that Satisfy Functional Equations, *Advances in Mathematics*, 44, **1982**, 180–225.
[259] R. Otter, The Number of Trees, *Annals of Mathematics*, 49, **1948**, 583–599.

[260] T. Ottmann and D. Wood, 1-2-Brother Trees or AVL Trees Revisited, *Computer Journal*, 23, **1980**, 248–255.

[261] J.J. Pansiot, Nombres d'Euler et inversions dans les arbres, *European Journal of Combinatorics*, 3, **1982**, 259–262.

[262] K. Pineau, *Une généralisation des séries indicatrices d'espèces de structures*, Ph.D. Dissertation, Université du Québec à Montréal, Publications du Laboratoire de combinatoire et d'informatique mathématique Vol. 21, **1995**.

[263] G. Pólya, Kombinatorische Anzahlbestimmungen für Gruppen, Graphen und Chemische Verbindungen, *Acta Mathematica*, 68, **1937**, 145–254.

[264] G. Pólya and R.C. Read, *Combinatorial Enumeration of Groups, Graphs and Chemical Compounds*, Springer–Verlag, Berlin, Heidelberg, and New York, **1987**.

[265] G. Pólya and G. Szego, *Problems and Theorems in Analysis*, Springer–Verlag, Berlin, Heidelberg, and New York, **1976**.

[266] M. Poulin, *Relations entre certaines fonctions arithmétiques et séries indicatrices en combinatoire*, M.Sc. Thesis, Université du Québec à Montréal, **1992**.

[267] D.S. Rajan, The Adjoints of the Derivative Functor on Species, *Journal of Combinatorial Theory, Series A*, 62, **1993**, 93–106.

[268] D.S. Rajan, The Equations $D^k Y = X^n$ in Combinatorial Species, *Discrete Mathematics*, 118, **1993**, 197–206.

[269] M.V.S. Ramanath and T.R. Walsh, Enumeration and Generation of a Class of Regular Digraphs, *Journal of Graph Theory*, 11, **1987**, 471–479.

[270] D. Rawlings, The ABC's of Classical Enumeration, *Annales des Sciences Mathématiques du Québec*, 10, **1986**, 207–235.

[271] D. Rawlings, A Binary Tree Decomposition Space of Permutation Statistics, *Journal of Combinatorial Theory, Series A*, 59, **1992**, 111–124.

[272] N. Ray, Umbral Calculus, Binomial Enumeration and Chromatic Polynomials, *Transactions of the American Mathematical Society*, 309, **1988**, 191–213.

[273] A.H. Read, The Solution of a Functional Equation, *Proceedings of the Royal Society*, Edinburgh, A 63, **1951–52**, 336–345.

[274] R.C. Read, The Use of S-Functions in Combinatorial Analysis, *Canadian Journal of Mathematics*, 20, **1968**, 808–841.

[275] J.M. Redfield, The Theory of Group-Reduced Distributions, *American Journal of Mathematics*, 49, **1927**, 433–455.

[276] C. Reutenauer, Séries formelles et algèbres syntactiques, *Journal of Algebra*, 66, **1980**, 448–483.

[277] C. Reutenauer, Propriétes arithmétiques de séries rationnelles et ensembles denses, *Acta Arithmetica*, 39, **1981**, 133–144.

[278] C. Reutenauer, Theorem of Poincaré-Birkhoff-Witt, Logarithm and Representations of the Symmetric Group Whose Orders are the Stirling Numbers, in: [196], **1986**, 267–293.

[279] C. Reutenauer, Mots circulaires et polynômes irréductibles, *Annales des Sciences Mathématiques du Québec*, 12, **1988**, 275–285.

[280] C. Reutenauer, *Free Lie Algebras*, Oxford University Press, Oxford, **1993**.

[281] J. Riordan, *An Introduction to Combinatorial Analysis*, Princeton University Press, Princeton, NJ, **1980**.

[282] R.W. Robinson, Enumeration of Non-Separable Graphs, *Journal of Combinatorial Theory, Series B*, 9, **1970**, 327–356.

[283] R.W. Robinson and T.R. Walsh, Inversion of Cycle Index Series Relations for 2- and 3-Connected Graphs, *Journal of Combinatorial Theory, Series B*, 57, **1993**, 289–308.

[284] G.C. Rota, On the Foundations of Combinatorial Theory, I: Theory of Möbius Functions, *Zeitschift für Wahrscheinlichkeitstheorie und Verwandte Gebiete*, 2, **1964**, 340–368.

[285] G.C. Rota, Baxter Algebras and Combinatorial Identities, II, *Bulletin of the American Mathematical Society*, 75, **1969**, 330–334.

[286] G.C. Rota, *Finite Operator Calculus*, Academic, New York, **1975**.

[287] G.C. Rota, Hopf Algebras in Combinatorics, in: *Problèmes combinatoires et théorie des graphes*, Colloques internationaux du centre national de recherche scientifique 260, Paris, **1978**, 363–365.

Bibliography

[288] G.C. Rota and B.E. Sagan, Congruences Derived from Group Action, *European Journal of Combinatorics*, 1, **1980**, 67–76.
[289] G.C. Rota and D.A. Smith, Enumeration Under Group Action, *Annali Scuola Normale Superiore Pisa*, Classe di Scienze, Serie IV, Vol. IV, **1977**, 637–646.
[290] B. Sagan, *The Symmetric Group*, Wadsworth Brooks/Cole, Pacific Grove CA, **1991**.
[291] B. Salvy, *Asymptotique, automatique et fonctions génératrices*, Ph.D. Dissertation, Ecole Polytechnique, Palaiseau, France, **1991**.
[292] W.R. Schmitt, Hopf Algebras of Combinatorial Structures, *Canadian Journal of Mathematics*, 45, **1993**, 412–428.
[293] M.-P. Schützenberger, Solutions non commutatives d'une équation différentielle classique, in: *New Concepts and Technologies in Parallel Information Processing*, editor: M. Caianello, NATO Advanced Studies Institute Series E: Applied Sciences, 9, **1975**, 381–401.
[294] H.I. Scoins, The Number of Trees With Nodes of Alternate Parity, *Proceedings of the Cambridge Philosophical Society*, 58, **1962**, 12–16.
[295] R. Sedgewick and P. Flajolet, *An Introduction to the Analysis of Algorithms*, Addison–Wesley, Reading, MA, **1996**.
[296] D. Senato and A.M. Venezia, Polynomial Species and Connections Among Bases of the Symmetric Polynomials, *Combinatorics '86*, Trento, Annals of Discrete Mathematics 37, **1988**, 405–412.
[297] J.P. Serre, *Linear Representation of Finite Groups*, Springer–Verlag, Berlin, Heidelberg, and New York, **1977**.
[298] N.J.A. Sloane, An On-Line Version of the Encyclopedia of Integer Sequences, *Electronic Journal of Combinatorics*, 1 (1), **1994**.
[299] N.J.A. Sloane and S. Plouffe, *The Encyclopedia of Integer Sequences*, Academic, New York, **1995**.
[300] R.P. Stanley, Binomial Posets, Möbius Inversion, and Permutation Enumeration, *Journal of Combinatorial Theory, Series A*, 20, **1976**, 336–356.
[301] R.P. Stanley, Exponential Structures, *Studies in Applied Mathematics*, 59, **1978**, 73–82.
[302] R.P. Stanley, Generating Functions, in: *MAA Studies in Combinatorics*, editor: G.C. Rota, Mathematical Association of America, Washington D.C., **1978**, 100–141.
[303] R.P. Stanley, Differentiably Finite Power Series, *European Journal of Combinatorics*, 1, **1980**, 175–188.
[304] R.P. Stanley, *Enumerative Combinatorics*, Volume 1, Wadsworth Brooks/Cole, Pacific Grove, CA, **1986**.
[305] D. Stanton, Recent Results for the q-Lagrange Inversion Formula, *Conference on Ramanujan revisited*, Urbana-Champaign, Ill., 1987, Academic, New York, Boston, **1988**, 525–536.
[306] D. Stanton and D. White, *Constructive Combinatorics*, Undergraduate Texts in Mathematics, Springer–Verlag, Berlin, Heidelberg, and New York, **1986**.
[307] P.K. Stockmeyer, *Enumeration of Graphs with Prescribed Automorphism Group*, Ph.D. Dissertation, University of Michigan, Ann Arbor, **1971**.
[308] V. Strehl, Combinatorics of Jacobi-configurations I: Complete Oriented Matching, in: [196], **1986**, 294–307.
[309] V. Strehl, Combinatorics of Jacobi-configurations II: A Rational Approximation via Matching Polynomials, in: *Actes du séminaire lotharingien de combinatoire, 13e session*, editor: G. Nicoletti, Publications del'Institut de recherche mathématiques, Strasbourg, France, **1986**, 112–133.
[310] V. Strehl, Combinatorics of Jacobi-Configurations III: The Srivastava–Singhal Generating Relation Revisited, *Discrete Mathematics*, 73, **1988**, 221–232.
[311] V. Strehl, *Zykel-Enumeration bei Lokal-Strukturierten Funktionen*, Habilitationsschrift, Institut für Mathematische Maschinen und Datenverarbeitung der Universität Erlangen-Nürnberg, Germany, **1989**.
[312] V. Strehl, Cycle Counting for Isomorphism Types of Endofunctions, *Bayreuther Mathematische Schriften*, 40, **1992**, 153–167.
[313] D.E. Taylor, A Natural Proof of the Cyclotomic Identity, *Bulletin of the Australian Mathematical Society*, 42, **1990**, 185–189.

[314] W.T. Tutte, *Connectivity in Graphs*, University of Toronto Press, Toronto, **1966**.
[315] B. Unger, *Asymptotics of Increasing Trees*, Ph.D. Dissertation, University of Sydney, **1993**.
[316] J.H. van Lint and R.M. Wilson, *A Course in Combinatorics*, Cambridge University Press, Cambridge, **1992**.
[317] G.X. Viennot, Une interprétation combinatoire des coefficients du développement en série entière des fonctions elliptiques de Jacobi, *Journal of Combinatorial Theory, Series A*, 29, **1980**, 121–133.
[318] G.X. Viennot, Interprétation combinatoire des nombres d'Euler et de Genocchi, *Séminaire de théorie des nombres* 11, Université de Bordeaux, Bordeaux, France, **1980–81**.
[319] G.X. Viennot, *Une théorie combinatoire des polynômes orthogonaux*, Publications du Laboratoire de combinatoire et d'informatique mathématique, Université du Québec à Montréal, **1983**.
[320] G.X. Viennot, Heaps of Pieces I: Basic Definitions and Combinatorial Lemmas, in: [196], **1986**, 321–350.
[321] J. Vuillemin, A Unifying Look at Data Structures, *Communications of the ACM*, 23, **1980**, 229–239.
[322] D.W. Walkup, The Number of Plane Trees, *Mathematika*, 19, **1972**, 200–204.
[323] T.R. Walsh, On the Size of Quadtrees Generalized to d-Dimensional Binary Pictures, *Computers and Mathematics with Applications*, 11, **1985**, 1089–1097.
[324] K.H. Wehrhahn, *Combinatorics, an Introduction*, 2nd edition, Carslaw Publications, Glebe, NSW, Australia, **1992**.
[325] D. White, Counting Patterns With a Given Automorphism Group, *Proceedings of the American Mathematical Society*, 47, **1975**, 41–44.
[326] H.S. Wilf, *Generatingfunctionology*. Academic, New York, **1990**.
[327] H.S. Wilf and D. Zeilberger, Towards Computerized Proofs of Identities, *Bulletin of the American Mathematical Society*, New Series 23, **1990**, 77–83.
[328] S. Wolfram, *Mathematica, A System for Doing Mathematics by Computer*, Addison–Wesley, Reading, MA, **1988**.
[329] S.G. Williamson, *Combinatorics for Computer Science*, Computer Science Press, Rockville, Maryland, **1985**.
[330] N. Worwald and E.M. Wright, The Exponential Generating Functon of Labelled Blocks, *Discrete Mathematics*, 25, **1979**, 93–96.
[331] J.S. Yang, Asymmetric Structures, Types and Indicator Polynomials, *Advances in Mathematics*, 100, **1993**, 133–153.
[332] J.S. Yang, The Plethystic Inverse of a Formal Power Series, *Discrete Mathematics*, 139, **1995**, 413–442.
[333] Y.N. Yeh, *On the Combinatorial Species of Joyal*, Ph.D. Dissertation, State University of New York at Buffalo, **1985**.
[334] Y.N. Yeh, The Calculus of Virtual Species and $\mathbb{K}$-Species, in: [196], **1986**, 351–369.
[335] Y.N. Yeh, Solutions to Some Problems in the Theory of Combinatorial Species, *Annales des Sciences Mathématiques du Québec*, 11, **1987**, 363–369.
[336] D. Zeilberger, A Combinatorial Approach to Matrix Algebra, *Discrete Mathematics*, 56, **1985**, 61–72.
[337] D. Zeilberger, Toward a Combinatorial Proof of the Jacobian Conjecture?, in: [196], **1986**, 370–380.
[338] D. Zeilberger, A Bijection From Ordered Trees to Binary Trees that Sends the Pruning Order to the Strahler Number, *Discrete Mathematics*, 82, **1990**, 89–92.
[339] J. Zeng, La β-extension de la formule d'inversion de Lagrange à plusieurs variables, in: *Actes du séminaire lotharingien de combinatoire, 20e session*, editors: L. Cerlienco and D. Foata, Publications de l'Institut de recherche mathématiques, Strasbourg, France, **1988**, 23–38.
[340] J. Zeng, Weighted Derangements and the Linearization Coefficients of Orthogonal Sheffer Plolynomials, *Proceedings of the London Mathematical Society*, 65, **1992**, 1–22.

Notation Index

GENERAL NOTATION

(F and G, generic species of structures)

aut(**n**)	number of automorphisms, 20
$[x^n]f(x)$	coefficient of formal power series, 13
coeff$_{\mathbf{n}}$	coefficient of index series, 20
$=_n$	contact of order n, 23
cyc(ψ)	connected components of ψ, 86
des	number of descents, 346
Des	descent set, 346
fix σ	number of fixed points of σ, 16
Fix σ	fixed points, 16
inv	number of inversions, 346
Inv	inversion set, 346
$\mathcal{O}(x)$	orbit of x, 395
$\mathcal{G}_x$	stabilizer of x, 396
$\mathcal{L}_n^{(\alpha)}(x)$	Laguerre polynomials, 95
p_n	power sum, 278
$r_n(\lambda)$	binomial type sequence, 169
rec(ψ)	recurrent points of ψ, 86
$\lambda^{\langle n \rangle}$	rising factorial, 170
$\lambda_{\langle n \rangle}$	falling factorial, 170
$\mathfrak{A}$	set of atomic species, 146
$\mathfrak{M}$	set of molecular species, 142
$F[n]$	F-structures on $[n]$, 13
$F[\sigma]$	transport function, 5
$F(x)$	generating series, 12
$\widetilde{F}(x)$	type generating series, 15
$P_{G;Y}$	cycle index polynomial, 397
Z_F	cycle index series, 17
Γ_F	asymmetry index series, 324
F'	derivative, 47
$F^{(k)}$	k^{th} derivative, 51
$F^{\bullet}$	pointing, 60
$F^{\bullet n}$	pointing n times, 69
$\int F$	integral, 348
F_w	weighted species, 83
$F_{w^{(\alpha)}}$	", with connected components counter, 97
F_α	F-structures with constant weight α, 86
F^c	connected F-structures, 46
$\kappa_n(F)$	average number of connected F-structures, 63
F_n	restriction to cardinality n, 30
F_{odd}	odd part, 38
F_{even}	even part, 38
Φ^-	negative part, 123
Φ^+	positive part, 125
$\overline{F}$	flat part, 322
T(F)	types of F-structures, 14
$F^{\text{inj}}\langle X_{\mathbf{t}} \rangle$	F-injective colorings, 325
$F \simeq G$	isomorphism (of species), 21
$F = G$	combinatorial equality (isomorphism), 21
$F \equiv G$	equipotence, 20
$F \circ G$	substitution, 41
$F^{\langle n \rangle}$	iterates for substitution, 46
$F \times G$	Cartesian product, 64
$F \square G$	functorial composition, 70
$F * G$	convolution product, 348
$F \cdot_{\mathcal{O}} G$	ordinal product, 348
$s(n,k)$	Stirling numbers of the first kind, 53
$c(n,k)$	signless ", 53
$S(n,k)$	Stirling numbers of the second kind, 53
$\ell_1 +_{\mathcal{O}} \ell_2$	ordinal sum, 343

PARTICULAR SPECIES

$\mathcal{A}_-$	rooted trees, 7
$\mathcal{A}_R$	R-enriched rooted trees, 165
$\mathcal{A}_L$	ordered rooted trees, 167
$\mathfrak{a}$	trees, 7
$\mathfrak{a}_R$	R-enriched trees, 178
Alt	alternating permutations, 345
$\mathcal{B}$	binary rooted trees, 8
$\mathcal{B}^\dagger$	increasing binary trees, 345
$\mathcal{B}_c^\dagger$	complete increasing binary trees, 345
Bal	ballots, 10

Notation Index

$\mathrm{Bal}^{[k]}$	k level ballots, 36	$\Lambda^{(\alpha)}$	connected components w... species, 98
Bij	bijections, 112		
$\mathcal{C}$	cyclic permutations, 7	L	linear orders, 7
$\mathcal{C}_{\mathrm{alt}}$	alternating cyclic permutations, 108	Lag	Laguerre configurations, 9...
Cha	chains, 11	n	$n \in \mathbb{N}$ as a species, 31
$C^{[m]}$	m-complexes, 78	Oct	octopuses, 12
Cov	coverings, 78	$\mathrm{Oct}_{\mathrm{alt}}$	alternating octopuses, 113...
$\mathrm{Cov}^{[m]}$	m parts coverings, 78	$\mathrm{Oct}_{\mathrm{reg}}$	regular octopuses, 56
$\mathcal{D}$	directed graphs, 7	Ord	order relations, 47
Der	derangements, 10	$\wp$	subsets, 7
E	sets, 8	$\wp^{[k]}$	k-element subsets, 34
$E^{\pm}$	oriented sets, 144	$\mathcal{P}$	commutative parenthesiza...
End	endofunctions, 7	$\mathfrak{p}$	plane trees, 178
End_R	R-enriched endofunctions, 172	$\mathcal{P}$	plane rooted trees, 168
$\mathrm{End}^{\wp}$	partial endofunctions, 172	Par	set partitions, 7
$\mathrm{End}_R^{\wp}$	R-enriched partial endofunctions, 172	$\mathrm{Par}^{[k]}$	partitions with k parts, 31
		Par_P	partial set partitions, 49
Φ	functions, 112	P	polygons, 10
$\mathcal{F}$	forests, 52	Preo	preorders, 47
$\mathcal{G}$	simple graphs, 4	$\mathrm{Rel}^{[m]}$	m-ary relations, 74
K_n	complete graphs, 297	Red	reduced posets, 57
$\mathcal{G}^{\mathrm{c}}$	connected simple graphs, 27	$\mathcal{S}$	permutations, 7
$\mathcal{G}^{\mathrm{d}}$	disconnected simple graphs, 27	$\mathcal{S}^{[k]}$	permutations with k cycle...
$\Gamma(X, Y)$	graphs on vertices and edges, 111	$\mathcal{S}^{<k>}$	permutations with all cycl... k, 40
$\mathcal{H}$	homeomorphically irreducible rooted trees, 283	$\mathcal{S}^{\mathrm{tric}}$	tricolored permutations, 1...
$\mathfrak{h}$	homeomorphically irreducible trees, 283	$\mathcal{S}^{\mathrm{mix}}$	mixed permutations, 114
		Sur	surjections, 112
$\mathcal{H}$	hedges of rooted trees, 10	$\mathcal{V}$	vertebrates, 61
Inv	involutions, 7	X	singletons, 8
Inj	injections, 112	1	empty set species, 8
Jac	Jacobi endofunctions, 203	0	empty species, 8
		ε	elements, 8

Index

...ed graphs, 299–303
...nnected graphs, 301, 307
...ted trees, 198, 211, 230
...s, 230, 244
...s, 243

...ed,
...
, 82
...tities, 189
...le formal power series, 257
...olecules, 287, 288
... identities (extension principle), 182
... singularity, 253
...olecules, 287, 288
...ig,
   105, 108
   144, 321
...ses, 113
...ations, 344, 345, 351
   55
...y of structures, 42
...tric,
...lies, 328
...r assemblies, 328
...d
...es, 93
...mutations, 93
...ted assemblies, 328
...rooted trees, 336
...trees, 336
... trees, 322, 329–330, 338
...s, 322
...res, 322, 324
...321, 329–330
...try,
...ting series, 324

  index series, 324, 327
Asymptotic, 247
  expansion, 255–256
  scale, 255
Asymptotically equivalent, 248
Atomic,
  decomposition, 147, 150
  k-sort species, 158
  species, 146, 154, 329, 430–431
AVL trees, 238, 245
Automorphism, 3, 6

$\mathbb{B}$-species, 341, 345
B-trees, 243
Ballots, 10, 36, 44, 57, 69, 124, 251, 254, 309, 412, 413, 415, 417
bc-tree of a graph, 300–301
Bell numbers, 63, 116
Bender's theorem, 261–262, 274
Bernoulli,
  numbers, 268, 269
  polynomials, 268
Bicolored,
  plane trees, 215–217
  rooted plane trees, 215
  rooted trees, 206
    R, S-enriched, 206
  trees, 206
    R, S-enriched, 206, 294
  vertebrates, 215
Billiard stroke, 237
Binary (plane) rooted trees, 8, 10, 162, 195, 235, 261, 373
  complete, 235
    increasing, 345, 358
  height of, 235
  increasing, 344, 345, 359, 373, 380

leftist, 246
pseudo leftist, 246
Strahler order of, 235
unary, 254
Binary search trees, 341
Binomial type sequences, 169, 181, 187, 188
Block-cut-point tree, 300
Block of a graph, 300
Bridge of a graph, 307
Buds, 361, 365
Burnside's lemma, 396

c-chromatic trees, 294
rooted, 294
Cancellation law for addition, 134
Canonical decomposition, 30, 39, 67, 109, 123
Cartesian product of,
species, 64, 410, 413, 414, 416, 418
virtual species, 121
(weighted) $\mathbb{L}$-species, 348
weighted species, 84
Catalan numbers, 163, 168, 252, 256, 291
Cauchy-Frobenius theorem (Burnside's lemma), 396
Cayley's,
formula, 167, 180
identity, 290
Center of a tree, 280
of mass, 290
Central trinomial coefficients, 183
Chains, 11, 56, 160, 297, 409, 412, 414, 415, 417
Characterization of atomic species, 154
Circular words, 92
Colorations, 404
Colored,
cycles, 93
derangements, 96
permutations, 93
Colorings, 399
F-injective, 325
G-reduced, 318, 399
Combinatorial,
differential equations, 358, 363
eclosions, 212, 365
equality, 21
exponential, 131
logarithm, 130, 149
separation of variables, 377
Commutative parenthesizations, 10, 198, 222, 241, 379, 410, 412, 414
Commuting group actions, 407
Compatible permutations, 218
Complementary subspecies, 134
Complete,

binary rooted trees, 235
increasing binary rooted trees, 345, 358, 373, 380
graphs, 297
Composite (partitional) of,
species, 41, 104
$\mathbb{L}$-species, 348, 353
Composite (functorial), 73
Composition of cycle index series, 43, 73, 84, 107
Concentrated (species) on a cardinality, 30, 140
Conjugate,
subgroups, 395
words, 92
Connected,
compatible permutations, 219
components weighting species, 99, 340
linear orders, lists, 131, 348
R-enriched endofunctions, 202
simple graphs, 7, 37, 46, 64, 297, 298
structures, 46, 55, 63, 96, 131, 159, 308, 410, 412, 414, 416, 418
Contact order, 22, 23, 194
Convolution, 348–349, 357
Coverings of sets, 78
Counter for,
arcs, 411
connected components, 97, 411, 413–418
cycles, 83, 86, 91, 184, 411
edges, 112, 297, 308, 405, 411
elements, 100, 411
fixed points, 89, 411
leaves, 80
parts, 86, 411
transpositions, 89, 411
Cubic anharmonic oscillator, 389
Cut-point of a graph, 299
Cycle counter, 83, 86, 91, 184, 411
Cycle index,
polynomial, 397
series, 17, 84, 106, 122, 207–209, 310
Cycle type of permutations, 16
Cyclic group, 406
Cyclic permutations, oriented cycles, 2, 7, 45, 48, 58, 86, 92, 109, 133, 144, 155, 190, 220, 293, 322, 346, 409, 411, 413, 415, 417
alternating, 105, 108
colored, 93
asymmetric, 93
Cyclic rooted trees, *see* mobiles
Cyclotomic identities, 98

Degenerate vertebrates, 62

Index   451

Derangements, 10, 31, 32, 33, 39, 220, 250, 409, 412, 413, 415, 417
 colored, 96
Derivation,
 generalized, 119
 of species, 47
 of virtual species, 125
 of (weighted) $\mathbb{L}$-species, 348
 of weighted species, 84
 partial, 105, 354
Descent of a permutation (list), 346
Differential equations,
 autonomous, 360
 combinatorial, 358, 363
 higher order, 371
 linear, 377
 non autonomous, 371, 377
 system of, 358, 371
Differential operators, 365, 372
 powers of, 368, 386
Dihedral group, 144, 321, 406
Directed,
 graphs, 7, 74, 76, 77, 409, 411, 415
 trees, 292
Disconnected simple graphs, 27
Discrete graphs, 297
Dissections, 104
Dissymmetry theorem, dissymmetry formula, for,
 asymmetric trees, 330
 c-chromatic trees, 294
 graphs, 301
 plane trees, 291
 R-enriched trees, 282, 291
 R, S-enriched trees, 207, 294
 R-enriched trees with leaves, 288
 trees, 277, 280, 296
Dobinski's formula, 116
Double cosets, 153, 328
Dominant singularities, 249
Duffing equation, 389

Eccentricity of a vertex, 280
Eclosion, 212
Eclosion points, 366
Edges, 72
Edge counter, 112, 297, 308, 405
Electric circuit with a quadratic resistance, 388
Elementary symmetric functions, 333
Elements, 8, 409, 412, 413, 415, 417
Empty set species, 8, 346, 409
Empty species, 8, 346, 409
endofunctions, 7, 41, 62, 63, 86, 209, 328, 412, 414

Jacobi, 172, 203, 245
 idempotent, 221
 partial, 172
 R-enriched, 172, 184, 200, 218, 220, 410
  connected, 202
  partial, 172, 200
Equipotence of species, 20
Euler,
 constant, 267
 gamma function, 266
 Euler–MacLaurin summation formula, 268
 numbers, 352, 383
Eulerian,
 numbers, 354
 polynomials, 354
Even,
 alternating permutations (lists), 351, 359
 complete increasing binary rooted trees, 359
 graphs, 351
 part of species, 38, 410, 412, 414, 416–417
 permutations, 144, 337
 sets, 29, 348, 409, 411, 413, 415, 416
Excedance, 355
Exponential,
 generating series, 13
 formula (weighted), 96
Extension principle of algebraic identities, 182

F-injective assemblies, 325
F-injective colorings, 325
F-Maclaurin expansion, 118
F-structures, 5, 102, 345
F-Taylor expansion, 118
Falling factorial, 170
Fatou's formula, 233
Fibers, 113, 165, 363
Fibonacci numbers, 249
Flat,
 part, 322–323
 species, 322
Fliess series, 389
Forests, 52, 58, 170, 260, 274, 293
Frobenius formula, 354
Functional digraph, 7
Functions with R-enriched fibers, 113, 170
Functorial composition of,
 species, 70, 71, 410, 413–414, 416, 418
 virtual species, 125
 weighted
  multisort species, 105
  species, 84
Functoriality, 5, 11
Fundamental transformation, 22, 174

G-eclosions, 212
G-rooted trees, 195, 197, 213, 229, 262, 275
    unlabelled, 263, 276
Gamma function, 266
Genealogical trees, 188
Generalized,
    differentiation, 119
    pointing, 117
    separation of variables, 387
Generating series, 13, 83, 106, 122, 346
    asymmetry type, 324
    type, 15, 84, 106, 122
Good's formula (one dimensional), 183
Good-Lagrange inversion formula, 200, 203, 220
Graphs,
    2-connected, 299
    2-edge-connected, 300, 307
    complete, 297
    directed, 7, 74, 76, 77, 409, 411, 415
    discrete, 297
    even, 351
    constructed on vertices and edges, 111, 114
    Husimi, 300
    multigraphs, 114, 308
    non-separable (or 2-connected), 299–304
    sagital, 7
    simple, 4, 7, 28, 46, 64, 71, 74, 76, 100, 146,
        151, 158, 297, 308, 351, 405, 409, 411,
        415
        connected, 7, 27, 46, 64, 297–298
        disconnected, 27
        tri-chromatic, 100, 116
        unlabelled, 405
        with loops, 77
        without end points, 304
Gröbner's formula, 365
Group,
    actions, 393
        commuting, 407
        isomorphism, 394
        transitive, 140, 395
    alternating, 144, 321
    cyclic, 144, 321, 406
    dihedral, 144, 321, 406
    of rotations of a cube, 397, 406
    symmetric, 11, 399

H-enriched rooted trees, 193, 210, 229
    increasing, 374–375
Hadamard product, 64, 65, 349
Hanlon's inversion method, 305, 306
Hayman's theorem, 258
Head vertex, 62
Heavy rooted trees, 224, 241, 390

Hedges, 10, 189
Height of a binary tree, 238
Hereditary finite sets, 338
Hermite polynomials, 89, 94, 184, 185
History of increasing rooted trees, 383
Homeomorphically irreducible,
    rooted trees, 283
    trees, 188, 283, 291, 337
Homogeneous (complete) symmetric f
    88
Hurwitz's inversion formula, 184
Husimi graphs, 300

Idempotent endofunctions, 221
Identity tree, 321
Implicit function theorem, 192, 209
Implicit species theorem, 195
    multisort species, 199
    virtual species, 212
Increasing,
    binary rooted trees, 344, 345, 359, 3
    complete, 345, 358, 373, 380
    functions, 343
    mobiles, 365, 391
    ordered rooted trees, 365, 379, 383
    m-ary, 365
    with even fibers, 381
    rooted trees, 344, 359, 367
        H-enriched, 374–375
        m-ary, 365
        plane, 363
        planted plane (or planar), 363
        R-enriched, 341, 361–363, 365, 3
            392
        ternary, 368, 376, 380
        with even fibers, 380
Indegree, 166
Induced order, 343
Injective,
    assemblies, 332
    coloring, 324
    functions, 91
Integral of (weighted) $\mathbb{L}$-species, 348, 3
Internal vertices, 361
Inventory, 81, 90
Inverse
    (multiplicative), 124
    under substitution, 130, 132
Inversion table, 355
Inversions,
    of permutations (lists), 346
    in trees (and rooted trees), 381
Involutions, 7, 16, 89, 94, 221, 258, 337
    411, 413

Index 453

8
c rooted trees, 5
sm
 actions, 394
cies, 345
es, 21
ures, 3, 6, 102
hted sets, 81
6, 14
ding to sorts, 110
tegrals, 386
rder, 232–234

ctions, 172, 203, 245
, 203, 219
tions, 374
nials, 172, 184, 203–205, 213
, 203
tidimensional inversion theorem, 221

pecies, 353
ies, 102
158
ar, 155

341, 344
rt, 353
d, 344
tructures, 12
, 399
ed, 317, 399
366
arson transform, 350
inversion formula, 164, 174, 183, 239
ing bidimensional, 215
d version, 189

ations, 95, 213
nials, 95, 172, 184, 214
ter, 80
56, 181

373
minima, 22
nary) trees, 246
ode, 355
 R-enriched endofunctions, 173
er-Taylor formula, 365
ed trees, 224, 241, 390
sequence of species, 23

tial equation, 377

functional equation, 195
Read Bajraktarevic equation, 233
Linear (or total) orders, lists, 7, 10, 12, 15, 50,
    131, 322, 346, 347, 354, 356, 359, 409,
    411, 413, 415, 417
connected, 131, 348
non-empty, 57, 138
of length k, 37, 346
Linearization coefficients, 95
Lucas numbers, 56
Lukasievich paths, 384
Lump of a graph, 307
Lyndon words, 92, 98

m-ary relations, 74, 76, 79
m-ary rooted trees, 211
Macmahon's master theorem, 217
Major index of a permutation (list), 346
Marked trees, 292
    rooted, 292
Meir and Moon theorem, 260, 274
Mixed,
    species, 380
    trees, 292
Mehler's formula, 94
Mobiles, 240, 261
    increasing, 365, 391
Molecules,
    alcohol, 287–288
    alkane, 287–288
Molecular,
    decomposition, 141, 148, 150, 226, 299, 410,
        421–424, 430, 431
    standard form, 142
k-sort species, 155
species, 140, 143–147, 154–155, 311, 314,
    317–318, 328, 333, 418–420, 431, 432
Morphism of $\mathbb{A}$-weighted sets, 81
Morphism of actions, 394
Motzkin paths, 267, 389
Multicardinality, 101
Multifunction, 101
Multigraphs, 114, 308
Multiplicable family of species, 39
Multiset, 101
Multisort,
    $\mathbb{L}$-species, 353
    species, 102, 199

Naturality condition, 21
Negative,
    part of a virtual species, 123
    virtual species, 123
Negligible, 247

Index

Neighbors of vertices, 178
Newton-Raphson iteration, 222–225
  higher order, 242
  for differential equations, 389
  for Read-Bajraktarevic equations, 246
Non-separable (or 2-connected) graphs, 299–304

Octopuses, 12, 55, 56, 57, 69, 221, 409, 412, 415, 417
  alternating, 113
  regular, 56
Odd,
  sets, 29, 348, 409, 411, 415, 417
  alternating permutations, 351, 359
  complete increasing binary rooted trees, 359
  part of a species, 38, 410, 412, 414, 416, 417
Orbit, 395
Order,
  induced, 343
  iterative, 232–234
  of a graph, 297
  of a formal power series, 264
  of an unlabelled structure, 14
  of contact, 22, 23, 194
  Strahler, 235, 245
  to be of, 247
Order relations, partial orders, 47, 138
  reduced, 57, 138
Ordered,
  partitions, *see* ballots
  rooted trees, 10, 167, 185–186, 190, 195, 261, 284, 379
    increasing, 363, 379, 383
Orderly labelling, 366
Ordinal,
  product, 348, 349
  sum, 343
Oriented,
  cycles, *see* cyclic permutations
  graphs, *see* directed graphs
  sets, 144, 152, 384
  rooted trees, 285–286, 292
  trees, 285–286, 292
Otter's formula, 277, 279

Paraffins, 287–288
Parenthesizations, 198, 379
  commutative, 10, 198, 222, 241, 379, 410, 412, 414
Partial,
  differentiation, 105, 354
  orders, 47, 138
  partitions, 49
Partitional composition, *see* substitution

Partitions, *see* set partitions
Permutation representation, 394
permutations, 7, 12, 15, 31, 40, 41, 45, 83, 91, 99, 114, 133, 139, 142, 220, 251, 266, 322, 346, 409, 411, 413, 415, 417
  alternating, 344, 345
    even, 351, 359
    odd, 351, 359
  colored, 93
    asymmetric, 93
  compatible, 218
  connected, 219
  even, 144, 337, 409, 411, 413
  Jacobi, 374
  R-enriched, 218
  random, 64
  standard form, 22
  tri-colored, 107
Plane trees, 178, 180, 182, 190
  bicolored, 215–217
  rooted, 168, 178, 186, 190, 284
Planted plane (or planar) rooted trees, 167
Plethystic substitution, 43, 84, 107
Pointed triangular cacti, 305
Pointing,
  generalized, 117
  species, 60, 410, 412, 414, 416, 418
  virtual species, 125
  weighted
    multisort species, 106
    species, 84
Pólya–Redfield theorem, 319, 403
polygons, 10, 30, 144, 161, 297
  n-gons, 30
Polynomial species, 111
Polynomials,
  Bernouilli, 268
  cycle index, 397
  Eulerian, 355
  Hermite, 89, 94, 184, 185
  Jacobi, 172, 184, 201–205, 213
  Laguerre, 95, 172, 184, 214
  tree inversion, 297, 382
Positive,
  species, 123
  part of virtual species, 123
Power sum symmetric functions, 83, 333
PQ-trees, 24
Preorders, 47
Product,
  of $\mathbb{A}$-weighted sets, 82
  of species, 32, 410, 413, 414, 416, 418
  of virtual species, 122
  of (weighted) $\mathbb{L}$-species, 347

of weighted
  multisort species, 104
  species, 84
  ordinal, 348, 349
Pseudo-leftist binary trees, 246
Pure simplicial complexes, 78

q-analogue, 339, 347
q-series associated to species, 339
Quotient species, 147, 159

R-enriched,
  endofunctions, 172–173, 184, 200, 218, 220
    connected, 202
    partial, 172, 200
  fibers, 113
  functions, 170
  permutations, 218
  rooted trees, 165, 169, 170, 181, 201, 207, 224–228, 239, 260, 273, 275, 281, 329, 383, 410, 424–429
    unlabelled, 263, 275
  trees, 178–182, 281, 335, 410
Radius of convergence, 263
Random,
  permutations, 64
  set partitions, 63
  simple graphs, 64
  structures, 63, 69
Read-Bajraktarevic (functional equation), 230, 238, 245
  linear, 233
  Newton-Raphson iteration, 246
Recurrent elements, 40, 86
Reduced form of virtual species, 122, 148, 151–152
Reduced order relations (partial orders), 57, 138
Regular octopuses, 56
Representative of a virtual species, 121
Restricted (species) to a cardinality, 30, 349
Restriction of a total order, 343
Rising factorial, 170
Robinson's iterative procedure, 299, 308
Rodrigues formula, 184
Root, 2
Rooted,
  c-chromatic trees, 294
  directed trees, 292
  marked trees, 292
  oriented trees, 285–286, 292
  plane trees, 168, 178, 180, 190, 284
  signed trees, 292
Rooted trees, 7, 10, 41, 43, 58, 60, 80, 100, 114, 127, 132, 141, 167, 178, 181, 185–186,
    209, 210, 211, 226, 260, 278, 303, 328, 409, 412, 415, 417
  (2, 3)-rooted trees, 198, 211, 230
  (2, 3)-trees, 230, 244
  (a, b)-trees, 243
  asymmetric, 322, 329, 338
    plane, 336
  AVL-trees, 245
  B-trees, 206
  bicolored, 206
  binary (plane), 8–10, 162, 195, 235, 238, 261, 373
    complete, 235, 358
    increasing, 344, 345, 358, 359, 373, 380
  c-chromatic, 294
  cyclic, see mobiles
  directed, 292
  G-rooted trees, 195, 197, 213, 229, 241, 262, 275
    unlabelled, 263, 276
    (2, 3)-rooted trees, 198, 211, 230
  H-enriched, 193, 210, 229
    increasing, 374–375
  heavy, 224, 241, 390
  homeomorphically irreducible, 283
  increasing, 344, 345, 359, 367
    with even fibers, 380
  light, 224
  m-ary, 211
  marked, 292
  mixed, 292
  oriented, 285–286, 292
  ordered, 10, 167, 185–186, 190, 195, 261, 284, 379
    increasing, 363, 374, 383
    with even fibers, 381
  plane, 168, 178, 185–186, 190, 191, 284
    asymmetric, 336
    bicolored, 215
    increasing, 363
  planted plane (or planar), 167
    increasing, 363
  R-enriched, 165, 169, 170, 181, 201, 207, 224–228, 239, 260, 273, 275, 281, 329, 383, 410, 424, 429
    increasing, 341, 361–363, 371, 389–392
    unlabelled, 263, 275, 425
  signed, 392
  ternary, 211
    increasing, 368, 375, 380
  unary-binary, 254
  unlabelled, 141, 290
Rotations of a cube, 397, 406

Sagittal graph of an endofunction, 7
Scoins formula, 206, 215
Secant,
   functions, 352, 373
   numbers, 352, 359, 381, 383
Separation of variables,
   combinatorial, 376
   generalized, 387
Sequence of binomial type, 169, 181, 187, 188
Set partitions, 7, 44, 49, 63, 86, 104, 409, 411, 413, 415, 417
Set successor, 57
Sets, 8, 50, 322, 346, 409, 411, 413, 415, 416
   with even cardinality, 29, 348, 409, 411, 413, 415, 416
   hereditary finite, 338
   multisets, 101
   non-empty, 10
   odd cardinality, 29, 348, 409, 411, 413, 415, 417
   of cardinality,
      2, 8
      n, 30
   oriented, 144, 152, 334, 409, 412, 416
   totally ordered, 342
Signed trees, 292
   rooted, 292
Simple graphs, 4, 7, 28, 46, 64, 71, 74, 76, 100, 116, 146, 151, 158, 221, 259, 273, 297, 308, 351, 405, 409, 411, 415
   unlabelled, 151, 297, 405
Simplicial complexes, 78
Singletons, 8, 87, 98, 322, 346, 409
   of sort i, 103
Species characteristic of the empty set, 8, 346, 409
Species of structures, 5
   atomic, 146, 154, 329, 430, 431
   flat, 322
   molecular, 140, 143–147, 154, 311, 314, 317–318, 328, 333, 418–420, 431, 432
   multisort, 102, 199
   weighted, 82
Stabilizer, 396
Standard form of,
   permutations, 22
   molecular decompositions, 142
Stereoisomers, 288
Stirling,
   asymptotic expansion, 272
   formula, 248, 258, 271
   numbers of first kind, 53, 69
   numbers of second kind, 53, 69, 116
Strahler order, 235, 245

Structure, 1
Structured words, 78
Subsets, 7, 34, 99, 409–411, 413, 415, 417
   non-empty, 78
   of cardinality k, 34, 70
Subspecies, 120, 134
Substitution of,
   index series, 43, 84, 107
   species, 40, 410, 413, 414, 416, 418
   virtual species, 127
   (weighted) $\mathbb{L}$-species, 348, 353
   weighted
      multisort species, 104
      species, 84
Subtraction of species, 120
Sum,
   of $\mathbb{A}$-weighted sets, 82
   of species, 28, 410, 412, 414, 416, 418
   of virtual species, 122
   of (weighted) $\mathbb{L}$-species, 347
   of weighted
      multisort species, 104
      species, 84
   ordinal, 343
Summable family of,
   series, 38
   species, 29, 37
   virtual species, 123
   weighted sets, 81, 90
Superposition of structures, 65
Surjections, 115
Symmetric,
   functions, 87
      power sum, 88, 333
      complete homogeneous, 88
      elementary, 333
   group, 11, 399
   square, 296
   square root, 296
Symmetry, 321

Tail vertex, 62
Tangent,
   functions, 352, 373
   numbers, 352, 359, 381, 383
Taylor's formula,
   generalized, 118
   with integral remainder, 358
Taylor–Maclaurin expansion, 119
Temperley's identity, 308
Ternary (plane) rooted trees, 211
   increasing, 368, 376, 380
Topological trees, 188, 285, 291, 337
Total cardinality, 101

Total orders, *see* linear orders
Totally ordered sets, 342
Touchard's formula, 97
Transitive group action, 115, 324
Transport of structures, 2, 5, 83, 102, 343
Tree inversion polynomial, 297, 382
Trees, 7, 51, 58, 127, 158, 178, 182, 278, 293, 300, 303, 410, 412, 414
  (2, 3)-trees, 230, 244
  (a, b)-trees, 243
  AVL, 238, 245
  B, 243
  bicolored, 206
    R, S-enriched, 206, 294
  binary search, 341
  block-cut-point, 380
  c-chromatic, 284
  directed, 292
  genealogical, 188
  homeomorphically irreducible, 188, 283, 291, 337
  identity, 321
  leftist, 246
  marked, 292
  mixed, 292
  oriented, 285–286
    unlabelled, 290
  plane, 178, 180, 182, 190, 284
    asymmetric, 336
    bicolored, 215–217
    unlabelled, 290
  PQ-trees, 242
  pseudo-leftist, 246
  R-enriched, 178–182, 281, 335, 410, 424–429
  rooted, *see* rooted trees
  signed, 292
  topological, 188, 283, 291, 337
  unlabelled, 17, 279, 290, 296
Tri-chromatic graphs, 100, 116
Tri-colored permutations, 107
Tri-colorings, 10
Tricomi identity, 96
Triangular cacti, 301, 305, 307
  pointed, 305

Type generating series, 15, 84, 106, 122
Types (isomorphism) according to a sort, 110, 112, 115, 364

Unlabelled,
  bicolored plane trees, 217
  G-rooted trees, 263, 276
  homeomorphically irreducible trees, 284
  oriented trees, 292
  plane rooted trees, 186, 191
  plane trees, 290
  R-enriched (rooted) trees, 263, 275, 410, 424–429
  rooted trees, 141, 290
  simple graphs, 151, 297, 405
  structures, 3, 12, 14, 102, 277
  trees, 10, 227, 236, 241
Unrelated species, 102, 124

Vertebral column, 61
Vertebrates, 61, 68
  bicolored, 215
  degenerate, 62
Virtual species, 121
  negative part, 123
  positive part, 123
  reduced form, 122, 148, 151–152
  strictly, 12

Walkup's formula, 186
Weight, 81
  preserving functions, 81
Weighted
  exponential formulas, 96
  $\mathbb{L}$-species, 344
  multisort species, 104
  species, 82
  sets, 81
Wreath of G-structures, 313
Wreath product, 145, 314
Words, 79
  abelian, 79
  circular, 79, 92
  Lyndon, 92, 98